Fire on Earth

Fire on Earth:

An Introduction

Andrew C. Scott
Royal Holloway, University of London, England

David M.J.S. Bowman
University of Tasmania, Australia

William J. Bond
University of Cape Town, South Africa

Stephen J. Pyne
Arizona State University, Tempe, Arizona, USA

Martin E. Alexander
University of Alberta, Edmonton, Alberta, Canada

WILEY Blackwell

This edition first published 2014 © 2014 by John Wiley & Sons, Ltd

Registered office: John Wiley & Sons, Ltd, The Atrium, Southern Gate, Chichester, West Sussex, PO19 8SQ, UK

Editorial offices: 9600 Garsington Road, Oxford, OX4 2DQ, UK
The Atrium, Southern Gate, Chichester, West Sussex, PO19 8SQ, UK
111 River Street, Hoboken, NJ 07030-5774, USA

For details of our global editorial offices, for customer services and for information about how to apply for permission to reuse the copyright material in this book please see our website at www.wiley.com/wiley-blackwell.

The right of the author to be identified as the author of this work has been asserted in accordance with the UK Copyright, Designs and Patents Act 1988.

Library of Congress Cataloging-in-Publication Data
Scott, Andrew C.
    Fire on earth : an introduction / Andrew C. Scott, David M.J.S.
Bowman, William J. Bond, Stephen J. Pyne, Martin E. Alexander.
        pages cm
    Includes bibliographical references and index.
    ISBN 978-1-119-95357-9 (cloth) – ISBN 978-1-119-95356-2 (pbk.)
1. Fire–History. 2. Fire management. 3. Fire ecology. 4.   Forest fires.
5.   Wildfires. I. Bowman, D. M. J. S. II. Bond, William J., 1948- III.
Pyne, Stephen J., 1949- IV. Alexander, Martin E. V. Title.
    GN416.S46 2013
    541'.361–dc23
                                                          2013018591

A catalogue record for this book is available from the British Library.

Wiley also publishes its books in a variety of electronic formats. Some content that appears in print may not be available in electronic books.

Cover images: Aykut ince, OGM-Turkey and NASA (http://lance-modis.eosdis.nasa.gov)

Cover design by Steve Thompson

Set in 10/12pt, Minion by Thomson Digital, Noida, India.

Printed and bound in Singapore by Markono Print Media Pte Ltd

1   2014

Contents

Preface

Earth is the only planet known to have fire. The reason is both simple and profound: fire exists because Earth is the only planet to possess life as we know it. Life created both the oxygen and the hydrocarbon fuel that combustion requires, it arranges those fuels according to processes of evolutionary selection and ecological dynamics and, in the form of humanity, it supplies the most abundant source of ignition. Fire is an expression of life on Earth and an index of life's history. Few processes are as integral, unique or ancient.

Yet, while the significance of fire can hardly be doubted, it rarely enters the discourse of relevant disciplines or appears in standard texts of geology, biology, human history, physics or global chemistry. Fifty years ago, the only organized inquiries lodged in applied contexts such as combustion engineering, urban fire services and, fitfully, forestry and range science. No journal exclusively reported on it; conferences, wholly national, might be held once a decade on how better to control it; and, even when it was examined, free-burning fire, which integrates everything around it, was narrowly confined within other disciplines. The outcome was an extraordinary disconnection. While fire was ubiquitous in various forms throughout Earth, it was absent from our formal inquiries about our world. It stood outside both science and scholarship.

Today, the literature has multiplied exponentially. Dedicated journals exist. Half a dozen international conferences might be held annually. A host of formal sciences, or programmes announcing interdisciplinary intentions, are willing to consider fire. Wildfire appears routinely in media reporting. What has not happened, however, is a synthesis of contemporary thinking that can bring together the most powerful concepts and disciplinary voices that have interested themselves in fire. There is no global survey that can convey why planetary fire exists, how it works and why it looks the way it does today. This volume intends to redress the problem.

The text consists of four parts. The choice of themes is not arbitrary. We wanted to:

- establish the autonomy and longevity of fire on Earth;
- centre its dynamics in the living world;
- accent the critical presence of fire for humanity, and of humanity for pyrogeography;
- have fire's behaviour serve more as an integration of factors, and hence a summary, than as a putative foundation to everything else.

No volume can hope to summarize everything that has been published on the subject, or convey fire's endlessly ramified expressions in the field. We have selected those organizing themes that we believe best introduce the subject.

Each part is intended to stand alone, yet allow for connections to the others. Instead of creating an artificial synthesis, an intellectual equivalent of Esperanto, we elected to let each author speak in his own disciplinary tongue, in the hope that the gains from fluency will overcome any losses from translation. Yet, as in any collaborative venture, we have had an influence on what each of us has written and, hence, this volume must be considered as a book with five authors rather than an edited volume. The result is not an encyclopaedia, but a studied description and explanation of how fire appears to a prominent cadre of fire researchers. As each discipline organizes the whole through its own disciplinary prism, so each author speaks in his own voice.

Inevitably, there are lapses and overlaps in the particulars of the four parts. Each of us, for example, sees the foundational fire differently. For someone interested in deep time, fire appears as an emerging property of an evolving planet – one that leaves a geologic record. For biologists, fire appears as a product of the living world – the substrate without which fire cannot exist. For a cultural historian, it appears as an informing and defining technology for humanity – a unique signature of our agency and identity. And for someone interested in fire behaviour, fire will appear as a chemical reaction shaped by its physical surroundings.

We felt it better to let each author follow his own vision and thematic arc than try to merge them into a common cauldron. In this way, each perspective:

- will understand the increasingly dominant presence of humanity differently;
- will see it as the latest in a long chronicle of fire eras;
- will see it as perturbation along all scales of Earth's biota;
- will see it as an index of humanity's changing power; or
- will see it an arena for the application of better understanding to protect ourselves from the fires we do not want and promote those we do.

Nor have we tried to describe field operations, as previous texts by some of the authors have. The reason is simple. Other technologies, notably video, can do that job much better; 30 seconds of film can convey more accurately and vividly how to scrape fireline or run a pump than 30 pages of text. We wanted to let a book do what it can do best, which is to explore our understanding of fire and our relationship to it. We have sought to explain what principles mean through ground-truthing details, selected examples and case studies.

Inevitably, we can include only a minuscule fraction of landscapes, events, information and published (and unpublished) studies. Our choices will reflect our own judgment of what is most useful within the setting of this text, what the fashion of the times prefers and, inevitably, our own personal experiences and tastes.

We have elected to hold in-text citations to a minimum and to supplement them – again selectively – in the rosters of references and further reading attached to each of the four parts. The published literature on fire now numbers in the tens of thousands (and is expanding exponentially) and, while it is densest in the more developed world, its topics range across the planet. The authors of this text alone have a collective bibliography that includes hundreds of citations. The fire literature since the early 1960s has multiplied exponentially so, just as only a handful of examples must stand for the whole, only a tiny fraction of this literature can enter our bibliography. A master bibliography belongs online, not on printed pages.

These choices will please those members of the fire community whose work has been selected and will doubtless irk those who work has not. To the many who may feel we have slighted important sources, we plead *nolo contendere* and repeat that our purpose has not been to summarize the entire state of the literature, but to demonstrate why fire matters and how we might better understand the complex ways it intertwines with Earth and humanity. As Plutarch famously put it, the mind is not a vessel to be filled, but a fire to be kindled.

Acknowledgements

This book resulted from two circumstances. The first was the formation of the International Pyrogeography Research Group, which was funded by the Kavli Institute for Theoretical Physics and NCEAS (National Center for Ecological Analysis and Synthesis) in Santa Barbara. This was led by David Bowman and Jennifer Balch and has led to a wide interchange of ideas and experiences. The second was the persistence of Ian Francis of Blackwell (now Wiley Blackwell) to persuade two of us (ACS and SJP) to undertake a book on Fire.

We thank the many researchers for allowing us to use their photos and diagrams and, in some cases, make new versions for us.

ACS: I am indebted to Bill Chaloner for introducing me to fire in deep time, and to both him and my colleague Margaret Collinson for their constant support over the past 40 years. I thank all of my research students for continuing to stimulate me and, in particular, I thank Ian Glasspool and Claire Belcher, who have helped in my fire research in so many ways. I thank Deborah Martin for getting me to look at modern fire systems and for her constant interest in my work, and members of the International Pyrogeography Research Group for their stimulating discussions. Finally, I thank the Department of Geology and Geophysics at Yale University, who hosted my sabbatical, which directly led to the expansion of my fire research, and am grateful to the late Leo Hickey, Bob Berner and Derek Briggs for organizing my visit and to the late Karl Turekian for many hours discussing fire on earth. I thank Guido van der Werf and Mark Wooster for commenting on Chapter 1. For photographs and diagrams, I am indebted to: Sally Archibald, Jennifer Balch, Chris Baisan, Claire Belcher, Sarah Brown, John Calder, Mark Cochrane, Margaret Collinson, J.H. Dieterich, Stefan Doerr, Ian Glasspool, Rick Halsey, John Keeley, Meg Krawchuk, Colin Long, Jen Marlon, Deborah Martin, John Moody, Max Moritz, Dan Neary, Gary Nichols, Susan Page, Roy Plotnick, Mitch Power, Sue Rimmer, Dave Scott, Greg Smith, Alan Spessa, Tom Swetnam, Kirsten Thonicke, Dieter Uhl, Guido van der Werf, Cathy Whitlock.

DMJSB: I thank the National Center for Ecological Analysis and Synthesis (NCEAS) and Australian Centre for Ecological Analysis and Synthesis (ACEAS) for enabling a focused, synoptic understanding of landscape fire. I acknowledge the generosity of the numerous members of the fire community who shared their original figures and photos, thereby enhancing the book and expediting its preparation. I am indebted to all my colleagues, students and collaborators who have taught me so much about fire over the last 30 years; you are so numerous the hazard of naming some is that I will risk neglecting many others. Nonetheless, I thank my lab team including Brett Murphy, Lynda Prior, Grant Williamson, Sam Wood, Andres Holz, David Tng, Clay Trauernicht and Jess O'Brien who have been fantastic sounding boards for thinking through ideas about fire. Over the years many people have shared their knowledge of fire with me including, in Australia, Ross Bradstock, Jeremy Russell-Smith, Dick Williams, Peter Clarke, Don Franklin, Mike Crisp, Simon Haberle, Rod Fensham, Barry Brook, Owen Price, Greg Jordan, Geoff Burrows, and Fay Johnston. Thanks also to my Aboriginal friends Joshua Rostron and Rahab Redford have welcomed me on their country in Arnhem Land and taught my family how to burn the bush. I have been blessed with wonderful overseas collaborators including Jennifer Balch, Meg Krawchuk, Sarah Henderson, Jon Keeley, Tom Veblen, Cathy Whitlock, Chris Roos, Jon Lloyd, David Janos, George Perry, Brad Marston, Matt McGlone, Tom Swetnam, Johann Goldammer and Mark Cochrane. Finally, I am indebted to Stephen Pyne, William Bond and Andrew Scott—this book is a testament to their drive, wisdom and generosity.

WJB: I thank the many people from different parts of the world who have hosted visits at various times and made my life as an ecologist so rewarding. For the preparation of this book, I am particularly grateful to Dave Bowman for hosting me at the University of Tasmania, Grant Wardell-Johnson for a spectacular introduction to the world of giant eucalypts, Caroline Lehmann, Dick Williams, Garry Cook, Mahesh Sankaran, Bill Hoffmann and others for global perspectives on savannas, Jon Keeley and Ross Bradstock for continuing insights into fire in woody systems, Sally Archibald, Carla Staver, Kath Parr, Ed February, Steve Higgins, and Jeremy Midgley for enrichment on fire ecology over the years and Andrew Scott for opening up the exciting window on fire in deep time. Steve Pyne put it all together in the first place, and I have enjoyed his historian's exhilarating breadth and depth. My thanks to Tracey Nowell and Tristan Charles-Dominique for assistance with figures.

SJP: I would like to gratefully acknowledge the following people for their assistance, particularly with regard to illustrations: Johann Goldammer, Ray Lovett, Olle Zackrisson, Christian Kull, Jennifer Balch, Rich Gullette, Chris Elvidge, Mark Melvin, Zach Prusak, Cliff White, Randy Bomar, Brian van Wilgen, Jeremy Russell-Smith, Mary Huffman, Anders Granström, David Rönnblom, Peter Frost and Jerry Williams.

MEA: My good friends and fellow colleagues Miguel Cruz and Dave Thomas kindly provided detailed comments on all three chapters comprising Part Four. Miguel Cruz and graphic illustrator at large Wanda Lindquist were instrumental in the production or reproduction of many of the illustrations. Appreciation is extended to my long-time friends and fellow wildland fire behaviourists, Rick Lanoville and Dennis Quintilio, for undertaking a broad review of the manuscript. Other reviews of selected chapters were performed by Greg Baxter, Neil Burrows, Paulo Fernandes, Paul Keller, Ralph Nelson, Jim Steele and Wade Wahrenbrock. Several people have assisted me in various ways during the course of my 40-plus year journey in wildland fire. In addition to Miguel, Dave, Rick, Dennis, and Ralph, I would especially like to acknowledge the following individuals: Jack Barrows, Al Beaver, Phil Cheney, Jim Davis, Jack Dieterich, Dennis Dube, Murray Dudfield, Bill Furman, Dave Kiil, Wally Lancaster, Bruce Lawson, Murray Maffey, John Mason, Lachie McCaw, Peter Murphy, Bob Mutch, Grant Pearce, Dick Rothermel, Dave Sandberg, Jim Shell, Brian Stocks, Steve Taylor, Rob Thorburn, Chris Trevitt, Domingos Viegas, Terry Van Nest, Charlie Van Wagner, Dale Wade and Ron Wakimoto. Steve Longacre was instrumental in securing the photo in Figure 14.26. Finally, I am particularly indebted to Steve Pyne for allowing me the opportunity to contribute to this book. Part Four is a contribution in part of Joint Fire Science Program Project JFSP 09-S-03-1.

FIRESCIENCE.GOV
Research Supporting Sound Decisions

About the Authors

Professor Andrew C. Scott is internationally recognized for his work in palaeobotany, palynology, coal geology, petrology and geochemistry and for his work on the geological history of wildfire. He is currently Professor of Applied Palaeobotany and a Distinguished Research Fellow in the Department of Earth Sciences at Royal Holloway University of London, England, where he has been since 1985. Professor Scott received his bachelor's degree in geology from Bedford College, University of London, England. He studied in the Botany Department of Birkbeck College, London where he received his PhD. He received a personal professorial chair in 1996. He was awarded a D.Sc. in 2002 by the University of London for his published work. Professor Scott is a Fellow of the Geological Societies of London, America, the Royal Society of Arts, and the Higher Education Academy. He received the Gilbert H. Cady award from the Geological Society of America in 2007.

Professor David M. J. S. Bowman holds a research chair in Environmental Change Biology in the School of Biological Sciences at the University of Tasmania. The primary motivation for his research is understanding the effects of global environmental change, natural climate variability and Aboriginal landscape burning on bushfire activity and landscape change across the Australian continent. He also studies the effect of bushfire smoke on human health. These research quests are truly transdisciplinary and involve numerous national and international collaborators and use a variety of techniques, including remote sensing and geographic information systems, epidemiology, historical ecology, palynology, dendrochronology, stable isotopes analyses, ecophysiology, mathematical modelling, biological survey, field experiments and molecular ecology. After completing his PhD in forest ecology and silviculture at the University of Tasmania in 1984, he spent two decades undertaking full time research in rainforest and savanna ecology throughout northern Australia. He received a DSc in 2002 from University of Tasmania, and has received travelling fellowships from the Australian Academy of Science, Harvard, Kyoto, Leeds and Arizona universities.

Professor William J. Bond holds the Harry Bolus Chair of Botany at the University of Cape Town. His interest in fire began while working for the Forestry Department of South Africa on understanding fynbos shrublands and how to manage them. From this beginning, he was able to forge links with researchers on fire ecology in the other Mediterranean-type ecosystems. In the 1990s, after he moved to UCT, he began working on the ecology of African savannas and the intriguing interactions of fire, large herbivores and physical forces in shaping these ecosystems. He has had the good fortune to work with colleagues in similar ecosystems elsewhere in the world helping to develop a global perspective. He has also made periodic excursions into the deep past to better understand the present. He is a Fellow of the Royal Society of South Africa and a foreign associate of the National Academy of the USA.

Professor Stephen J. Pyne is a historian and Regents Professor in the School of Life Sciences, Arizona State University, Tempe, Arizona, USA. He has written over a score of books, including fire histories of the U.S. (*Fire in America; Between Two Fires*), Canada (*Awful Splendour*), Australia (*Burning Bush*), Europe (*Vestal Fire*), and the world generally; two editions of a textbook, *Introduction to Wildland Fire*; and numerous articles about fire elsewhere in the world. Among his other interests is the history of exploration, to which he has contributed *The Ice: A Journey to Antarctica*, *How the Canyon Became Grand*, and *Voyager: Exploration, Space, and the Third Great Age of Discovery*. He spent 18 seasons in fire management with the National Park Service. He is a MacArthur Fellow, a member of the American Academy of Arts and Sciences, and twice a fellow at the National Humanities Center.

Dr. Martin E. Alexander, a forester by training, but began specializing in wildland fire with the 1972 and 1973 fire seasons when he worked as a U.S. Forest Service hotshot crew member. He obtained his B.Sc.F. (1974) and M.Sc.F. (1979) degrees from Colorado State University and Ph.D. degree in forestry from the Australian National University (1998). Marty retired in late 2010 as a Senior Fire Behaviour Research Officer with the Canadian Forest Service stationed at the Northern Forestry Centre in Edmonton, Alberta, after nearly 35 years of public service. He presently serves as an Adjunct Professor of wildland fire science and management at the University of Alberta and Utah State University. His research and technology transfer efforts have focused on practical applications of wildland fire behaviour knowledge, including firefighter and public safety. In 2003, Dr. Alexander received the International Wildland Fire Safety Award from the International Association of Wildland Fire and the Canadian Forestry Achievement Award from the Canadian Institute of Forestry in 2010. His work has taken him to all the provinces and territories of Canada, and to many parts of the world, including the continental USA and Alaska, Australia, New Zealand, Portugal, Greece, Italy, Turkey, and Fiji.

About the Companion Website

This book is accompanied by a companion website:

www.wiley.com/go/scott/fireonearth

The website includes:

- Powerpoints of all figures from the book for downloading
- PDFs of all tables from the book for downloading
- Links to key fire websites
- Links to videos and podcasts
- Additional teaching material

PART ONE
Fire in the Earth System

Photo

Recent research using satellite data has revolutionized our understanding on the distribution of fire on Earth. This image shows smoke plumes from Californian fires between Los Angeles and San Francisco in October 2007 billowing out over the Pacific Ocean. Red spots indicate active fires. (Image from Modis Rapid Response Project at NASA/GSFC, image 1163886).

Fire on Earth: An Introduction, First Edition. Andrew C. Scott, David M.J.S. Bowman, William J. Bond, Stephen J. Pyne and Martin E. Alexander.
© 2014 John Wiley & Sons, Ltd. Published 2014 by John Wiley & Sons, Ltd.

Preface to part one

The first part of this book is an introduction to fire that not only considers fundamentals of fire as a physical/chemical process but also includes methods for the study of fire, an appreciation of the geological history of fire and its importance in the Earth System.

For some, fire is an every day part of life; for others, it is a remote phenomenon and is unimportant; for still others, it evades consciousness altogether. This may be said not only for individuals, but also for entire subject areas where fire has yet to be given its rightful place.

In this section, we discuss the nature and occurrence of fire and illustrate ways by which it can be recognized and studied. The past ten years has seen a revolution in our perception of fire, and news of major wildfires may now be instantly broadcast through a wide range of media. In addition, the increase in the ways we can observe fire through the use of satellites and the ability to view maps of the positions of active fires – even from our mobile phones –has brought a phenomenon unfamiliar to many to the forefront of current debate on human impact on the planet.

What is less well known or appreciated is the long geological history of fire on our planet and the role that fire has played in deep time in shaping our Earth. In this section, we demonstrate the methods we can use to unravel the history of fire – not just in terms of thousands of years, but in terms of hundreds of millions of years. In only the past few years, we have begun to unravel the relationship between fire and atmospheric change, especially with oxygen in the fossil record. This has led to a reassessment of the relationship between fire and vegetation, both from an ecological as well as from an evolutionary perspective. Part One sets up, therefore, the role of fire as an Earth System process and its special role in the evolution of life on land.

Chapter 1

What is fire?

This chapter serves as an introduction not only to Part One but also to the book as a whole. It considers many of the fundamentals of fire. We introduce here a number of concepts that are developed throughout the text and, where relevant, the chapter numbers or parts are given for reference. In addition, some areas are dealt with here because there is no space to develop them more fully within this book, as to do so would make it too long and unwieldy. Due to this, we have tried to provide a wide range of illustrative material here, as well as more extensive references for further reading.

1.1 How fire starts and initially spreads

Simply put, fire – generally called combustion – is a rapid chemical oxidative reaction that generates heat, light and produces a range of chemical products (Torero, 2013). However, in the context of vegetation fires, it is important to consider not only the range of materials that may be combusted, but also the conditions under which fire may occur and even be ignited.

It is obvious, therefore, that the basis of a fire is the nature of the fuel that will be combusted and the type of ignition source. The general principle for vegetation fires is that there is an initial high-temperature heat source. This may be produced by lightning, volcanic activity, a spark from a rock fall or, of course, by humans. Plants contain a range of organic compounds that include cellulose, a carbohydrate that is a linear polysaccharide polymer found in many cell walls. The high initial temperature causes a breakdown of the cellulose molecule and produces a range of gaseous components that include ammonia (NH_3), carbon dioxide (CO_2) and methane (CH_4). These gases mix with atmospheric oxygen and undergo a rapid exothermic reaction – combustion. This rapid increase in heat, together with the readily available oxygen, allows the reaction to continue and a fire is started (Cochrane and Ryan, 2009). These features may be characterized by the use of a fire triangle (Figure 1.1, Fire fundamentals).

Each element will be discussed in more detail below, but it is worth making a few general points at the outset.

- First, the fuel needs to be as dry as possible. This is because the initial heat may be dissipated by the need to evaporate water. If dry, then the heat can begin to break down the cellulose in the plant material. The moisture value of the fuel will depend on whether the plant is alive or dead. If alive, then the plant may contain moisture in the leaves, branches and trunk. If dead, the plant may be more prone to drying out.
- The second element is the fuel itself. For a fire to spread, it is necessary to have sufficient fuel to burn. Extreme build-up of litter that is dry would obviously be conducive to the spread of fire. However, how the fuel is arrayed and how quickly it is combusted is also important (Van Wagtendonk, 2006). There are also differences in the ways in which woody and non-woody vegetation burn, as well as other features such

Fire on Earth: An Introduction, First Edition. Andrew C. Scott, David M.J.S. Bowman, William J. Bond, Stephen J. Pyne and Martin E. Alexander.
© 2014 John Wiley & Sons, Ltd. Published 2014 by John Wiley & Sons, Ltd.

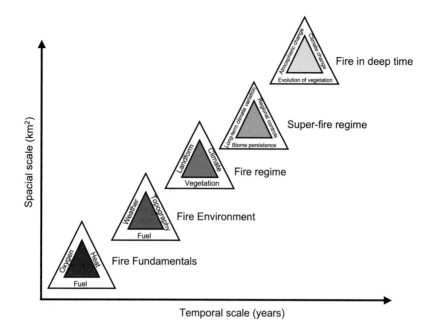

Figure 1.1 Fire triangles. The importance of different elements of fire is shown in relation to different scales, from the initial starting of a fire to the controls on fire in deep time. (This figure is compiled from a range of different authors' work including S. Pyne, M. Oritz, C. Whitlock, A. C. Scott).

as calorific value, the rate of fire spread and its intensity (see Chapter 14, Part Four).

- Third, a key element is readily available oxygen. In today's atmosphere, where the air contains 21% O_2, then combustion and fire spread is possible. For fire to be maintained, oxygen must continue to arrive at the burning point or the fire will be exhausted. This is why wind is so dangerous, as it not only drives the fire, but also replenishes the oxygen at a faster rate.

The implications of the above are also that to put a fire out, water may be added to the fuel to stop flame spread; or, in a confined space, oxygen may be excluded by smothering the fire by the use of inorganic materials such as sand or CO_2 to replace the oxygen-rich air.

1.2 Lightning and other ignition sources

Of all the natural ignition sources for a wildfire, lightning, volcanic eruptions and sparks from rock falls, it is lightning that is the most important. Human sources of ignition will be considered elsewhere in this book (see Part Three).

Lightning occurs when there is electrostatic discharge from the atmosphere. The most significant is sky-to-ground lightning (Figure 1.2). Here, a strong electrical charge is transferred from a cloud to the ground. Where the lightning hits the ground, there is a sudden increase in temperature, creating temperatures sometimes in excess of 30 000 °C. Lightning may or may not occur associated with rainfall.

Lightning may strike across many parts of the Earth's surface, but it is found concentrated in particular regions (see map, Figure 1.3). One problem with lightning maps, however, is that they show all lightning, including cloud-to-cloud lightning, not just cloud-to-ground lightning. It is significant that there may be as many as eight million lightning strikes every day.

When not associated with rain, the lightning may be referred to as 'dry lightning' and may occur in cumulonimbus clouds, which then may produce pyrocumulus clouds that create more lightning as a result of a warming ground surface from fire and is, therefore, a result of part of a positive feedback mechanism.

Not all lightning gives rise to a wildfire. In many cases, when trees are struck, this may result merely in scorching. However, if the tree is dead or dry because of drought, the great heat allows combustion to occur. This is equally the case with herbaceous vegetation,

Figure 1.2 Lightning strike. Dry lightning (not associated with rain) is one of the major ignition sources for fire (Courtesy valdezrl/Fotolia).

but sufficient fuel also needs to be available for a fire to spread.

The occurrence of fire may, therefore, be limited because of the amount or nature of the fuel (fuel limited) or because of moisture content of that fuel (moisture limited). In the tropics, this can lead to a single tree on fire, as it is unable to spread because of fuel moisture (Figure 1.4).

In regions of grassland, however, such as in savannas in Africa, fire may start just hours following a

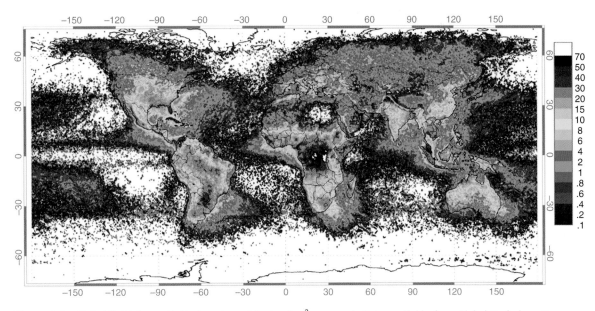

Figure 1.3 Global lightning activity (number of flashes/km^2 per year). Data available from Global Hydrology Resource Center (http://ghrc.msfc.nasa.gov). (From Bowman, 2005).

Figure 1.4 Fire burning a solitary tree in the Amazon rainforest. (Photo: M. Cochrane).

rainstorm, as the atmosphere is warm and dry, which allows the fine fuels to dry out very quickly. All of these facts are of particular significance to those producing fire potential maps (Figure 1.5).

1.3   The charring process

Most plant material comprises of a range of organic compounds, including a variety of macromolecules. For example, wood is composed of cellulose and lignin, but also includes hemicelluloses. Leaf coatings contain cutin, whereas spores and pollen are composed of the inert macromolecule sporopollenin. All of these compounds, including those from other organic sources (e.g. chitin from fungi), will break down upon heating. Of particular significance are aliphatic compounds such as cellulose, a carbohydrate, and lignin, which is an aromatic compound that is heavily cross-linked.

When heated in the absence of air this pyrolysis process results in the decomposition of the biomacromolecules to produce liquid and gaseous materials. The resultant residue is termed charcoal, and this is highly aromatic, with an increased

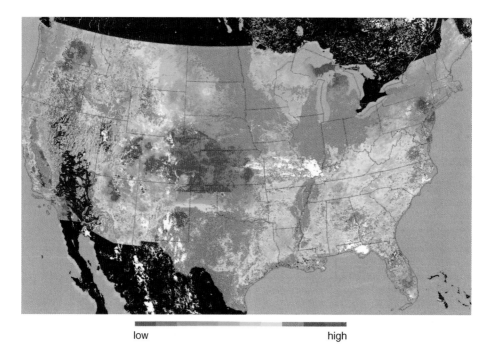

low   high

Figure 1.5   Fire potential index map of the United States for August 13, 1998. White areas are cloud cover; grey are agricultural lands (not rated). (From USGS factsheet 125–98, 1998).

proportion of carbon over the starting material. Some tar-like liquids may be produced, along with other volatile components and other gases. These materials may be mixed with oxygen in the air and combustion results, which, in turn, generates more heat for the process to continue.

The burning of plants is, therefore, a two-stage process in which:

1 pyrolysis occurs in the absence of oxygen, whereby the bio-macromolecules are charcoalified (Scott, 2010), releasing volatile components;
2 combustion then follows, representing an oxidative process whereby these components mix with oxygen in the air to allow burning.

Temperatures in the process are variable. For charcoalification to begin, the temperatures generally of 275 °C are required but higher temperatures will create a broad range of pyrolysis products that may be combusted.

1.4 Pyrolysis products

The result of the pyrolysis and combustion process leads to the production of a number of materials and compounds from charcoal: inorganic ash, volatile gases and compounds (aerosols), and soot (Santín *et al.*, 2012). Most fires generate this full range of material, and this may have a significant impact on the environment. Many of these products may be incorporated into the smoke plume of the fire and be transported considerable distances (Figure 1.6). Each of these products will be briefly examined in turn.

The products of a fire can be divided into two main groups: the material that remains in place after a fire

Figure 1.6 Smoke from fires.

A. White smoke columns from the Las Conchas Fire, New Mexico, USA, 2011 (see also Figure 1.42).
B. Grey smoke from the Norton Point Fire, Wyoming 2011.

(Photos from National Interagency Fire Center, https://picasaweb.google.com).

has passed (including mineral ash and charcoal); and the material that is transported away from a fire within a smoke plume. White or grey smoke (Figure 1.6) will depend on a range of factors, from fuel type to moisture content to temperature.

Usually the first visible evidence of a wildfire is its smoke plume. This may include water vapour which will be dependent on the water content of the fuel, small charcoal particles, usually less than 125 μm, but in some fires, significantly larger pieces may be lofted with the plume. Perhaps more important is the presence of soot, volatile components and aerosols (Artaxo *et al.*, 2009).

1.4.1 Soot

Soot, together with charcoal, is often referred to as black carbon. There is considerable disagreement among researchers on the nomenclature of these products. Some use the term 'black carbon' to mean only soot, whereas others include any combustion product that is recalcitrant in the biosphere (see Chapter 2 and Glasspool and Scott (2013)). Soot is formed by the recombination of vaporized organic molecules to form a new carbon material. Chemically, it is nearly pure carbon, and it is morphologically distinctive. Under the scanning electron microscope, it can be seen to have a range of morphologies, with a particle size less than 1 μm (see Figure 2.5g). Soot may also be produced by a range of other combustion processes (including petroleum), but that from vegetation fires may have this particular morphology.

Small cenospheres may also be produced from the burning of fossil fuels such as peat and coal. This soot may be widely dispersed into the atmosphere and may subsequently be deposited across the globe, even into deep-sea sediments (see Chapter 2). The soot may also be associated with micro-sized charcoal particles.

1.4.2 Volatile gases and compounds

A range of gases and aerosols may also be incorporated into the smoke plume. These include CO_2, carbon monoxide (CO), CH_4 and oxides of nitrogen (NO_x). Fire is therefore a significant producer of greenhouse gases. Most of the CO_2 is recaptured from the atmosphere by the re-growth of vegetation. If a fire results in the burning of peat, however, then this may become a significant issue for climate forcing.

Other important compounds include complex organic molecules such as pyrolitic polycyclic aromatic hydrocarbons (PAHs). These compounds may be produced in large quantities and their composition may depend of the type of vegetation being burned and the temperature involved. The higher the temperature, the larger the number of carbon rings found in the molecule. Table 1.1 shows a list of these compounds and their origin. The most common of these are cadanene and retene, but they also include phenanthracene, fluoromethene and chrysene, pyrene and coronene. Laevoglucosan derived from cellulose is widely used as a biomarker for vegetation fires. These compounds may also stay

Table 1.1 Major biomarker tracers in smoke from biomass burning. (From Simoneit, 2002).

Compound	Structure	Composition	Indicator for source
Anisic acid (p-methoxy-benzoic acid)	V	$C_8H_8O_3$	Gramineae lignin
Vanillic acid	II	$C_8H_8O_4$	Lignin
Syringic acid	IV	$C_9H_{10}O_5$	Angiosperm lignin
Matairesinol	VII, R = O	$C_{20}H_{22}O_6$	Conifer lignin[a]
Shonanin	VII, R = H$_2$	$C_{20}H_{24}O_5$	Conifer lignin[a]
Divanillyl	VIII	$C_{16}H_{18}O_4$	Lignin dimer
Divimillylmethane	IX	$C_{17}H_{20}O_4$	Lignin dimer
Divanillylelhane	X	$C_{18}H_{22}O_4$	Lignin dimer
Vanillylsyringyl	XI	$C_{17}H_{20}O_5$	Angiosperm lignin dimer
Disyringyl	XII	$C_{18}H_{22}O_6$	Angiospyrm lignin dimer
Dianisyl	XIII	$C_{16}H_{18}O_2$	Gramineae lignin dimer

Table 1.1 (*Continued*)

Compound	Structure	Composition	Indicator for source
Levoglucosan	XIV	$C_6H_{10}O_5$	Cellulose
Mannosan	–	$C_6H_{10}O_5$	Hemicellulose
Galactosan	–	$C_6H_{10}O_5$	Hemicellulosc
1,4:3,6- Dianhydro-β-D-glucopyranose	–	$C_6H_8O_4$	Cellulose
Dehydroabielic acid	XV	$C_{20}H_{28}O_2$	Conifer resin
n-Nonacosan-10-ol	–	$C_{29}H_{60}O$	Wax[a]
3- Methoxyfriedelane	XVI	$C_{31}H_{54}O$	Angiosperm[a]
Abietic acid	XVII	$C_{20}H_{30}O_2$	Conifer resin[a]
Pimaric acid	XVIII	$C_{20}H_{30}O_2$	Conifer resin[a]
iso-Pimaric acid	XIX	$C_{20}H_{30}O_2$	Conifer resin[a]
Sandaracopimaric acid	XX	$C_{20}H_{30}O_2$	Conifer resin[a]
Cyclopenta[c,djpyrene	XXI	$C_{18}H_{10}$	PAH all burning
Retene	XXII	$C_{18}H_{18}$	Conifer
Pimanthrene	XXIII	$C_{16}H_{14}$	Conifer
Simonellite	XXIV	$C_{19}H_{24}$	Conifer
Acetosyringone	XXV, R = C_2H_3O	$C_{10}H_{12}O4$	Angiosperm lignin
Syringyl acetone	XXV, R = C_3H_5O	$C_{11}H_{14}O_4$	Angiosperm lignin
Oleana-2,l2-diene	XXVI, R = CH_3	$C_{30}H_{48}$	Angiosperm
Uraina-2,12-diene	XXVII, R = CH_3	$C_{30}H_{48}$	Angiosperm
Oleana-2,12-dien-18-oic acid	XXVI, R = CO_2H	$C_{30}H_{46}O_2$	Angiosperm
Ursana-2,12-dien-18-oic acid	XXVII, R = CO_2H	$C_{30}H_{46}O_2$	Angiosperm
Allobclul-2-ene	XXVII	$C_{30}H_{48}O$	Birch (*Betula*)
β-Sitosterol	XXLX, R = $βC_2H_5$	$C_{29}H_{50}O$	Vegetation[a]
1-Palmitin	XXXV	$C_{19}H_{38}O_4$	Fauna (flora)
1-Stearin	–	$C_{21}H_{42}O_4$	Fauna (flora)
Cholesterol	XXIX, R = H	$C_{27}H_{46}O$	Fauna algae
Campesterol	XXIX, R = $αCH_3$	$C_{28}H_{48}O$	Gramineae
Stigmasta-3,5-diene	XXX	$C_{29}H_{48}$	Vegetation[a]
Lupa-2,22-diene	XXXI	$C_{30}H_{48}$	Angiosperm
Stigmaslerol	XXXII	$C_{29}H_{48}O$	Vegetation[a]
Ferruginol	XXXIII	$C_{20}H_{30}O$	Some conifers
1,6-Anhydro-2-acetamido-2-deoxyglucose	XXXVI	$C_8H_{13}NO_5$	Chitin
17α(H),21 β(H)-Hopanes	XXXIV	e.g., $C_{30}H_{52}$	Petroleum, lignite
Moretanes (βα-hopanes)	XXXVII	e.g., $C_{30}H_{52}$	Petroleum, lignite
α-Amyrone	XXXVIII, R = 0	$C_{30}H_{48}O$	Angiosperm
β-Amyrone	XXXIX, R = 0	$C_{30}H_{48}O$	Angiosperm
α-Amyrin	XXXVIII, R = βOH	$C_{30}H_{50}O$	Angiosperm[a]
β-Amyrin	XXXIX, R = βOH	$C_{30}H_{50}O$	Angiosperm[a]
β-Oxodeliydioabietic acid	XL	$C_{20}H_{26}O_3$	Conifer resin
7-Oxodehydroabictit acid	XLI	$C_{20}H_{26}O_3$	Conifer resin

Source: Reproduced by permission of Elsevier.
[a]Natural product, compound (unaltered).

Figure 1.7   Photos of different types of fire.

A. Surface fire in grassland at edge of forest in Gibbon Meadows, Yellowstone National Park 1988 (photo: Jim Peaco, 12048 National Park Service (www.nps.gov/features/yell/slidefikle/fire/wildfire88)).
B. Surface fire in grassland. (http://cedarcreek.umn.edu/high res.savanna-fire.jpg). Reproduced with permission of Cedar Creek Ecosystem Science Reserve.
C. Surface to crown fire in conifer forest. Trees torching at Grant village junction, July 1988, Yellowstone National Park (Jeff Henrey: Slide 12064 (www.nps.gov/features/yell/slidefikle/fire/wildfire88)).
D. Large crown fire in conifer forest (Arrow fire, 1976, Yellowstone National Park. Slide 11818 (www.nps.gov/features/yell/slidefikle/fire/wildfire)).

in the atmosphere for a considerable time and result in the prevention of rain formation, hence prolonging a wildfire event. The compounds may be washed out of the atmosphere and be incorporated into sediments (Simoneit, 2002).

Other gases that may occur in the smoke plume are the hazardous nitrous oxides that have an impact upon human health. Studies have shown that human populations regularly subjected to smoke from wildfires have a susceptibility to a range of diseases, especially lung diseases (Johnston *et al.*, 2012). Significantly, the toxicology of wildfire smoke is different, and more harmful, than the vehicle emissions that cause smoke in urban airsheds.

1.5   Fire types

Fire may occur where there is sufficient build-up of fuel that is dry enough to burn. Most often, a fire starts on the surface, where litter and duff and herbaceous plants and shrubs may occur (Figure 1.7). With forest systems, such surface fires (Figure 1.8A) may only burn the fuel on the forest floor (Figure 1.7A). These fires tend to burn relatively coolly – often less than 400 °C – and some relatively slowly. However, some shrubby vegetation, such as chaparral in California, may burn much hotter – up to 900 °C at the ground surface (Figure 1.7F) and spread faster (Figure 1.7C). Grass fires (Figure 1.7A,B) may also burn more at a

E. Crown fire in coniferous forest, Castle Rock fire, Ketchum, Idaho, USA, August 2007 (National Interagency Fire Center).
F. Crown fires in Chaperral, East Basin Complex, California, USA, July 2008 (National Interagency Fire Center).
G. Peat Fire in Indonesia. Although the fire above ground is out, the peat underground continues to burn (photo: S Page).

much faster rate. In the case of grass or scrub fires, wind may drive the fire to move faster.

If there is a thick humic layer within the soil and it is sufficiently dry, a fire may spread to burn this layer. In such a case, with restricted oxygen supply, the fire may then smoulder. The movement of such a fire may be quite slow, as much fuel may be consumed. Such a fire is termed a ground fire (Figure 1.8C). Ground fires may also burn thick peat layers (Figure 1.7G), where the water table has been lowered and the peat has dried out. In such circumstances, water may not be sufficient to extinguish such a fire, as this may introduce

Figure 1.8 Types of wildfire and fuel sources. (From Scott, 2000, modified from Davis, 1959).

oxygen to the system and the fire may actually flare up in response.

If there is a significant build-up of surface fuel within a forested ecosystem, then fire temperature may increase. Here, the fire may spread up the trunks of the tree and into the crowns of the forest trees (Figure 1.7C). This type of fire is referred to as a crown fire (Figure 1.8B). The energy release of a

Figure 1.9   Ponderosa Pine forests of Colorado, USA after fire. (All from Scott, 2010).

A. Standing trees remaining after fire. Most of the trees remain (Buffalo Creek (1996) fire, Colorado, USA). Reproduced with permission from Elsevier.
B. Cut charred stump from the Hayman (2002), Colorado, USA fire, showing that only the outside of the trunk was charred. Reproduced with permission from Elsevier.
C. Leaves and fine twigs have been removed in the fire, leaving even small branches intact. (Hayman, Colorado, USA fire, 2002). Reproduced with permission from Elsevier.

fire and its spread relates to fire intensity, and it is this factor that partly controls the spread of a fire from the surface to the crown. The leaves and fine branches appear to be the major source of fuel for the fire (as well as dead dry trees), as large upright trunks may still be visible after such a fire (Figure 1.9). When cut down, often only the outermost bark and trunk will show signs of burning (Figure 1.9B).

Crown fires (Figure 1.7D-F) may burn much hotter. While many crown fires produce temperatures only around 800–900 °C, in some cases where there is abundant dry fuel and wind to feed into the fire with oxygen, temperatures may rise to 1200 °C. It is important, however, to distinguish the temperature at the flame tip and that a metre above the tip, which may be higher. Most often it is the Fire Radiative Power (FRP) that is measured, using infrared data from satellites (see section 1.13). Crown fires may separate from surface fires and move much more quickly (Figure 1.8B). Glowing embers may be lofted up into the atmosphere (Figure 1.10) and set fire to other vegetation, causing the spread of the fire and developing many fire fronts. The result of this may be a mosaic burn pattern in the forest (Figure 1.11). In extreme fire, gas balls may explode above the fire front.

Figure 1.10   Glowing embers from burning trees.

A. Fire at tops of palm trees, Orange County, California (Courtesy LA Times. Photographer Glenn Koenig).
B. Glowing embers from burning crown of palm tree, Santa Barbara California (Photo: Ray Ford – ray@sb-outdoors.com).

1.6   Peat fires

The extreme ends of ground fires are peat fires. In such wetland mire systems, natural wildfire may be relatively rare. Studies have shown that, globally, less than 5% of the peat is charcoal formed by fire. However, when there are very dry periods, even in these ever-wet systems, a surface fire may burn away the peat layer. This may have a dramatic effect upon the environment and will be discussed in a later chapter (see Part Two). Two of the effects may be to cause a switch in the vegetation cover or, in some cases, to develop into a lake when the water table is restored to its original level.

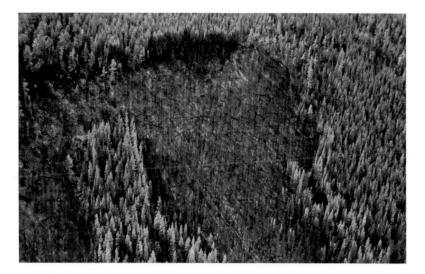

Figure 1.11   Mosaic burn in Yellowstone National Park, USA after fire. This shows the patchy nature of a fire. North side of Willow Creek Fire, August 1974. (www.nps.gov/features/yell/slidefikle/fire/wildfire 1358).

While peat fires may occur naturally in some ecosystems, it is the action of humans that creates particular problems. Peatland drainage, such as is seen in Kalimantan, Indonesia, may create the conditions for peat fires (unintentional or intentional) to take hold and may produce significant emissions (Figure 1.7 G) (Page *et al.*, 2002, 2009; see also Parts Two and Three).

Recent studies of biomass burning across peat areas in Indonesia have shown the release of up to 2.5 Gt of carbon per year, which is significant to the global carbon budget. Unlike in other systems, where perturbations in the system can be reversed, the habitat may be permanently changed within these tropical peat systems. These peat-burning fires may also create a significant smoke hazard that is deleterious to human health (Johnston *et al.*, 2012; see Figure 1.12).

Temperatures within peat fires may be very variable. Most of these are smouldering fires, and temperatures reached may be up to 900 °C (Rein, 2013). Often, combustion is near complete, with a total conversion of the biomass to CO_2. This is important, as combustion completeness is a significant factor in the calculation of carbon emissions from wildfire.

1.7   Fire effects on soils

A significant impact on the environment relates to the temperatures that occur within a fire. It is important to stress that how and when a temperature is measured is particularly significant. Temperature figures are often quoted that have very little relevance to the issue being discussed. For example, temperatures within crown fires may bear little relation to the effect of surface fires upon soil systems. In addition, the length of exposure to a particular temperature may also be important (Table 1.2), as it may create increased temperatures at different depths (Figure 1.13). In the context, therefore, of the fire effects on soil, the temperature reached in surface fires is particularly significant (Ubeda and Outeiro, 2009; see Table 1.2).

There are three important concepts that need a brief introduction here:

- First, the term 'fire intensity' refers to the amount of energy released from a fire, usually expressed as the amount of energy per unit flame length (in kW/m).
- Second, the term 'fire severity' is related to the loss or damage to vegetation that may be determined through either ground observations or by satellite data.
- Third, related to this is 'burn severity', which is often measured by the loss of organic matter in the soil and is, therefore, related to the impacts of the fire on a range of properties of the ecosystem that are concerned with the land surface.

We have already seen that both the nature of the fuel (wood/grass/dry litter, living trees), as well as its

Figure 1.12 Smoke from peat fires in Indonesia, seen from NASA's Terra satellite on September 15 2009. Individual fire areas shown with red spots (http://earthobservatory.nasa.gov/NaturalHazards/view.php?id=40182).

structure (closely or sparsely packed) and biomass, may influence fire temperature. Instead of considering the fire temperature above the burning fuel (the flame tip temperature and that measured above 1–3 m above the flames may also be significantly different), the temperature at the fuel burn-soil interface and within the soil layer needs to be considered.

The temperature reached at the soil-burn interface will depend upon a number of factors. Importantly, these include fuel load, which has implication for forest management and prescribed burns (see Chapter 16, Part Four). A significantly high surface fuel load may create the conditions for a fire not only to burn hotter

but also to remain at a particular point longer, which can impact upon soil temperatures (Figure 1.13A).

The condition of the fuel within a soil may also be important, as combustion occurs differently in wet fuels, as opposed to dry fuels. The type and distribution of the fuel load should also be considered – whether it comprises fine fuels or slash logs, for example. Finally, different species of plants have different calorific values and mineral contents, and some contain resins or other volatile compounds that may cause a fire to burn hotter (Figure 1.13B). The ecological consequence of burn temperature will be considered in a later chapter (Chapter 8, Part Two). Two

Table 1.2   Soil temperatures measured during fires. (From Ubeda and Outeiro, 2009).

Temp (°C)	Depth (cm)	Vegetation
135	0.32–0.64	Pines
550	Surface	Grassland
250	2.5	Dense forest
105	7.5	
60	15	
538	Surface	Scrubland
149	3.8	
700	Surface	Savanna
438	Surface	Conifers
27	3	
17	7	
590	Surface	Scrubland
399	1	
177	Surface	Grassland
93	1.3	
1150	Surface	Pines
500	3	
900	Surface	Eucalyptus
100	5	
400–200	Surface	Shrubs
510	Surface	Dense forest
44		
666	Surface	Eucalyptus
112		
245	Surface	Grassland
68	1.3	
716	Surface	Dense forest
166	2.5	(by afternoon)
66	5	
316	Surface	Dense forest
66	2.5	(by night)
43	7.6	
93	Surface	Pines
800	Surface	Scrubland
500	1	
250	Surface	Scrubland
125	2.5	
50	5	
700–250	Surface	Scrubland
200–90	2.5	
250	Surface	Different kinds
100	2	Black ashes
500–750	Surface	Different kinds

Table 1.2   (*Continued*)

Temp (°C)	Depth (cm)	Vegetation
350–450	2	
150–300	3	
< 100	5	
388–442	Surface	Masticated fuel
170–330	Surface	Wheat
700	Surface	Pines
300	15	
340	Surface	Pines
740	Surface	*Cistus*
280	Surface	No vegetation
180	Surface	Scrubland
50	2.5	
475	Surface	Low dense
90	2.5	Scrubland
40	5	
600	Surface	Grassland
50	1	
702	Surface	*Ulex parviflorus*
22	5	

Source: Reproduced by permission of Taylor & Francis.

areas will be discussed here: the conversion of the plant biomass to ash (i.e. both mineral ash, charcoal and other pyrolysis products – see Santín *et al.*, 2012); and temperature reached and its impact within the mineral soil profile.

An important feature of soils is what is termed soil hydrophobicity (DeBano, 2000). In simplest terms, this relates to how quickly or slowly water infiltrates a soil. Hydrophobic soils are ones where drops of water remain on the soil surface for a considerable time. Techniques have been developed to measure this by timing the absorption of a measured water droplet into a soil. When water falls upon a soil surface, it may be absorbed into the top surface of the soil or flow off the surface by overland flow. Equally, the water may be absorbed into the upper layers of a soil but then flow laterally within the soil profile.

The burning of the surface fuels by a surface fire has two major impacts. First, there is the temperature rise within the topmost layer of the mineral soil. Organic matter may combust and the soil structure altered. This may be seen by a change in colour, from browns and blacks to yellows, oranges and reds (Figure 1.18A). In some extreme cases, some clays may be baked into red

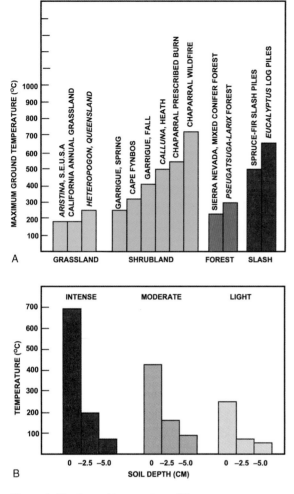

Figure 1.13   Ground temperature of fires.

A. Maximum ground temperatures reached during natural wildfires (modified from Rundel, (1981)). Reproduced with permission from Springer Science+Business Media B.V.
B. Profiles of maximum temperature with soil depth for three intensities of chaparral fires (after DeBano *et al.*, 1977). USDA.

brick-like fragments. The pyrolysis of the organic matter may generate liquids that may be freed into the soil layers coating the soil particles.

The effect of these changes is to change or introduce a hydrophobic layer. The strength and position of the hydrophobic layer plays a major role in post-fire erosion (Figures 1.14, 1.15). Clearly, the impact of a surface fire upon soil properties depends on a large array of variables and these are subject to considerable on-going research (see Doerr and Shakesby, 2013).

1.8   Post-fire erosion-deposition

The impact of a surface fire upon the landscape may be severe. This will depend on three principal factors:

- the amount of fuel built-up on the land surface;
- the intensity and severity of the fire; and
- the timing between the fire and the first rainstorm.

Surface fires with minimal fuels may burn at relatively low temperatures. Their speed will depend on a range of factors, including the wind speed. A fast-moving low-temperature surface fire may have little effect upon the underlying soil or plant roots. A build-up of fuel may mean that the fire burns hotter, remaining at a single site for longer, despite the fire front moving and have a greater effect on the soil.

The fire may play two major functions. It may impact on the binding of soil particles by destruction of their organic content and, more importantly, it may have an impact on soil hydrophobicity. We have seen how a fire may enhance soil hydrophobicity by primarily precipitating volatile compounds within the soil. The effect is that the uppermost layer of the soil becomes more porous, but a more water-impenetrable layer forms beneath (Figure 1.14, 1.15).

The impact of a fire upon the soil may also depend on the nature of the soil itself, so that immature granite soils that contain rock fragments and little organic matter, for example, may behave differently from mature sandy or clay soils or those with a high organic content. Fire can remove organic chemicals leached from some vegetation types that make soils hydrophobic.

If there is a delay between the fire and significant rainfall, this may allow some plants to re-sprout. This may be helped by slight rainfall where, in some settings, ferns and grasses may re-sprout within a few days or weeks of fire (Figure 1.16). This will help in the stabilization of the soil and will prevent its movement and removal from the environment. Even if there is significant ash production (including mineral ash and charcoal), growth of plants may stabilize the sediment.

If, however, there is a significant rainstorm soon after the fire, this may have a devastating effect upon the environment (Figure 1.17). This is particularly the case if there is any significant topography. The rain in such circumstances will not fall on living vegetation

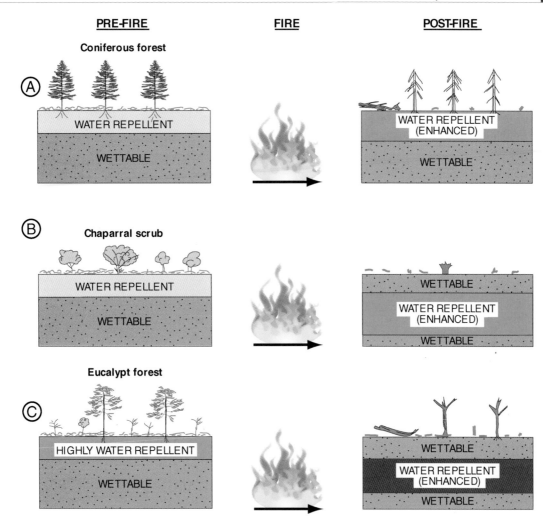

PRE-FIRE	FIRE	POST-FIRE

Coniferous forest

(A)
WATER REPELLENT
WETTABLE

WATER REPELLENT (ENHANCED)
WETTABLE

(B) **Chaparral scrub**
WATER REPELLENT
WETTABLE

WETTABLE
WATER REPELLENT (ENHANCED)
WETTABLE

Eucalypt forest

(C)
HIGHLY WATER REPELLENT
WETTABLE

WETTABLE
WATER REPELLENT (ENHANCED)
WETTABLE

Figure 1.14 Soil water repellency changes following fire of moderate or high severity for:

A. coniferous forest in the north-western United States;
B. Californian chaparral;
C. Australian eucalypt forest.

Darker shading represents more severe repellency.
(After Doerr *et al.*, 2009). Reproduced by permission of Stefan Doerr.

and be absorbed by the plants, but will fall instead upon a bare landscape. We can consider if this landscape comprises bare soil, where there is little ash residue, or if there is a significant layer of ash and charcoal.

In zones of high burn severity, there may be little residue of ash/charcoal on the soil surface. The rain will tend to pound on the soil surfaces, as the presence of a strong hydrophobic layer may prevent rapid infiltration of the water (Figure 1.18).

The water will, therefore, move by overland flow. The hydrophobic layer, however, may be discontinuous, so that water may penetrate beneath the layer. In such circumstances, small rills are produced that may widen into very large erosional channels (Figure 1.19). Sediment may move very quickly and in large quantity (Moody and Martin, 2009).

In cases where there is significant topography, this may not only trigger the erosion of large catastrophic channels, but sediment may come to rest at a change

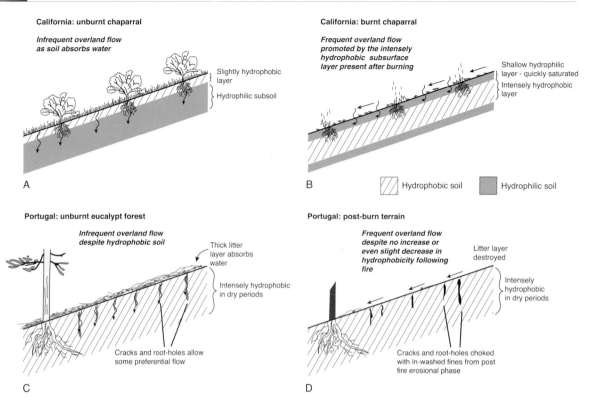

California: unb, unburnt chaparral

*Infrequent overland flow
as soil absorbs water*

Slightly hydrophobic layer

Hydrophilic subsoil

A

California: burnt chaparral

*Frequent overland flow
promoted by the intensely
hydrophobic subsurface
layer present after burning*

Shallow hydrophilic layer - quickly saturated

Intensely hydrophobic layer

B

Hydrophobic soil   Hydrophilic soil

Portugal: unburnt eucalypt forest

*Infrequent overland flow
despite hydrophobic soil*

Thick litter layer absorbs water

Intensely hydrophobic in dry periods

Cracks and root-holes allow some preferential flow

C

Portugal: post-burn terrain

*Frequent overland flow
despite no increase or
even slight decrease in
hydrophobicity following
fire*

Litter layer destroyed

Intensely hydrophobic in dry periods

Cracks and root-holes choked with in-washed fines from post fire erosional phase

D

Figure 1.15 Impact of rainfall on a burnt soil. Hydrological response of forested terrain with high natural levels of coil water repellency for (A) unburnt and (B) burnt conditions following fire in California chaparral and (C, D) in eucalypt forest in Portugal. (From Doerr *et al.*, 2000). Reproduced with permission from Elsevier.

Figure 1.16 Burning and recovery of heather heathland in southern England (see Scott *et al.*, 2000).

A. Frensham, Surrey after fire showing re-growth of ferns and grass after two weeks (photo: A. C. Scott). Reproduced with permission from Elsevier.

B. Same area after ten years of re-growth of heather (*Calluna*) (photo: A. C. Scott). Reproduced with permission from Elsevier.

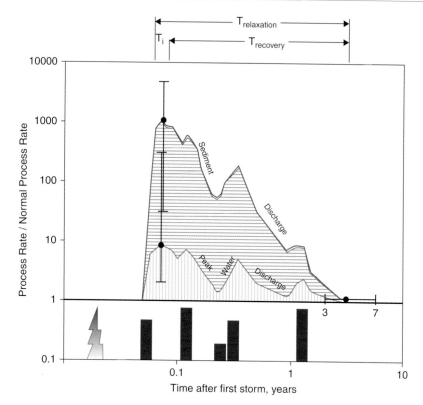

Figure 1.17 Conceptionalization of the response for one fire-flood cycle. The pre-fire process rates have a magnitude of 1.0, so that the process rates for peak water discharge and sediment transport are relative rates. The time of the fire is shown by the 'lightning bolt' (modified from Moody and Martin, 2009). Reproduced with permission from Taylor & Francis.

of slope in the form of an alluvial fan (Figures 1.19, 1.20). Fan thicknesses of more than one metre may form following single thunderstorms events.

This sediment, now released into the depositional system, may be fluvially reworked and be deposited many kilometres downstream, or it may find its way into lakes or even on to flood plains (Figure 1.21). In some cases, it may even reach the sea.

If there is a significant ash/charcoal layer, this may also have an impact upon the erosion depositional cycle. It has been suggested recently that a thick layer of ash may be effective in reducing hill-slope responses to surface run-off and erosion following a fire. The ash may develop a significant storage capacity. In some cases, the charcoal in the ash will float and may be transported both within the burnt area and out of the burnt area by overland flow (Figure 1.22).

The nature of the ash produced by the fire may also be significant in a range of geomorphic processes (Doerr and Shakesby, 2013). Ash may include various

types, from black to grey to white ash that may be composed of a range of soluble and insoluble materials, from pure mineral ash (i.e. the organic materials have been largely combusted, but inorganic compounds that were present in the plant remain after combustion, leaving a residue that is, therefore, white) to ash that includes a greater amount of organic matter, often in the form of charcoal, and hence is more grey or black in colour. On some burnt surfaces, a layer of grey or black ash may underlie a coating of white ash (Figure 1.23).

Within the burn site, charcoal may be transported into local hollows or be dammed up against fallen trees (Figure 1.24).

Extensive water run-off may transport sediment, charcoal and even uncharred litter, as a slurry, out of the burn area (Figure 1.22). This charcoal-rich sediment may be deposited many kilometres from the burn site, in some cases filling channels or even lakes (Figure 1.22). A range of studies has shown

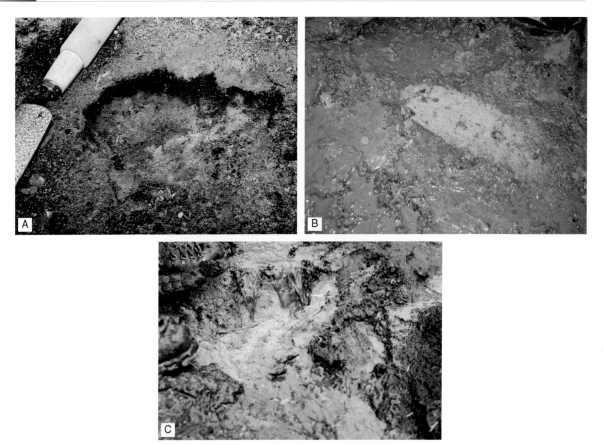

Figure 1.18 A burnt soil surface and the impact of rain after fire in Canadian Conifer forest. (Photos: D. F. Scott). Photos were taken in Okanagan Mountain Park, Kelowna, BC during a light rain event in October 2003, roughly six weeks after the passage of a late summer wildfire through old mountain pine forest (predominantly *Pinus contorta*). Dead and down fuel loads were high as the area was long unburnt and unharvested.

A. Burnt soil surface.
B. Water saturated upper layer with un-wettable hydrophobic layer beneath.
C. Water moving by overland flow eroding gulley.

that depositional rates, in some deposits, and setting may increase more than 30 fold (Swanson, 1981; Moody and Martin, 2009). As yet, we have rather limited understanding of this transport depositional process or how to recognize how it has occurred from the nature of the sediments. The occurrence of charcoal, especially in sediments that have been rapidly deposited, may provide a significant clue.

The significance of fire as a trigger for alluvial fan deposition was first studied in sediments from Yellowstone National Park in the USA. Further investigations in the historic records there and elsewhere, has demonstrated the widespread nature of this phenomenon

(Meyer and Pierce, 2003; Cannon *et al.*, 2001). Most studies have involved forested ecosystems, and there is a need for more studies in grassy ecosystems. There is a growing understanding that not only the vegetation, but also the hydrology and hill slope, will play a part in the erosion deposition process (Figure 1.21).

1.9   Fire and vegetation

Another key element in wildfire is the nature of the vegetation and fuel. There are fundamental differences in fires within forests, for example, and those in grasslands. The different vegetation types result not

Figure 1.19 Geomorphic impacts of fire.

A. Channel incised after the Cerro Grande Fire near Los Alamos, New Mexico, USA, 2003.
B. Wide gully eroded after Buffalo Creek Fire, Colorado, USA, 1996.
C. Alluvial fan created following Buffalo Creek Fire and first rainstorm, Colorado, USA, 1996.

(All photos: John Moody, USGS).

only in different fuel loads and flammability, but also in the influence of wind and speed of fire spread. Even fires within similar types of vegetation (e.g. shrubs) will differ because of both fuel structure and calorific value of the fuel.

Shrubs and grasses are burnt predominantly by surface fires (Figure 1.7). However, it is useful to know that Mediterranean-type shrublands (e.g. chaparral, fynbos, matorral) are usually described as crown fire systems (Figures 1.7, 1.8). The build-up and moisture content of the fuel is critical in how they burn. In grasslands, fuel may grow rapidly and dry dead fuel may accumulate quickly in significant quantities. Fires in such vegetation may be of cool to moderate temperatures (less than 600 °C) and move rapidly through the vegetation. Because of this, the roots of the plants may not be killed, nor the seeds in the soil bank destroyed. A key for grassland fires relates to the fire return interval (FRI) (see below), whereby any small tree sapling could be killed if the FRI is less than ten years. If there is too long between fires, it is possible that the nature of the vegetation will change to growth of shrubs and trees, thereby changing the nature of subsequent fires (Bond *et al.*, 2005).

Figure 1.20 Impact of the Buffalo Creek Fire, Colorado, USA, 1996.

A. Area after fire before first rain storm (June 1996).
B. Area as above but after first rain storms, showing large area of sediment accumulation (16 July 1996).

(All photos: John Moody, USGS).

Ground cover vegetation, ranging from ferns through to shrubs, will also burn by surface fires. Heather heathland in Europe is a prime example of this (Scott *et al.*, 2000, Figure 1.16). In such cases, much of the fuel comes from the living vegetation rather than a build-up of litter. In other cases, such as chaparral, the fuel load comprises a combination of dead and fine fuel. This vegetation, however, burns particularly hot and may be spread by stormy winds.

As we have seen, though, the important result of a fire relates to its intensity and severity. Fire must be considered in the context of the ecosystem, as pointed out by Keeley *et al.* (2012) and, hence, the factors involved may be thought of in terms of a fire diamond (Figure 1.25).

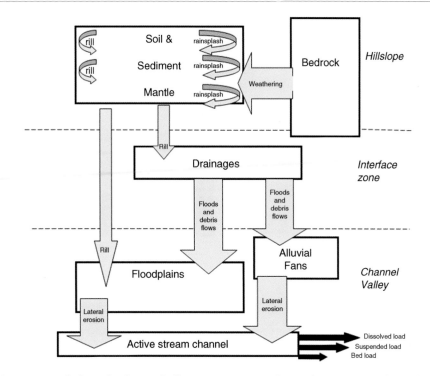

Figure 1.21 Components of a burnt landscape. Sediment storage reservoirs are shown as rectangles and transfer processes as grey-green arrows. (Modified from Moody and Martin, 2009). Reproduced by permission of Taylor & Francis.

Here, not only is the amount and distribution of the fuel important, but also climate seasonality. Within forest systems, the fires may be more complex. The majority of wildfires in forests start as surface fires that consume the on-ground fuel load, typically comprised of dead litter, logs and duff, as well as living herbaceous plants and shrubs (Figure 1.8). This fuel build-up is key. If surface fires burn regularly through many types of forest ecosystems (e.g. ponderosa pine forests of western North America), then the temperature and flame height may be insufficient for the fire to either kill the roots of the plants or spread into the canopy layer as a crown fire.

Key elements in the nature of the surface fuels will be not only the moisture content of the plants but also the size, distribution and type of fuel available to burn. Both from field and laboratory observations, it can be shown that logs, twigs and leaves have different flammability and different rates of drying. Even different shapes and sizes of leaves can ignite differently and can also produce different fire spread rates (Belcher et al., 2010a). Obviously, there are some surface fires within woodlands or forests that may

have a significant impact upon that ecosystem, but in forests where fires are a regular feature, such as in ponderosa pine forests of western North America, regular burning of surface fuel is important in preventing fuel load build-up. Consequently, a surface fire may spread into the tree canopy as a crown fire if the surface fuel loads are very high (Figure 1.7; Roos and Swetnam, 2012).

Where there is a significant build-up of soil litter, an organic-rich soil layer may form. In extreme circumstances, such as in a normally waterlogged environment (e.g. mire), this organic-rich layer may dry up and a fire may not only burn surface litter but also begin to burn humus or peat layers in a ground fire. In some cases, these fires may change from quick-moving flaming surface fires to slow-moving smouldering fires that are relatively starved of oxygen. These fires may last hours, days, months or even years after the surface fire has moved away and been extinguished (Rein, 2013). Putting out such fires may be problematic, as exposing the fire to air may induce flaming and, in some cases, water may only act to introduce oxygen into the system, making the fire worse rather than

Figure 1.22   Moss movement and deposition of charcoal following fire and rain.

- A. Charcoal-rich flows after rainstorm, from a forest fire by overland flow from the Rodeo-Chediski Fire, Apache-Sitgreaves National Forest, Arizona, USA, 2002 (photo: D. Neary, US Forest Service).
- B. Charcoal-rich flows after rainstorm, from a forest fire by overland flow from the Rodeo-Chediski Fire, Apache-Sitgreaves National Forest, Arizona, USA, 2002, and into a nearby river (photo: D. Neary, US Forest Service).
- C. Charcoal filling channel from the Hayman fire, Colorado, USA, 2002 (photo: Greg Smith, USGS).

(From Scott, 2010).

putting it out. An extreme example of such fires are underground lignite and coal fires, which can smoulder for decades and cover thousands of square kilometres (Figure 1.26).

As will be shown later, the frequent occurrence of surface fires in some forest ecosystems has led to the evolution of some fire-resistant traits in some plants and even, in some cases, the need for fire as part of their reproductive strategy. With some plants, features of their growth and composition may even promote fire. Most eucalypts, for example, have evolved a re-sprouting ability from root stocks, so even if all of the main above-ground biomass is consumed in the fire,

the plant will still survive (see Chapter 7, Part Two for a discussion of flammable traits).

Different tree shapes, densities and even wood type may all affect the nature and type of a fire. This wide range of variables makes the prediction of wildfire occurrence and behaviour very difficult (Chapter 15, Part Four).

1.10   Fire and climate

However much fuel build-up there is, it is climate that provides one of the ultimate controls on wildfire

Figure 1.23   Charcoal and ash from wildfire in forest, Colorado, USA (from Scott, 2010).

A. Ash on forest floor from the Overland fire (2003). Reproduced by permission of Elsevier.
B. Pale upper ash layer and thick dark charcoal layer beneath (photo: A. C. Scott).
C. Area with only charcoal-rich ash on forest floor from the Hayman fire, 2002. Quadrat is 20 × 20 cm. Reproduced by permission of Elsevier.

(Figure 1.1), both in the ignition and spread, as well as in the extinguishing of, a major fire (see Chapter 9, section 9.16, Part Two). As we have seen, fuel moisture content is critical for fire ignition through a lightning strike (or even through a human agency). Simply wet plants do not burn (at least under the present oxygen composition of the atmosphere). A lengthy hot-dry period has two effects: to dry out the dead fuel making it available to burn and to reduce the moisture content of living vegetation to allow at least the finer leaf/twig fraction to more easily combust (Figure 1.19).

However, one of the observations from the Yellowstone fires of 1988 was that living plants were fully hydrated, despite long preceding drought. Fire risk will therefore be a combination of the condition of fuel, the moisture content of the air and also the temperature. There is no doubt that hotter temperatures promote increased fire. Aspects of fire weather are detailed later in this book (Chapter 14, Part Four), but it is important to grasp that climate (and hence climate change) plays a particularly significant role in wildfire activity. Small changes in the timing of snow melt or summer length may have a profound affect upon the occurrence or length of a fire season. Such small climate shifts have been noted in western North America, which has been claimed to lead to a significant increase in wildfire activity (Westerling et al., 2006; Figure 1.27).

In parts of the world where fire is not a common element in the natural vegetation, periods of drought can lead to an increase in wildfire activity (e.g. southern England), even if a number of the fires are started

Figure 1.24   Charcoal from surface fires in heathland and conifer forest.

A. Charcoal after fire across heathland in southern England (from Scott *et al.*, 2000). Reproduced by permission of Elsevier.
B. Charcoal from Hayman fire, Colorado, USA (2002). Scale is 1 cm (from Scott, 2010). Reproduced by permission of Elsevier.
C. Charcoal washed up against downed tree trunk that is acting as a trap. Hayman fire, Colorado, USA, 2002 (from Scott, 2010). Reproduced by permission of Elsevier.

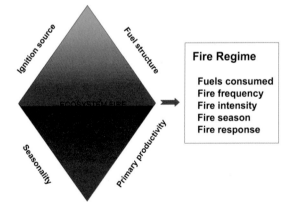

Figure 1.25   Ecosystem fire. Fire diamond schematic of factors necessary and sufficient for predicting the distribution of fire as an ecosystem process. Fire regimes are controlled strongly by ignition frequency and fuel structure, with important feedback loops between all four factors shown on the sides of the diamond. (Modified by J. Keeley from Keeley *et al.*, 2012. Courtesy USGS)

by human activity. Strong winds can fan fires, even under mild temperatures.

An additional impact of climate are the southern oscillations, such as El Niño/La Niña cycles. This leads to periods of more drought and increased fire activity, such as seen in the south-western USA (Swetnam and Betancourt, 1990). Such weather cycles also affect other regions, such as Indonesia, where a combination of the human draining of peat lands and exceptionally dry years led to extensive peat fires such as those seen in 1983 and 1992 (Page *et al.*, 2002, 2009).

Together, therefore, climate and weather play a major role in wildfire activity. Fires, however, unlike other disturbances such as floods, cyclones, but very like herbivores in 'eating' vegetation, play a major role in selecting for particular plant traits (Bond and Keeley, 2005). Without the organic matter, fires would not exist. We emphasize the importance of the biology in the fire ecology chapters later in this book (see Part Two).

Figure 1.26 Russian coal fires. (Photos A. C. Scott).

A. Coal fire, Permian coals, Kusnetsk Basin, Siberia, Russia.
B. Burnt coal after coal fire, Permian coals, Kusnetsk Basin, Siberia, Russia.

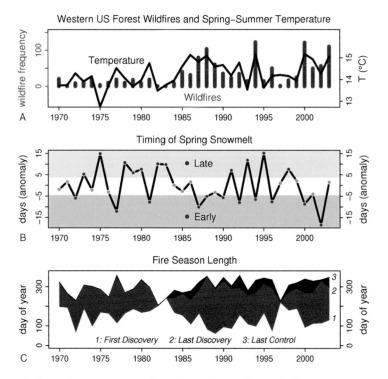

Figure 1.27 Western USA forest fires 1970–2005. (From Westerling *et al.*, 2006). Courtesy TW Swetnam.

A. Annual frequency of large (9400 ha) western US forest wildfires (bars) and mean March through August temperature for the western United States (line).
B. First principal component of centre timing of streamflow in snowmelt-dominated streams (line). Low (pink shading), middle (no shading) and high (light blue shading) tercile values indicate early, mid- and late timing of spring snowmelt, respectively.
C. Annual time between first and last large fire ignition, and last large fire control.

1.11   Fire triangles

Traditionally, a range of characteristics of fires has been expressed in the form of fire triangles (Figure 1.1). These have the advantages of helping to simplify the fundamental aspects of fire, but the disadvantage in some cases is restriction of other elements. In this chapter, so far, we have considered two of these triangles. The first considers plant combustion, with fuel, oxygen and heat being the key elements. Lightning is a source of heat and the fuel is ignited, but oxygen availability is key to fire spread. The second triangle fire considers fundamental aspects of fire spread. This considers the fuel, its condition as determined by weather, and also topographic variation.

Yet these triangles are not sufficient to fully define all of the variables of fire (see, for example, the fire diamond in Figure 1.25). As will be discussed later, there is also a time dimension to consider (see Chapters 4 and 5). In today's world there is increasing interest in modelling fire both on a local and global scale. New approaches of fire science add aspects of scale and a consideration of the 'mega-fire' (Figure 1.1). These approaches have been driven by the need to understand the longer-term aspect of fire suppression. An important additional element in our discussion concerns the fire return interval.

1.12   Fire return intervals

A key element in understanding fire is the concept of fire return interval. This is the re-burning of vegetation that has been burnt at some time in the past. The significance of this factor is manifold, but it has particular importance with regard to fire intensity or severity, and also in ecological maintenance or succession. There is a complete spectrum between fire return intervals, from less than ten years in many grassland savannas to several hundreds, or even thousands, of years in some forested ecosystems (Table 1.3).

The fire return interval may be influenced both by the vegetation type and climate regime. In addition, the resistance of vegetation to fire may also play an

Table 1.3   Fire return intervals. FRI has been classified using a number of scales (see Pyne *et al.*, 1996) and here that of Davis and Mutch (1994) is followed. (Modified from Pyne *et al.*, 1996).

Class A	Natural fires rare or absent.
	Examples: Wetter regions of Eastern USA deciduous forest; Southern USA deserts; Inland cypress communities of Florida.
Class B	Infrequent, low intensity surface fires with more than a 25-year return interval. Most fires are small.
	Examples: Eastern USA deciduous forest; Subalpine forests; Pinyon-juniper woodlands of USA and some montane meadows in western USA.
Class C	Frequent low intensity surface fires with 1- to 25-year return intervals for small areas. They are often combined with sporadic, small-scale fires with long or very long return fires and/or high intensity surface fires with a 200- to 1000–year interval. Typical burns are a few hundred to a few thousand acres.
	Examples: Sierra mixed conifer forests with giant sequoia, ponderosa pines etc, western USA. Prairies of USA; Sawgrass everglades of Florida.
Class D	Infrequent, severe surface fires with more than a 25 -year interval. These are usually combined with long return intervals of 300 years, sporadic crown fires, and/or higher intensity surface fires killing most but not all stand elements. Many fires can cover large areas from 1000 to 10 000 acres with some portions of still larger crown fires that belong to other regimes.
	Examples: White and red pine forests of Canada; Lodgepole pine forests of Rockies; Redwoods in northern California; Pinyon-pine and mixed juniper communities in USA.
Class E	Shorter to medium length return interval crown fires and/or stand-killing, high-intensity surface and ground fires with 25- 100- yearr return intervals.
	Most stand elements are killed over large areas. Ecologically significant fires are generally 5000 to 10 000 acres or more in area.
	Examples: Boreal forest of Eastern Canada; Coastal chaparral in California.
Class F	Long and very long return intervals sporadic crown fires and high intensity surface fires that kill most but not all stand trees, often over large areas. Return intervals are often 100–300 years or more and probably are longest at higher elevations. Fire areas are often 5000 to 50,000 acres, but smaller in subalpine forests.
	Examples: Most Douglas fir, and red cedar forests in western USA and Canada; High sub-alpine forest of the Rockies; Rainforest of Hawaii; Coastal redwood forests.

important role. In a grassland savanna ecosystem, for example, rapid fuel build-up, together with periods of drought, encourages frequent fires. These fires, however, generally burn rapidly and do not affect the soil seed bank, so that the vegetation may quickly regenerate. The fire may also have a flame height that may kill any tree saplings. In such circumstances, if there is no fire at the same location for more than ten years, the trees may grow to a height where they are not killed by the fire. In such a case, the vegetation may be transformed from a savanna to a woodland/forest (Bond *et al.*, 2005; see Chapters 7, 8, Part Two).

Frequent burning of some shrubby vegetation, such as heather heathland in Europe, may also help in the maintenance of the ecosystem. In some ponderosa pine forests in western North America, the return of surface fires at regular intervals may maintain the ecosystem, as these fires are often of low intensity and severity, burning only dead surface fuels. In such systems, if the fire return interval is too long, the build-up of fuel will allow for both a more intense surface fire and possibly a stand-replacing crown fire. In many temperate forest ecosystems, fire return intervals may vary between 40–400 years (Roos and Swetnam, 2012).

Some ecosystems may become vulnerable to devastating mega-fires on the sites of recently burnt areas, such as in California. In such cases, the initial fire may have killed, but not consumed, all of the woody vegetation. The dead trees and shrubs will then become a large fuel load for a fire if the return interval is too short. Table 1.3 shows typical fire return intervals for a range of vegetation types.

Climate change may play an important role in altering fire intervals. However, humans may also exert an influence, both in setting fires in vegetation that rarely experiences fire naturally (e.g. Indonesian peats) or in suppressing fire (e.g. the forests of western USA) (Parts Two and Three) in which case a future fire may be much larger and more damaging. A significant new threat is that of plant invasives. This is where plants (often, but not always, grasses such as Gamba and Cheat grass) has been introduced and spread (see Part Two). The spread of flammable grasses into a vegetation that rarely burns, or with a long fire return interval, can be devastating. Equally, these plants may change the fire frequency or timing of a fire, which also may have a major impact on the native ecosystem (Balch *et al.*, 2013).

1.13 How we study fire: satellites

For much of the past two centuries, our understanding of fire has come from direct on-the-ground observations and manage reports following wildfires. In such cases, observations may be limited because of both access and safety considerations. In some cases, awareness of a fire has come only after a fire has been extinguished. Over the past 50 years, there has been an increase in aerial observations of wildfire, often as a result of both aerial dropping of fire retardants and from firefighters in the air. However, even in such cases, direct observations of fires may be made much more difficult because of smoke plumes (see Figure 1.6).

There are several important drivers for the need for a range of broader scale observations of fires. These include the need to access fire damage and impact upon the land environment, and also to access the atmospheric impact of a fire, including smoke, aerosol and as emissions. For many years, aerial photography of burn sites, together with surface observations, have provided much valuable data, especially concerning burn severity. Burn severity assessments are, for example, regularly carried out by the US Forest Service (Figure 1.28).

In recent years, satellite monitoring of fires and effects, not just locally but also on a global scale, has proved invaluable in advancing the understanding of wildfire (Figure 1.29). Early assessments of global wildfire activity were undertaken using Landsat imaging techniques (Figure 1.30). These were undertaken at a variety of scales, using both normal light photography and imagery using other bands, such as infrared. Such techniques have proven very useful in assessing not only the extent of a wildfire, but also the severity of the fire, as the occurrence of dead vegetation may become particularly visible using non-visible spectra (Graham, 2003).

Satellite images of fires over parts of the globe have become commonplace and dramatic (Figure 1.31). The 1980s saw the development of the Advanced Very High Resolution Radiometer (AVHRR) that is used to scan the Earth's surface (Table 1.4).

There are a number of channels that can be used to measure different wavelengths, too – for example, determining smoke plumes from fires, as well as the occurrence of fire, using thermal infrared data from which temperatures can be derived. The latest instrument version is AVHRR/3, with six channels, which

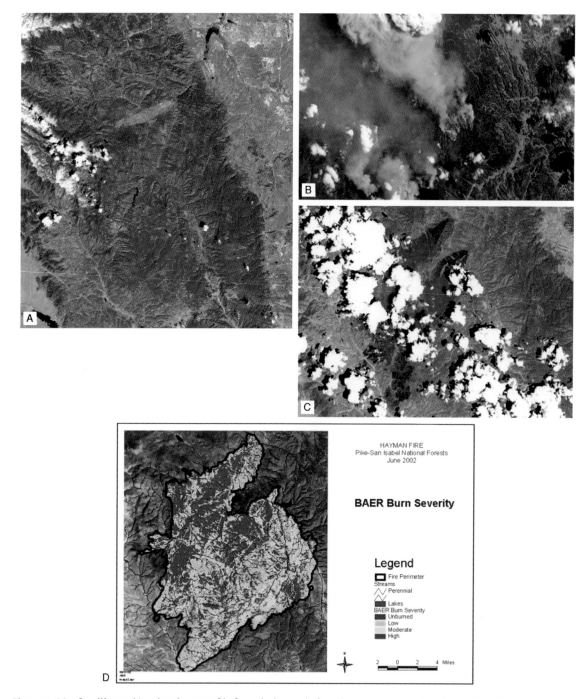

Figure 1.28 Satellite and Landsat images of before, during and after the 2002 Hayman Fire, Colorado, USA.

A. Landsat image of the Hayman Fire area in early 2002 before the fire. Vegetation appears red in the false-colour image (http://earthobservatory.nasa.gov). NASA.

B. Photo taken by in International Space Station crew on Tuesday 19 June 2002, showing the eastern flank of the Hayman fire (http://earthobservatory.nasa.gov). NASA.

C. Landsat image of burn scar. Clouds are shown in white. The burnt area shows in black, June 16 2002. (http://earthobservatory.nasa.gov). NASA.

D. Burned Area Emergency Response (BAER) burn severity map of the Hayman fire (from Graham, 2003). USDA.

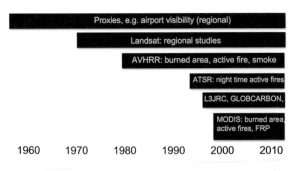

Figure 1.29 Development of the remote sensing of fire showing satellites and products. (Data from G. van der Werf).

was launched in 1998. In all such data gathering equipment, it is not simply the obtaining of the data that is important but also its processing into usable products (Roy *et al.*, 2013). For example, a database of fire activity in Russia was derived from 1 km resolution remote sensory imaging using AVHRR that was itself largely derived from active fire observations.

There has been an increasing need to monitor active fires, both during the day and at night, and to build databases to examine fire frequency within

Table 1.4 Satellites and sensors used for monitoring fires

Satellite type	Satellite	Sensor
Polar orbiting	NASA Terra and Aqua	MODIS (Moderate Resolution Imaging Spectroradiometer)
Polar orbiting	NOAA POES	AVHRR (Advanced Very High Resolution Radiometer)
Polar orbiting	ERS	ATSR (Along Track Scanning Radiometer)
Polar orbiting	ENVISAT	AATSR (Advanced Along Track Scanning Radiometer)
Polar orbiting	TRMM	VIRS (Visible and Infrared Scanner)
Polar orbiting	NASA Terra	ASTERR (Advanced Spacebourne Thermal Emission and Reflection Radiometer)
Geostationary	Meteosat SG	SEVIRI (Spinning Enhanced Visible and Infrared Imager)
Geostationary	GOES East and West	GOES (Geostationay Operational Environmental Satellite Imager)

different areas and different types of vegetation, in order to help develop an understanding of how fire may be related both to climate and to human activities (Figure 1.32).

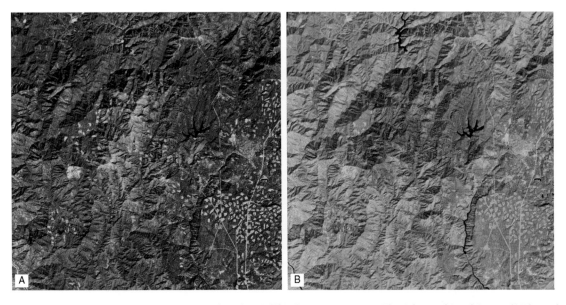

Figure 1.30 Landsat images of fire scar, Bagley Fire, Californian, August 2012. The Advanced Land Image (ALI) on the Earth Observing-1 (EO-1) satellite acquired these images on September 11, 2011.

A. Showing natural colour with the burn scar in brown.
B. Showing the burnt vegetation in red.

(http://earthobservatory.nasa.gov). NASA.

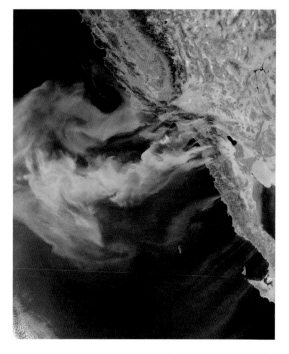

Figure 1.31 Smoke plumes from Californian fires between Los Angeles and San Francisco in October 2007 billowing out over the Pacific Ocean. Red spots indicate active fires. (Modis Rapid Response Project at NASA/GSFC, image 1163886). NASA.

Another important development during the 1990s was the ability to monitor night-time active fires (Figure 1.33).

The European Space Agency launched its satellite ENVISAT in 2002 to continue the work of ERS satellites to monitor environmental and climate changes. ENVISAT flies in a sun-synchronous polar orbit of about 800 km altitude, and the repeat cycle is 35 days. The satellite carries the Advanced Along-Track Scanning Radiometer (AATSR), following on the ATSR-1 and ATSR-2 on board ERS-1 and ERS-2. The AATSR has a resolution of 1 km at nadir and measurements are derived from reflected and emitted radiation: 0.55 μm; 0.66 μm; 0.87 μm; 3.7 μm, 11 μm and 12 μm.

The data from this satellite has been used, for example, to develop a fire atlas. Each year, the fire atlas requires the processing of 80,000 images. All hotspots with a temperature higher than 312 °K at night are precisely located (better than 1 km). Information concerning the fire atlas is available online. The satellite also allows monitoring of smoke base in addition to fire (Figure 1.34). The orbiting space station has also provided both large and smaller scale images of smoke plumes, such as those from human-set peat fires across Indonesia (Figure 1.12).

Other developments were L3JRC and Glob carbon. These are products developed by a consortium of universities and the European Commission to image and classify burnt areas. This data was derived from the Earth Observation system SPOT VEGETATION.

Figure 1.32 Fires across the USA from January 1 to October 31, 2012, as detected by MODIS instruments (http://earthobsevatory.nasa.gov). NASA.

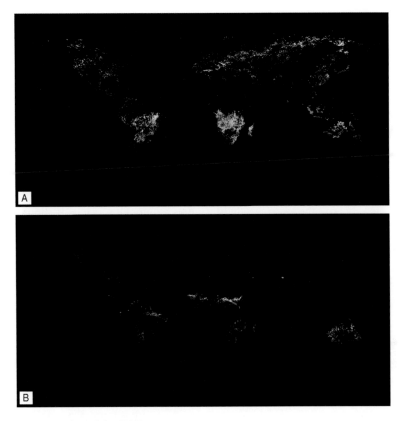

Figure 1.33   Active fires imaged at night. NASA.
A. This image, using MODIS on the Terra satellite, shows the fires seen from Aug 1 - Sept 1 2012.
B. This images shows fires from Nov 1 to Dec 1, 2012 (http://neo.sci.gsfc.nasa.gov).

Figure 1.34   Smoke plumes from fires across Russia on 29 July, 2010, with a resolution of 300 m. This image was taken by the European Space Agency's ENVISAT Earth-observing satellite. Moscow is in the lower left hand corner. (http://esamultimedia.esa.int/images/EarthObservation/images_of_the_week/forest_fires_MoscowMER_FR_20100729_43977.jpg (Aug 10, 2010)). European Space Agency.

Perhaps the most significant development over the past 10–15 years is the development of MODIS products (and FRP). Fire Radiative Power (FRP) is estimated from active fire observations and is the rate at which a fire emits radiant energy. Fire Radiative Energy (FRE) is a temporal integer of FRP over the period of observations. The units are Joules. This data can be used to retrieve fire temperature. In general, the greater the amount of fuel, the greater the amount of energy that is released, and the heat released is also related to both the fuel consumption and the heat content of the fuel.

The Moderate Resolution Imaging Spectra-radiometer (MODIS) is aboard the Terra satellite launched by NASA. The amount of data collected by this satellite is astonishing, and a wide range of products has been developed to utilize this vast quantity of data. For example, the Fire Information for Resource Management System (FIRMS) integrates remote sensing and GIS technologies to deliver global MODIS hotspot fire locations and burnt area information to a variety of end users (Figures 1.35, 1.36, 1.37). This builds upon a mapping interface, *Web Fire Mapper*, which delivers near real-time hotspot-fire information and monthly burnt area information (see van der Werf *et al.*, 2006).

It is possible to receive a fire alert from the NASA data site. It should be noted that this fire data relates to fires within 1 km pixels. It should also be noted that this fire data might be viewed using Google Earth, but size restrictions may mean that not all the data is

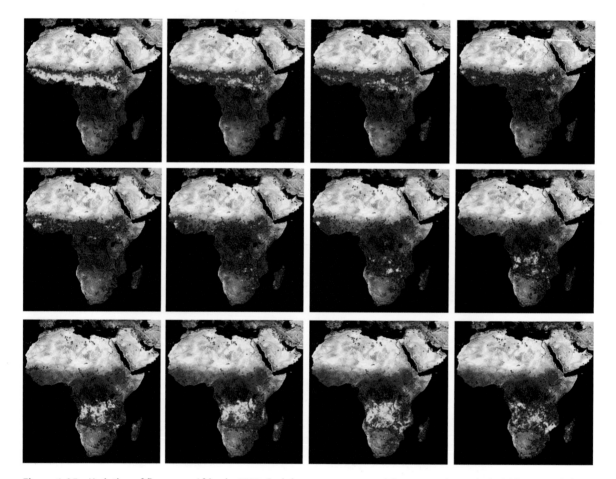

Figure 1.35 Variation of fire across Africa in 2005. Each image was processed from a ten-day period within a month from Jan to August (Jan 1–10; Jan 21–30; Feb 10–19; Mar 2–11; Mar 22–31; April 11–20; May 1–10; May 20–30; Jun 10–19; Jun 30–Jul 9; Jul 20–29; Aug 9–18). Successive images show the shift of fire through the year. Red indicates a single fire in an area over the period and yellow indicates more. These images were processed using data from MODIS on NASA's Terra and Aqua satellites. (http://earthobsevatory.nasa.gov). NASA.

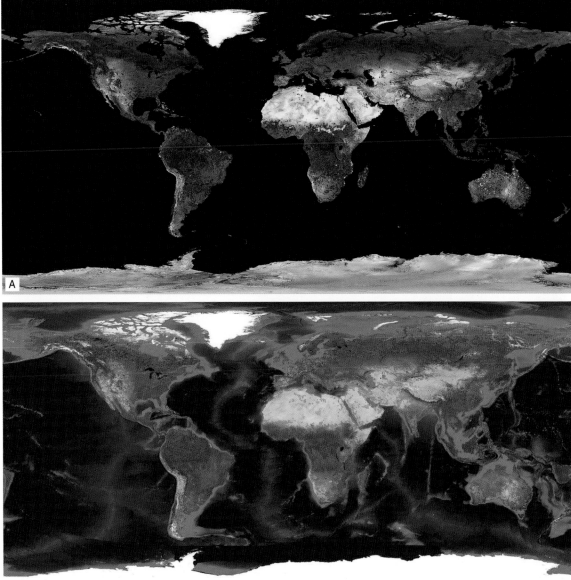

Figure 1.36 Fire across the world.
A. This image is a composite of active fires in part of a single year. Areas of red show a single fire in a ten-day period and yellow shows multiple fires, Jan 1–10, 2013). This image was collected using MODIS on NASA's Terra and Aqua satellites (http://lance-modis.eodis.gov).
B. This map has all fires collated through the year for 2012 (NASA FIRMS 2012. MODIS Active Fires Detections Data Set. Available online http://earthdata.nasa.gov/firms). (Map made and provided by Min Minnie Wong, Department of Geographical Sciences, University of Maryland, USA).

displayed (http://activefiremaps.fs.fed.us). These new data also allow information such as burnt area as well as carbon emissions to be calculated (see Chapter 6, Part Two).

It is not only the fires themselves that can be imaged. The effects of not only smoke (e.g. using the Calipso satellite; Figure 1.38) but also other gases and aerosols derived from fires (Figures 1.39, 1.40),

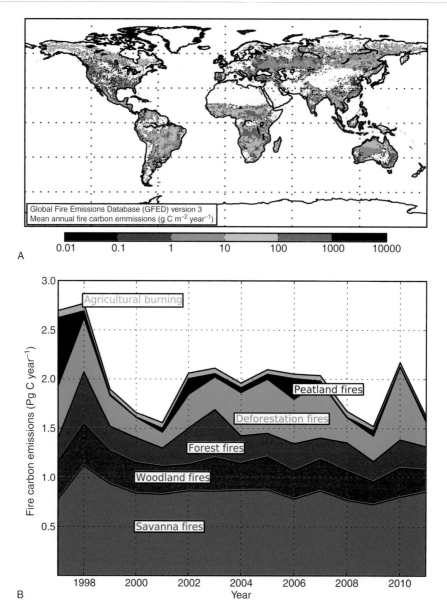

Figure 1.37 Emissions from fires from mid 1997 - mid 2010 shown on a global map (A) and through time related to vegetation type (B). (Diagrams supplied by G. van der Werf).

such as carbon monoxide (CO; Figure 1.41) and oxides of nitrogen (NO_x; Figure 1.42) can be seen spreading into the upper atmosphere, and this emphasizes the global impact of fire (Part Three).

The new fire data available using satellite monitoring techniques (Roy *et al.*, 2013) has revolutionized our understanding of global fire and also of the impact that humans play in both starting and suppressing fires (Part Three).

1.14 Modelling fire occurrence

The ability to map fires at an increasingly fine resolution in both space and time has led to the development of fire models that simulate the observed patterns. The fire occurrence data can be linked together with fuel data and also meteorological data, including

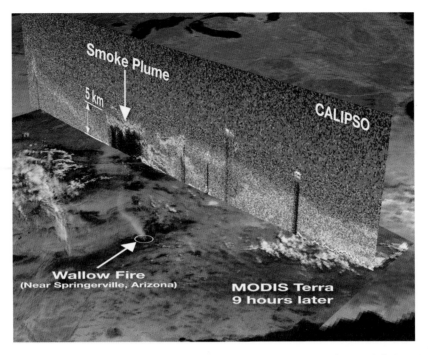

Figure 1.38   Smoke plume from the Wallow fire (Arizona, USA), June 2011, as seen from NASA's CALIPSO satellite. This vertical profile from space shows the smoke plume reached heights of 5 km (3 miles) high. Photo: Jason Tacett and Calipso team (www.nasa.gov). NASA.

Figure 1.39   Aerosol emissions from fires in Indonesia, October 1997. White represents smoke near the fire and colours indicate smog being carried in to the Troposphere by high altitude winds. This was taken using NASA's Earth Probe Total Ozone Mapping Spectrometer (TOMS) satellite instrument (http://visibleearth.nasa.gov). NASA.

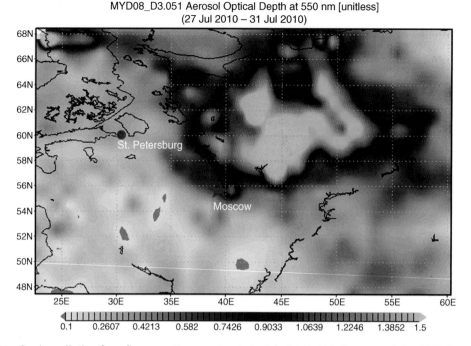

Figure 1.40 Smoke pollution from fires near Moscow, Russia in July 2010. This image used the MODIS sensor on the EOS Aqua satellite and was averaged over a five-day period (July 27–31, 2010) (see Keywood *et al.*, 2013). (Image from www.earthdata.nasa.gov). NASA.

temperature, humidity and wind speeds. For the short term, this may lead to the prediction of fire occurrence through fire danger maps that may consider overall risk to an area through to daily forecasts (Figure 1.5).

The development of powerful climate models also has allowed the integration of fire prediction to consider not only the relationship of fire and climate (also fire, vegetation and climate) in the present but also in the future. This approach is at an early stage, and several research groups have developed computer models allowing prediction of fire and climate. The computing power needed for such models is considerable, so most of these models tend to be global and fairly large scale.

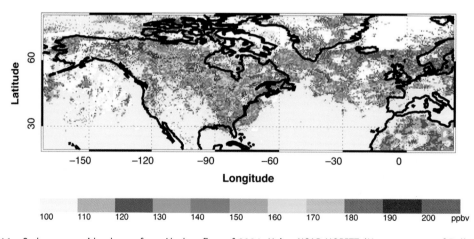

Figure 1.41 Carbon monoxide plumes from Alaskan fires of 2004. Using NCAR MOPITT (Measurements of Pollution in the Troposphere) (UCAR, University Corporation for Atmospheric Research (D101890).

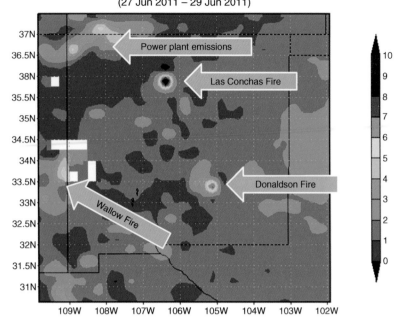

OMNO2e.003 NO2 Tropospheric Column (Cloud–Screened at 30%) [10^15 molec/cm^2]
(27 Jun 2011 – 29 Jun 2011)

Figure 1.42   Levels of nitrogen dioxide (NO_2) levels from fires across New Mexico and Arizona in June, 2011, using Ozone Measuring Instrument (OMI) on NASA's Aura Satellite (NASA and James Acker: www.nasa.gov.20110701). NASA.

There are two main approaches to the construction of fire-climate models. One is correlative and, therefore, based on statistics of past fire events. The other is an attempt to build a mechanistic model analogous to the GCMS for predicting entirely new combinations of vegetation and climate of the future. SPITFIRE is currently the most advanced attempt to build a mechanistically based simulation model for application at a global scale.

One correlative model (Krawchuk and Moritz, 2011) constructed multivariate statistical, generalized additive models (GAMS), combining existing fire occurrence, climate, net primary productivity (NPP) and ignition data (see Chapter 6, Part Two). In addition, data such as global vegetation distribution was included. This data used spacial data at a spacial resolution of 100 km (10 000 km) and mapped global vegetation fire using the ATSR fire atlas (Figure 1.36). Such an approach allows for a consideration of a range of climate scenarios, from the most conservative estimates of climate change to the most extreme (Figure 1.43).

The results of such analyses may allow analysis of whether fire will become more or less likely within a region, or if fire may retreat or invade an area. These

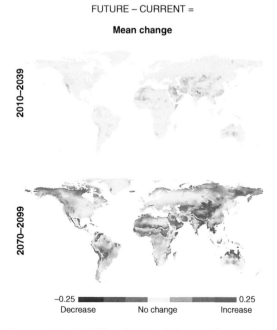

Figure 1.43   Modelling fire in relation to climate change showing particularly strong increases in fire in some areas in the later part of the 21st century. Note: in some areas, fires decrease. (From Moritz *et al.*, 2012, with permission from the authors).

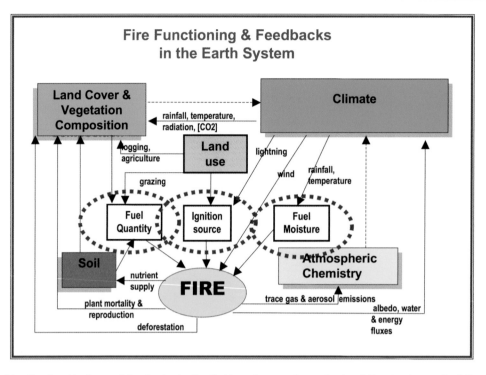

Figure 1.44 Fire functioning and feedbacks in the Earth system used as a basis of the development of the SPITFIRE model. The three key factors of fuel quantity, moisture and ignition are highlighted. (Supplied by Spessa *et al.*, 2012). Reproduced with permission from Springer Science+Business Media B.V.

models should alert planners to future fire risk and they should also link into major considerations of discussions concerning future global biodiversity assessments (Shlisky *et al.*, 2007).

The other recent approach using a mechanistically based simulation model has also involved simulating climate-vegetation-fire interactions using regional applications of the LPJ-SPITFIRE model (Figures 1.44, 1.45). This research is concerned with the long-term changes in vegetation and global carbon storage, as well as quantifying different trace gasses from biomass burning (CO_2, etc.) at regional and global scales. The model also allows the effect of regional climate phenomena such as El Niño on fire activity, vegetation and emissions, and it tries to distinguish between patterns of natural versus human fire ignition patterns (Thonicke *et al.*, 2010).

1.15 Climate forcing

An increased understanding of the distribution of fire on Earth has led to a consideration not only of how

fire is influenced by climate change, but also how fire may be involved in climate forcing. Fire may produce increased CO_2 in the atmosphere, adding to the total of greenhouse gases and contributing significantly to global warming. In contrast, smoke emissions may have an effect upon the atmosphere that creates a cooling effect. Burning of a region may also change albedo, so that green vegetation may become black. Fire suppression may change a reflective grassland to an absorbent forest, thereby heating the planet (while supposedly sequestering carbon). In addition, soot particles may cover a snowfield, again adding to that effect. Such positive and negative impacts make the overall contribution of fire to climate change difficult to calculate.

Recent global analyses of fire and its effects on climate change have been published. Some of these publications consider the effect of regional fire activity on global emission data. For example, the major peat fires in Indonesia are shown to have had a significant effect upon the global carbon cycle. A particular problem is separating natural and human-caused fire effects.

A. **MODIS**

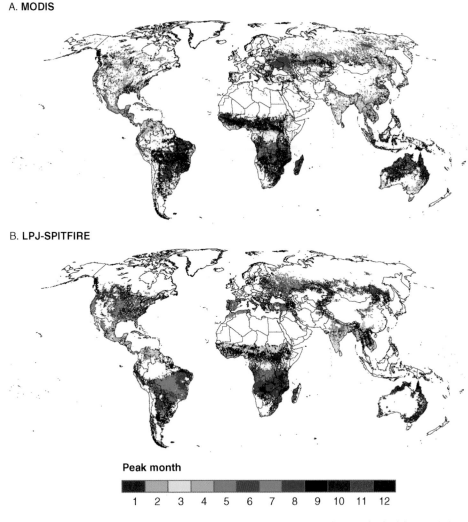

B. **LPJ-SPITFIRE**

Peak month

1 2 3 4 5 6 7 8 9 10 11 12

Figure 1.45 Simulating fire using the SPITFIRE model. These simulations can be matched with actual data (MODIS) to indicate the power of the model. (From Thonicke *et al.,* 2010).

Bowman *et al.* (2009) provides a detailed analysis of climate forcing and fire (see Chapter 6, Part Two). Radiative forcing is defined by the Intergovernmental Panel on Climate Change (IPCC), as the change in stratospherically adjusted radiative influx at the tropopause, compared with 1750 AD (Forster *et al.,* 2007). Positive forcing will increase global mean surface temperature, while negative forcing will decrease it. Fires change radiative forcing through altered atmospheric composition and/or changes in surface albedo.

The calculations involve a number of assumptions to be made. These include an understanding of the extent and frequency of fire in the pre-industrial era.

Some assumptions also consider that fire emissions (e.g. from tropical forest regions) are directly related to deforestation, and that the peat fires in south-east Asia represent a new anthropogenic emission source. All of the data combined provide an estimate that fires have contributed to about 19% of the anthropogenic radiative forcing since the pre-industrial era.

Unravelling CO_2 emissions from natural fires is complex. It is considered that fires contribute a constant 50% of total carbon emissions through time from deforestation. Fires represent an important source of ozone precursors, such as NO_x, especially in tropical regions. There are, however, many uncertainties in the calculations. Basically we can calculate:

$$\begin{aligned}
\text{Emissions} = {} & \text{Burnt area} \times \text{Fuel(biomass)} \\
& \times \text{Combustion completeness} \\
& \times \text{Emission factor}
\end{aligned}$$

As has been indicated, surface albedo can be changed by fire. Burning may cause short-term warming of the surface due to blackening or cooling of the surface due to especially increased exposure of snow at high latitudes. Black carbon on snow warms the surface by decreasing albedo. However, only 20% of this effect may be attributed to fires.

The direct aerosol effect is both positive and negative. A small cooling effect ($< 2\%$) is associated with light-scattering sulphate aerosols. Black carbon, however, has a tropospheric warming effect. Aerosol particles emitted by fires also have a significant impact on clouds, which has been termed an 'indirect aerosol effect'. There are two opposing effects of aerosols on clouds: the cloud condensation (microphysical) and black carbon (radiative). Where there is heavy smoke, there is an increase in cloud cover but a decrease in cloud droplet size. In contrast, large bouts of black carbon increase the susceptibility of low clouds to evaporation, which inhibits cloud formation and development. Quantification of these effects is difficult (see Chapter 9, Part Two).

One aspect of fire that is often forgotten is the production of charcoal (Figure 1.24B) As this is relatively inert, it may be buried and survive not only for thousands of years but for millions of years (Scott, 2010). In such cases, this may contribute to the long-term draw down of CO_2 in the atmosphere. It is this feature that has encouraged the production and use of bio-char to help in the reduction of current carbon dioxide levels (Lehmann *et al.*, 2006; Chapter 2).

1.16 Scales of fire occurrence

While the majority of fires are small, there has been an increasing concern over mega-fires ($> 20\,000$ ha) in the past few years. This is particularly significant, not only in terms of the impact to the ecosystem or on humans, but especially in relation to emissions as discussed earlier. Most fires burn areas of a few to a few hundred hectares. In some regions, there may be many fires that are burning at the same time, as was the case during the fires in Yellowstone National Park (USA) in 1988. In total, more than 570 000 ha burnt, of which 400 000 ha were in the National Park.

There is also a relationship between fire size and ecosystem impact. Many large fires burn areas that have been previously burnt, and this may have a devastating ecological effect. Large fires may burn for a much longer period and are difficult to

Figure 1.46 Fires threatening the Jet Propulsion Laboratory. The fires raged in the foothills around NASA's Jet Propulsion Lab (JPL). The 'Station Fire' broke out on August 26, 2009, in La Cañada Flintridge, California, just a few miles from the JPL. It was started by a combination of factors: triple-digit temperatures; extremely low humidity; dense vegetation that has not burned for several decades; and years of extended drought. The blaze burnt 145 000 acres (227 square miles) of the Angeles National Forest, destroyed 64 houses, forced tens of thousands of people to evacuate their homes and caused the deaths of two firemen who were involved in a crash while trying to escape rapidly advancing flames. (Photo: Brent Buffington, JPL, NASA).

extinguish, and they also may have a much greater atmospheric and, hence, global impact. These impacts may also include human health (Johnston, 2009; Johnston *et al.*, 2012).

In California, the Santa Ana wind may contribute to the size of a fire, in addition to human fire suppression policies, so that some fires may expand from 100 000 ha to over one million ha in events that may last several days and even weeks (Keeley *et al.*, 2011). In populated regions, fire size may be limited by habitat fragmentation and effective fire suppression measures. However, the problem of fire at the wildland-urban interface is becoming increasingly important (Figure 1.46; see Parts Two and Three).

An understanding of the scale of fire is important for fire modelling and the assessment of future fire with climate change. Recent studies in the western USA have found that mega-fires did not occur in the past, as non-catastrophic small fires reduced fuel load and, over the past century, fire suppression has reduced fire size. Climate change is, however, leading to an increase in the occurrence of mega-fires, and this trend may continue in the future.

Further reading

Artaxo, P., Luciana V. Rizzo, L.V., Paixão, M., de Lucca, S., Oliveira, P.H., Lara, L.L., Wiedemann, K.T., Andreae, M.O., Holben, B., Schafer, J., Correia, A.L., Pauliquevis, T.M. (2009). Aerosol Particles in Amazonia: Their Composition, Role in the Radiation Balance, Cloud Formation, and Nutrient Cycles associated with deposition of trace gases and aerosol particles. *Amazonia and Global Change*. Geophysical Monograph Series **186**, 233–250.

Bowman, D.M.J.S., Balch, J.K., Artaxo, P., Bond, W.J., Carlson, J.M., Cochrane, M.A., D'Antonio, C.M., DeFries, R.S., Doyle, J.C., Harrison, S.P., Johnston, F.H., Keeley, J.E., Krawchuk, M.A., Kull, C.A., Marston, J.B., Moritz, M.A., Prentice, I.C., Roos, C.I., Scott, A.C., Swetnam, T.W., van der Werf, G.R., Pyne, S.J. (2009). Fire in the Earth System. *Science* **324**, 481–484.

Cerdà, A. and Robichaud, P. (eds) (2009). *Fire Effects on Soils and Restoration Strategies*. Science Publishers Inc., New Hampshire.

Cochrane, M.A. and Ryan, K.C. (2009). Fire and fire ecology: concepts and principles. In: Cochrane, M.A. (ed). *Tropical fire ecology: Climate change, land use and ecosystem dynamics*, pp. 24–62. Springer, Berlin.

Doerr, S.H., Shakesby, R.H. (2013). Fire and the Land Surface. In: Belcher, C.M. (ed). *Fire phenomena and the Earth System: An interdisciplinary guide to fire science. 1st edition, pp* 135–155. J. Wiley & Sons, Ltd, Chichester.

Fites-Kaufman, J., Bradley, A.F., Merril, A.G. (2006). Fire and plant interactions. In: Sugihara, N.G., Van Wagtendonk, J.W., Shaffer, K.E., Fites-Kaufman, J., Thode, A.E. (eds). *Fire in California's ecosystems*, pp. 94–117. University of California Press, Berkley, CA.

Keeley, J.E., Bond, W.J., Bradstock, R.A., Pausas, J.G., Rundel, P.W. 2011. *Fire in Mediterranean Climate Ecosystems: Ecology, evolution and management*. Cambridge University Press.

Roy, D.P., Boschetti, L., Smith, A.M.S. (2013). Satellite remote sensing of fires. In: Belcher, C.M. (ed). *Fire phenomena and the Earth System: An interdisciplinary guide to fire science. 1st edition*, pp. 97–124. J. Wiley & Sons, Ltd, Chichester.

Rundel, P.W. (1981). Fire as an ecological factor. In: Lange, O.L., Nobel, P.S., Osmond, C.B., Ziegler, H. (eds.), *Physiological plant ecology I, Response to the physical environment*, pp. 501–538.

Shakesby, R.A., Doerr, S.H. (2006). Wildfire as a hydrological and geomorphological agent. *Earth Science Reviews* **74**, 269–307.

Spessa, A., van der Werf, G., Thonicke, K., Gomez-Dans, J., Lehsten, V., Fisher, R. & Forrest, M. (2012). Modelling Vegetation Fires and Emissions. In: Goldammer, J. (Ed.) *Fire and Global Change*, Chapter XIV.

Sugihara, N.G., Van Wagtendonk, J.W., Fites, Kaufman, J. (2006). Fire as an ecological process. In: Sugihara, N.G., Van Wagtendonk, J.W., Shaffer, K.E., Fites-Kaufman, J., Thode, A.E. (eds). *Fire in California's ecosystems*, pp. 58–74. University of California Press, Berkley, CA.

Van Wagtendonk, J.W. (2006). Fire as a physical process. In: Sugihara, N.G., Van Wagtendonk, J.W., Shaffer, K.E., Fites-Kaufman, J., Thode, A.E. (eds). *Fire in California's ecosystems*, pp. 38–57. University of California Press, Berkley, CA.

Chapter 2

Fire in the fossil record: recognition

2.1 Fire proxies: fire scars and charcoal

While fire may be considered a destructive force, there are a number of ways through which fire history may be interpreted. In the more recent past, an increasingly important method has been to study fire scars as part of a tree ring analysis (Figure 2.1). When a surface fire passes through a forest, the outer part of the trunk may be partially burnt or destroyed in part, but not sufficiently to kill the tree. The tree then will resume its normal growth. In temperate environments where trees exhibit growth rings, this may be very useful. If a tree is felled and a cross-section of a tree made, then the age of the tree may be calculated through counting the rings. However, it is possible to calculate the age of any fire affecting the tree through the occurrence of a fire scar. A single tree may have many scars and, hence, a fire return interval may also be calculated.

The occurrence of fire scars provides researchers with three important pieces of data (Roos and Swetnam, 2012):

1 The recognition of a fire event that touched the tree.
2 The fire record during the life of the tree.
3 The extent of fires over distinctive time intervals, if several trees have been examined over a wide geographical area.

The spatial extent of a fire can be considered at two main scales. If trees are studied within a forested area – perhaps a few square kilometres – then the size of an individual fire might be calculated. If trees are studied over a much larger region – perhaps over hundreds or thousands of square kilometres – then it is possible to interpret high and low fire years (Figure 2.2). Such an approach has been found useful in the study of the relationship between fire and climate (Figure 2.3).

In climate studies, the fire scars and tree ring data can be used to link, for example, the occurrence of fire in relation to El Niño and La Niña. This relationship has been proven to be very strong, with more fires occurring in the south-west of the USA during the La Niña years (Swetnam and Betancourt, 1990). In addition, fire records can also be combined with a range of climate records, so that it is possible to demonstrate that more fires occur in warmer years. This relationship between increasing temperature and increasing fire is of particular significance and concern. Recent studies in the western USA have demonstrated an increasing occurrence of larger fires with increasing temperature (Figure 1.27). It should be noted that the effects of El Niño and La Niña would be different in different regions.

Fire on Earth: An Introduction, First Edition. Andrew C. Scott, David M.J.S. Bowman, William J. Bond, Stephen J. Pyne and Martin E. Alexander.
© 2014 John Wiley & Sons, Ltd. Published 2014 by John Wiley & Sons, Ltd.

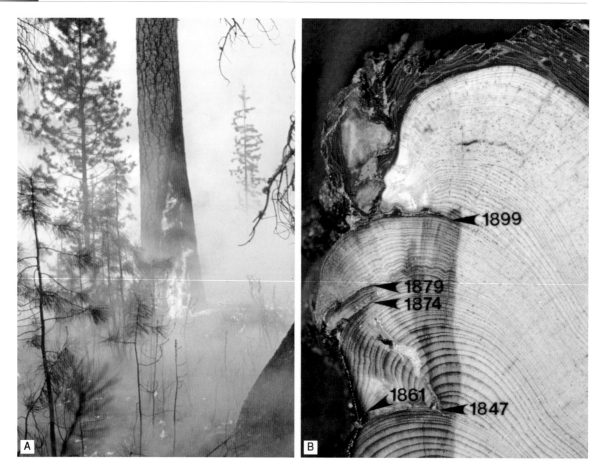

Figure 2.1   Tree rings and fire scars.

A. Low severity surface fire, re-igniting within existing fire scar cavity (photo: J. H. Dieterich, reproduced by permission of T. W. Swetnam).
B. Fire-scarred ponderosa pine (photo: Chris Baisan, Laboratory of Tree-Ring Research, University of Arizona). Courtesy Chris Baisan.

Tree-ring fire scar analysis is a very powerful tool in unravelling climate and human impacts. In a detailed study of fires in forests in the south-western USA, using data from trees up to 1500 years old, Roos and Swetnam (2012) were able to show that recent mega-fires in the region were truly unusual. The data from the tree rings provides important information about the climate, and particularly about the occurrence of wet and dry years.

The fire scar data has been used to test whether today's hot-dry climate alone is causing mega-fires that destroy large tracts of forest. The researchers compared fire frequency during hotter and drier periods with that during cooler and wetter periods. Interestingly, they showed that the fire frequencies were similar. However, over the past 100 years, they noted a reduction in fire as a result of human fire suppression.

The historic data has demonstrated that frequent surface fires were important in keeping fuel loads low.

The fires were predominantly low-temperature surface fires. The suppression of fire in such forests, however, leads to a build-up of fuel so that, during the increasingly warm arid phase of climate, any fire becomes a devastating high-temperature crown fire that destroys the forest (Figure 2.3).

Many tree-ring/fire scar studies are limited to data from less than 1500 years. In some vegetation types, much longer fire records can be obtained. In the *Sequoia* redwood forests of north-west USA, fire scar records have been obtained from trees more than 3000 years old (Swetnam, 1993). In addition, a link can be made between data from fire scars and data from charcoal in varved lake sediments (Figure 2.4). In the lake records, there may be less precision of exact fire years, as some charcoal may be reworked. The linkage of fire scar data and charcoal data has proven very important in

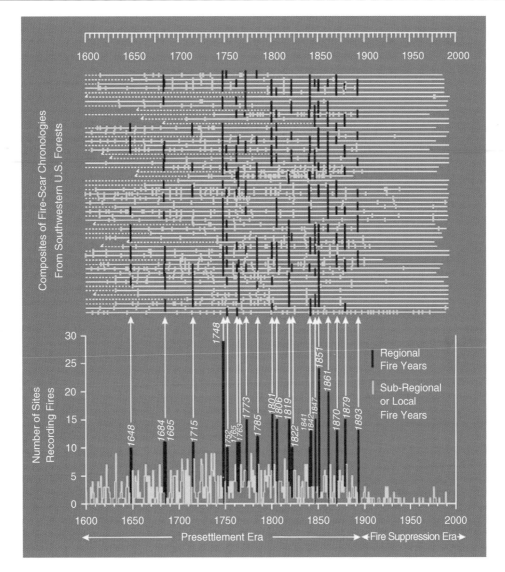

Figure 2.2 Fire scars across an area; south-western US fire scar network. This is a composite of fire scar chronologies from 55 forest and woodland sites in Arizona, New Mexico and northern Mexico, AD 1600 to Present. (From Swetnam *et al.*, 1999). Reproduced by permission of Elsevier.

the interpretation of local and regional fire events (Swetnam *et al.*, 2009).

2.2 The problem of nomenclature: black carbon, char, charcoal, soot and elemental carbon

As discussed in Chapter 1, a range of products is formed as a result of a fire, and one significant problem is that of nomenclature. When plant material is subjected to fire, it undergoes both pyrolysis and combustion. During the initial heat increase, pyrolysis takes place in the absence of oxygen and, for example, wood may be converted to charcoal. This, however, is not a single stage process. The process of charcoalification has been called 'carbonization' by some authors. This may cause some misunderstanding, as this term has also been used to describe the thermal upgrading of coal.

Some authors prefer to refer to a combustion continuum model (Figure 2.5). This considers a continuum ranging from slightly charred biomass through to soot (see Hammes and Abiven (2013), for example).

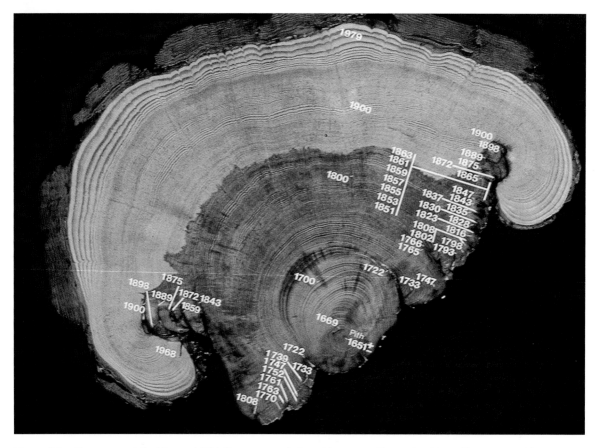

Figure 2.3   Fire scars and fire history; LMF-3 fire scar sample from Limestones Flats, AZ, USA. 42 fire scars on this one tree. (From Dieterich and Swetnam, 1984). Reproduced by permission of the Society of American Foresters.

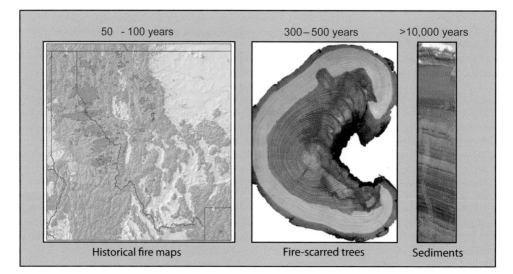

Figure 2.4   Types of fire data from maps to fire scars to charcoal in varved lake sediments. (from M. Power). Courtesy H. D. Grissino-Mayer.

This concept is basically a chemical one, reflecting an increase in aromaticity as the charring temperature increases. Plant material may be charred. Some authors distinguish a low temperature char to a higher temperature charcoal, but there is disagreement as to the identification of this transition. A significant problem results from the different methods used to study such material. Geochemists often refer to the spectrum of char to soot as black carbon (BC), but concentrations of such material in sediments may differ considerably, depending on the methods used for their detection (Hammes and Abiven, 2013). In some cases, discrepancies are a result of differing nomenclature. In others, it is the failure to detect some or part of the fire-derived material. In still other cases, over-detection by some methods means that some non-fire derived material is also detected (Masiello, 2004).

Soot (Figure 2.5) has often been included in some definitions of black carbon. It differs from charcoal, for example, in that it represents a condensation product. It may form from the combustion of any fuel – not only vegetation, but also from coal, oil and gas. The abundance of soot in sediment may not, therefore, be solely a function of biomass burning. In other cases (e.g. Bond *et al.*, 2013), the two terms are considered to be synonymous.

Charcoal is the main residue from biomass burning (Figure 2.5; and see Scott, 2010). It is clear that chemical changes occur during the charring process. It has been shown that charcoal may retain the anatomy of the plant, so that it can be recognized visually (Figure 2.5). It has often been considered that charcoal is chemically inert and is not easily broken down, allowing it to accumulate in a range of carbon pools, from soils to ocean sediments. This has led to a chemical approach to the quantification of black carbon, using an oxidation process.

Recent experiments have shown that low-temperature charcoals (those formed by temperatures less than 450 °C) may be partly or completely oxidized. It is, therefore, likely that analysis of black carbon using this technique may underestimate the quantity of plant material that has been affected by fire. Other techniques use a range of physical methods, such as 13C NMR, or the use of molecular markers, such as the measurement of benzene polycarboxylic acids (BPCA). A range of other chemical, spectroscopic and optical methods also exist. Likewise, a range of microscopic methods exist that may be used to identify fire residues; these will be considered next.

It is clear that it is important, when considering the quantity of wildfire residue in any sample, to clarify what is being identified and to take in to account the method used in its quantification. For some, the importance of charcoal quantification is to interpret fire history, including the link between fire and climate. For others, the purpose for quantification relates to unravelling the global carbon cycle and, hence, the role of fire in biogeochemical cycles. For some, therefore, identification of what, and how much, is being burnt may be important. For others, however, the main concern may be how recalcitrant is the buried carbon.

Alongside the black carbon it is possible to also find other chemical compounds that are related to combustion. Pyrolitic polyaromatic hydrocarbons (PAHs) are often found associated. Many of these compounds may have tracers to allow the nature of the vegetation that was burnt to be established (see Simoneit, 2002 for a discussion). These compounds are also found in the fossil record (see Chapter One).

2.3   How we study charcoal: microscopical and chemical techniques

One of the most common methods of quantifying charcoal in sediment is the use of acid oxidation. This uses the fact of the recalcitrance of the charcoal carbon. In such studies, non-charred plant material is oxidized away, leaving a charred residue. There is a range of thermal oxidation methods used, and by far the most widely used is that of chemical oxidation using dichromate/sulphuric acid (H_2CrO_4/H_2SO_4). As has recently been shown, low temperature charcoals are destroyed using this process (Ascough et al., 2010, 2011).

This research also indicates that, in some cases in the natural environment, low temperature charcoals may decay preferentially. Some studies, therefore, use a combination of approaches to black carbon quantification.

Some of the most useful methods study a range of fire residues are microscopical (Glasspool and Scott, 2013). We consider here a range of methods, from simple low-powered light microscopy through to the most advanced use of synchrotron radiation X-ray microtomography (SRXTM).

As charcoal is found in a wide range of sizes (from micron to centimetre), there are several approaches to light microscopy. Fine charcoal fragments, usually less

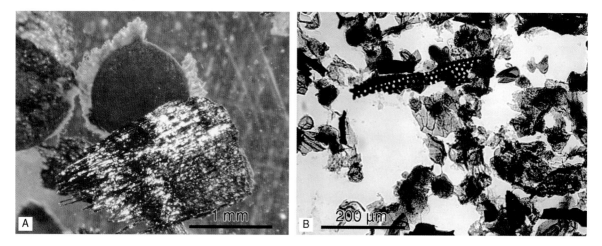

Figure 2.5 Plate of SEM images – soot to charcoal.

A. Macerated coal in water using dark field microscopy. Note brown megaspores and black charcoal with its silky sheen and fibrous appearance (Pennsylvanian (Carboniferous 320 ma) from coal from Poland). Scale bar 1 mm. (From Scott, 2010). Reproduced by permission of Elsevier.

B. Microcharcoal in palynology slide (Pennsylvanian (Carboniferous 315 ma) coal measures of England). (From Glasspool and Scott, 2013).

than 125 μm, may be encountered in palynological preparations. These may be found as black angular or lath-shaped fragments when studied under transmitted light (Figure 2.5B). In such cases, however, it is not possible to identify the origin of the charcoal, except in a few exceptional cases. Small charred grass cuticle fragments have a distinctive epidermal structure that can be used for broad identification. Quantification analysis of charcoal in palynological slides has been routinely used to interpret fire history but there are numerous drawbacks to this approach, and these will be considered in Chapter 5.

Charcoal will commonly be sieved from sediment. With unconsolidated sediment, some researchers float off the charcoal. This is a common technique in archaeological studies (Figure 2.7). Some charcoal does not easily float and may become part of the sink residue. It has been shown that charcoal formed at different temperatures will behave differently in regard to floating or sinking. In addition, sediment may infiltrate the charcoal, causing it to sink. In some cases, mineral matter may precipitate within the charcoal to prevent floatation.

The most common approach is to remove charcoal from the sediment using sieves. Commonly, a lower size of 125 μm or 180 μm is used. Although dry sieving can be used, wet sieving is much more effective and may prevent charcoal disintegration. 125 μm has routinely been used for quantification studies where

anatomical data is required for plant identification, then 180 μm is often employed.

Charcoal may be sieved from unconsolidated sediment. However, not all sediment breaks down simply in water. For weakly lithified sediment, a 10% solution of hydrogen peroxide may be used to help disaggregate the sediment. It may be necessary, however, to use a range of stronger acids. For sediments with carbonates 10% hydrochloric acid (HCl) is commonly used. For siliceous sediment, including mudstones, 40% hydrofluoric acid (HF) is used; extreme care should be employed in its use. Details of techniques can be found in Jones and Rowe (1999).

Where there is an organic matrix, the charcoal may be released using concentrated nitric acid (HNO_3) followed (after neutralization by water) by dilute 10% ammonia solution. With such a method, charcoal may be released even from a moderate rank bituminous coal (Figure 2.6).

The charcoal (mesoscopic (125 μm – 1 mm) and macroscopic (> 1 mm)) may first be examined using a low-power binocular microscope. Although transmitted light can be employed, top lighting is more successful. In addition, it may often be easier to study and pick the charcoal in a Petri dish under water, using a fine 00 or 000 brush and a mounted needle.

Identification of charcoal may not always be easy. Employing dark field microscopy has proven very useful

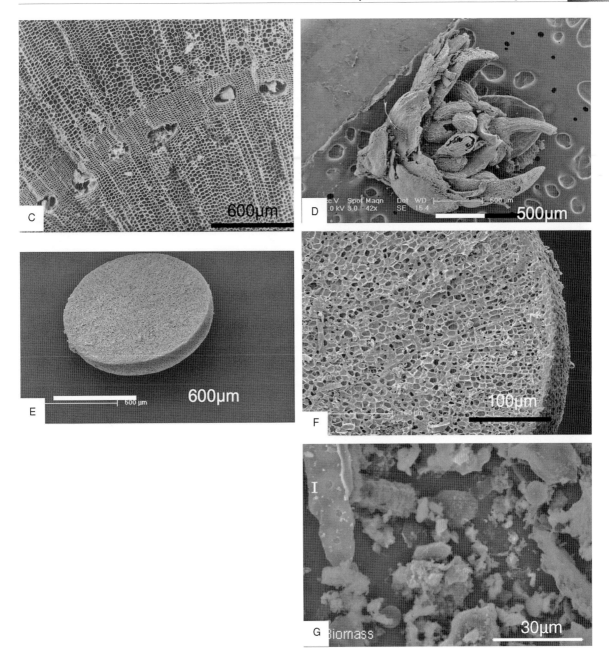

C. Scanning electron micrograph of pine charcoal illustrated in Fig 1.12b. Scale bar 500 μm. (From Scott, 2000). Reproduced by permission of Elsevier.

D. Scanning electron micrograph of charred flower from the Thursley (2006) fire, Surrey, England. Scale bar 500 μm. (From Scott, 2010). Reproduced by permission of Elsevier.

E,F. Scanning electron micrograph of fungal sclerotium from the Thursley (2006) fire, Surrey, England. Scale bar 100 μm (Photos A. C. Scott).

G. Windblown charcoal and soot from the 2002 Hayman wildfire, Colorado. (From Belcher et al., 2005). Reproduced by permission of The Geological Society.

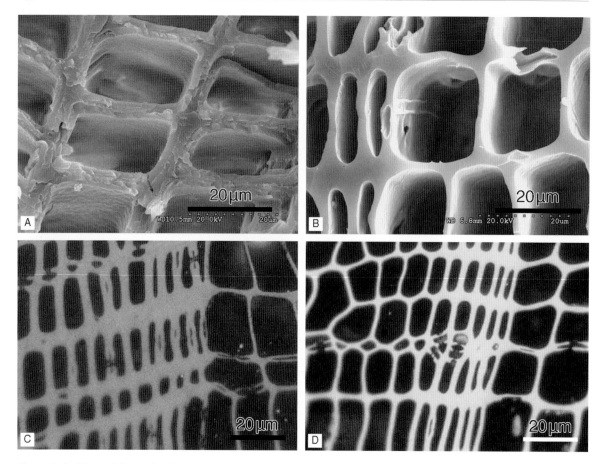

Figure 2.6   Plate of charcoal – SEM, SRXTM and reflectance. Micrographs of artificially charred sequoia wood (From Scott, 2010). Reproduced by permission of Elsevier.

A. and B. Scanning electron micrographs. A: Wood charred at 300°C for one hour. Cells still show middle lamella (arrow). B: Wood charred at 500 °C for one hour. Cell walls are homogenized.

C. and D. Reflectance micrographs of polished blocks under oil. C: Wood charred at 300 °C for one hour. D: Wood charred at 500 °C for one hour.

to help distinguish charcoal particles (Scott 2010; Glasspool and Scott, 2013). Some researchers simply wish to count charcoal particles to give data that can be used in determining fire history (see Chapter 5).

As documented below, charcoal is an information-rich material, so a number of other microscopical techniques may be employed in its study. Charcoal fragments may be picked into small vials/tubes or into cavity slides for later study. Commonly, charcoal fragments are mounted onto aluminium stubs and gold coated for study using the scanning electron microscope (SEM). The use of SEM has two main purposes:

• to observe anatomical details of the charcoal to help taxonomic assignment;

• to observe details of preservation, such as cell wall homogenization, that can be used to help identify the plants as being charcoal (Scott, 2010; see Figure 2.6).

In addition, information such as cracks formed by the burning of green wood or fungal infection indicating decayed wood can also be seen. It is important to note here that many charcoal fragments (even sub-mm or, indeed, whole organs) may be preserved, and SEM is needed for their recognition and identification (Figure 2.5). A good case point is the identification of small organic spheres that may have a range of origins. Some of these spheres, 200 mm to 1 mm in size, may represent fungal sclerotia (Scott *et al.*, 2010;

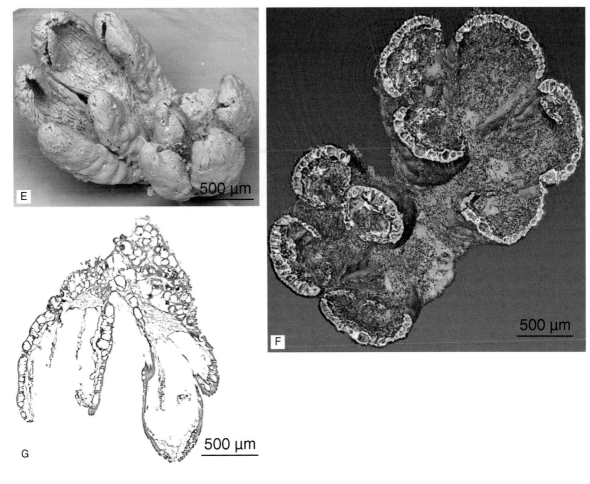

E-G. Micrograph of Mississippian (335 ma) charcoalified pollen organ from Kingswood, Fife, Scotland.
E. Scanning electron micrographs of pollen organ showing clusters of sporangia.
F. Digital section using synchrotron radiation X-ray tomographic microscopy (SRXTM). longitudinal section.
G. Cut reconstructed image with anatomical data. (All images from Scott, 2010).

Figure 2.5E,F). The best images, even of charcoal that has high carbon content, are taken using standard SEM techniques on coated specimens using secondary electrons. It is possible, however, to examine uncoated specimens using a low pressure SEM using backscattered electrons.

Charcoal may be very brittle, and may shatter when attempting to break a specimen to reveal internal anatomy. However, new techniques such as X-ray tomography may reveal internal anatomy without the need of physical sectioning (Figure 2. E-5G). In normal X-ray micro-tomography, the resolution may be relatively low, in the region of microns or

tens of microns. For small specimens (usually less than 5 mm), synchrotron radiation X-ray micro-tomography (SRXTM) may be used to examine details of internal anatomy without the need of physical destruction. The resolution of this technique is sub-micron, so fine structures may be determined. Three-dimensional images may be rendered and virtual sections in any orientation may be obtained. Details of the internal anatomy of fossil charcoalified fertile organs have been obtained using this method (Figure 2.5E-G).

These techniques have been used almost exclusively in fossil charcoal specimens. It is time-consuming and

Figure 2.7 Floating charcoal from archaeological dig. (Photo A. C. Scott)

costly, so the need for its use in studying modern charcoal may be limited, but it may have a use in the study of unusual or difficult to identify specimens in charcoal residues from wildfires.

Reflectance microscopy is a powerful tool for the study of both modern and ancient charcoal. It should not be confused with a technique developed for those studying plant anatomy in archaeological charcoal assemblages; in such studies, charcoal is examined simply in reflected light in air, viewing broken surfaces. There is a mistaken belief that an increase in the reflectance in air indicates an increase in temperature at which the material burned, and that glassy carbon or vitrified charcoal indicates high formation temperatures.

True reflectance microscopy is a coal petrographic technique, whereby a specimen is embedded in a resin and a surface polished. The specimen may be examined using a reflectance microscope, but under oil (Figure 2.5C,D). This not only allows an image to be viewed, but also allows measurement of the reflectance that may provide information on the temperature of formation (see below). This technique has become used increasingly in the study of ancient charcoals, and has now also found a use in archaeological studies and in studies of charcoal residues from wildfires.

2.4 Charcoal as an information-rich source

Charcoal is an under-utilized resource, not only for fire scientists but also for ecologists and evolutionary biologists, as well as for Earth scientists (Scott, 2010). The preservation of plant anatomy in most charcoal fragments – even pieces and organs less than 1 mm in size – offers the opportunity to identify what was being burnt (Figure 2.5). It can be argued that charcoal from deep time (i.e. many millions of years old) can be recognized from the features of anatomical preservation, in addition to the black colour of the material and the characteristic sheen.

Examination of charcoal from modern fires rarely includes an analysis of what plants are being preserved. This, however, may be significant. Such a study may allow the distinction of ground cover and plants found as litter from those plants that are burnt in crown fires.

A study on charcoal assemblage from heather-dominated (*Calluna*) heathland and surrounding mixed forest vegetation in England demonstrated that a large number of the taxa in the living vegetation were preserved in the charcoal (Scott *et al.*, 2000). This included not only woody taxa such as *Calluna* and *Pinus* but also ferns such as *Pteridium* and mosses. It is possible to study charcoal assemblages not only from recent fire sites, but also from assemblages from throughout geological history, to determine the identity of the vegetation being burnt (See Chapter 4).

Another feature of charcoal concerns its reflectance, as seen in polished blocks under oil (see Figure 2.6C,D and section 2.5). It can be demonstrated that, for an individual charcoal particle, there is increasing reflectance with temperature. Reflectance data from charcoal assemblages may provide evidence of the minimum fire temperature and help in the assessment of fire intensity.

2.5 Charcoal reflectance and temperature

Charring experiments have been undertaken on a wide range of plant material. In these experiments, air is excluded by a range of methods, including burial in sand, placing in specially designed containers or in ovens with an inert nitrogen atmosphere. Experiments

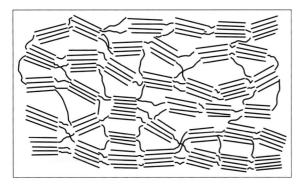

Figure 2.8 Charcoal graphitic domains. (From Cohen-Ofri *et al.*, 2006). Reproduced by permission of Elsevier.

have involved a wide range of temperatures and time, from 200–1200 °C and from minutes to days to weeks. Shorter time scales have been designed to mimic wildfire environments, and very long time scales were designed to look at the effect of long-term burial in hot volcanic pyroclastic block and ash deposits.

These experiments, by various research groups, all demonstrate a number of similar features. With increasing temperature and time, there is a reduction in both mass and density. In addition, some shrinkage may occur. The anatomy of the plant is not necessarily affected by increasing temperature. It is possible that, if oxygen leaks into the experimental system, some destruction of the plants may occur. With increasing temperature, the samples may become more brittle; this is particularly so above 800 °C. Charcoals

produced around 450 °C may, however, be fairly robust and resist fragmentation.

The carbon content of the charcoal increases with increased temperature and the aromatic rings become aligned into graphitic domains (Figure 2.8). The carbon isotopic composition of the charcoal may change little as 2 per mil (Figure 2.9).

When studied in polished blocks under oil, it was initially shown that wood increased reflectance with increasing temperature (Figure 2.10). Early experiments chose a one-hour charring time (Jones *et al.*, 1991). Later experiments showed that reflectance also increased with the time of heat exposure (Scott and Glasspool, 2005). It was demonstrated that a maximum reflectance was established after about 4–5 hours. The implications of this finding are significant, as it allows the establishment of a minimum charring temperature when heat exposure is less than five hours. If a charcoal fragment has a reflectance of 2%, then the temperature must have been more than 400 °C. It can be demonstrated, for example, that if the charcoal was exposed to less that 400 °C (e.g. 350 °C) for even 192 hours, it would never reach 2% reflectance (Figure 2.11).

Additional studies on a range of plant and other organic tissues (e.g. ferns and fungi) have demonstrated that this is a widespread relationship. It opens up the potential of obtaining fire temperatures from charcoal assemblages in a wide range of settings: wildfire, archaeological and volcanic.

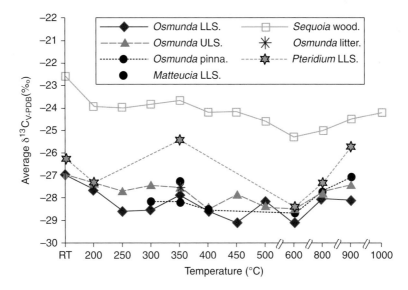

Figure 2.9 Changes in ∂13 during charring of conifer and ferns. (From McParland *et al.*, 2007). Reproduced by permission of the Society for Sedimentary Geology(SEPM).

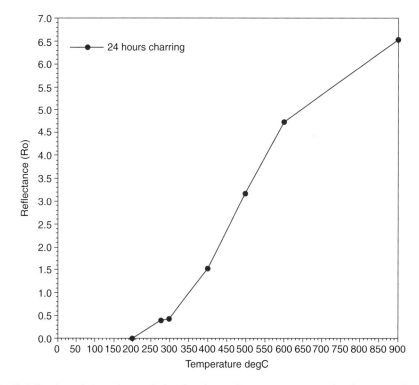

Figure 2.10 Artificially charred *Sequoia* wood showing increasing temperature and reflectance. (From Scott, 2010). Reproduced by permission of Elsevier.

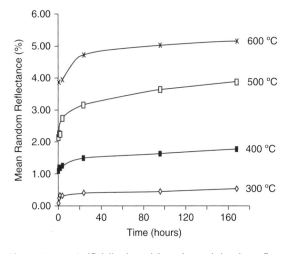

Figure 2.11 Artificially charred *Sequoia* wood showing reflectance change with time. (From Scott and Glasspool, 2005). Reproduced by permission of the Geological Society of America.

2.6 Uses of charcoal

The above discussion demonstrates the considerable information content of charcoal. Charcoal has a wide range of uses across a number of sciences in addition to fire science (see Scott and Damblon (2010) and Scott (2010) for details). Many of the uses of charcoal are shown in Figure 2.12.

The ability to obtain both taxonomic data and temperature data from charcoal residues from recent fires has implications for the discussion of fire type, fire intensity and severity. Charcoal may be readily transported from the site of a fire and through water transport (air transport for very small particles) into a depositional site (Figure 2.13). Studies of such charcoal may be used to interpret fire history. The fact that most charcoal preserves anatomical data leads to its use in the interpretation of the vegetation that was subjected to fire.

In deep time studies, not only can anatomical data from charcoal be used to identify what was burnt, but also many plants may only be found in the fossil record as charcoal, so that charcoal provides important information for plant evolution. The fact that organs such as flowers may be preserved as charcoal is particularly significant, as it is the study of such material that has helped shape our understanding of early flowering plant evolution (Friis *et al.*, 2006).

In the archaeological field, studies of charcoal are important. Traditionally, archaeological charcoal

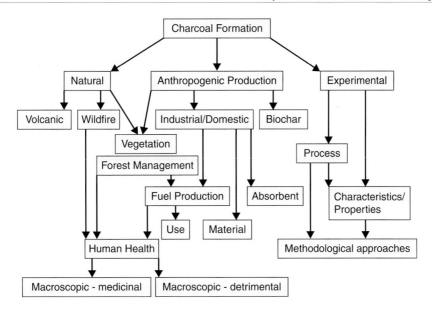

Figure 2.12   The study and uses of charcoal and their interrelationships (From Scott and Damblon, 2010). Reproduced by permission of Elsevier.

studies have been dominated by taxonomic works, often related to wood economy. More recently, studies have used reflectance data to interpret anthropogenic processes, including metalworking (McParland *et al.*, 2009a), and also possible temperatures of Roman hypocausts that have significance in deciding whether wood or charcoal was used as a fuel (McParland *et al.*, 2009b).

In geological studies, charcoal in pyroclastic flow deposits have been used to interpret flow temperature,

and the data has been used to help model the movement of density flows following volcanic eruptions.

2.7 Fire intensity/severity

There has been considerable debate over the use of the terms 'fire intensity', 'burn intensity' and 'fire severity'. A useful discussion may be found in Keeley (2009)

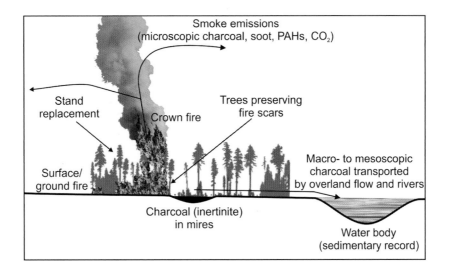

Figure 2.13   Transport of charcoal during and after a fire. (From Glasspool and Scott, 2013).

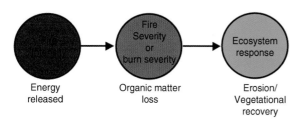

Energy   Organic matter   Erosion/
released   loss   Vegetational
                                      recovery

Figure 2.14 The relationship of fire intensity, severity and ecosystem response. (After Keeley, 2009). Reproduced by permission of CSIRO.

(see Chapter 8, Part Two). In essence, fire intensity is a measure of the energy released, while fire severity measures organic matter loss (Figure 2.14).

It is not possible to measure fire severity in the fossil record, although there may be some indication from post-fire erosion events. It is possible that some element of fire intensity is preserved in the charcoal assemblage. Intense fires are likely to be hot and, therefore, may be expected to produce higher-temperature charcoals. It is possible, therefore, that reflectance values of charcoal assemblages may provide evidence of hot versus cool fires.

2.8   Deep time studies

The appreciation of a widespread fire history in deep time is relatively recent (Scott, 2000). There has always been an appreciation of fire in the Quaternary and the more recent past, but studies on the pre-Quaternary have been much scarcer. In part, this was related to debates concerning the origin of fusain, the term used (and introduced by Marie Stopes as one of her coal lithotypes) to describe 'charcoal-like' plant fragments in the fossil record. Some authors were unable to accept that fusain represented fossil charcoal and was the result of wildfire.

Over the past 30-plus years, extensive research has demonstrated conclusively that fusain represents fossil charcoal (see Scott, 1989 for a discussion of this topic). Scott (2000) collated a comprehensive record of known charcoal records in deep time and documented a near continuous record of wildfire from the Devonian period onwards. Since then, publication records have been added at a rapid rate, so that we now have evidence of fire from the late Silurian and a

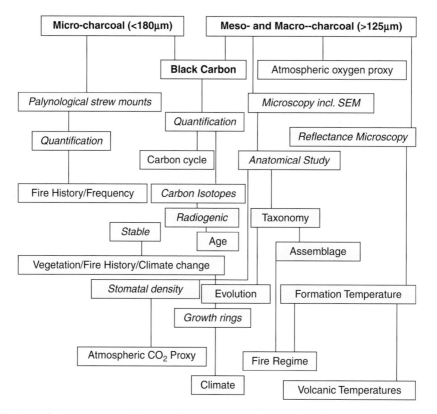

Figure 2.15   Methods of study and uses of charcoal (From Scott, 2010). Reproduced by permission of Elsevier.

much broader assessment of the impact of wildfire in deep time can be made (see Chapter 3, section 3.4). Most of these studies are qualitative, providing data on fire occurrence and also indicating, using data on plant anatomy, what vegetation was being burnt (Figure 2.15).

2.9 Pre-requisite for fire: fuel – the evolution of plants

When unravelling the geological history of wildfire, it is necessary to consider the evolution of plants. The

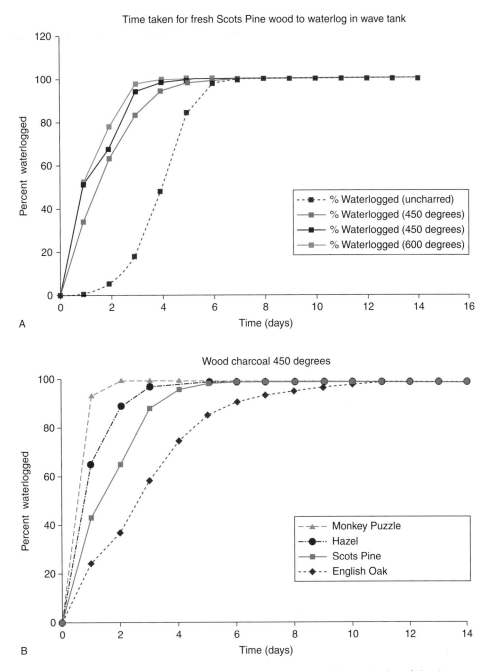

Figure 2.16 The settling of charcoal in water. (After Scott, 2000). Reproduced by permission of Elsevier.

terrestrialization of the continents is an area of intense current research. It is clear that there would not have been any significant wildfire before the advent of plant life on land. The first non-vascular plants probably spread onto land in the Ordovician (460 ma), and the earliest vascular land plants are found in the late Silurian (420 ma). These plants would have been small – less than 10 cm high – and would not have accumulated a significant fuel load. Despite this, there is evidence of charcoalified plants from the latest Silurian period (419 ma), indicating the oldest known wildfires.

Trees did not evolve until the mid-late Devonian period (385 ma) so there could be no forest fires until that time. Fuels changed as plants evolved. Upland forests did not appear until the late Carboniferous (320 ma).

At different times in Earth's history, there is evidence of changing fire regimes, and this is documented further in Chapter 4.

2.10   Charcoal in sedimentary systems

While some charcoal may be incorporated into soils at the site of the wildfire, some may be transported in the smoke plume during the fire (Clark *et al.*, 1998). A significant quantity of charcoal may be transported

away from the fire site by water. This may occur in only a few days after the fire. Charcoal may be transported by overland flow and into lakes, rivers and, in some cases, into the sea (Figure 2.13).

Charcoal floats, but different plant organs, sizes of charcoal and charcoals formed at different temperatures differ in the way that they float and settle out of a water column (Nichols *et al.*, 2000; Figure 2.16). It is often thought that the occurrence of large charcoal particles indicates a local fire. Settling experiments have shown that larger wood charcoals may be transported further by water than small charcoal fragments. In contrast, with wind-blown small charcoal fragments, very small pieces may remain in the air for much longer and may be transported hundreds or thousands of kilometres away from the fire site, even with deep ocean sediments (Forbes *et al.*, 2006).

Charcoal may be incorporated into almost any depositional setting and be associated with a wide range of lithologies. Charcoal is commonly found in fluvial overbank fine sediments, as well as in bed-load sandstones (for an introduction to sedimentology, see Nichols (2009)). Within sandstones, it may be found preferentially associated with cross-beds (Figure 2.17A), indicating flows of around 35 cm/sec (Figure 2.17B). Charcoal may be found in debris flows and flood deposits associated with deposition following post-fire erosion.

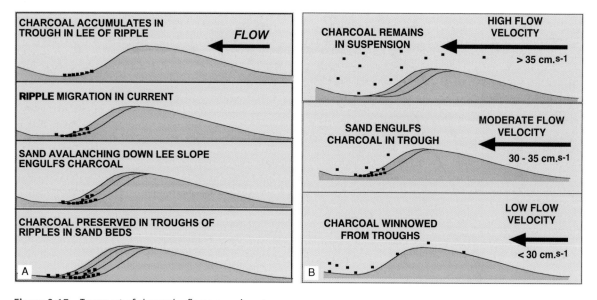

Figure 2.17   Transport of charcoal – flume experiments.
A. The process of deposition of charcoal in the A. The process of deposition of charcoal in the troughs in front of migrating ripples forming a charcoal 'lag' (From Nichols et al., 2000). Reproduced by permission of Elsevier.
B. The deposition of charcoal depending on current flow (Data from G. Nichols, J. Cripps, M.E.Collinson, A.C.Scott).

Commonly, charcoal may accumulate within lake basins and be found in varved sediments (Figure 2.4). This has proven useful for the interpretation of recent fire history. Charcoal may also be found in peats and coals, representing fires within mire environments (Figure 2.18). In modern peats, charcoal may represent less than 4% biomass but, in some ancient coal deposits, charcoal may represent more than 70% of some coal seams, such as those from the Permian of Australia or India.

Charcoal may be found in near-shore estuarine sediments or as isolated fragments in a range of shelf sediments, both clastics (e.g. sands, silts, clays) and carbonates (e.g. limestones). Finally, small charcoal fragments and soot particles may accumulate in deep ocean sediments, where they may be used to interpret fire history (see, for example, Herring (1985)).

Further reading

Antal Jr., M.J., Grønli, M. (2003) The art, science, and technology of charcoal production. *Industrial and Engineering Chemistry Research* **42**, 1619–1640.

Carcaillet, C., Bouvier, M., Fréchette, B., Larouche, A.C., Richard, P.J.H. (2001) Comparison of pollen-slide and sieving methods in lacustrine charcoal analyses for local and regional fire history. *The Holocene* **11**, 467–476.

Clark, J.S. (1988) Particle motion and the theory of charcoal analysis: source area, transport, deposition and sampling. *Quaternary Research* **30**, 67–80.

Glasspool, I.J. and Scott, A.C. (2013). Identifying past fire events. In: Belcher, C.M. (ed). *Fire phenomena and the Earth System: An interdisciplinary guide to fire science.* 1st edition , pp. 179–205. J. Wiley & Sons, Ltd, Chichester, UK.

Jones, T.P., Rowe, N.P. (eds.) (1999). *Fossil Plants and Spores: Modern Techniques.* Geological Society, London.

Scott, A.C. (2010). Charcoal recognition, taphonomy and uses in palaeoenvironmental analysis. *Palaeogeography, Palaeoclimatology, Palaeoecology* **291**, 11–39.

Schmidt, M.W.I., Noack, A.G. (2000). Black carbon in soils and sediments: Analysis, distribution, implications, and current challenges. *Global Biogeochemical Cycles* **14**, 777–793.

Swetnam, T.W., Baisan, C.H., Capiro, A.C., Brown, P.M., Touchan, R., Anderson, R.S., Hallett, D.J. (2009). Multi-millennial fire history of the giant forest, Sequoia National Park, California, *USA. Fire Ecology* **5**(3), 120–150.

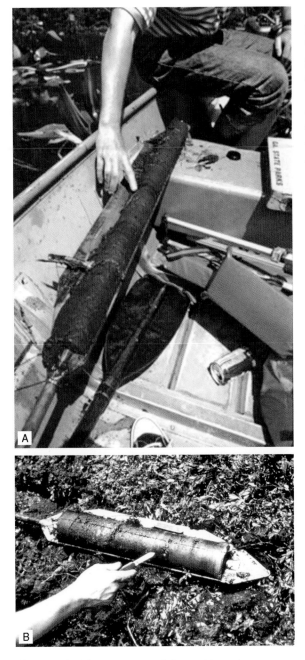

Figure 2.18   The occurrence of charcoal in peat.

A. Charcoal layer in peat core, Okeefenokee swamp, USA (photo: M. E. Collinson).
B. Core through the MacDonald Bog, NS, Canada, showing position of charcoal layer (photo: J. Calder).

Chapter 3

Fire in the fossil record: earth system processes

3.1 Fire and oxygen

We have already seen that oxygen is an important part of one of the fire triangles. Heat causes the breakdown of cellulose and lignin in plants to give rise to flammable gases. These gases need to mix with atmospheric oxygen for combustion to take place. Combustion, therefore, is an exothermic chemical oxidative reaction, and removal of oxygen leads to extinguishing a fire. Today, the Earth's atmosphere contains around 21% oxygen, which is quite sufficient to sustain large fires. In part, this is why wind is so dangerous, not only driving a fire but also replenishing the oxygen.

It has long been appreciated that the composition of the atmosphere has varied considerably over geological time. This is clear from an understanding of the evolution of photosynthetic bacteria. Another key event was the evolution of plants on land and the consequent rise of oxygen (Figure 3.1).

Biogeochemical models, however, show that oxygen levels have not remained constant through time. At some periods, it has been claimed, oxygen levels were less than 15%, yet at other times they may have been as high as 33%, such as in the late Paleozoic (late Carboniferous-Permian, 300–250 ma). There are, however, wide variations in the different models used to estimate these concentrations and, in some geological periods, there is little agreement about oxygen values (Figure 3.2). The most widely cited models are those of Berner (e.g. Berner et al., 2003; Berner, 2009) and Bergman et al. (2004) (see Lenton (2013) for detailed discussion).

There has, however, been a wide range of experiments to help understand the relationship between oxygen levels and fire in the fossil record. These experiments have involved combustion both with oxygen levels below the current figure of 21% and above that level. There has been disagreement as to the significance of some of the experimental procedures. For example, the minimum oxygen threshold for fire differs, depending on whether the data collated relates to the oxygen level at which a fire can be sustained or the level at which a fire can be extinguished. There is also disagreement on the possible maximum level of atmospheric oxygen that can be maintained before all plants (including wet plants) are destroyed.

The results can be broadly summarized as follows:

- Below 15% oxygen, fires will be suppressed.
- 16% oxygen is the minimum concentration in order for plants to ignite and for fire to be self-sustaining.
- Burn probabilities remain low until 19% oxygen.
- At 23% oxygen, fire is sustained in wetter vegetation.
- As oxygen levels increase up to 30%, wetter plants may burn.

It has been claimed that above 35% O_2, a fire could not be put out, as saturated plants can still burn. This upper limit is disputed, but undertaking experiments

Fire on Earth: An Introduction, First Edition. Andrew C. Scott, David M.J.S. Bowman, William J. Bond, Stephen J. Pyne and Martin E. Alexander.
© 2014 John Wiley & Sons, Ltd. Published 2014 by John Wiley & Sons, Ltd.

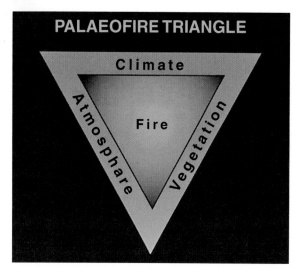

Figure 3.1 Palaeofire triangle. (From Scott, 2000). Reproduced by permission of Elsevier.

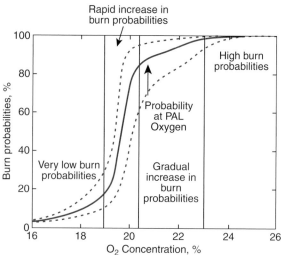

Figure 3.3 Estimated burn probabilities as a function of atmospheric oxygen concentration according to the FIREOX model, PAL, present atmospheric level (From Belcher et al., 2013, modified from Belcher et al., 2010).

with this concentration of oxygen is hazardous. Lenton (2013) argues that feedbacks in the system means that fire could not be sustained in very high oxygen concentrations, as there would be a significant reduction in fuel available to burn. He argues that feedbacks make levels above 25% more difficult to sustain.

The FIREOX model of Belcher et al. (2010) predicts burn probability (Figure 3.3) as a function of atmospheric oxygen. This indicates that there is a rapid increase in burning from 19–22%, after which it begins to plateau. This may prove significant for

arguing that fire itself may regulate many aspects of the Earth's system.

We have already seen that modelled atmospheric oxygen levels over the past 400 million years have been claimed to range from below 15% to 30%. The implications of this are profound in not only the evolution of fire systems, but also for a range of Earth System processes.

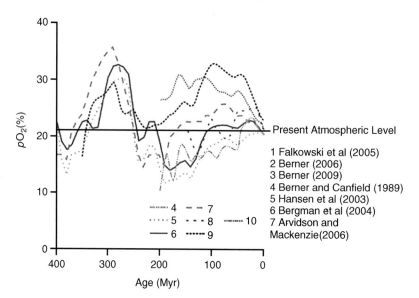

Figure 3.2 Modelled oxygen through time. (After Glasspool and Scott, 2010). Reproduced by permission of Nature Publications.

3.2 Fire feedbacks

The consequences of wildfire go far beyond simple destruction of vegetation or ecological disturbance. Impacts may have global importance. We can consider these in terms of both positive and negative feedbacks. Here we shall look at individual effects of wildfire, and in the next section we shall examine how these can be linked together to estimate overall impact.

Fire can be considered as reverse photosynthesis. In photosynthesis, sunlight, carbon dioxide and water are all linked to produce a carbon-rich plant skeleton, with oxygen as a waste product. This can be represented as a reversible chemical equation:

$$H_2O + CO_2 \leftrightarrow CH_2O + O_2.$$

Going from left to right, this represents net photosynthesis (photosynthesis minus respiration). The reverse process is representative of the oxidation of organic matter (Berner *et al.*, 2003).

We know that the burning of vegetation provides carbon dioxide, so that the carbon from the wildfire will be reused during plant growth. Extensive biomass burning may, however, produce more CO_2 than can be used by new plant growth. The increase in net CO_2 from biomass burning represents a contribution to overall rising CO_2 values that, in turn, have a significant greenhouse effect. Increasing global temperature may lead to increased fire, so this then becomes a significant feedback. Quantification of these effects are currently being undertaken by a number of authors (e.g. Van der Werf *et al.*, 2010).

We also know that charcoal may be produced from biomass burning and, as this may be relatively inert, it may be buried and act as a long-term carbon sink. This is the theory behind the use of biochar to help reduce long-term CO_2 values and thereby combat global warming via the greenhouse effect (Lehmann *et al.*, 2006).

The consequence of fire may also be more complex. Rainstorms following fires may produce post-fire erosion, leading to a significant increase in the sediment deposition rate. This sediment may incorporate not only charcoal but also uncharred plant material.

Increased burial of carbon will have two effects. The first is to cause an increase in atmospheric oxygen. The second is to affect the phosphorous cycle. Phosphorous is an important element for plant growth, and wildfire may release phosphorous into the environment. If phosphorous is released into the ocean, then this may encourage plankton blooms, which will have an impact in reducing atmospheric CO_2. If there are significant plankton blooms, the oceans may become anoxic, with the result of long-term deposition and burial of organic matter. This, too, will have significance for atmospheric oxygen levels (Berner *et al.*, 2003; Lenton and Watson, 2000; Kump, 1988; and see Lenton (2013) for detailed discussion of this topic).

Fires, therefore, play an important role, not only in effecting vegetation but also upon atmospheric evolution and climate.

3.3 Systems diagrams

We have shown that atmospheric oxygen levels play an important role in the occurrence of wildfire in deep time. The level of atmospheric oxygen can be altered or forced by a range of both geological and biological processes. In addition, there are feedbacks to consider (Lenton, 2013). These consider relationships that are positive or negative. If state A increases, then this leads to an increase in state B. This is positive feedback. If, however, state B increases and leads to a decrease in state A, this is considered to be a negative feedback.

An effective way of examining relationships between variables is to undertake a systems analysis with the aid of a system diagrams. In such a diagram, a plain arrow indicates a direct response. An arrow with an attached 'bulls-eye' represents an inverse response. If arrows are followed around the loop and the sum of the bulls-eye is zero or an even number, then there is a net positive feedback. If, however, there is an odd number of bulls-eyes, this indicates that there is a negative feedback.

Here we show the use of two such systems diagrams (Figures 3.4, 3.5). The first, developed by Berner *et al.* (2003); concerns the relationship between fire and atmospheric oxygen (Figure 3.4). We have seen that oxygen stimulates fire, and this results in multiple feedbacks. If there were an increase in oxygen, we would then expect there to be an increase in fire, which would cause a decrease in terrestrial biomass. This would lead to a reduction in oxygen, so it is a negative feedback. However, as fire also produces charcoal, this enhances organic burial as it is relatively resistant to decay and, hence, this would lead to increased oxygen. Through time, evolution may modify these effects, so that some plants may become resistant to fire, while some may produce tissues

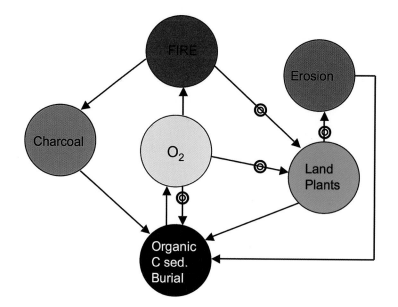

Figure 3.4 Systems diagram of fire and oxygen. Systems analysis showing the feedbacks between fire and atmospheric oxygen: arrows originate with causes and end at effects. Plain arrows indicate direct responses, while arrows marked with bulls-eyes show inverse responses. Closed loops with an even number of bulls-eyed arrows or solely plain arrows are positive feedbacks, and those with an odd number of arrows with bulls-eyes are negative feedbacks. Straight arrows lead to a positive response (e.g. oxygen increases, fires increase) and arrows with bulls-eyes are negative responses (e.g. fires increase, vegetation decreases). A closed loop with an odd number of bulls-eyes leads to negative feedback and stability. An even number or no bulls-eyes leads to positive feedback and enhancement (but not always destabilization, as can be shown mathematically). (After Berner *et al.*, 2003). Reproduced by permission of Annual Reviews.

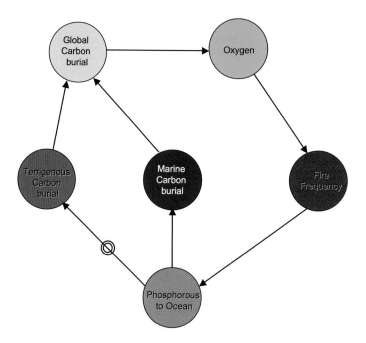

Figure 3.5 Systems diagram of fire-phosphorous feedback cycle. Positive feedbacks are shown by arrows and negative feedbacks are shown by bulls-eyes. In this case, increasing fire leads to an increase in phosphorous to the ocean which, in turn, promotes widespread anoxia and marine carbon burial. (From Brown *et al.*, 2012, after Kump, 1988). Reproduced by permission of Elsevier and Nature Publications.

that will tend to char rather than combust. The effects of fire also need to be considered, so that fire may destroy vegetation and enhance erosion. This may increase sediment yields and also organic burial.

The second concerns the phosphorous cycle (Figure 3.5). There have been two different attempts to look at this cycle (Kump, 1988; Lenton and Watson, 2000; see Lenton, 2013). Here we will consider that of Kump (1988). In this model, the disturbance of the vegetation by fire causes a transfer of phosphorous to the ocean. Marine carbon increases and may be buried. However, this is balanced by the change of organic burial on the land. This effect, Kump argues, provides a general mechanism for the regulation of atmospheric oxygen. A detailed analysis of fire and the phosphorous cycle is given in Lenton (2013).

3.4 Charcoal as proxy for atmospheric oxygen

Although there are several biogeochemical models that predict atmospheric oxygen through time, these do not agree with each other (Figure 3.2). It is only in the late Paleozoic (300–250 ma) that all models agree on a high O_2 content, perhaps 30–35% in comparison to the present at atmospheric level (PAL) of 21%. This is particularly significant, in that experiments have indicated (see section 3.1) that oxygen content over 23% means that wetter plants can burn, and over 30–35% very wet plants may burn and, indeed, rainfall may fail to extinguish a fire.

It was noted that in modern mire systems (peat-forming wetlands), fire does occur. Analysis of recent peats indicated that charcoal makes up around 4% on average, globally. However, coals of the late Palaeozoic age often contain up to 70% charcoal (Figure 3.6). It seems probable that, as oxygen levels increase, wetter plants can burn, and fires become more common and widespread during such periods. We can, therefore, use the global average percentage charcoal in coal as a potential oxygen proxy.

Glasspool and Scott (2010) used the following assumptions, based upon experimental observations:

- No charcoal should be present if the atmospheric oxygen was less than 15%.
- At 21% O_2 (PAL), global average in peats was 4%.
- The maximum level of O_2 possible for the Earth to sustain vegetation is 30 or 35%.

Using these assumptions, a mathematical equation was developed to convert percentage charcoal in coal into an atmospheric oxygen content. The resulting curve indicates a number of periods in Earth's history (Figure 3.7) where O_2 content may have played a major role in suppressing or enhancing fire (Figure 3.8).

The rise of fire in the Paleozoic (420–350 ma) may be coupled not only to the spread and diversification of plants, but also to the rise of the atmospheric oxygen, reaching a maximum in the late Palaeozoic (320–250 ma) (see Scott and Glasspool (2006) for a discussion of the significance of this).

The O_2 content appears to have varied considerably through the Jurassic (200–145 ma), but again rose significantly in the Cretaceous (145–66 ma). Both the Permian and the Cretaceous, therefore, may be considered as high fire worlds. The impact of fire in the Cretaceous is of particular significance. Recent research has suggested that the frequent occurrence of fire had an important ecological and evolutionary consequence (see Bond and Scott, 2010; Brown *et al.*, 2012; and Chapter 7, Part Two), creating the conditions for the rapid spread and diversification of early flowering plants, as well as having an impact upon dinosaur communities (Brown *et al.*, 2013).

Oxygen levels appear to have fallen from a high level in the mid-Cretaceous (100 ma) to the present level of 21% at the beginning of the Eocene (around 55 ma), and have remained at about that level until the present (Figure 3.7). Why this happened is difficult to determine, but it may be related to changing rainfall patterns and the evolution and spread of broad-leaved tropical rain forests in the Eocene (56–34 ma).

3.5 Burning experiments – fire spread

We have seen, at the start of this chapter, that experiments were performed to determine the oxygen concentration required for a fire to start or be extinguished. This data has proved very useful for the construction of atmospheric oxygen curves as indicated above.

Another approach has been to examine both fire ignition and fire spread in relation to changing atmospheric oxygen. The experiments of Belcher *et al.* (2010) showed that fire activity would be greatly suppressed below 18.5% O_2, switched off below 16% and

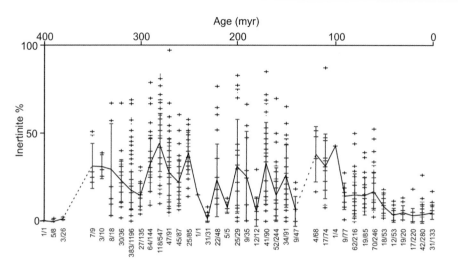

Figure 3.6 Charcoal (inertinite) in coal through time. (After Glasspool and Scott, 2010). Reproduced by permission of Nature Publications.

rapidly enhanced between 19–22% O_2. These authors used their experimental data to produce a mathematical model of fire spread in relation to different atmospheric oxygen conditions. This model, termed the FIREOX model, used oxygen outputs from Bergman *et al.* (2004) to predict the occurrence of fire in the fossil record (Figure 3.3). This model output was then compared to the fossil record of charcoal. The authors also concluded that the relationship between fire and oxygen was significant in the geological past and should inform our discussion about fire as an Earth System Process.

3.6   Fire and the terrestrial system

We have now established that there is a significant record of fire in deep time. This fact is important not only for the discussion of biogeochemical cycles, but also for a range of terrestrial ecological and evolutionary aspects. The integration of fire science into discussions of environmental and climate change are only just beginning, and there is as yet little appreciation by geologists of the significance of wildfire in the

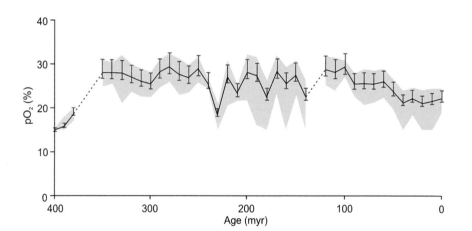

Figure 3.7 Calculated atmospheric oxygen through time using charcoal proxy. (From Glasspool and Scott, 2010). Reproduced by permission of Nature Publications.

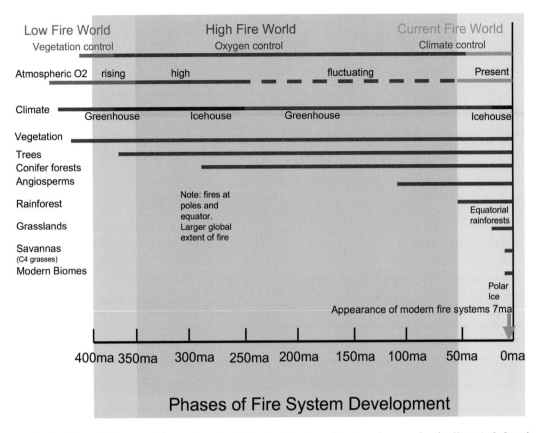

Figure 3.8 Evolution of plants, and fire systems in relation to changing climate and oxygen levels. (Data A. C. Scott).

pre-Quaternary geological record. Fire may influence a number of features of the terrestrial environment through time.

As we have seen, there is a strong relationship between fire and atmospheric composition. Not only is fire affected by changing atmospheric composition, it may also influence atmospheric composition through a range of feedbacks. In relation to atmospheric oxygen levels, we have shown that this dramatically influences not only the occurrence of fire, but also its frequency and spread. It is also probable that there is a range of significant feedbacks which both enhance and suppress fire.

Likewise, there is a relationship between fire and carbon dioxide, which may have both a long-term and a short-term effect. In the short term, extensive burning may result in increased CO_2 levels which may, in turn, lead to an increase in warming that may affect fire activity. However, the accumulation of abundant charcoal may result in a long-term reduction of CO_2 and cause global cooling. This effect may be overwhelmed by tectonic processes and weathering. Modelling can, however, allow different effects to be quantified (e.g. Beerling *et al.*, 2012).

If fire is a significant force in deep time, this is important when considering terrestrial ecology. Fire should be considered a significant disturbance factor not only for plants, but also for animals. We may see changes in plant communities as a direct result of fire, and we can also demonstrate that these fires may have had an impact on animal communities (see Scott and Jones, 1994; Brown *et al.*, 2012; and also Chapter 7, Part Two).

In many deep time studies, there is clearly a linkage between plant and animal communities and climate. As we have seen, however, increasing atmospheric oxygen levels suppress the influence of climate on fire. The development of the concept of low and high-fire worlds is very new, but it may play an important role in the evolution of several plant groups.

Recent research has demonstrated that a number of biological traits related to fire first rose in the

Cretaceous (140–65 ma), one of the fieriest Periods of earth history. It has also been claimed that the frequency and occurrence of fire may have played a role in the spread of flowering plants during the period (Chapters 4 and 7, Part Two).

The linkage between plant evolution, fire and animals is also seen with the spread of flammable C4 grasslands in the late Cenozoic (e.g. Scheiter *et al.*, 2012; see Chapters 4 and 7, Part Two). Having established the significance of fire in deep time, we can now look more closely at the evidence and interpretation of the fossil data.

Further reading

Belcher, C.M., Yearsley, J.M., Hadden, R.M., McElwain, J.C., Rein, G. (2010). Baseline intrinsic flammability of Earths' ecosystems estimated from paleoatmospheric oxygen over the past 350 million years. *Proceedings of the National Academy of Sciences, USA* **107**, 22448–22453.

Berner R.A., Beerling, D.J., Dudley, R., Robinson, J.M., Wildman, R.A. (2003). Phanerozoic atmospheric oxygen. *Annual Review of Earth and Planetary Sciences* **31**, 105–134.

Glasspool, I.J., Scott, A.C. (2010). Phanerozoic atmospheric oxygen concentrations reconstructed from sedimentary charcoal. *Nature Geoscience* **3**, 627–630.

Kump, L.R. (2010). Earth's second wind. *Science* **330**, 1490–1491.

Lenton, T.M. (2013). Fire feedbacks on atmospheric oxygen. In: Belcher, C.M. (ed). *Fire phenomena and the Earth System: An interdisciplinary guide to fire science*. 1st edition, Pp. 289–308. J. Wiley & Sons, Ltd, Chichester.

Lenton, T.M., Watson, A.J. (2000). Redfield revisited: 2. What regulates the oxygen content of the atmosphere? *Global Biogeochemical Cycles* **14**, 249–268.

Masiello, CA. (2004). New directions in black carbon organic geochemistry. *Marine Chemistry* **92**, 201–213.

Simoneit, B.R.T. (2002). Biomass burning – a review of organic tracers for smoke from incomplete combustion. *Applied Geochemistry* **17**, 129–162.

Chapter 4

The geological history of fire in deep time: 420 million years to 2 million years ago

4.1 Periods of high and low fire, and implications

We have seen that there are a number of fundamental controls on the occurrence of wildfire, not only in the modern world, but also in the geological record. There are, however, additional issues to consider. The first is the occurrence of plants as a fuel source (see Chapter 3, Figure 3.1). Until the Ordovician, and even the Silurian (480–420 ma), land surfaces are likely to have been covered by non-vascular plants living in damp habitats. It is unlikely that these plants would have provided sufficient fuel to sustain a wildfire, but they may have had a role in increasing atmospheric oxygen levels. Fossil evidence suggests the vascular plants had spread onto the terrestrial landscape by the late Silurian, and underwent rapid diversification in the Devonian (Gensel, 2008).

The oxygen content of the atmosphere also influences the occurrence of fire in deep time. Biogeochemical models have indicated that oxygen levels have varied considerably over the past 450 million years. Oxygen levels play a key role in the ignition and spread of fire. As discussed in Chapter 3.1, fire is unlikely to have been initiated with oxygen levels below 15%, and fire spread experiments suggest that at levels 17–19%, fire would have been much less likely and that over 23%, fire would have been more likely than at the present day. These experiments have suggested that as atmospheric oxygen contents increase above 25%, the moisture content of the fuel becomes less of a constraint than in the modern Earth System.

These experiments, models and proxy data all go to suggest that throughout Earth's history we have had periods of time when oxygen levels were low and we had a low fire world, periods where oxygen levels were around contemporary levels of 21%, giving us the modern fire world, and periods when oxygen levels were high and we had a high fire world (see Figure 3.8). It is in this context that we will discuss the evolution of fire systems and consider their impact on the evolution of plants and vegetation.

4.2 The first fires

Our geological history of wildfire depends predominantly on the charcoal fossil record. The debate on the occurrence of charcoal in sedimentary rocks in deep time has considerably hampered our understanding of ancient fire systems. However, now that

Fire on Earth: An Introduction, First Edition. Andrew C. Scott, David M.J.S. Bowman, William J. Bond, Stephen J. Pyne and Martin E. Alexander.
© 2014 John Wiley & Sons, Ltd. Published 2014 by John Wiley & Sons, Ltd.

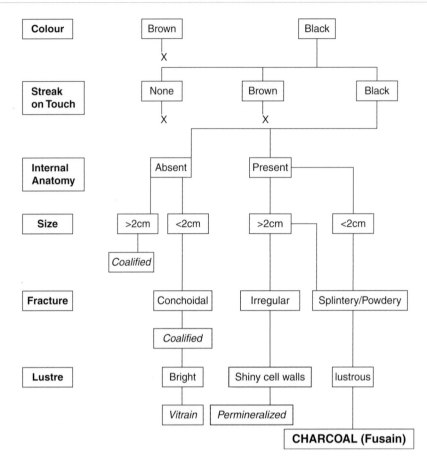

Figure 4.1   Method to identify charcoal in hand specimen. (From Scott, 2010). Reproduced by permission of Elsevier.

we have a more widely understood set of criteria to distinguish coalified plants from charcoalified plants (Scott, 2010; Figure 4.1), we can be confident that the occurrence of charcoal is more likely to be identified and published, and that the absence of charcoal is real.

This is particularly the case with the oldest charcoals. Until 2000, there were very few examples of pre-Carboniferous charcoals (older than 350 ma). This is particularly the case in early land floras, where the plants did not produce secondary wood. All of the earliest plants from the late Silurian (420–400 ma) were small – less than 10 cm tall – and they only produced primary tissues. In such circumstances, they were unlikely to have built up significant quantities of fuel to sustain a large fire. In addition, as all the plants were spore bearing, they tended to live near water, which again are areas unlikely to burn.

Despite this, we now have evidence that some of these small 'Lilliputian' plants did experience fire and are preserved as charcoal. Since that discovery, a number of early Devonian plants have also been found as charcoal (Glasspool et al., 2006). These occurrences are very rare, but they do imply that the oxygen content of the atmosphere is likely to have been over 17%.

We know that plants rapidly evolved through the Devonian, with the first trees appearing by the mid-Devonian (375 ma) and widespread fires appearing at the end of the Devonian (350 ma). Despite the spread of land plants and the potential build up of fuel, we have little evidence of wildfire from the early to late Devonian. It has been suggested that this may because of low atmospheric oxygen levels during this period (models suggest as low as 15%), and it was only when levels rose above 17%, towards modern levels, that

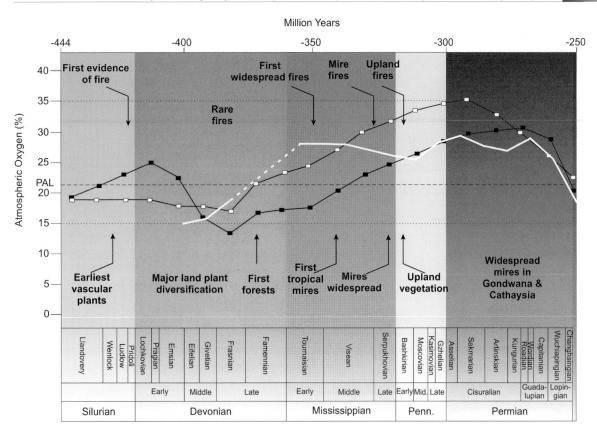

Figure 4.2 Rise of fire in the Paleozoic. Atmospheric oxygen levels taken from the model of Berner *et al.* (2003 – open circles), Berner (2006 – closed squares) and the proxy of Glasspool and Scott (2010 – white line). (Modified after Scott and Glasspool, 2006, with data from Glasspool and Scott, 2010).

extensive wildfires were recorded (Scott and Glasspool, 2006; Figure 4.2).

4.3   The rise of fire

The beginning of more widespread fires is difficult to document in the fossil record. Relatively few charcoal assemblages have been reported from the late Devonian but, as their importance is appreciated, more are being published. The reason for this is less to document fire history than to report new plants and on the anatomy of known plant fossils. The fact that fossil charcoal preserves plant anatomy is of fundamental importance in helping us to unravel ancient fire systems.

We might have expected that the spread of forests (mainly from the progymnosperm *Archaeopteris* – wood known as *Callixylon*) would have coincided with the spread of forest fires. Macroscopic charcoal

may appear to look like pieces of burnt wood (e.g. Figure 4.3A) but, until detailed microscopy is undertaken, this cannot be demonstrated. In the late Devonian charcoal assemblages of North America, some of the larger pieces of charcoal have been demonstrated to have come from ferns, not from woody trees (Figure 4.3B–D). Charcoal has been found in increasing quantity, both in terrestrial and clastic and marine clastic sediments in the very latest Devonian (Famennian, 360 ma). Assemblages have now been reported from both North America and Europe, and a wide range of plants has been found as charcoal. This data suggests that fire systems had become much more widespread by the Devonian/Carboniferous boundary (350 ma).

We can document this 'rise of fire' by examining microscopic charcoal in marine sediments. Sediment cores in the Ohio Basin, USA, during the late Devonian to early Carboniferous, show a progressive rise in

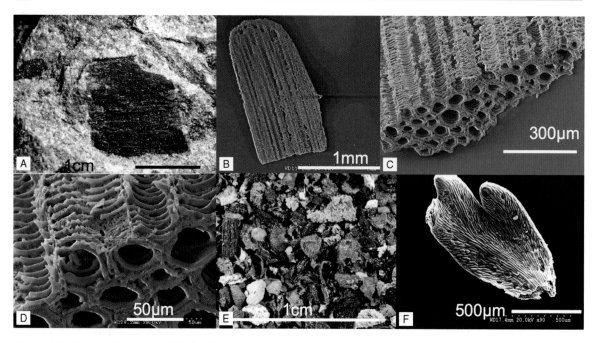

Figure 4.3 Plate of Devonian and Carboniferous charcoal.

A. Devonian charcoal fragment (*Rhacophyton*) New Hampshire, USA.
B. Scanning electron micrograph of Devonian charcoal fragment (*Rhacophyton*) New Hampshire, USA.
C. Detail of B. Tracheids of the fern *Rhacophyton*, Late Devonian, New Hampshire, USA. Scale 300 μm.
D. Detail of C
E. Charcoal-rich residue from silty limestones, Mississippian, Scotland, showing uncharred brown megaspores. Scale bar 1 cm. Charcoal fragments are shown in Figures G and H.
F. Charcoalifed pteridosperm leaf showing stomata (arrows). Scale 200 μm.

the occurrence of microscopic charcoal in the sediment (Figure 4.4).

This charcoal is recognized petrographically as inerinite (inertodetrinite), and it probably represents either wind-blown micro-charcoal or even charcoal washed into the sea from widespread wildfires on the land. It is particularly significant, as this record of increasing occurrences of wildfire seen in these marine areas coincides with an increasing number of occurrences of macroscopic charcoal found in terrestrial and near-shore marine environments. This rise of fire may well be linked to the rise of atmospheric oxygen at this time (see Figure 4.2).

The latest Devonian saw the evolution of several key biological innovations. The evolution of secondary growth was key to plants being able to grow taller and, as a result, allowing the build-up of fuel loads. The development of secondary wood was, therefore, key for the appearance of trees and forests. In addition, the evolution of secondary wood also has chemical consequences. This material, although made up of 70% cellulose, has around 30% lignin as a strengthening agent. This material decayed more slowly and also produced significant amounts of charcoal following a fire. This led to an increasing burial of organic matter in the soil and this, together with increased weathering because of the evolution of deeper rooting structures, had a profound effect upon the atmosphere (Berner *et al.*, 2003).

Even most of the trees of the latest Devonian Period reproduced using spores, so that the plants tended to be restricted to damper environments. However, in the late Devonian, some plants evolved the seed as a reproductive strategy (see Chapter 3, Figure 3.7). This freed plants from needing to be in permanently damp environments, and allowed their spread into drier environments. The combination of increased oxygen in the atmosphere, with increased fuel loads and the spread of plants into drier environments, all led to the widespread occurrence of fire by the earliest Carboniferous period (350 ma) (Figure 4.2).

G. Charcoalifed lycopsid stele from the Mississippian of Foulden, Berwickshire, Scotland. A. Whole stele, scale 500 μm.

H. Detail of tracheids showing typical thickenings with striations. Scale 30 μm.

I–L. Scanning electron micrographs of charcoalified fertile organs from the Mississippian of Kingswood, Fife, Scotland.

I. Pteridosperm pollen organ.

J. Charcoalified pteridosperm ovule with spirally arranged glandular hairs. Scale bar 500 μm.

K. Detail of J from above, showing free lobes and micropyle. Scale 500 μm.

L. Detail of J, showing glandular tips to hairs. Scale bar 100 μm.

M–N. Charcoalifed ferns and pteridosperms from the Mississippian of Pettycur, Fife, Scotland. Scanning electron micrographs of stele of fern petiole (*Metaclepsydropsis*) macerated from the rock.

M. Whole stele. Scale 1 mm.

N. Detail of tracheids. Scale 100 μm. E. detail of tracheids showing pitting in walls. Scale 50 μm.

(Figures C, D, G and N from Scott, 2010; K, L from Scott *et al.*, 2009, all others from A. C. Scott).

4.4 Fire in the high-oxygen Paleozoic world

It is in the earliest Carboniferous that we see the first really extensive charcoal deposits. Abundant charcoal is found in coaly shales from the Devonian/Carboniferous boundary of Bear Island, Spitsbergen. Some of these sediments were described as coals. When viewed petrographically, they can be shown to represent dark shales with abundant charcoal. Charcoal then becomes common in many early Carboniferous terrestrial deposits (Mississippian, 350–325 my). It is only recently that such charcoal has

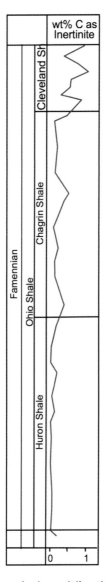

Figure 4.4 Microscopic charcoal (inertinite) in late Devonian marine sediments and the rise of fire. (Diagram from S. Rimmer, Data from Rimmer et al., 2004).

been recognized and described. SEM studies of isolated charcoal fragments have shown a wide variety of plants that were found as charcoal, including herbaceous lycopsids and seed plants (Figure 4.5).

Very thick (> 1 m) sedimentary units containing abundant charcoal were discovered at several localities in western Ireland in sequences dating from the middle part of the Mississippian (Viséan) (310 ma). The charcoal in the Shalwy Beds in Donegal is found in near-shore marine estuarine sandstones (Figure 4.5 a, b). The sediments are widespread in the Donegal

area and the rocks are comprised of over 10% charcoal. It has been suggested by Nichols and Jones (1992) that the charcoal is the result of major wildfires across the region, and that extensive post-fire erosion, transport and deposition led to the formation of these charcoal-rich deposits. These authors calculated that the fire covered an area greater than 70 000 km².

SEM studies of some of the charcoal have also shown that it comprises both woody plants as well as herbaceous lycopsids, such as the small *Oxroadia* (Figure 4.5 C,D; Scott, 2000). Other localities in western Ireland with rocks of similar age (e.g. Mayo) also contain abundant charcoal. A similar scenario existed here to that seen at Donegal, whereby major wildfire produced abundant charcoal which, together with sediments produced by post-fire erosion, was transported into estuaries. This, it has been proposed, had a catastrophic effect upon the environment, killing fish and other animals. It is the earliest example of environmental catastrophe caused by a wildfire in the fossil record (Falcon-Lang, 1998) (Figure 4.6).

Charcoal became increasingly common in the Carboniferous in a wide range of settings (Figure 4.2). It has been claimed from both models and proxies that atmospheric oxygen levels were elevated above the present atmospheric value of 21% during the Carboniferous, reaching peaks of 30% or 35%. The timing of this peak is uncertain, with early models suggesting that it occurred in the late Carboniferous (315–300 ma), while later models and proxies suggest a Permian high (300–250 ma) (Figure 4.2). What is clear is that charcoal becomes increasingly common in coal (fossil peat) deposits.

We have seen that, in modern peats, charcoal may be only 4% of the total volume. In the Carboniferous, charcoal was often more than 20%, and this has been used to suggest that there were increased fires during this period, due to high atmospheric oxygen (see Chapter 3, Figure 3.6). Fires were abundant in many of the tropical wetland systems at this time. Recognition of these ancient fires has led to an increasing appreciation of fire as an ecological and evolutionary force (Scott and Jones, 1994; Falcon-Lang, 2000). It is possible that some plants evolved thick bark and periderm as a fire-adapted trait.

The late Carboniferous (Pennsylvanian 315–300 ma) was a period of rapid plant diversification (Figure 4.7). As we have noted, some plant groups evolved seeds that allowed them to live in drier

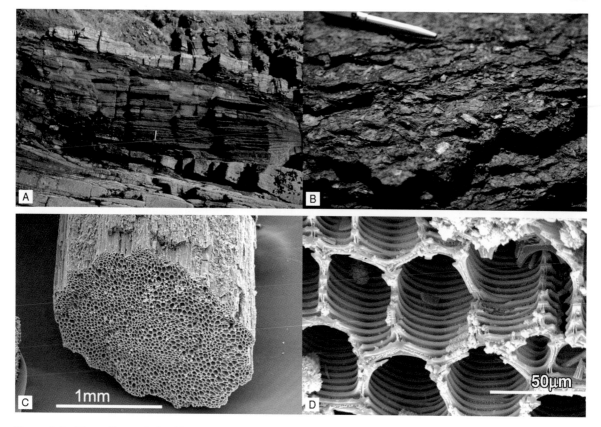

Figure 4.5   Plate of Lower Carboniferous charcoal from Shalwy, Ireland.

A. Outcrop of charcoal-rich near-shore marine sediments (Photo A. C. Scott).
B. Detail of sediment surface, showing abundant charcoal (Photo G. Nichols).
C. Stele of the herbaceous lycopsid, *Oxroadia*, from B. stele. Scale 1 mm (from Scott, 2010). Reproduced by permission of Elsevier.
D. Detail of tracheids, scale 50 μm (from Scott, 2010). Reproduced by permission of Elsevier.

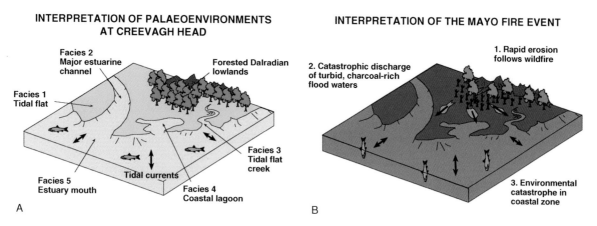

Figure 4.6   Carboniferous fires in Mayo – a catastrophe in the estuarine zone. (After Falcon-Lang, 1998). Reproduced by permission of Elsevier

A. Before fire.
B. Post fire.

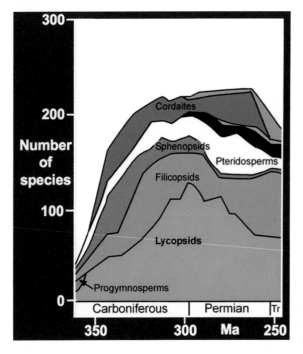

Figure 4.7 Evolution of plants in the Carboniferous. Data from Niklas *et al.*, 1985.

environments. One group – the large tree-like cordites that had long strap-shaped leaves – is found in a wide range of environments, from coastal plains to upland sites. Charcoalified wood of these plants is common in a wide range of sediments. This is often associated with very large silicified logs, and it is possible that some of these represent log-jams that resulted from major forest fires in the Carboniferous hinterlands (Falcon-Lang and Scott, 2000).

The earliest conifers also evolved at this time, and many of the oldest records in the Carboniferous are from charcoal. It has been suggested that these plants evolved in an extra-basinal setting, possibly in drier uplands where they were subjected to regular wildfires. A recent discovery of Carboniferous cave deposits to the north of Illinois basin has yielded both charcoal and geochemical evidence of fires in conifer-domi-nated vegetation (Scott *et al*, 2010; Figure 4.8).

Our understanding of Permian (300–250 ma) fire systems has developed rapidly over the past few years. The paucity of Permian charcoal records is apparent rather than real, and an increasing number of localities have been discovered, both in the northern and southern hemispheres. The significance of the more recent studies is that not only is charcoal reported, but also the plants

that are preserved are being studied and identified. A wide variety of plants (dominated by tree-like conifers and other seed plants) have been found as charcoal.

Both biogeochemical models and oxygen proxies suggest that atmospheric oxygen levels were high through the Permian, always above 25% but reaching a peak of 30% or 35% at some stage during this period (see Chapter 3, Figure 3.6). This is well demonstrated by the abun-dance of charcoal in Permian coals (Figure 3.6), where it may reach a level as high as 70% and is often over 40%. This is the case both in the widespread Gondwana coals of the southern hemisphere, as well as the northern tem-perate coal deposits of Angaraland (Siberia) (Figure 4.9).

It is not only the total quantity of charcoal in Permian coals that is significant, but also its distribu-tion. Most coals are studied using crushed coal pellets that are examined in reflected light under oil. This method, which is an industrial standard for the study of coal, destroys the spatial data of charcoal occur-rence. Polished blocks of *in situ* coal pillars, however, offer the opportunity to establish the value and dis-tribution of charcoal in coal. Studies from the late Permian coals of the Kusnetsk Basin (Siberia, Russia) have allowed the fire return interval to be calculated in these fossil peats. The vegetation was dominated by arborescent gymnosperms that are similar to the cordaites of the Carboniferous. The coals contain both horizons of charcoal, as well as scattered charcoal in some cases (Figure 4.10). This charcoal has been interpreted as having come predominantly from sur-face fires. Calculations on fire return intervals suggest a much higher frequency than in peats of the present day, and this supports the view of higher atmospheric oxygen levels at this time (Hudspith *et al.*, 2012).

Frequent wildfires, both from the charcoal record and occurrence of pyrolitic PAHs are reported in the late Permian up to the Permo-Triassic boundary in China.

4.5 Collapse of fire systems

The Permian-Triassic boundary (250 ma) provides the most important extinction event of all time, affecting both marine and terrestrial ecosystems. Perhaps up to 90% of all living species became extinct. This fundamen-tal collapse of the Earth System, when 'life almost died', may have been brought about by a number of events. The formation of the supercontinent Pangea in the Permian played a significant role in altering the climate and movement of the oceans. During the late Permian, the

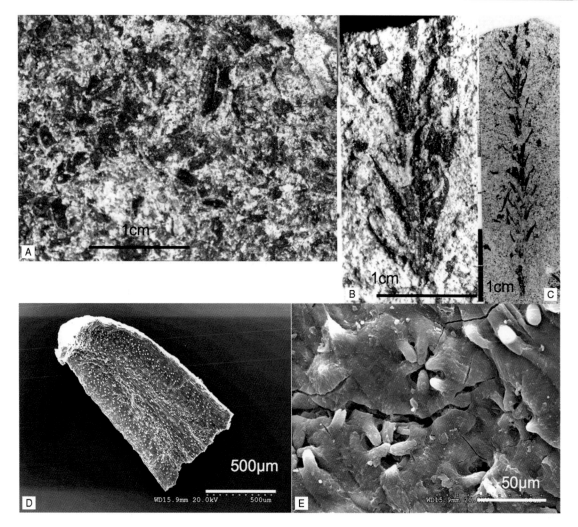

Figure 4.8   Plate of charred Carboniferous conifers. Charcoalifed conifer needles from mid-Pennsylvanian of cave deposits of Illinois, USA.

A. Charred needles in sediment.
B. Charred leafy shoot.
C. Charred leafy shoot.
D. Scanning electron micrograph of whole needle, scale 1 mm.
E. Detail of stomataliferous surface, showing two rows of stomata with papillae. Scale 30 μm.

(Figures A, C, D, E from Scott *et al.*, 2010, B. photo from R. Plotnick). Reproduced by permission of Elsevier.

glaciers, such an important part of the late Paleozoic, receded and methane may have been released, creating conditions for global warming.

It was probably the eruption of the Siberian trap volcanoes that was the 'nail in the coffin' for most life. A sudden massive injection of CO_2 and SO_2 into the atmosphere created a potent mix. Temperatures rose, possibly as much as 8 °C, and the sulphur resulted in

acid rain. This would have devastated life, both on land and in the sea.

Conditions would have remained difficult for several million years after the events at the boundary, and biogeochemical models indicate a rapid fall of oxygen. The burial of carbon was restricted, leading to a gap in our coal record in the early Triassic. Under such conditions, we may not expect to see a fossil record

Figure 4.9 Charcoal in Permian coal (Photos A. C. Scott).

A. Thick Permian coal seam, Kusnetsk Basin, Siberia, Russia that contains abundant charcoal.
B. Bedding surface of coal with wood charcoal fragments.

of fire. Models suggest a Phanerozoic low for O_2 levels, perhaps only 15% and, under such circumstances, we should not expect to find charcoal in the earliest Triassic rocks. Unfortunately, the record of terrestrial rocks of this age is scarce. Few researchers have studied the early Triassic marine sediments for charcoal, and there is always the danger that rare scattered records may represent charcoal eroded from older rocks.

Our earliest post-fire collapse charcoal comes from the early middle Triassic and becomes more common from that period (Figure 4.11). By the late Triassic, significant fire systems have been re-established and it is likely that O_2 levels were back to PAL or even higher, as late Triassic coals contain significant quantities of charcoal.

4.6   Fire at the Triassic-Jurassic boundary

The Triassic-Jurassic boundary (200 my) is another important period in Earth's history. There are significant extinctions, both in marine and terrestrial ecosystems. Studies on the distribution of plants and charcoal in sediments that span the Triassic-Jurassic boundary in Greenland indicate major changes in the fire system (Belcher *et al.*, 2010). It is probable that there was a significant rise in atmospheric CO_2 at this time, leading to a significant global temperature rise – as much as 4 °C. It has also been suggested that, at this

time, oxygen levels were above the PAL of 21%. It is claimed by the authors that increasing temperature would have led to an increase in lightning strikes and, hence, the ignition of fires.

Charcoal is shown to have increased in abundance in the early Jurassic (up to five times that of the late Triassic) and indicates a sharp rise in fire following the boundary (see Chapter 3, Figures 3.6 and 3.7). In addition, the nature of the vegetation changed, so that there was a climate-driven shift from broad-leaved taxa to predominantly narrow leaved assemblages. Belcher *et al.* (2010) showed from experiments that narrow leaved forms are more flammable than broad-leaved forms; this, coupled with the temperature rise, created a feedback that saw the increase in fire after the boundary.

4.7   Jurassic variation

In a classic paper, Harris (1958) documented the occurrence of charcoal in Jurassic sediments and noted for the first time the significance of fire in the fossil record. Despite this early recognition, the number of charcoal records in Jurassic sediments is relatively low, despite their sediments representing more than 50 ma of time. A range of atmospheric oxygen models suggested that the Jurassic was predominantly a period of low oxygen. However, recent oxygen proxy data suggests that O_2 levels remained at, or above, the PAL of 21% throughout this period (see Chapter 3, Figure 3.3). When the level was at PAL, or even

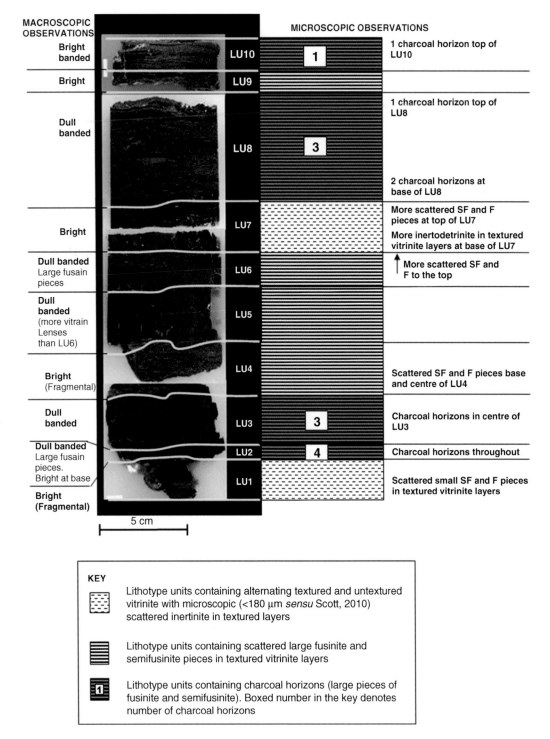

Figure 4.10 Charcoal in Russian Permian coal showing numbers of charcoal layers in several lithological units (LU). (From Hudspith *et al.*, 2012). Reproduced by permission of Elsevier.

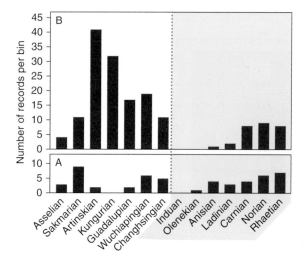

Figure 4.11 Records of charcoal from the Permian and Triassic (From Abu Hamad *et al.*, 2012). Reproduced by permission of Elsevier.

A. Reported occurrences of macroscopic charcoal.
B. Reported occurrences of inertinites/black carbon.

perhaps slightly below, changes in climate may have had a major impact on the distribution of wildfires. Harris (1958) described charred conifers found within caves or crevices within a limestone landscape, and these charcoals may be the remains of more upland fire regimes.

There has been an increasing number of charcoal records of fire found within the Jurassic, as well as records based upon the co-occurrence of charcoal and PAHs, both in the northern and southern hemispheres (Marynowski *et al.*, 2011).

Many assemblages, including ones from the middle Jurassic (170 ma) of Yorkshire, England, are dominated by gymnospermous wood charcoals. These include not only conifers but also gymnospermous taxa such as *Ginkgo*. It should be noted, however, that both here and elsewhere, the associated flora is rare in conifers and abundant in cycads and cycadeoids, indicating that perhaps not all vegetation types were subjected to frequent wildfire.

4.8 Cretaceous fires

The Cretaceous is a particularly important period of Earth history (145–65 ma). It is not only a period when dinosaurs dominated terrestrial faunas, but also one when land plants underwent one of the most dramatic

changes since they evolved on land. It is through this period when the flowering plants (angiosperms) first evolved and spread and, by the end of the period, they dominated many regions of the Earth (Figure 4.12). We need to consider the Cretaceous as a very different world from the one that we are familiar with today. First, it was a greenhouse world, with CO_2 levels and global temperatures much higher than those of today. No only was there no polar ice, but fossil evidence shows there were extensive high latitude (and even polar) coniferous forests. Many of the vegetational zones that have been identified are broader than those in the present day.

The vegetation changed considerably through the Cretaceous (Figure 4.12). Floras were dominated by conifers, ferns, cycads, bennettitaleans and ginkgophytes. It was in the early Cretaceous that angiosperms (flowering plants) first evolved (Figure 4.12). They are thought to have been understorey clonal plants living in 'damp, dark and disturbed' woodland habitats (Feild and Arens, 2005). By the mid-Cretaceous, angiosperms had diversified into open, sunlit habitats and drier uplands. They are thought to have had weedy attributes, such as small seeds, rapid reproduction and rapid growth, so it is thought that these were weedy plants that thrived well in disturbed environments.

Both palynological and paleobotanical evidence suggest that these plants evolved and spread rapidly through the Cretaceous but that is was not until the late Cretaceous that angiosperms became to dominate some vegetation, and that angiosperm-dominated forests evolved. By the end of the Cretaceous, angiosperms had increased their global distribution from high to low latitudes. The genera of conifers differ, depending on palaeolatitude.

There has been considerable debate as to the levels of atmospheric oxygen in the Cretaceous. Early models suggested this was a period of high O_2, but subsequently it was even been suggested that it was a period where oxygen was low. The widespread occurrence of charcoal in Cretaceous sediments, however, suggested that model estimates at PAL, or even above, were likely. O_2 estimates from the charcoal in coal proxy (Figure 4.12) suggest that O_2 levels were above PAL throughout the Cretaceous and reached a peak in the mid-Cretaceous (100 ma), which may have been as high as 30%. The significance of fire in the Cretaceous needs to be considered in this context (see Brown *et al.* (2012) for discussion).

Charcoal has been widely described from the early Cretaceous deposits (Figure 4.13) since charcoalified ferns were first described from the Wealden rocks of

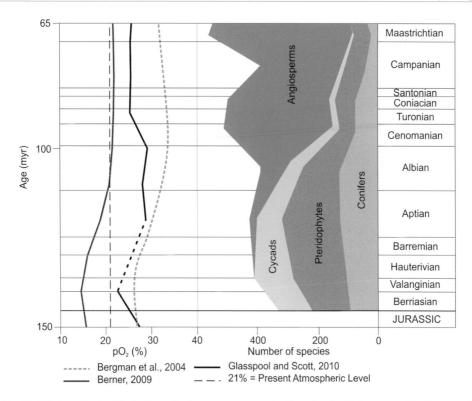

Figure 4.12 Modelled and calculated atmospheric oxygen concentrations in the Cretaceous with the global change in vegetational composition. (From Brown *et al.*, 2012). Reproduced by permission of Elsevier.

Palaeogeography Map- Albian to Santonian

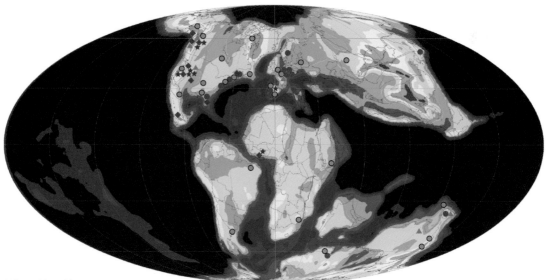

◉ Charcoal Assemblage
● Charcoal Assemblage with charred angiosperm reproductive organs
★ Coal with inertinite values above 20%
◆ Coal with inertinite values of 10%–20%
▲ Coal with inertinite values below 10%

Figure 4.13 Geographical distribution of charcoal assemblages, with and without angiosperm fertile organs, in the Cretaceous. Plate reconstructions by C.R. Scotese (Paleomap project), (figure by S. Brown after Brown *et al.*, 2012). Reproduced by permission of Elsevier.

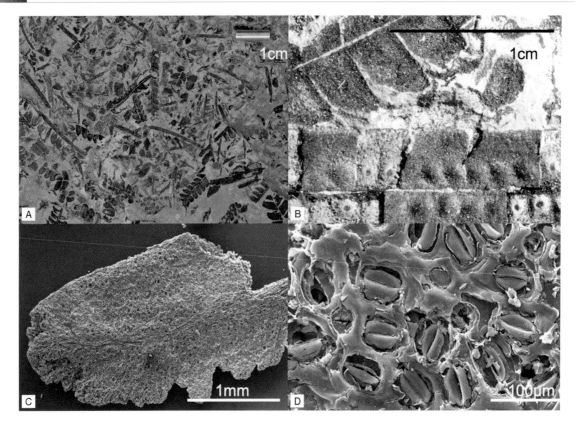

Figure 4.14 Plate of Cretaceous charred plants. Charred plants from the Wealden of Shepherds Chine, Isle of Wight, England.

A. Siltstone with charred ferns, mainly *Weichselia*.
B. Detail showing charred ferns of *Weichselia* and *Phlebopteris*.
C. Scanning electron micrograph of charred pinnule of the fern *Weichselia*.
D. Detail of C. showing abundant stomata.

Southern England (Figure 4.14). Charred ferns have been reported throughout the early Cretaceous, and several charcoalified assemblages have been described. Many of these assemblages are also found in terrestrial sequences that yield charcoalified conifers. These assemblages appear to show that, during the early Cretaceous (145–125 ma), there was widespread fire both in conifer forests and also in 'fern-prairies'. Frequent fires would have helped to maintain this fern-prairie habitat.

It is in rocks of this age, predominantly in Europe (but also slightly later in North America), that the first angiosperms occur. Significantly, many of the small flowers are preserved as charcoal (Figure 4.14H,I). In Portugal, the plant assemblages comprise of a range of taxa, including conifers and angiosperms, with abundant charcoalified flowers. It is suggested, however, that all the early angiosperms were predominantly weedy and thrived in disturbed environments. Large angiosperm trees did not evolve until later in the Cretaceous.

Fire represents a major disturbance mechanism, and it has been suggested by Bond and Scott (2010) that fire may have accelerated the diversification and spread of early angiosperms. The Cretaceous should be considered as a 'high-fire' world, where the moisture content of fuels may have played a less significant role than today. Recent molecular studies have indicated that some groups with fire-adapted traits may have evolved during this time period (He *et al.*, 2012) (see Chapter 7 for discussion on fire traits).

The full impact of fire in the Cretaceous has yet to be established but, as noted earlier, this was a period when dinosaurs dominated the landscape. Clearly, extensive and frequent wildfires would have had a major impact upon their food sources and habitat.

E. Coarse sandstone with conifer wood charcoal.

F. Charcoal from the Campanian sediments of Dinosaur Provincial Park, Alberta, Canada. Scanning electron micrograph of charred gymnosperm wood showing distinctive growth rings.

G-I. Scanning electron micrographs of charred plants from the Santonian of the Allon flora, Georgia, USA.

G. Charred fern.

H. and I. Charred flowers.

(Figures A, B, E, F–I from Brown *et al.*, 2012, C,D photos from A. C. Scott). Reproduced by permission of Elsevier.

Further, extensive post-fire erosion-deposition may have played a significant role in the formation of some dinosaur bone beds (Brown *et al.*, 2012).

4.9   Fire at the Cretaceous-Paleogene (K-P or K-T) boundary

The Cretaceous-Paleogene (K-P) boundary (also known as the Cretaceous-Tertiary or K/T boundary) represents one of the most important periods in Earth history. It has long been known that wide ranges of animals, both marine and terrestrial, went extinct across this boundary. In popular imagination, this is when the dinosaurs become extinct, and the cause of this has been the subject of considerable controversy. Some authors have pointed to the major volcanism of the Deccan traps in India. However, many now accept the occurrence of a major asteroid impact at this time as being a likely cause of the extinction event. In the context of the excitement of this discovery, some soots recovered from marine sections across the boundary were claimed to represent evidence of a global fire at the boundary following the impact (Wolbach *et al.*, 1980).

Over the subsequent 30 years, the extensive analysis of Cretaceous sequences has indicated that the period was one of high natural wildfire activity. Studies on a large number of terrestrial sites have now demonstrated that there was no global fire at this boundary and, even now, models of the impact do not suggest such a scenario (see Belcher (2009) for summary).

4.10   Paleocene fires

Oxygen proxy data suggests that atmospheric oxygen levels remained above PAL through the Paleocene (65–55 ma) (see Chapter 3, Figure 3.6). Charcoal is abundant in many coal seams, which suggests that oxygen levels may have been as high as 24% through the period but beginning to decline. Charcoal occurs commonly in Paleogene sediments, but few assemblages have yet been described. It should be noted that, however, in contrast to the Cretaceous, where charcoalified flowers have been widely found, none have been reported or described from the Paleocene. The reasons for this are currently not known.

4.11   Fires across the Paleocene-Eocene thermal maximum (PETM)

The end of the Paleocene (55 ma) shows a period of rapid global warming. There has been considerable interest in this event, both in terms of the causes and impacts upon the environment, as it is considered to provide us with lessons concerning future global warming. At this time, there was a sudden increase in global CO_2 levels that resulted in a rapid and short-lived period of elevated temperatures (estimates range from 2–4 °C). The reasons for this sudden CO_2 rise is hotly debated, and ideas range from sudden methane release from methane clathrates (unstable methane deposits in the uppermost sediments on the sea floor) to CO_2 emissions from volcanic activity.

There are relatively few sequences that show this boundary in terrestrial sediments. One such locality is in southern England, which can be linked to a fire story. The Cobham Lignite was exposed by the extension of the Channel Tunnel rail link into London. Carbon isotopic data and palynology demonstrated that the sequence through the Cobham Lignite showed the beginning of this rapid warming event. It was noted by Collinson *et al.* (2007) that the lower part of the lignite sequence was dominated by ferns. These authors suggested that there were frequent fires through this fern-dominated vegetation (Figure 4.15). However, they also showed that there was a sudden change to a period of no fire. The reason for this is unclear, but it may relate to an increase in rainfall.

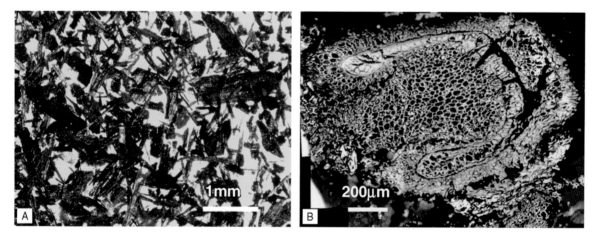

Figure 4.15   Plate of Cobham Lignite across the start of the Paleocene-Eocene Thermal Maximum. The base of the section contains abundant charred ferns.

A. Charcoal macerated from the lower laminated lignite. Many of these fragments are of ferns. (Photo M. E. Collinson).
B. Reflectance image under oil made from a montage of 56 images of charred fern axis (photo from Collinson *et al.*, 2007). Reproduced by permission of the Geological Society of London.

4.12   Dampening of fire systems

The oxygen proxy data suggests that, at this time, atmospheric O_2 stabilized at around the present level of 21% and remained there from 50 my ago to the present time (see Chapter 3, Figure 3.7). Oxygen can no longer be considered a significant driver of wildfire from this time (Figure 3.8). It may also be noted that charcoal is very rare in the Eocene (55–35 ma), suggesting a relatively low fire or modern fire world in contrast to the earlier high fire world (Figure 3.8). We also note that this was the period that saw the evolution and spread of angiosperm-dominated tropical rainforests, which are not naturally high-fire vegetation. The detail and causes of this major transition on fire systems has yet to be established.

4.13   Rise of the grass-fire cycle

Until now, there has been little interest and few studies on the evolution of late Cenozoic fire systems (35 ma–1 ma). This period, however, represents a time when grasslands first appeared and spread, and their significance in relation to fire has been appreciated only over the past few years. It has now been firmly established that C4 savanna grasslands are maintained by fire, and that there is a significant grass-fire cycle (see Chapter 8, Part Two). The timing of this event is interesting, and can be traced using carbon isotope techniques in addition to the fossil record of grasses (see Osborne and Beerling (2006) for discussion).

It has long been recognized that there has been an increase in charcoal occurrence in deep marine sediments (Figure 4.16) over the past 7 my, and this coincides with the spread of C4 savanna, supporting the idea of a coupled fire-grass system (Bond and Keeley, 2005). Grass charcoal has been reported from late Cenozoic deposits

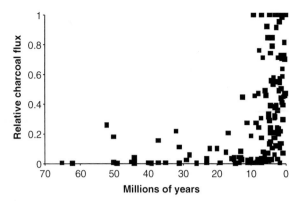

Figure 4.16   Charcoal in deep sea sediments over the past 55 million years. (From Bond and Scott, 2010, after Herring, 1985).

in Africa. It is now important to develop Cenozoic charcoal studies, in order to help flesh out ideas on the evolution of flammable grassland systems.

Further reading

Belcher, C.M. (2009). Reigniting the Cretaceous-Paleogene firestorm debate. *Geology* **37**, 1147–1148.

Belcher, C.M., Collinson, M.E., Scott, A.C. (2012). A 450 million year record of fire. In: Belcher, C.M. (Ed), *Fire Phenomena and the Earth System – An Interdisciplinary guide to Fire Science*, pp. 229–249. Wiley-Blackwell, Oxford.

Bond, W.J., Midgley, G.F. (2012). Fire and the angiosperm revolutions. *International Journal of Plant Sciences* **173**, 569–583.

Pausas, J.G., Keeley, J.E. (2009). A burning story: the role of fire in the history of life. *Bioscience* **59**, 593–601.

Scott, A.C. (2000). The pre-Quaternary history of fire. *Palaeogeography, Palaeoclimatology, Palaeoecology* **164**, 281–329.

Scott, A.C., Glasspool, I.J. (2006). The diversification of Paleozoic fire systems and fluctuations in atmospheric oxygen concentration. *Proceedings of the National Academy of Sciences, USA* **103**, 10861–10865.

Chapter 5

The geological history of fire – the last two million years

5.1 Problems of Quaternary fire history

In the last chapter, we considered the evolution of fire in the natural environment. Humans, however, are characterized as a species that uses fire (see Part Three). With the use of fire, humans are able to change a number of aspects of fire occurrence and spread, and so complicate any analysis of future fire. The most recent evidence from caves in South Africa is that humans were using fire, at least to keep warm and cook, as far back as one million years ago (Berna *et al.*, 2012). While the early use of fire by humans may not have had a significant impact upon natural fire systems, it is clear that by at least the evolution of *Homo sapiens* and their spread across the world, a number of aspects of fire in the natural environment may have been affected. In this context, it is important to examine not only the occurrence of fire, but also how the interwoven strands of fire, climate and human modification can be unpicked (Figures 5.1, 5.2; see chapters 10, 11, Part Two).

In Chapter 2, we examined the way in which we can obtain evidence concerning the history of fire. In deep time, we are constrained by our inability to date sequences with sufficient resolution to interpret a human influence on fire systems. However, as we move towards the present, an increasing number of techniques are available to us to help unravel fire history (Figure 5.3).

Traditionally, the occurrence of fire through the Quaternary (from 2.6 my) has been documented through the study of charcoal occurrence in lake and peat sequences. The dating of these sequences can be difficult, as either scattered dated points or broad biostratigraphic or climatic events are used for correlation. The lack of resolution allows only a limited number of questions to be asked of the data.

In addition, a wide range of methods has been used to collect the data. In many cases, the record of fire has been the result of identifying – and sometimes counting – charcoal fragments on palynological slides, which have often been prepared for vegetational or biostratigraphical analysis. In such cases, most charcoal will be less than 180 μm and sometimes less than 125 μm, depending on the preparation technique that was used. Using this technique, broad correlations may be made relating fire occurrence to specific vegetational regimes or climate states (Figure 5.4).

In rock sequences younger than 70 000 years old, the ability to use carbon dating with increasing precision allows a much more accurate fire history to be developed (see Power (2013) for a detailed analysis). When this is combined with the counting of varves in lake sediments, it is sometimes possible to examine relationships between climate history and fire return interval, for example.

Fire on Earth: An Introduction, First Edition. Andrew C. Scott, David M.J.S. Bowman, William J. Bond, Stephen J. Pyne and Martin E. Alexander.
© 2014 John Wiley & Sons, Ltd. Published 2014 by John Wiley & Sons, Ltd.

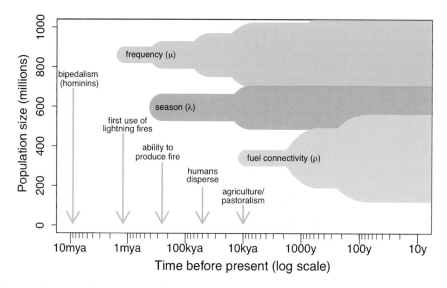

Figure 5.1 Fire and human activity. Stages of human evolution defined by their ability to manipulate μ (frequency of ignition events), λ (timing of ignition events), and ρ (the connectivity of the fuel bed). The thickness of the coloured lines gives a rough representation of the magnitude of the effect. (From Archibald *et al.*, 2012).

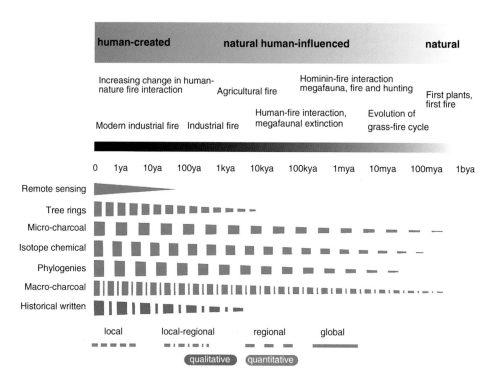

Figure 5.2 The evolution of fire and humans. Summary of the available historical sources and palaeoecological proxies to reconstruct fire regimes, spanning the period from the advent of fire on Earth in deep time to the modern industrial period characterized by fossil fuel combustion. The spatial and temporal resolution of all these approaches varies and decays with increasing time depth, constraining our understanding of fire regimes, especially before the Industrial Revolution. (From Bowman *et al.*, 2011).

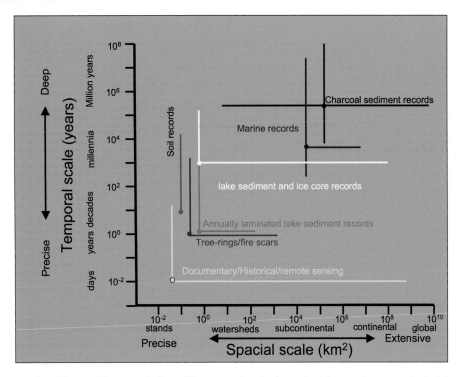

Figure 5.3 Records of fire activity recorded at different scales. Each method is complementary in helping to make us understand fire on Earth. (Modified from a number of authors including T. Swetnam, D. Gavin, C. Whitlock and others).

Our oldest records of fire based upon fire scars in trees can be traced back to 3000 years ago. Increasingly from this time, and especially over the past 500–1000 years, it has been possible to link fire data derived from lake and peat sequence with fire scar data (see Chapter 2). This is a powerful tool that allows the development of a clearer understanding of the relationship between fire and climate. The data from fire scars, together with tree-ring data, also helps in seeing human modification of fire (both on a yearly basis and within a year), and also an increase in fire above the natural background or the suppression of fire.

In historic time, both of these records can be linked to verbal, written or other historical collections of data. This may be particularly useful in more recent times in helping to validate the interpretation of charcoal records in lakes, for example. The scattered records and range of methodologies has often hampered a wide use of fire data to unravel aspects of climate change. This has led to a number of initiatives to help the standardization of charcoal data, in particular, so that this can then be examined, both on its own and also in conjunction to climate models.

5.2 The Paleofire working group: techniques and analysis

There are several basic problems in the interpretation of fire from recent (<70 000 years) sedimentary records:

- How to record the charcoal data from individual sequences?
- How to compare data from different areas and using different collection techniques?
- How to date the sequence to a significant resolution to make fire frequency meaningful?
- How to be sure that the charcoal record is recording actual fire events?

These questions cover a broad range of fundamental issues that are still being debated in the literature. Perhaps one of the most important issues is how to record charcoal occurrence and how to

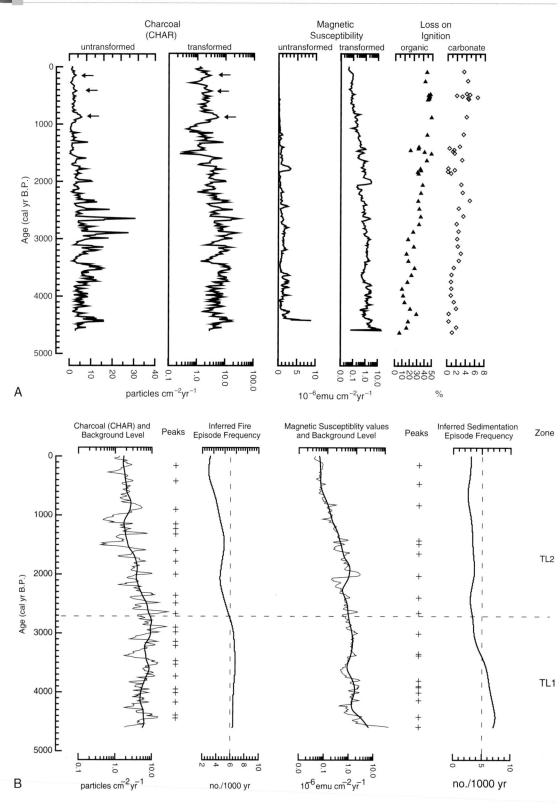

Charcoal (CHAR)
untransformed   transformed

Magnetic Susceptibility
untransformed   transformed

Loss on Ignition
organic   carbonate

particles cm⁻²yr⁻¹

10^{-6} emu cm^{-2}yr^{-1}

%

A

Charcoal (CHAR) and Background Level

Peaks

Inferred Fire Episode Frequency

Magnetic Susceptibility values and Background Level

Peaks

Inferred Sedimentation Episode Frequency

Zone

particles cm⁻²yr⁻¹

no./1000 yr

10^{-6} emu cm^{-2}yr^{-1}

no./1000 yr

TL2

TL1

B

interpret the data in terms of fire occurrence. The most serious issues concern charcoal taphonomy, i.e. how the charcoal is produced, transported and fragmented during several phases of its journey to a depositional environment, and how it may change even during the collecting and processing procedure.

As we have already seen, the amount and nature of charcoal (including which organ is charred, size, fragmentation, physical and chemical properties) is controlled by a wide range of variables, including the nature of the original vegetation, surface or crown fire, fire temperature, etc. Experiments have shown that the transport and fragmentation behaviour of charcoal may be influenced by the temperature at which it was formed (Scott, 2010). This leads to a fundamental question: how should charcoal be recorded in a sedimentary sequence (and *what* should be recorded)?

First, the size of charcoal fragments needs to be considered. Charcoal may be produced both by crown and surface fires. Crown fires tend to produce smaller charcoal fragments that may be predominantly wind-blown. Observations on recent crown fires have shown that small charcoal fragments may be lifted and distributed by wind. It has further been observed that larger fragments fall before smaller fragments (Clark, 1988).

It has also been demonstrated that large quantities of larger (>1 mm) charcoal fragments may be pro-duced from surface fires (Scott, 2010). In such cases, the charcoal may be removed from the site of the fire by water, often by overland flow initially (see Chap-ter 2). Experiments have shown that if charcoal enters a water body, then larger fragments take longer to sink and, hence, may be transported further than smaller fragments. Organ type, size and temperature of forma-tion, may all affect transport of the charcoal (Nichols *et al.*, 2000), so that the simple observation of large charcoal fragments does not necessarily imply a local fire.

Attempts at standardization of the recording of charcoal in lake sediments have involved using data from the fraction above 125 μm charcoal size. The next consideration is of how the charcoal may be quantified. Ranges of optical and chemical methods have been used, each giving quite different data sets. Comparing volume or numbers of fragments are also possibilities, and attempts have been made to compare the two data sets. A large volume of charcoal may represent one fragment, but a large number of frag-ments may have less volume.

Equally problematic is the possibility of charcoal fragmentation during collection and/or processing, so that one large fragment in the rock may produce a number of small fragments in the sample being counted. Some authors fail to appreciate this problem, so data should always be carefully analyzed by taking these possibilities into consideration.

In recent years, there has been a clear attempt to compile datasets that allow meaningful consideration, including the development of publicly available software called *CharAnalysis* (http://sites.google.com/site/charanalysis). This software requires the input of the number of charcoal fragments of greater than 125 mm from a sample. It also requires a detailed age model for the sample set, based on a number of dated points in the sequence. This approach allows data from each time interval (e.g. year) to be compared quantitatively with samples above and below.

The data (for example, Figure 5.5) shows that the quantitative data can be 'decomposed' with two com-ponents: a slowly varying background component that is considered to represent the average amount of charcoal through time, and the high frequency varia-tions or peaks component that might be associated with distinctive fire episodes. These peaks, when com-bined with the age data, can provide estimates of fire frequency and/or fire return intervals by establishing how often fire occurs in a system.

Figure 5.4 Charcoal from a Quaternary lake sequence. (After Long and Whitlock, 2002). Reproduced by permission of Elsevier.

A. Untransformed and log-transformed charcoal (CHAR), magnetic susceptibility accumulation rates, and data from loss on ignition analysis plotted against age of the long core. CHAR peaks assumed to be watershed fires in the last 1000 yr are marked with arrows (←).

B. Log-transformed charcoal (CHAR) and magnetic susceptibility accumulation values, background levels, peaks, and inferred fire frequency and sedimentation event frequency for the Taylor Lake long core. A 600-yr background window width and a 1.25 threshold-ratio value were used to determine background and peak series. The horizontal line marks the boundary between pollen zones TL1 (ca. 4600–2700 cal yr B.P.) and TL2 (ca. 2700 cal yr B.P.–present).

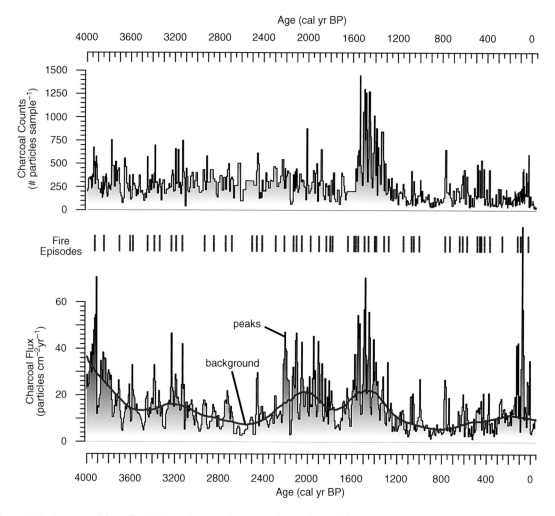

Figure 5.5 Decomposition of a 4000 year long sedimentary charcoal record from Foy Lake, Montana, USA. Charcoal counts (top) were converted to charcoal accumulation rates (CHAR = particles cm^{-2} year^{-1}) (bottom) by multiplying the charcoal concentration by accretion rate. CHAR values were interpolated to five-year intervals and smoothed using a 500-year window width, shown as background (blue). Fire episodes are indicated by vertical (red) symbols when CHAR exceeds background by calculating residuals and using a locally defined threshold and Gaussian mixture model. (The data used in this figure was provided by M. Power).

For a broader analysis, for example looking at fire and climate change, a more sophisticated approach is required. The standardization of records was undertaken by the Palaeofire Group (http://www/ncdc. noaa.gov/paleo/impd/paleofire.html), who then integrated these records into the Global Charcoal Database (http://www.gpwg.org) (Power *et al.*, 2010). The key was to standardize more than 120 different approaches for classifying and reporting charcoal. They input data from the >125 µm size fraction as pieces cm^{-3} or as charcoal influx or accumulation rate pieces^{-2} year^{-1}.

The standardization protocol included:

- rescaling the values using a min-max transformation (Figure 5.6);
- transforming and homogenizing the variance using Box-Cox transformation;
- rescaling values once more to z scores (Power *et al.*, 2010).

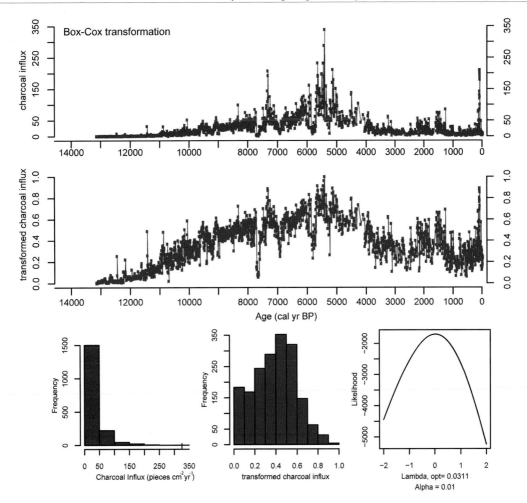

Figure 5.6 Data standardization of charcoal records (from Power, 2013). *Top and middle panels*: A comparison of transformed and original charcoal data shows the impact of the Box-Cox transformation on the charcoal influx data (particles cm^{-2} yr^{-1}) from Foy Lake (Power *et al.*, 2011). *Bottom panel*: A comparison of the distribution of the original data (blue) and transformed data (green) show how the optimal value of λ (e.g. 0.03 in this example) normalizes the distribution of the data. Plotting both the raw charcoal influx and transformed charcoal influx (using 100–4000 yrs BP as the base period, as in Power *et al.*, 2008) shows the curvilinear relationship in the Box-Cox transformation.

The impact of such a transformation is shown in Figure 5.6. Overall, data is compared to a baseline and fire is expressed as a positive or negative number relative to the base. Ranges of base lines are possible (Figure 5.7). There is considerable power in this approach to charcoal data but care is needed in interpretation (Power, 2013).

5.3 Fire and climate cycles

There has been increasing interest in the relationship between fire and climate, especially in relation to the current debate on global warming. The long record of fire and climate, as seen in combining the Charcoal Database with climate data from polar ice cores, is of particular significance. The integration of these data into global climate models provides another important tool and approach (Power *et al.*, 2008). We can use these tools to try to answer the question of the relationship between fire and climate, and also to consider the role of human activity in the story.

Analysis of the charcoal and climate records in North America from the past 20 000 years indicated a number of important conclusions. Marlon *et al.* (2009) showed that periods of cooling led to lower fire and that periods of warming led to higher fire in North

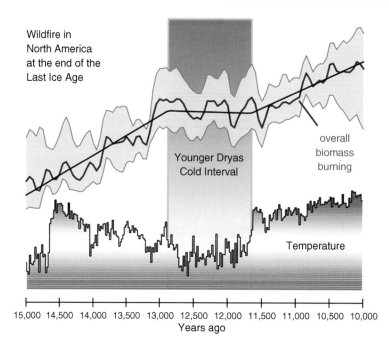

Figure 5.7 Fire in North America 15–10 000 years. (From J. Marlon with data from Marlon *et al.*, 2009). Reconstruction of biomass burning in relation to the north Atlantic temperature record based upon the GISP2 ∂ 0_{ice} record as a proxy for north Atlantic temperatures.

America (Figure 5.6). More significantly, they demonstrated the response of fire to abrupt climate change, noting a distinct shift in the average level of burning at the beginning (12.9 ka) and end (11.7 ka) of the Younger Dryas cold period. In particular, they noted a specific rise in biomass burning and fire frequency when the cold period ended, and a continued increase as temperatures rose.

Likewise, in a study of fire over the past 2000 years, Marlon *et al.* (2008) noted a link to fire and changing climate (Figures 5.7, 5.8). Here, the authors observed a reduction of fire as the climate cooled. During this interval, however, the impact of human activity can be increasingly discerned.

A more detailed analysis of charcoal records is possible over the past 21 000 years, as the number of sites increases and the dating becomes more precise (Power, 2013). This has allowed a more detailed discussion of fire in four distinctive periods:

- The glacial period, 21–16 ka (note that dates are often expressed as calendar years before present).
- The late glacial, 15–12 ka.
- The early Holocene (11–7 ka).
- The mid to late Holocene (6 ka to present).

A series of recent analyses has indicated that during the glacial interval, 21–16 ka, fire activity was reduced in general, although there is regional variation (e.g. with fire in parts of Australia and South America increasing). Fire activity remained low during the late glacial period (15–12 ka) and even continued to decrease, especially in Europe and North America. The switch from glacial to inter-glacial conditions (11 ka onwards) also shows an increase in fire activity. This is particularly evident in the higher latitudes and this continues over the past 6000 years. Over the last two millennia, there is a match between long-term cooling and fire (see Marlon *et al.*, 2008), but significant impact of humans on global fire has been detected over the last century.

5.4 Fire and humans: the fossil evidence

As indicated at the beginning of this chapter, our earliest evidence of the relationship between fire and humans comes from evidence, in sediments 1 ma, from a cave in South Africa. It is frustratingly difficult to be sure of the impact of humans on wider natural

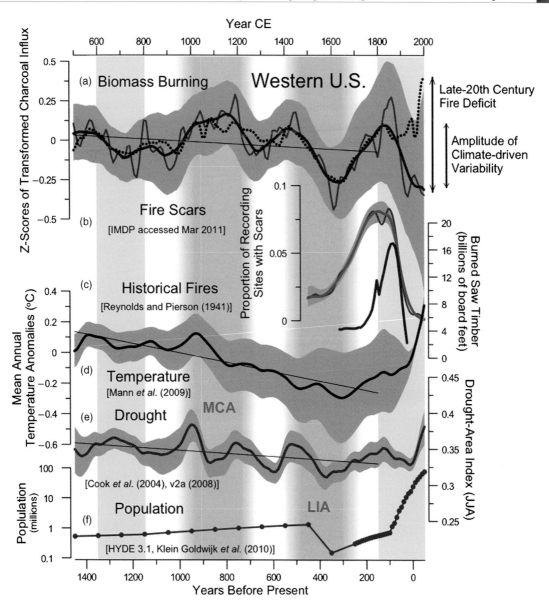

Figure 5.8   Fire in North America, 1 500 years to present. (Modified by J. Marlon from Marlon *et al.*, 2012).

A. Smoothed and standardized 25-year (grey) and 100-year (red) trend line through standardized biomass burning records (n = 69) along with predicted biomass burning based on a GAM (black dashed line) fit to the 100-year biomass burning records.
B. Smoothed proportions of dendrochronological sites recording fire scars (the green curve is based on locally fitting nearest-neighbor parameter of 0.25, while the gray curve is based on a parameter value of 0.10.
C. Estimated historical saw timber affected by fires.
D. Smoothed gridded temperature anomalies for the western United States.
E. Smoothed Palmer Drought Severity Index for the western United States.
F. Population estimates for the western United States.

fire systems. It is possible to consider the following scenarios, where human influence may be more distinctly recognized:

- In areas where there is no natural fire, humans may introduce fire as part of their agricultural practice.
- In areas where there is already fire, humans may alter the extent, timing and frequency of fire through their activities.

- Finally, humans may change fire through agricultural practice, changing the natural vegetation and by fire suppression (see Figures 5.1, 5.2).

There are a number of examples given later in this book (Part Three), but we may briefly consider several here (e.g. Chapter 10). The introduction of fire into non-flammable landscapes may be exemplified by a study of the Brazilian rain forest (Cochrane, 2003).

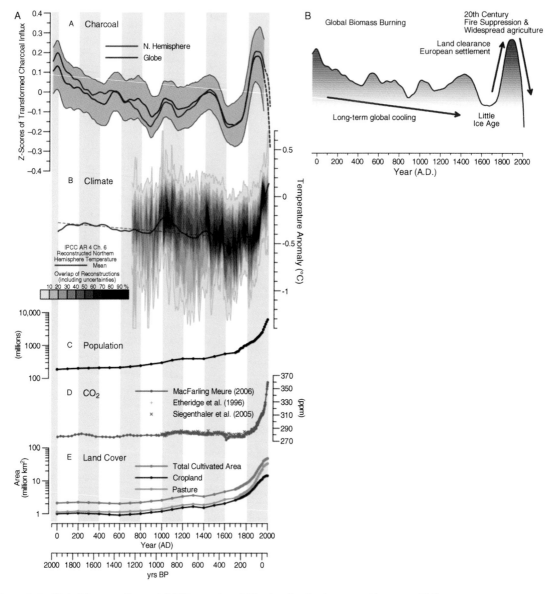

Figure 5.9   Global fire over the past 2 000 years in relation to climate change and human activity.

A. Reconstruction of biomass burning, climate, population and land cover. (After Marlon *et al.*, 2008). Reproduced by permission of Nature Publications.

B. Summary of key findings showing fall in biomass burning because of cooling, followed by extensive land clearance fires and subsequent fire suppression. (From J. Marlon).

This is a system without natural fire, yet fire is used in slash-and-burn agriculture to clear the forest. Increasing attention is being given to how this has happened, using sedimentary charcoal records, and these data can be linked to the history of indigenous peoples and their practices and to the invasion of Europeans, who brought with them their own agricultural practices. As always, these data must be considered in light of a changing climate record (see Power, 2013).

The relationship between fires and humans in Australia has always been controversial. Australia has many flammable biomes, and fossil data suggests a very long history of fire on the continent. However, some have claimed an increase in charcoal content of cores at around 70 000 years ago to be evidence of the spread of humans into Australia. The unravelling of natural fire systems, and the use of fire by humans and its recognition in the fossil record continues to create debate. What is clear is that there is an Aboriginal tradition of setting frequent patchy fires in Northern Australian savannas that create a fine-scale mosaic of burnt and unburnt areas. Changes in land use practice brought in by European settlers created a major change to large frequent fires (see Bowman *et al.* (2011) and Chapter 10 (Part Two) for discussion).

The European story is even more complex, as we see thousands of years of change in agricultural practice. Fire occurrence may be influential, not only by vegetational change, but also by changes in land management and even human population growth.

In some cases, humans may have used fire to maintain a particular vegetation type, to clear a vegetation and replace it by another that may be more economically advantageous or even, in some cases, to use fire to manage natural harvests, such as manipulating the acorn supply in oak forests (see Mason, 2000 as an example of this). This range of potential impacts of fire (in addition to the use of fire to drive large mammals to areas where they could be killed) shows that there may be an over-emphasis on the use of fire solely as a mechanism for forest clearance (see Chapter 13 (Part Three)).

5.5   Fire and the industrial society

There has been much recent debate on the existence and importance of the Anthropocene – the period during which humans have created major lasting changes to the environment (see Zalasiewicz *et al.*, 2011). This is certainly a diachronous boundary, but it relates to the past 150 or so years of Earth history. When examining the palaeofire records of the past 2000 years, Marlon *et al.* (2008) found a major change in the pattern of fire with a major decline relating to fire suppression (Figure 5.9).

This feature is entangled with population growth, the growth and spread of urban culture and a cleavage between some populations with natural and managed fire systems. This has been termed the pyric transition, whereby fire has been 'boxed up' to provide heat and energy (and fuel) for the urban population, and there has been a progressive tension between fire and people at the wildland-urban interface.

Many cultures have forgotten both the need of fire and also the use of fire by humans, and have begun to demonize any fire in the natural environment (see Chapter 12, Part Three). However, as we have seen in the past two chapters, fire has been an important part of the Earth System for 400 million years, and we must learn to understand fire and its place, not simply consider it to be something that should be suppressed. We shall deal with the tension between humans and fires in later chapters (Parts 2 and 3), but humans are fire creatures of what has been increasingly shown to be a fire planet.

Further reading

Bowman, D.J.M.S., Balch, J., Artaxo, P., Bond, W.J., Cochrane, M.A., D'Antonio, C.M., DeFries, R., Johnston, F.H. Keeley, J.E., Krawchuk, M.A., Kull, C.A., Mack, M., Moritz, M.A., Pyne, S.J., Roos, C.I., Scott, A.C., Sodhi, N.S., Swetnam, T.W. (2011). The human dimension of fire regimes on Earth. *Journal of Biogeography* **38**, 2223–2236.

Marlon, J.R., Bartlein, P.J., Carcaillet, C., Gavin, D. G., Harrison, S. P., Higuera, P. E., Joos, F., Power, M. J., Prentice, I. C. (2008). Climate and human influences on global biomass burning over the past two millennia. *Nature Geoscience* **1**, 607–702.

Marlon, J.R., Bartlein, P.J., Walsh, M.K., Harrison, S.P., Brown, K.J., Edwards, M.E., Higuera, P.E., Power, M.J., Anderson, R.S., Briles, C., Brunelle, A., Carcaillet, C., Daniels, M., Hu, F.S., Lavoie, M., Long, C., Minckley, T., Richard, P.J.H., Scott, A.C., Shafer, D.S., Tinner, W., Umbanhowar, C.E. Jr., Whitlock, C. (2009). Wildfire responses to abrupt climate change in North America. *Proceedings of the National Academy of Sciences, USA* **106**, 2519–2524.

References for part one

Abu Hamad, A.M.B., Jasper, A., Uhl, D. (2012). The record of Triassic charcoal and other evidence for palaeo-wildfires: Signal for atmospheric oxygen levels, taphonomic biases or lack of fuel? *International Journal of Coal Geology* **96–97**, 60–71.

Antal Jr., M.J., Grønli, M. (2003) The art, science, and technology of charcoal production. *Industrial and Engineering Chemistry Research* **42**, 1619–1640.

Archibald, S., Staver, A.C. and Levin, S.A. (2012). Evolution of human driven fire regimes in Africa. *Proceedings of the National Academy of Sciences* **109**, 847–852.

Artaxo, P., Luciana V. Rizzo, L.V., Paixão, M., de Lucca, S., Oliveira, P.H., Lara, L.L., Wiedemann, K.T., Andreae, M. O., Holben, B., Schafer, J., Correia, A.L., Pauliquevis, T.M. (2009). Aerosol Particles in Amazonia: Their Composition, Role in the Radiation Balance, Cloud Formation, and Nutrient Cycles associated with deposition of trace gases and aerosol particles. Amazonia and Global Change. *Geophysical Monograph Series* **186**, 233–250.

Ascough, P.L., Bird, M.I., Scott, A.C., Collinson, M.E., Weiner, S., Cohen-Ofri, I., Snape, C.E., Le Manquais, K. (2010). Charcoal reflectance: implications for structural characterization. *Journal of Archaeological Science* **37**, 1590–1599.

Ascough, P.L., Bird, M.I., Francis, S.M., Thornton, B., Midwood, A.J., Scott, A.C., Apperley, D. (2011). Variability in oxidative degradation of charcoal: influence of production variables and environmental exposure. *Geochimica et Cosmochimica Acta* **75**, 2361–2378.

Balch, J.K., Bradley, B.A., D'Antonio, C.M., Gomez-Dans, J. (2013). Introduced annual grass increases regional fire activity across the arid western USA (1980–2009). *Global Change Biology* **19**, 173–183.

Beerling, D.J., Taylor, L.L., Bradshaw, C.D.C., Lunt, D.J., Valdes, P.J., Banwart, S.A., Pagani, M., Leake, J.R. (2012). Ecosystem CO_2 starvation and terrestrial silicate weathering: mechanisms and global scale quantification during the late Miocene. *Journal of Ecology* **100**, 31–41.

Belcher, C.M. (2009). Reigniting the Cretaceous-Palaeogene firestorm debate. *Geology* **37**, 1147–1148.

Belcher, C.M., Collinson, M.E., Scott, A.C. (2005). Constraints on the thermal energy released from the Chicxulub impactor: New evidence from multi-method charcoal analysis. *Journal of the Geological Society, London* **162**, 591–602.

Belcher, C.M., Collinson, M.E., Scott, A.C. (2013). A 450 million year record of fire. In: Belcher, C.M. (ed). *Fire phenomena and the Earth System: An interdisciplinary guide to fire science*. 1st edition, pp. 229–249. J. Wiley & Sons, Ltd, Chichester.

Belcher, C.M., Mander, L., Rein, G., Jervis, F.X., Haworth, M., Hesselbo, S.P., Glasspool, I.J., McElwain, J.C. (2010a). Increased fire activity at the Triassic/Jurassic boundary in Greenland due to climate-driven floral change. *Nature Geoscience* **3**, 426–429.

Belcher, C.M., Yearsley, J.M., Hadden, R.M., McElwain, J.C., Rein, G. (2010b). Baseline intrinsic flammability of Earths' ecosystems estimated from paleoatmospheric oxygen over the past 350 million years. *Proceedings of the National Academy of Sciences, USA* **107**, 22448–22453.

Bergman, N.M., Lenton, T.M., Watson, A.J. (2004). COPSE: A new model of biogeochemical cycling over Phanerozoic time. *American Journal of Science* **304**, 397–437.

Berna, F., Goldberg, P., Horwitz, J., Holt, S., Bamford, M., Chazan, M. (2012). Microstratigraphic evidence of *in situ* fire in Archeulean strata of Wonderwerk Cave, Northern Cape province, South Africa. *Proceedings of the National Academy of Sciences, USA* **109**(20), E1215–E1220.

Berner, R.A. (2006). A combined model for Phanerozoic atmospheric O_2 and CO_2. *Geochemica et Cosmochimica Acta* **70**, 5653–5664.

Fire on Earth: An Introduction, First Edition. Andrew C. Scott, David M.J.S. Bowman, William J. Bond, Stephen J. Pyne and Martin E. Alexander.
© 2014 John Wiley & Sons, Ltd. Published 2014 by John Wiley & Sons, Ltd.

Berner, R.A. (2009). Phanerozoic atmospheric oxygen: new results using the GEOCARBSULF model. *American Journal of Science* **309**, 603–606.

Berner, R.A., Beerling, D.J., Dudley, R., Robinson, J.M., Wildman, R.A. (2003). Phanerozoic atmospheric oxygen. *Annual Review of Earth and Planetary Sciences* **31**, 105–134.

Bird, M.I., Cali, J.A. (1998). A million-year record of fire in sub-Saharan Africa. *Nature* **349**, 767–769.

Bond, T.C., *et al.* (2013). Bounding the role of black carbon in the climate system: a scientific assessment. *Journal of Geophysical Research: Atmospheres.* http://onlinelibrary.wiley.com/doi/10.1002/jgrd.50171/pdf

Bond, W.J., Keeley, J.E. (2005). Fire as global 'herbivore': the ecology and evolution of flammable ecosystems. *Trends in Ecology and Evolution* **20**, 387–394.

Bond, W.J., Midgley, G.F. (2012). Fire and the angiosperm revolutions. *International Journal of Plant Sciences* **173**, 569–583.

Bond, W.J., Scott, A.C. (2010). Fire and the spread of flowering plants in the Cretaceous. *New Phytologist* **188**, 1137–1150.

Bond, W.J., Woodward F.I., Midgley G.F. (2005). The global distribution of ecosystems in a world without fire. *New Phytologist* **165**, 525–538.

Bowman, D. (2005). Understanding a flammable planet – climate, fire and global vegetation patterns. *New Phytologist* **165**, 341–345.

Bowman, D.M.J.S., Balch, J.K., Artaxo, P., Bond, W.J., Carlson, J.M., Cochrane, M.A., D'Antonio, C.M., DeFries, R.S., Doyle, J.C., Harrison, S.P., Johnston, F.H., Keeley, J.E., Krawchuk, M.A., Kull, C.A., Marston, J.B., Moritz, M.A., Prentice, I.C., Roos, C.I., Scott, A.C., Swetnam, T.W., van der Werf, G.R., Pyne, S.J. (2009). Fire in the Earth System. *Science* **324**, 481–484.

Bowman, D.J.M.S., Balch, J., Artaxo, P., Bond, W.J., Cochrane, M.A., D'Antonio, C.M., DeFries, R., Johnston, F.H. Keeley, J.E., Krawchuk, M.A., Kull, C.A., Mack, M., Moritz, M.A., Pyne, S.J., Roos, C.I., Scott, A.C., Sodhi, N.S., Swetnam, T.W. (2011). The human dimension of fire regimes on Earth. *Journal of Biogeography* **38**, 2223–2236.

Brown, S.A.E., Collinson, M.E., Scott, A.C. (2013). Did fire play a role in formation of dinosaur-rich deposits? An example from the Late Cretaceous of Canada. *Palaeobiodiversity and Palaeoenvironments.* Doi: 10.1007/s12549-013-123-y

Brown, S.A.E., Scott, A.C., Glasspool, I.J., Collinson, M.E. (2012). Cretaceous wildfires and their impact on the Earth system. *Cretaceous Research* **36**, 162–190.

Caldararo, N. (2002). Human ecological intervention and the role of forest fires in human ecology. *The Science of the Total Environment* **292**, 141–165.

Cannon, S.H., Bigio, E.R., Mine, E. (2001). A process for fire-related debris flow initiation, Cerro Grande fire, *New Mexico. Hydrological Processes* **15**, 3011–3023.

Cannon, S.H., Gartner, J.E., Rupert, M.G., Michael, J.A. (2010). Predicting the probability and volume of postwildfire debris flows in the intermountain western United States. *Geological Society of America Bulletin* **122**, 127–144.

Carcaillet, C., Bouvier, M., Fréchette, B., Larouche, A., Richard, P. (2001). Comparison of pollen-slide and sieving methods in lacustrine charcoal analyses for local and regional fire history. *The Holocene* **11**, 467–476.

Cerdà, A., Robichaud, P. (eds.) (2009). *Fire Effects on Soils and Restoration Strategies.* Science Publishers Inc., Enfield, NH.

Clark, J.S. (1988) Particle motion and the theory of charcoal analysis: source area, transport, deposition and sampling. *Quaternary Research* **30**, 67–80.

Clark, J.S., Lynch, Stocks, B.J., Goldammer, J.G. (1998). Relationships between charcoal particles in air and sediments in west-central Siberia. *The Holocene* **8**, 19–29.

Cochrane, M.A. (2003). Fire science for rainforests. *Nature* **421**, 913–919.

Cochrane, M.A., Ryan, K.C. (2009). Fire and fire ecology: concepts and principles. In: Cochrane, M.A. (ed). *Tropical fire ecology: Climate change, land use and ecosystem dynamics*, pp. 24–62. Springer, Berlin, Germany.

Cohen-Ofri, I., Weiner, L., Boaretto, E., Mintz, G., Weiner, S. (2006). Modern and fossil charcoal: aspects of structure and diagenesis. *Journal of Archaeological Science* **33**, 428–439.

Collinson, M.E., Steart, D., Scott, A.C., Glasspool, I.J. and Hooker, J.J. (2007). Fire and episodic runoff and deposition at the Paleocene-Eocene boundary. *Journal of the Geological Society, London* **164**, 87–97.

Davis, K.M., Mutch, R.W. (1994). Applying ecological principles to manage wildland fire. In: *Fire in Ecosystems Management.* Training course. National Advanced Resources Technology Centre.

DeBano, L.F. (2000). The role of fire and soil heating on water repellency in wldland environments: a review. *Journal of Hydrology* **231–2**, 195–206.

DeBano, C.F., Dunn, P.H., Conrad, C.E. (1977). Fire's effects on physical and chemical properties of chaperral soils. In: Mooney, H. A. and Conrad, C. E., *Proceedings of Symposium of Environmental Consequences of Fire and Fuel Management in Mediterranean Ecosystems*, pp. 65–74. USDA Forest service General Technical Report WO-3.

Dieterich, J.H., Swetnam, T.W. (1984). Dendrochronology of a fire scarred ponderosa pine. *Forest Science* **30**(1): 238–247.

Doerr, S.H., Shakesby R.H. (2013). Fire and the Land Surface. In: Belcher, C.M. (ed). *Fire phenomena and the Earth System: An interdisciplinary guide to fire science.* 1st edition, pp 135–155. J. Wiley & Sons, Ltd, Chichester.

Doerr, S.H., Shakesby, R.A., MacDonald, L.H. (2009). Soil water repellency: a key factor in post-fire erosion. In Cerdà, A., Robichaud P. (eds.), *Fire Effects on Soils and Restoration Strategies*, pp. 197–224. Science Publishers, Inc., Enfield, NH.

Doerr, S.H., Shakesby, R.A., Walsh, R.P.D. (2000). Soil water repellency: its characteristics, causes and hydro-geo-morphological consequences. *Earth Science Reviews* **51**, 33–65.

Duffin, K.I. (2008). The representation of rainfall and fire intensity in fossil pollen and charcoal records from a South African savanna. *Review of Palaeobotany and Palynology* **151**, 59–71.

Duffin, K.I., Gillson, L., Willis, K.J. (2008). Testing the sensitivity of charcoal as an indicator of fire events in savanna environments: quantitative predictions of fire proximity, area and intensity. *The Holocene* **18**, 279–291.

Falcon-Lang, H.J. (1998). The impact of wildfire on an Early Carboniferous coastal system, North Mayo, Ireland. *Palaeogeography, Palaeoclimatology, Palaeoecology* **139**, 121–138.

Falcon-Lang, H.J. (2000). Fire ecology of the Carboniferous tropical zone. *Palaeogeography, Palaeoclimatology, Palaeoecology* **164**, 339–355.

Falcon-Lang, H., Scott, A.C. (2000). Upland ecology of some Late Carboniferous Cordaitalean Trees from Nova Scotia and England. *Palaeogeography, Palaeoclimatology, Palaeoecology* **156**, 225–242.

Field, T.S., Arens, N.C. (2005). Form, function and environments of the early angiosperms: merging extant phylogeny and ecophysiology with fossils. *New Phytologist* **166**, 383–408.

Fites-Kaufman, J., Bradley, A.F., Merril, A.G. (2006). Fire and plant interactions. In: Sugihara, N.G., Van Wagtendonk, J.W., Shaffer, K.E., Fites, Kaufman, J., Thode, A.E. (eds.). *Fire in California's ecosystems*, pp. 94–117. University of California Press, Berkley, CA.

Flannigan, M.D., Stocks, B.J., Wotton, B.M. (2000). Climate change and forest fires. *The Science of the Total Environment* **262**, 221–229.

Forbes, M.S., Raison, R.J., Skjemstad, J.O. (2006). Formation, transformation and transport of black carbon (charcoal) in terrestrial and aquatic ecosystems. *Science of the Total Environment* **370**, 190–296.

Forster, P., Ramaswamy, V., Artaxo, P. et al. (2007). Changes in atmospheric constituents and in radiative forcing. In: Solomon, S. et al. (eds.) *Climate Change 2007: The Physical Science Basis. Contribution of Working Group I to the Fourth Assessment Report of the Intergovernmental Panel on Climate Change*, pp. 129–234. Cambridge University Press, Cambridge, UK.

Friis, E.M., Pedersen, K.R., Crane, P.R. (2006). Cretaceous angiosperm flowers: innovation and evolution in plant reproduction. *Palaeogeography, Palaeoclimatology, Palaeoecology* **232**, 251–293.

Gensel, P.G. (2008). The earliest land plants. *Annual Reviews of Ecology, Evolution and Systematics* **39**, 459–477.

Giglio, L., Csiszar, I. & Justice, C. O. (2006). Global distribution and seasonality of active fires as observed with the Terra and Aqua Moderate Resolution Imaging Spectroradiometer (MODIS) sensors. *Journal of Geophysical Research* **111**, 1–12.

Glasspool, I.J., Edwards, D., Axe, L. (2006). Charcoal in the Early Devonian: A wildfire-derived Konservat-Lagerstatte. *Review of Palaeobotany and Palynology* **142**, 131–136.

Glasspool, I.J., Scott, A.C. (2010). Phanerozoic atmospheric oxygen concentrations reconstructed from sedimentary charcoal. *Nature Geoscience* **3**, 627–630.

Glasspool, I.J., Scott, A.C. (2013). Identifying past fire events. In: Belcher, C.M. (ed.) *Fire phenomena and the Earth System: An interdisciplinary guide to fire science*. 1st edition, pp. 179–205. J. Wiley & Sons, Ltd, Chichester, UK.

Goldberg, E.G. (1985). *Black carbon in the environment*. J. Wiley and Sons, Chichester.

Graham, R.T. (Ed.) (2003). *Hayman Fire Case Study*. Gen. Tech. Rep. RMRS-GTR-114. Ogden, UT: U.S. Department of Agriculture, Forest Service, Rocky Mountain Research Station. 396 p.

Hammes, K, and Abiven, S. (2013). Identification of black carbon in the Earth System. In: Belcher, C.M. (ed.) *Fire phenomena and the Earth System: An interdisciplinary guide to fire science*. 1st edition, pp. 157–176. J. Wiley & Sons, Ltd, Chichester.

Harris, T.M. (1958). Forest fire in the Mesozoic. *Journal of Ecology* **46**, 447–453.

Harrison, S.P., Marlon, J., Bartlein, P.J. (2010). Fire in the Earth system. In: Dodson, J. (Ed.), *Changing Climates, Earth Systems and Society*, pp. 21–48. Springer-Verlag, Berlin.

He, T., Pausas, J.G., Belcher, C.M., Schwilk, D.W., Lamont, B.B. (2012). Fire-adapted traits of *Pinus* arose in the fiery Cretaceous. *New Phytologist* **194**, 751–759.

Herring, J.R. (1985). Charcoal fluxes into sediments of the North Pacific Ocean: the Cenozoic record of burning. In: The carbon cycle and atmospheric CO2: natural variations Archean to Present. *Geophysical Mongraphs* **32**, 419–442.

Hudspith, V., Scott, A.C., Collinson, M.E., Pronina, N., Beeley, T. (2012). Evaluating the extent to which wildfire history can be interpreted from inertinite distribution in coal pillars: an example from the late Permian, Kuznetsk Basin, Russia. *International Journal of Coal Geology* **89**, 13–25.

Johnston, F. H. (2009). Bushfires and human health in a changing environment. *Australian Family Physician* **38**, 720–725.

Johnston, F. H. Henderson, S.B., Chen, Y., Randerson, J.T., Marlier, M., DeFries, R.S., Kinney, P., Bowman, D.M.S., Brauer, M. (2012). Estimated global mortality attributable to smoke from landscape fires. *Environmental Health Perspectives* **120**, 695–701.

Jones, T.P., Rowe, N.P. (eds.) (1999). *Fossil Plants and Spores: Modern Techniques*. Geological Society, London.

Jones, T., Scott, A.C., Cope, M. (1991). Reflectance measurements against temperature of formation for modern charcoals and their implications for the study of fusain. *Bulletin de La Société Géologique de France* **162**, 193–200.

Keeley, J.E. (2009). Fire intensity, fire severity and burn severity: a brief review and suggested usage. *International Journal of Wildland Fire* **18**, 116–126.

Keeley, J.E., Bond, W.J., Bradstock, R.A., Pausas, J.G., Rundel, P.W. (2012). *Fire in Mediterranean Climate Ecosystems: Ecology, evolution and management.* Cambridge University Press, Cambridge, UK.

Keeley, J.E., Pausas, J.G., Rundel, P.W., Bond, W.J., Bradstock, R.A. (2011). Fire as an evolutionary pressure shaping plant traits. *Trends in Plant Science* **16**, 406–411.

Keeley, J. E., Rundel, P.W. (2005). Fire and the Miocene expansion of C4 grasslands. *Ecology Letters* **8**, 683–690.

Keywood, M., Kanakidou, M., Stohl, A., Dentener, F., Grassi, G., Meyer, C.P., Torseth, K., Edwards, D., Thompson, A.M., Lohmann, U., Burrows, J. (2013). Fire in the Air: Biomass burning impacts in a changing climate. *Critical Reviews in Environmental Science and Technology* **43**, 40–83.

Kitzberger, T., Brown, P.M., Heyerdahl, E.K., Swetnam, T.W., Veblen, T.T. (2007). Contingent Pacific-Atlantic Ocean inflence on multicentury wildfire synchrony over western North America. *Proceedings of the National Academy of Sciences, USA* **104**, 532–448.

Krawchuk, M.A., Moritz, M.A. (2011). Constraints on global fire activity vary across a resource gradient. *Ecology* **92**, 121–132.

Krawchuk, M.A., Moritz, M.A., Parisien, M-A., Van Dorn, J., Hayhoe, K. (2009). Global pyrogeography: the current and future distribution of wildfire. *PloS One* **4**(4), e5102.

Kump, L. (1988). Terrestrial feedback in atmospheric oxygen regulation by fire and phosphorous. *Nature* **335**, 152–154.

Kump, L.R. (2010). Earth's second wind. *Science* **330**, 1490–1491.

Lehmann, J., Gaunt, J., Rondon, M. (2006). Bio-char sequestration in terrestrial ecosystems – a review. *Mitigation and Adaptation strategies for Global Change* **11**, 403–427.

Lentile, L.B., Holden, Z.A., Smith, A.M.S., Falkowski, M.J., Hudak, A.T., Morgan, P., Lewis, S.A., Gessler, P.E., Benson, N.C. (2006). Remote sensing techniques to assess active fire characteristics and post-fire effects. *International Journal of Wildland Fire* **15**, 319–345.

Lenton, T.M. (2013). Fire feedbacks on atmospheric oxygen. In: Belcher, C.M. (ed) *Fire phenomena and the Earth System: An interdisciplinary guide to fire science.* 1st edition, pp. 289–308. J. Wiley & Sons, Ltd, Chichester.

Lenton, T.M., Watson, A.J. (2000). Redfield revisited: 2. What regulates the oxygen content of the atmosphere? *Global Biogeochemical Cycles* **14**, 249–268.

Long, C.J., Whitlock, C. (2002). Fire and vegetation history from the Coastal rain forest of the Western Oregon Coast Range. *Quaternary Research* **58**, 215–225.

Marlon, J.R., Bartlein, P.J., Carcaillet, C., Gavin, D. G., Harrison, S. P., Higuera, P. E., Joos, F., Power, M. J., Prentice, I. C. (2008). Climate and human influences on global biomass burning over the past two millennia. *Nature Geoscience* **1**, 607–702.

Marlon, J.R., Bartlein, P.J., Walsh, M.K., Harrison, S.P., Brown, K.J., Edwards, M.E., Higuera, P.E., Power, M.J., Anderson, R.S., Briles, C., Brunelle, A., Carcaillet, C., Daniels, M., Hu, F.S., Lavoie, M., Long, C., Minckley, T., Richard, P.J.H., Scott, A.C. Shafer, D.S., Tinner, W., Umbanhowar, C.E., Jr., Whitlock, C. (2009). Wildfire responses to abrupt climate change in North America. *Proceedings of the National Academy of Sciences, USA* **106**, 2519–2524.

Marlon, J.R., Bartlein, P.J., Gavin, D.G., Long, C.J., Anderson, R.S., Briles, C.E., Brown, K.J., Colombaroli, D., Hallett, D.J., Power, M.J., Scharf, E.A., Walsh, M.K. (2012). Long-term perspective on wildfires in the western USA. *Proceedings of the National Academy of Sciences, USA* **109**(9), E535–E543.

Marynowski, L., Scott, A.C., Zatoń, M., Parent, H., Garrido, A.C. (2011). First multi-proxy record of Jurassic wildfires from Gondwana: evidence from the Middle Jurassic of the Neuquén Basin, Argentina. *Palaeogeography, Palaeoclimatology, Palaeoecology* **299**, 129–136.

Masiello, CA. (2004). New directions in black carbon organic geochemistry. *Marine Chemistry* **92**, 201–213.

Mason, S.L.R. (2000). Fire and Mesolithic subsistence – managing oaks for acorns in northwest Europe? *Palaeogeography, Palaeoclimatology, Palaeoecology* **164**, 139–150.

McParland, L.C., Collinson, M.E., Scott, A.C., Steart, D.C., Grassineau, N.J. and Gibbons, S.J. (2007). Ferns and fires: Experimental charring of ferns compared to wood and implications for paleobiology, coal petrology, and isotope geochemistry. *PALAIOS* **22**, 528–538.

McParland, L.C., Collinson, M.E., Scott, A.C., Campbell, G. (2009a). The use of reflectance for the interpretation of natural and anthropogenic charcoal assemblages. *Archaeological and Anthrolopological Sciences* **1**, 249–261.

McParland, L.C. Hazell, Z. Campbell, G., Collinson, M.E., Scott, A.C. (2009b). How the Romans got themselves into hot water: Temperatures and fuel types of a Roman hypocaust fire. *Environmental Archaeology* **14**, 172–179.

Meyer, G.A., Pierce, J.L. (2003). Climatic controls on fire-induced sediment punses in Yellowstone National Park and central Idaho: a long term perspective. *Forest Ecology and Management* **178**, 89–104.

Meyn, A., White, P.S., Buhk, C., Jentsch, A. (2007). Environmental drivers of large, infrequent wildfires: the emerging conceptual model. *Progress in Physical Geography* **31**, 287–312.

Moody, J.A., Martin, D.A. (2009). Forest fire effects on geomorphic processes. In: Cerdá, A., Robichaud, P.

(Eds.), *Fire Effects on Soils and Restoration Strategies*, pp. 41–79. Science Publishers, Inc, Enfield, NH.

Moody, J.A., Martin, D.A., Cannon, S.H. (2008). Post-wildfire erosion response in two geologic terrains in the western USA. *Geomorphology* **95**, 103–118.

Moritz, M.A., Morais, M.E., Summerell, L.A., Carlson, J.M., Doyle, J. (2005). Wildfires, complexity, and highly optimized tolerance. *Proceedings of the National Academy of Sciences, USA* **102**, 17912–17917.

Moritz, M.A., Parisien, M-A., Batllori, E., Krawchuk, M.A., Van Dorn, J., Ganz, D.J., Hayhoe, K. (2012). Climate change and disruptions to global fire activity. *Ecosphere* **3** (6) A49, 1–22.

Nichols, G.J. (2009). *Sedimentology and Stratigraphy*. Wiley-Blackwell, West Sussex.

Nichols, G.J., Cripps, J.A., Collinson, M.E., Scott, A.C. (2000). Experiments in waterlogging and sedimentology of charcoal: results and implications. *Palaeogeography, Palaeoclimatology, Palaeoecology* **164**, 43–56.

Nichols, G.J., Jones, T.P. (1992). Fusain in Carboniferous shallow marine sediments, Donegal, Ireland: The sedimentological affects of wildfire. *Sedimentology* **39**, 487–502.

Niklas, K.J., Tiffany, B.H., Knoll, A.H. (1985). Patterns in vascular land plant diversification: An analysis at the species level. In Valentine, J.W. (ed). *Phanerozoic diversity patterns: profiles in macroevolution*. Princeton University Press, Princeton, NJ.

Osborne, C.P., Beerling, D.J. (2006). Nature's green revolution: the remarkable evolutionary rise of C-4 plants. *Philosophical Transactions of the Royal Society B* **361**, 173–194.

Page, S.E., Hoscilo, A., Langner, A., Tansey, K., Siegert, F., Limin, S. & Rieley, J. (2009). Tropical Peatland Fires in Southeast Asia. In: Cochrane, M.A. (ed.), *Tropical Fire Ecology: Climate Change, Land Use and Ecosystem Dynamics*, pp. 263–287. Springer-Praxis, Heidelberg, Germany.

Page, S.E., Siegert, F., Rieley, J.O., Boehm, H-D.V., Adi, J., Limin, S. (2002). The amount of carbon released from peat and forest fires in Indonesia during 1997. *Nature* **420**, 61–65.

Pausas, J.G., Keeley, J.E. (2009). A burning story: the role of fire in the history of life. *Bioscience* **59**, 593–601.

Pfeifer, M., Disney, M., Quaife, T., Marchant, R. (2011). Terrestrial ecosystems from space: a review of earth observation products for macroecology applications. *Global Ecology and Biogeography* **21**, 603–624.

Pierce, J.L., Meyer, G.A. and Jull, A.J.T. (2004). Fire-induced erosion and millennial-scale climate change in northern ponderosa pine forests. *Nature* **432**, 87–90.

Power, M.J. (2013). A 21,000-year history of fire. In: Belcher, C.M. (ed). *Fire phenomena and the Earth System: An interdisciplinary guide to fire science*. 1st edition, pp. 207–227. J. Wiley & Sons, Ltd, Chichester, UK.

Power, M.J., Marlon, J.R., Bartlein, P.J., Harrison, S.P. (2010). Fire history and the Global Charcoal Database: a new tool for hypothesis testing and data exploration. *Palaeogeography, Palaeoclimatology, Palaeoecology* **291**, 52–59.

Power, M.J., Marlon, J., Ortiz, N., *et al.* (2008). Changes in fire regimes since the Last Glacial Maximum: an assessment based on a global synthesis and analysis of charcoal data. *Climate Dynamics* **30**, 887–907.

Pyne, S.J., Andrews, P.L., Laven, R.D. (1996). *Introduction to wildland fire*. J. Wiley and Sons, New York, NY.

Radzi bin Abas, M., Oros, D.R., Simoneit, B.R.T. (2004). Biomass burning as the main source of organic aerosol particulate matter in Malaysia haze episodes. *Chemosphere* **55**, 1089–1095.

Rein, G. (2013). Smouldering fires and natural fuels. In: Belcher, C.M. (ed.). *Fire phenomena and the Earth System: An interdisciplinary guide to fire science*. 1st edition, pp. 15–33. J. Wiley & Sons, Ltd, Chichester.

Rimmer, S.M., Thompson, J.A., Goodnight, S.A., Robl, T. (2004), Multiple controls on the preservation of organic matter in Devonian-Mississippian marine black shales: Geochemical and petrographic evidence. *Palaeogeography, Palaeoclimatology, Palaeoecology* **215**, 125–154.

Roos, C.I., Swetnam, T.W. (2012). A 1414-year reconstruction of annual, multidecadal, and centennial variability in area burned for ponderosa pine forests of southern Colorado Plateau region, Southwest USA. *The Holocene* **22**, 281–290.

Roy, D.P., Boschetti, L., Smith, A.M.S. (2013). Satellite romote sensing of fires. In: Belcher, C.M. (ed.). *Fire phenomena and the Earth System: An interdisciplinary guide to fire science. 1st edition*, pp. 97–124. J. Wiley & Sons, Ltd, Chichester.

Rundel, P.W. (1981). Fire as an ecological factor. In: Lange, O.L., Nobel, P.S., Osmond, C.B., Ziegler, H. (eds.), *Physiological plant ecology I, Response to the physical environment*, pp 501–538. Springer Verlag, Berlin, Germany.

Santín, C., Doerr, S.H., Shakesby, R.A., Bryant, R., Sheridan, G.J., Lane, P.N.J., Smith, H.G., Bell, T.L. (2012). Carbon loads, forms and sequestration potential within ash deposits produced by wildfire: new insights from the 2009 'Black Saturday' fires, *Australia. European Journal of Forest Research* **131**, 1245–1253.

Scheiter, S., Higgins, S.I., Osborne, C.P., Bradshaw, C., Lunt, D., Ripley, B.S., Taylor, L.L., Beerling, D.J. (2012). Fire and fire-adapted vegetation promoted C4 expansion in the late Miocene. *New Phytologist* **195**, 653–666.

Schmidt, M.W.I. and Noack, A.G. (2000). Black carbon in soils and sediments: Analysis, distribution, implications, and current challenges. *Global Biogeochemical Cycles* **14**, 777–793.

Schroeder, W., Prins, E., Giglio, L., Csiszar, I., Schmidt, C., Morisette, J. and Morton, D. (2008). Validation of GOES and MODIS active fire detection products using ASTER and ETM+ data. *Remote Sensing in the Environment* **112**, 2711–2726.

Scott, A.C. (1989). Observations on the nature and origin of fusain. *International Journal of Coal Geology* **12**, 443–475.

Scott, A.C. (2000). The pre-Quaternary history of fire. *Palaeogeography, Palaeoclimatology, Palaeoecology* **164**, 281–329.

Scott, A.C. (2010). Charcoal recognition, taphonomy and uses in palaeoenvironmental analysis. *Palaeogeography, Palaeoclimatology, Palaeoecology* **291**, 11–39.

Scott, A.C., Collinson, M.E., Pinter, N., Hardiman, M. Anderson, R.S., Brain, A.P.R., Smith, S.Y., Marone, F., Stampanoni, M. (2010). Fungus, not comet or catastrophe, accounts for carbonaceous spherules in the Younger Dryas 'impact layer'. *Geophysical Research Letters* **37**, L14302.

Scott, A.C., Cripps, J., Nichols, G., Collinson, M.E. (2000). The taphonomy of charcoal following a recent heathland fire and some implications for the interpretation of fossil charcoal deposits. *Palaeogeography, Palaeoclimatology, Palaeoecology* **164**, 1–31.

Scott, A.C., Damblon, F. (2010). Charcoal: taphonomy and significance in geology, botany and archaeology. *Palaeogeography, Palaeoclimatology, Palaeoecology* **291**, 1–10.

Scott, A.C., Galtier, J., Gostling, N.J., Smith, S.Y., Collinson, M.E., Stampanoni, M., Marone, F., Donoghue, P.C.J., Bengtson, S. (2009). Scanning Electron Microscopy and Synchrotron Radiation X-Ray Tomographic Microscopy of 330 million year old charcoalified seed fern fertile organs. *Microscopy and Microanalysis* **15**, 166–173.

Scott, A.C., Glasspool, I.J. (2005). Charcoal reflectance as a proxy for the emplacement temperature of pyroclastic flow deposits. *Geology* **33**, 589–592.

Scott, A.C., Glasspool, I.J. (2006). The diversification of Paleozoic fire systems and fluctuations in atmospheric oxygen concentration. *Proceedings of the National Academy of Sciences, USA* **103**, 10861–10865.

Scott, A.C., Jones, T.J. (1994). The nature and influence of fires in Carboniferous ecosystems. *Palaeogeography, Palaeoclimatology, Palaeoecology* **106**, 91–112.

Scott, A.C., Kenig, F, Plotnick, R.E., Glasspool, I.J., Chaloner, W.G., Eble, C.F. (2010). Evidence of multiple Late Bashkirian to Early Moscovian (Pennsylvanian) fire events preserved in contemporaneous cave fills. *Palaeogeography, Palaeoclimatology, Palaeoecology* **291**, 72–84.

Shakesby, R.A., Doerr, S.H. (2006) Wildfire as a hydrological and geomorphological agent. *Earth Science Reviews* **74**, 269–307.

Shlisky, A.J., Hickey, S., Bragg, T.B. (2007). Using models to assess fire regime conditions and to develop restoration strategies in grassland systems at landscape and global scales. *Tall Timbers Fire Ecology Conference Proceedings* **23**, 84–84.

Simoneit, B.R.T. (2002). Biomass burning — a review of organic tracers for smoke from incomplete combustion. *Applied Geochemistry* **17**, 129–162.

Spessa A., van der Werf, G., Thonicke, K., Gomez-Dans, J., Lehsten, V., Fisher, R. & Forrest, M. (2012). Modelling Vegetation Fires and Emissions. In: Goldammer, J. (Ed.) *Fire and Global Change*, Chapter XIV. Springer publishers.

Spracklen, D.V., Mickley, L.J., Logan, J.A., Hudman, R.C., Yevich, R., Flannigan, M.D., Westerling, A.L. (2009). Impacts of climate change from 2000 to 2050 on wildfire activity and carbonaceous aerosol concentrations in the western United States. *Journal Geophysical Research* **114**, D20301.

Sugihara, N.G., Van Wagtendonk, J.W., Fites, Kaufman, J. (2006). Fire as an ecological process. In: Sugihara, N.G., Van Wagtendonk, J.W., Shaffer, K.E., Fites, Kaufman, J., Thode, A.E. (eds). *Fire in California's ecosystems*, pp. 58–74. University of California Press, Berkley, CA.

Swanson, F.J. (1981). Fire and geomorphic processes. In: Mooney, H.A., Bonnicksen, T.H., Christensen, N.L., Lotan, J.E., Reiners, W.A. (Eds.). *Fire Regimes and Ecosystem Properties*, pp. 401–420. USDA For. Serv. Gen. Tech. Rep., WO-26.

Swetnam, T.W. (1993). Fire history and climate change in giant sequoia groves. *Science* **262**, 885–889.

Swetnam, T.W., Allen, C.D., Betancourt, J.L. (1999). Applied historical ecology: Using the past to manage for the future. *Ecological Applications* **9**(4): 1189–1206.

Swetnam, T.W., Baisan, C.H., Capiro, A.C., Brown, P.M., Touchan, R., Anderson, R.S., Hallett, D.J. (2009). Multimillennial fire history of the giant forest, Sequoia National Park, California, USA. *Fire Ecology* **5**(3), 120–150.

Swetnam, T.W., Betancourt, J.L. (1990). Fire-southern oscillation relations in the Southwestern United States. *Science* **249**, 961–1076.

Thonike, K., Spessa, A., Prentice, I.C., Harrison, S.P., Dong, L., Carmona-Moreno, C. (2010). The influence of vegetation, fire spread and fire behaviour on biomass burning and trace gas emissions: results from a process-based model. *Biogeosciences* **7**, 1991–2011.

Torero, J.L. (2013). An itroduction to combustion in organic materials. In: Belcher, C.M. (ed.). *Fire phenomena and the Earth System: An interdisciplinary guide to fire science*. 1st edition, pp 3–13. J. Wiley & Sons, Ltd, Chichester.

Ubeda, X., Outerio, L.R. (2009). Physical and Chemical Effects of Fire on Soil. In: Cerdà, A., Robichaud, P. (eds.), *Fire Effects on Soils and Restoration Strategies*, pp. 105–133. Science Publishers Inc., NH.

van der Werf, G.R., Randerson, J.T., Collatz, G.J., Giglio, L. (2003). Carbon emissions from fires in tropical and subtropical ecosystems. *Global Change Biology* **9**, 437–562.

van der Werf, G.R., Randerson, J.T., Giglio, L., *et al.* (2006). Interannual variability in global biomass burning emissions from 1997 to 2004. *Atmospheric Chemistry and Physics* **6**, 3423–3441.

van der Werf, G.R., Randerson, J.T., Giglio, L., Collatz, G.J., Mu, M., Kasibhatla, P.S., Morton, D.C., DeFries, R.S., Jin, Y., van Leeuwen, T.T. (2010). Global fire emissions and the contribution of deforestation, savanna, forest, agricultural, and peat fires (1997–2009). *Atmospheric Chemistry and Physics* **10**, 11707–11735.

Van Wagtendonk, J.W. (2006). Fire as a physical process. In: Sugihara, N.G., Van Wagtendonk, J.W., Shaffer, K.E., Fites, Kaufman, J., Thode, A.E. (eds). *Fire in California's ecosystems*, pp. 38–57. University of California Press, Berkley, CA.

Westerling, A.L., Hidalgo, H.G., Cayan, D.R., Swetnam, T.W. (2006). Warming and Earlier Spring Increase Western U.S. Forest Wildfire Activity. *Science* **313**, 940–943.

Whtilock, C., Higuera, P.E., McWethy, D.B., Briles, C.E., 2010. Paleoecological Perspectives on Fire Ecology: Revisiting the Fire-Regime Concept. *The Open Ecology Journal*, 2010, **3**, 6–23.

Whitlock, C., Tinner. W. 2010. Editorial: Fire in the Earth System. *PAGES news* **18**, 55–57.

Wolbach, W.S., Gilmour I., Anders, E. (1990). Major wildfires at the Cretaceous/Tertiary Geological Society of America. *Special Paper* **247**, 391–400.

Zalasiewicz, J., Williams, M., Haywood, A. & Ellis, M. (2011). The Anthropocene: a new epoch of geological time? *Philosophical Transactions of the Royal Society A: Mathematical, Physical and Engineering Sciences* **369**, 835–841.

PART TWO
Biology of fire

Photo

Crown fire in an Australian *Banksia* shrubland (Proteaceae). Recent research using data from molecular phylogenies calibrated with the fossil record suggests flammable traits in several distinct plant lineages evolved in the Cretaceous, a fiery period in Earth history (photo from B. Lamont).

Fire on Earth: An Introduction, First Edition. Andrew C. Scott, David M.J.S. Bowman, William J. Bond, Stephen J. Pyne and Martin E. Alexander.
© 2014 John Wiley & Sons, Ltd. Published 2014 by John Wiley & Sons, Ltd.

Preface to part two

The idea that fire and biological life are paired, or at least strongly coupled, is intellectually challenging and begs numerous other questions about the ecology and evolution of fire. In Part Two of this book, we explore these ideas, which necessarily involves revisiting fundamental ideas in biogeography, ecology and evolution. Orthodox accounts of life on Earth still treat fire like other external disturbances from which biological systems have had to periodically survive and recover. However, we will show here that such treatments misrepresent critical features of life on Earth.

Fire, like herbivores, requires organic matter to propagate, whereas disturbances such as windstorms, floods and landslides do not. Fire has caused a decoupling of climatic patterns and vegetation – a world without fire would have more forest cover because savannahs would not exist and grasslands would be restricted to climatically or edaphically extreme environments. Through evolutionary adjustment, fire and life have shaped each other, creating characteristic syndromes of fire activity known as 'fire regimes'.

Fire regimes influence ecological processes affecting soil formation, nutrient cycling, wildlife habitats, carbon dynamics and regional climates.

Understanding the biology of fire is a rapidly progressing field, demanding integrative and synthetic thinking because fire cuts across traditional disciplinary boundaries such as ecology, climatology and hydrology. Fire feedbacks challenge the assumption of simple cause and effect relationships. Further, considering fire as a major Earth System process has become an urgent research priority, because of the evidence that rapid global environmental changes caused by humans – particularly climate change – are affecting global fire activity which, in turn, causes further global environmental change.

In this section, we treat humans as biological actors in this unfolding ecological and evolutionary drama, acknowledging that this perspective cannot capture the complexity of different cultures and the motivations and concerns of individual humans. These issues are explored in the next section of the book.

Chapter 6

Pyrogeography – temporal and spatial patterns of fire

6.1 Fire and life

Biogeography is concerned with the spatial and temporal variation of life on Earth. In this chapter, we show that landscape fire also has a biogeography, because of the strong interactions amongst vegetation, climate, terrain and fire. This realization is important in understanding the biology of fire, and it is a powerful organizing principle in understanding the role of landscape fire in the Earth System, a subject we discuss in subsequent chapters.

First, we consider how climate shapes global vegetation patterns and how fire decouples this relationship. We then review global patterns of landscape fire activity, sometimes called 'pyrogeography'. Considering the more local scale, we then explore controls on vegetation boundaries within the same climate zone, using abrupt forest-savanna boundaries as a case study. We develop the fire regime concept (see Chapter 1, Figure 1.25) and conclude with a sketch of fire ecology, outlining how this discipline grapples with problems of correlation and causality in association with numerous direct and indirect feedbacks between climate, vegetation and fire.

6.2 Global climate, vegetation patterns and fire

When European explorers commenced undertaking oceanic voyages the changes in climate and vegetation were strikingly hard to ignore. Even so, it took towering intellects, like Alexander von Humbolt in the early 18th century and Andreas Schimper in the late 19th century, to provide a globally coherent explanation of these patterns: climate and vegetation were closely coupled (Shugart and Woodward, 2011).

Despite continents having different floras, it was realized that the same climate favoured similar life forms – for example, hard-leaved (sclerophyllous) shrubs in climates with mild moist winters and hot, dry summers (Mediterranean climates). Schimper is credited with the first modern synthesis of global biomes, and it was he who coined many names for vegetation types that are still in use today, such as tropical rain forest, savanna, steppe and tundra; because of the strong relationship between vegetation and climate, some climatic zones have been named by dominant vegetation, such as 'tundra' or 'savanna' climates (Figure 6.1).

Fire on Earth: An Introduction, First Edition. Andrew C. Scott, David M.J.S. Bowman, William J. Bond, Stephen J. Pyne and Martin E. Alexander.
© 2014 John Wiley & Sons, Ltd. Published 2014 by John Wiley & Sons, Ltd.

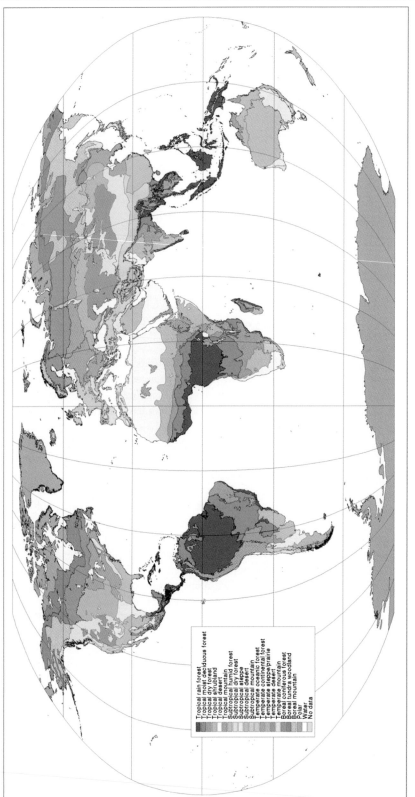

Figure 6.1 A map of global ecological zones, as categorized by the Food and Agriculture Organization of the United Nations (FAO 2001). This classification is based on the Koppen-Trewartha climate system, in combination with natural vegetation characteristics. This map typifies the belief that global biomes are the product of climate and vegetation, and that fire is an effect of this climate-vegetation coupling. (Reproduced with permission from Food and Agriculture Organization, 2001.).

Legend:

- Tropical rain forest
- Tropical moist deciduous forest
- Tropical dry forest
- Tropical shrubland
- Tropical desert
- Tropical mountain
- Subtropical humid forest
- Subtropical dry forest
- Subtropical steppe
- Subtropical desert
- Subtropical mountain
- Temperate oceanic forest
- Temperate continental forest
- Temperate steppe/prairie
- Temperate desert
- Temperate mountain
- Boreal coniferous forest
- Boreal tundra woodland
- Boreal mountain
- Polar
- Water
- No data

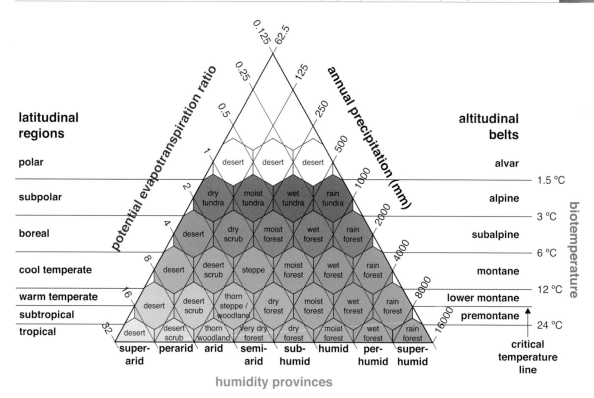

Figure 6.2 Holdridge Life Zones defined biomes on Earth as a function of 'biotemperature' (the mean of the annual temperature between 0° and 30° C.), mean total annual precipitation and potential evapotranspiration ratio. Such a climatic approach has been pivotal for both the interpretation of satellite imagery of the globe and in the development of predictive process models to global biomes. With such a perspective, fire is an effect of climate determined vegetation patterns rather than a causal factor. (After Holdridge, 1967).

The idea that climate determines vegetation has been subsequently elaborated. For example, Holdridge (1967) developed a classification that defines the climate space of different 'life zones', or biomes, using a lattice with three axes: annual temperature (influenced by latitude and altitude); annual rainfall; and annual potential water loss through evapotranspiration (Figure 6.2).

The role of climate in determining global vegetation is entrenched in biogeographic thinking, although the capacity to analyze vegetation patterns using globally consistent data from satellites (see Chapter 1.13, Part One) is now forcing a revision of this generalization. For example, recent analysis of satellite estimates of tree cover have identified large areas of the tropics that are climatically suitable for the development of either savanna or forests. These 'bi-stable areas' are sandwiched between areas of forest and savanna that are apparently climatically determined (Figure 6.3).

Computer models have also investigated the control of global vegetation patterns by climate and fire (see Chapter 1.14, Part One). Just as general circulation models (GCMs) have been built based on physical principles to explore future climate change (see Chapter 1.14), biogeographic models based on physiological principles have been developed to explore the dynamics of future vegetation. These 'dynamic global vegetation models' (DGVMs) provide predictions of physiologically constrained potential vegetation and, therefore, provide us with tools to separate correlation from causality in climate-vegetation relationships. DGVMs are built by simulating physiological response of plants to climate and carbon dioxide (CO_2), using a small suite of generalized plant life forms. Various combinations of these plant life forms can be used to recognize broad vegetation types or 'biomes'.

The first generation of DGVMs represented landscape fire activity from empirical relationships between fire probability and weather conditions. Bond *et al.*

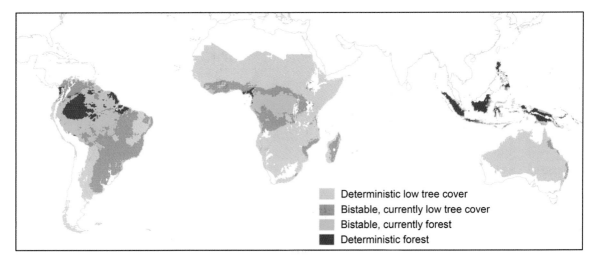

Figure 6.3 Global distribution of woody vegetation cover, classified according to whether it is climatically determined ('deterministic') or not ('bi-stable'). Low tree cover is predicted by low rainfall and high seasonality, and high tree cover is predicted by high rainfall. However, areas with intermediate rainfall and seasonality may have low tree cover (savanna) or high tree cover (forest). The cause of the bi-stability is thought to be due to high fire frequency. (From Staver *et al.*, 2011). Reproduced by permission of the American Academy for the Advancement of Science.

(2005) investigated what happens if these simulated 'fires' were switched off. They found that a 'world without fire' would have twice as much forest and half the extent of grassland that occurs on Earth (Figure 6.4). These results underestimate the global importance of fire, since the DGVM used by Bond *et al.* (2005) could not represent fire-dependent forest vegetation types such as pine forests of the western United States and eucalypt forests in Australia, being primarily focused on grasses and shrubs versus tree life forms.

More sophisticated DGVMs that include physically-based simulations of fire activity and greater range of plant life forms, including types that favour flammable environments, are currently in development (Chapter 1). The next generation of DGVMs will, therefore, further illuminate how fire shapes global vegetation patterns.

6.3 Pyrogeography

The routine availability of satellite imagery (Chapter 1.13) has transformed understanding of the biogeography of fire by highlighting the ubiquity of fire on Earth (Figure 6.5). This was a truly unexpected spin-off of remote sensing, as the primary mission of the early satellite sensors launched in the late 1970s (e.g. National Oceanic and Atmospheric Administration's (NOAA) Advanced Very High Resolution Radiometer (AVHRR)

sensor) was to observe meteorological conditions ('weather satellite') and were not designed to map fire activity.

Over the last 20 years, refinements in satellite monitoring have seen a revolution in understanding of the seasonal and geographical patterns of area burnt by fire, leading to the new science of 'pyrogeography'. The most important generalization to emerge from this has been the recognition of the relationship between area burnt and net primary productivity (which is controlled by climate) (Figure 6.6).

In environments with very low levels of primary productivity, such as deserts, area burnt is constrained by available fuel. In highly productive environments, such as tropical rainforests, fire is constrained by suitably dry weather conditions. The 'sweet spots' for fire activity are environments with intermediate levels of net primary productivity and climates with a high frequency of suitable weather conditions to sustain fires in seasonally abundant fuels. These conditions are met in tropical savannas, which are the most widespread flammable environments on Earth. Savannas have hot, wet seasons that promote rapid plant growth, followed by dry seasons that desiccate grass fuels and provide weather conditions suitable for fire. The savanna environment highlights the direct (fire weather) and indirect (plant growth) effects of climate on fire activity (see Chapter 1, Figure 1.35).

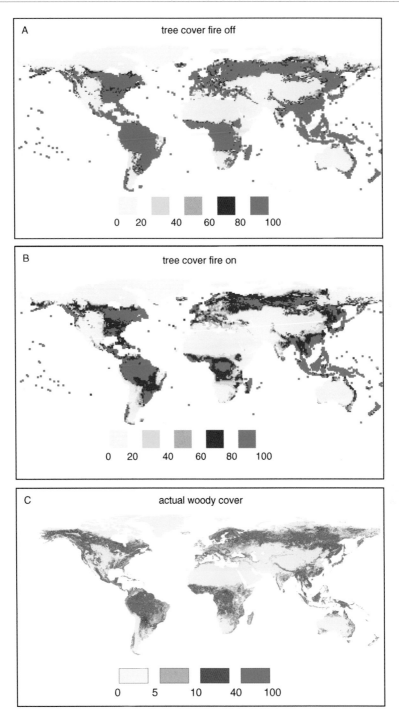

Figure 6.4   Change in predicted percentage tree canopy cover for a dynamic global vegetation model with 'fire' switched off, (A) fire switch 'on' and (B) compared to satellite estimates of woody cover from both natural forests and plantations. (C). The cover classes correspond to broad woody vegetation types where 40–100% canopy cover corresponds to closed canopy with no grass; 10–40% densely treed savanna and sorts of 'forest'; 5–10% canopy cover are savannas with scattered trees. The key message of this analysis is the disequilibrium of woody vegetation with climate, especially in the tropics, due to the effect of fire. (From Bond *et al.*, 2005).

A

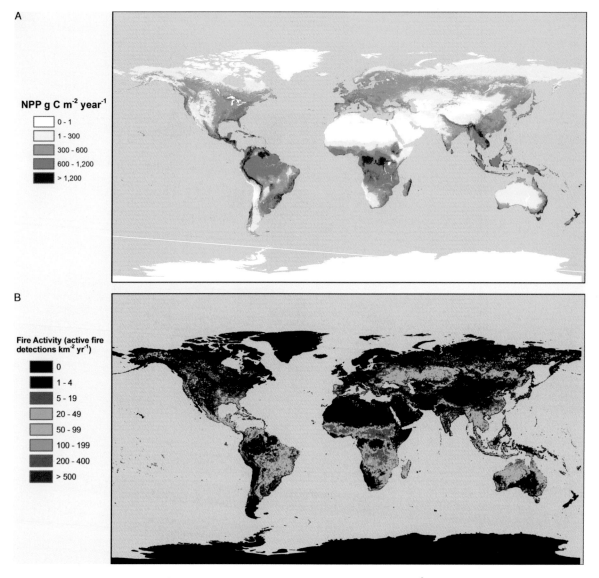

Figure 6.5   Satellite estimates of net primary productivity (NPP, grams of carbon per m² per year) and the annual average number of fires observed. There is a striking association between intermediate levels of NPP and fire activity. Human-caused fires are also responsible for the fire activity observed in the high productivity Amazonian rainforests of South America. (From Bowman *et al.*, 2009). Reproduced by permission of the American Association for the Advancement of Science.

The influence of productivity on fire activity is also apparent in the 'charcoal record' that records fire activity through time (Chapter 2, Part One; Glasspool and Scott, 2013). Lakes and bogs trap both macro- and microscopic charcoal particles in sediments, making it possible to reconstruct past fire activity reaching back into the ice age (Pleistocene) (see Chapter 5, Part One). Recent global syntheses of numerous charcoal records by Power *et al.* (2008) showed that, during the end of the last glacial (21 000 to 16 000 years ago), there was much lower fire activity in extra-tropical regions of North America, Europe and South America than in the current interglacial that commenced 10 000 years ago (see also Power, 2013). This pattern is consistent with the lower productivity of the glacial Earth due to cold, dry climates combined with low atmospheric

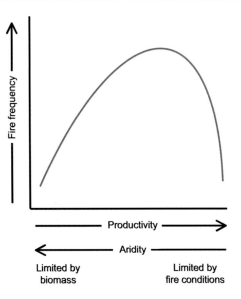

Figure 6.6 A conceptual diagram showing the humped response of fire frequency to primary productivity (controlled by available moisture). In arid environments, fire activity is typically limited by available biomass, the exception being preceding wet periods that produce heavy fuel loads. In moist environments, fire activity is limited to climate conditions sufficiently dry to allow fires to propagate. In environments with intermediate primary productivity, there is often the right combination of fuel and suitable fire weather. (From Murphy et al., 2011).

CO_2 levels, and possibly reduced lightning activity due to fewer convection storms.

The fundamental 'fire triangle' representing fuel, heat and oxygen (Chapter 1, Figure 1.1) is sometimes scaled up to landscapes to represent weather/seasonality, primary productivity and fuel structure/type and ignitions sources. Underlying these variables are direct and indirect effects of climate which manifest at a variety of different temporal scales. To deal with this, Bradstock (2010) has developed a conceptual model which integrates these direct and indirect effects of climate that control pyrogeography, known as the four-switch model. This model shows that fire occurrence can only occur when four conditions are met (Figure 6.7).

- First, there must be sufficient biomass to provide fuel for a fire. In arid environments, biomass is strongly controlled by preceding wet years (also known as antecedent rainfall). In mesic vegetation, the time since the last fire is an important constraint on fuel availability.
- The second switch is the availability of the biomass to burn. This is controlled by seasonal conditions or weather patterns conducive for fires, such as droughts. In mesic climates, most years may be too moist to sustain large fires, whereas, in arid

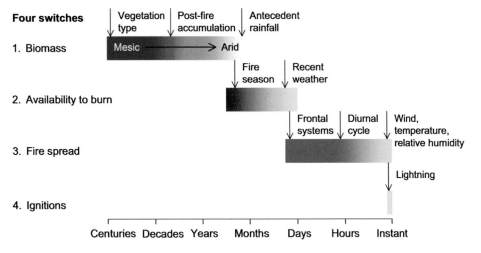

Figure 6.7 Conceptual diagram of Bradstock (2010) – the four-switch model that shows how four key 'switches' control fire activity organized on a time axis. In mesic environments, biomass can take centuries to accumulate, and years in intermediate productive environments, but only months in arid environments. Conditions suitable to making biomass flammable vary seasonally and are also affected by antecedent weather. Fire spread is controlled by meteorological conditions that operate over day to hour scales. Ignitions are instantaneous processes, but propagation of fire can only occur if the other three 'switches' are activated. Unlike lightning, humans can exploit a greater number of 'windows' to set self-sustaining fires. (After Murphy et al., 2011).

environments, climate conditions are nearly always suitable for burning.

- The third switch is the capacity of the fire to spread, which is controlled by immediate weather conditions such as strong wind speeds, low humidity and high temperatures.
- Finally, the fourth switch is the presence of ignitions from either lightning or humans.

It is important to note that this model works well for flammable systems, but it is not able to explain the coexistence of vegetation with strongly contrasting flammabilities, such as savanna and rainforests – a topic we address in the next section.

The four-switch model is also useful to understand how humans can directly and indirectly override constraints imposed by gradients in productivity. For example, pastoralism can substantially reduce fuel availability in savannas, the introduction of flammable plants increase fuel loads and fuel continuity in arid environments, and humans can choose to set fires in tropical rain forests during severe drought (Table 6.1). Further, humans can use fire to change fuel loads and create firebreaks, thereby influencing the extent of lightning-started fires (see Part 4).

The effect of human fire management interfering with climate driven 'natural' fire activity is apparent in sedimentary records. The synthesis of charcoal records

Table 6.1 How humans influence landscape fire parameters by modifying key variables that affect fire activity (from Bowman *et al.*, 2011)

Fire variable	Natural influences	Human influences	Fire parameters
Wind speed	• Season • Weather • Topography • Land cover	• Climate change • Land cover change	*Fire spread*
Fuel continuity	• Terrain type (slope, rockiness, aspect) • Rivers and water bodies • Season • Vegetation (type, age, phenology)	• Artificial barriers (Roads, fuel breaks) • Habitat fragmentation (fields) • Exotic grasses • Land management (patch burning, fuel treatments) • Fire suppression	
Fuel loads	• Tree, shrub and grass cover • Natural disturbances (e.g. insects or frost damage, windthrow,) • Herbivory • Soil fertility • Season	• Grazing • Timber harvests • Exotic species establishment • Fire suppression • Fuel treatments • Land use and land cover (deforestation, agriculture, plantations)	*Fire intensity and severity*
Fuel moisture	• Season • Antecedent precipitation • Relative humidity • Air temperature • Soil moisture	• Climate change • Land management (logging, grazing, patch burning) • Vegetation type and structure (species composition, cover, stem density)	
Ignitions	• Lightning • Volcanoes • Season	• Human population size • Land management • Road networks • Arson • Time of day • Season • Weather conditions	*Number and spatial and temporal pattern of fires*

from around the world, but with a concentration of records from north America and Europe, revealed that global biomass burning declined from AD 1 to 1750, then abruptly increased between 1750 and 1870, before sharply declining. Marlon *et al.* (2008) explained these patterns as reflecting the effect of a global cooling trend that was overridden by expanding human land clearance in the 19th century. There appears to be a subsequent decline in the late 19th and 20th century, attributed to reduced landscape fire use by humans, despite an overall warming trend and elevated atmospheric CO_2.

6.4 Fire and the control of biome boundaries

Studies of boundaries between highly flammable (pyrophilic) and less flammable (pyrophobic) vegetation provide powerful insights into how fire can control plants within one climatic zone and how vegetation can control fires. Of particular importance have been studies on the controls of boundaries between closed canopy forest (henceforth forest) and open tree canopies with a ground cover of light demanding grasses or treeless vegetation (henceforth savanna).

In less than 100 m, forests can switch to savannas that have open tree canopies (Figures 6.8 and 6.9). These abrupt boundaries cause equally abrupt changes in fire activity, where savannas can be regarded as fire promoting and forests fire retarding. The causes of these striking ecological transitions have fascinated ecologists, in part because they demonstrate that climate does not necessarily control vegetation patterns, contrary to the assumptions of many biogeographic classifications such as the Holdridge scheme (see Figure 6.2).

Some ecologists have argued that the forest boundaries are controlled by environmental factors other than climate, e.g. changes in soil fertility. They assert that fire is not a cause of the vegetation boundary, but rather that the difference in fire activity is an effect of the environmental control of the vegetation boundary. However, this view is contradicted by field observations, analysis of aerial photography and experiments that demonstrate that forests invade open vegetation when it is protected from fire.

Figure 6.8 Tiny rainforest patch embedded in a savanna in the Kimberley, northern Australia. In the space of a few metres, there is a switch from eucalypt savanna trees and a dense perennial grass to a closed canopy forest with numerous species from a wide variety of plant families. (See Bowman, 1992). (photo: David Bowman).

Figure 6.9 Aerial photograph of a rainforest-savanna boundary on the Atherton Tablelands in the humid tropics of north Queensland. (photo: Reproduced with permission of Wet Tropics Management Authority, 2008.).

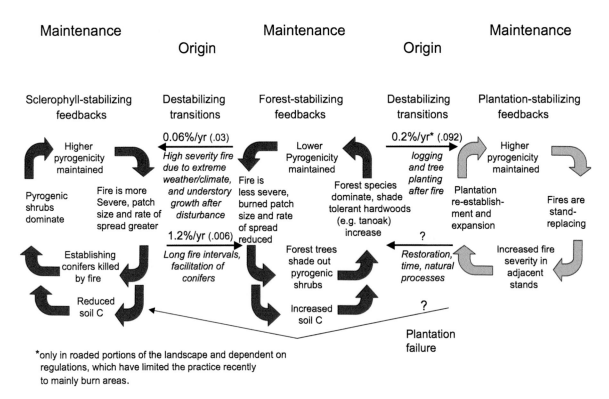

Figure 6.10 Conceptual model of the alternative stable system that enables the coexistence of open sclerophyll vegetation with natural and plantation conifer forests in the Klamath Mountains, north-western California, USA. The transition from forest to sclerophyll woodland is related to fire activity. The sclerophyll vegetation is also maintained by frequent burning. (After Odion *et al.*, 2010).

Nonetheless, it is important to acknowledge cases where forest boundaries have been shown by palaeoecological research to be very stable over hundreds of years. In such cases, soil differences across boundaries may also develop as a consequence of contrasting ecosystem processes in the pyrophilic vs. pyrophobic vegetation states. There have been a number of conceptual models, discussed below, that are able to explain why some boundaries are unstable while others are not.

Classical ecological successional theory asserted that, for each climate zone, there was a corresponding perfectly suited vegetation type known as the 'climatic climax'. For instance, in north-eastern USA it was assumed that deciduous beech (*Fagus*) forests were the climax vegetation of the temperate zone. It was recognized that disturbance could upset the equilibrium between climax vegetation and climate. It was believed that the equilibrium, however, could be restored through the orderly sequential replacement of different 'pioneer' vegetation types – a process known as 'succession'. Fire was recognized as an important disturbance that could initiate 'pyric' succession. The pyric succession model of vegetation dynamics dominated thinking about the effects of fire, and has entrenched the idea that climate was the driver of vegetation patterns.

By contrast, alternative stable state (ASS) theory can explain how flammable open and less flammable closed vegetation types are able to coexist in the same climate. This is well illustrated by Odion *et al.* (2010) who applied alternative stable states to the Klamath Mountains, north-western California, USA (Figure 6.10). They showed that frequently burnt shrub-dominated sclerophyllous (chaparral) vegetation exists side-by-side with less frequently burnt conifer forests.

The shrublands are highly flammable and recover quickly following fire. By contrast, the forests are much less pyrogenic, and flammability further decreases with time between fires. The forests are infrequently burnt by high-severity stand-replacing fires and, if the regeneration is burnt again (a short fire free interval), then shrubland can establish. Thus fire and vegetation are involved in feedback loops that provide stabilizing effects of vegetation patterns. These self-reinforcing feedbacks can be overridden by extreme fire disturbances, in the case of forests, and by long periods of fire protection in the case of shrublands.

Soil fertility can also provide powerful stabilizing feedbacks. For example, reduced fire activity changes nutrient cycles and increases nutrient capital, and it can improve soil drainage through increased transpiration and the development of deep roots. Conversely, increased fire activity can reduce site productivity due to increased nutrient losses. These edaphic feedbacks have been used to explain the juxtaposition in the southwest of Tasmania of temperate rainforest with fertile soils alongside treeless flammable vegetation on infertile areas with uniform bedrock (Figure 6.11).

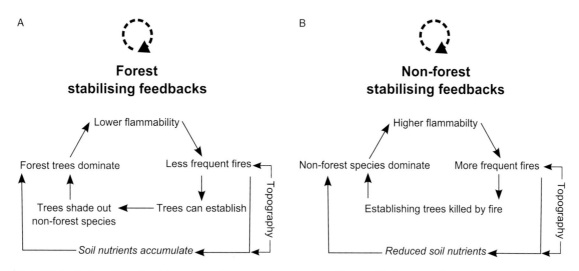

Figure 6.11 Model of how the interaction of topography and soil nutrients with fire cause the maintenance of forest (A) and non-forest (B) in south-west Tasmania. Accumulation of nutrients due to reduced burning favours increased forests that have a lower frequency of burning than non-forest communities. (From Wood *et al.*, 2011).

Figure 6.12 Temperate rain forest in topographic fire refugia in southwest Tasmania. Note the evidence of extensive burning surrounding the forest. (photo: S.W. Wood).

Topography is also important in this landscape, shaping both vegetation and fire patterns. Infrequently burnt rainforests occur in topographic settings less likely to be burnt, such as enclosed valleys, with the converse being true for flammable treeless vegetation (Figure 6.12).

Differences in the tolerance of forest and savanna trees to fire and their growth rates are also stabilizing feedbacks. Closed forest cover has a suppressive effect on fire activity, because of cooler and moister microclimate and shading out of grass fuels. Forest trees have limited capacity to withstand fire because, for instance, they have thin bark. By contrast, savanna trees are comparatively tolerant of frequent fire, due to thick bark and the ability of small plants to re-sprout from the base (and in some cases, their stems).

If forests are burnt, then they are vulnerable to being replaced by savanna, because the loss of tree cover favours the invasion of light-demanding grasses that increase fire risk and, in turn, perpetuate the loss of tree cover. Conversely, unburnt savannas are vulnerable to invasion by forest trees that have strategies to establish in low-light conditions. The likelihood of a closed forest canopy forming is determined by the interplay between fire frequency and tree growth rates. Rapid growth of savanna

sapling trees enables them to grow tall enough, and to have thick enough bark, to escape the 'fire trap' and be recruited into the canopy. This has been described as the fire resistance threshold. With enough recruitment canopy, closure can occur, which causes a decrease in the occurrence of savanna fires. This has been described as the fire suppression threshold (Figure 6.13).

It is important to note that any factor that increases tree growth is likely to increase the chance of forest formation, including high rainfall, abundant soil moisture, high nutrient availability, sheltered topographic positions and increased CO_2. These will influence the rate at which savannas can be converted to forest (Figure 6.14). Thus, on productive sites, high frequencies of burning are required to cause a switch from forest to savannas, whereas low frequencies of burning are required on unproductive sites.

Flammable invasive grasses can transform forest and woodland ecosystems through a process known as the 'grass-fire cycle'. This is a feedback where invasive grasses promote frequent, intense fires, because these grasses produce high biomasses of quick-drying and well-aerated fine fuels. Frequent fires defoliate and eventually kill trees, while the grasses are able to

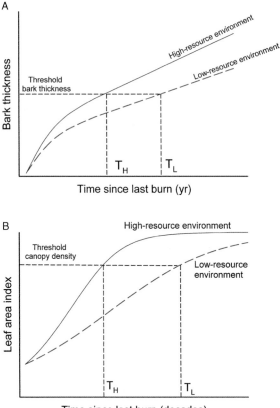

Figure 6.13 Graphical representation of how resource availability influences the 'fire resistance threshold' for re-sprouting trees (A) and 'fire threshold model' for savannas and forests (B). There is a threshold where bark thickness (that also scales to tree height) provides trees with the capacity to survive recurrent fires. There is a threshold where sufficient canopy cover reduces the severity of fires. Resource availability determines the time since the last fire when these thresholds are crossed. Once they are crossed, savanna switches to forests. (From Hoffmann *et al.*, 2012).

recover rapidly from fire because their buds are protected beneath the soil surface.

The loss of woody biomass can also result in drier microclimates, further adding momentum to the grass-fire cycle. Frequent burning disrupts nutrient cycles, making nutrients more available for the quick growing grasses, while reducing the overall nutrient stocks in the soils and, hence, slowing the growth of any surviving woody plants. The grass-fire cycle can convert species from diverse woody vegetation to exotic grasslands that resist invasion from native species even when protected from fire.

Aggressive invasive grasses that are driving grass-fire cycles are Cheatgrass (*Bromus tectorum*) in the Great Basin of the United States (Figure 6.15), tufted beard-grass (*Schizachyrium condensatum*) and molasses (*Melinis minutiflora*) in Hawaii, gamba grass (*Andropogon gayanus*) in north Australian savannas, and Buffel Grass (*Cenchrus ciliaris*) in arid Australian and the Sonoran desert of the western USA (see Chapter 10).

6.5   The fire regime concept

An important concept in fire ecology that captures many interactions between fire and vegetation types, landscape settings and climate zones is the fire regime (see also Chapter 1). A fire regime can be thought of as a recipe made up of various interrelated components shaped by both vegetation and climate; it can be defined as 'the pattern of repeated fires expressed as frequency, season, type, severity and areal extent in a landscape'. Six examples of fire regimes are provided in Table 6.2.

It is important to note that the biological effect of fires is not simply related to fire intensity because, in some environments (such as tropical rain forest), low-intensity, infrequent fires can be very damaging (high severity), due to the very limited capacity of the vegetation to tolerate fires. Conversely, some vegetation types are very resilient to frequent fires because of a range of biological traits to enable rapid recovery and, hence, the fires are typically of low severity. Key plant traits that provide fire tolerance and recovery are listed in Table 6.2 and will be discussed in detail in the following chapter (Chapter 7).

Fire regimes cannot be predicted by either climate or vegetation alone because of numerous interactions (see Chapter 1, Figure 1.25). For example, vegetation is also influenced by site productivity, terrain and human management practices, and vegetation can modify local climate. These feedbacks blur simple predictive relationships between climate, fire and vegetation. Indeed, as we have discussed, it is possible to have fundamentally different vegetation types growing side by side under the same climate.

Over evolutionary time, the appearance of new growth forms, such as C4 grasses, has changed fire regimes (see Chapter 4, Part One). Likewise, the introduction of non-native plants has also altered vegetation by changing fire regimes – notoriously

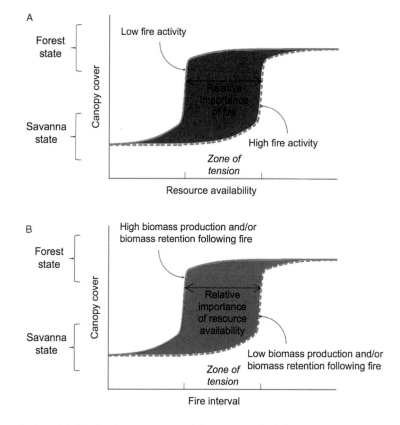

Figure 6.14 Conceptual model showing how savanna and forest can switch in response to resource availability (A) and fire interval (B). In low productive environments, comparatively low fire activity can maintain savannas, but much higher fire activity is required in productive environments. (From Murphy and Bowman, 2012).

Figure 6.15 A continuous cover of invasive cheatgrass fuels an intense fire in sagebrush in the Great Basin of the western USA. Such invasive grasses cause the loss of native vegetation through increase frequency and severity of fires. (See Balch *et al.*, 2013). (photo: Mike Pellant, Bureau of Land Management).

Table 6.2 Representative examples of fire regimes that occur in different woody vegetation types and associated plant life history strategies and traits (adapted from Bowman *et al.*, 2009). Reproduced by permission of the American Association for the Advancement of Science

Fire type	Surface	Surface	Surface	Crown	Crown	Crown
Fuel type	Leaf litter and soil organic matter	Stratified fuels including litter, twigs, shrubs	Stratified fuels including litter, twigs, shrubs	Above-ground woody biomass	Above-ground woody biomass and organic soil layers	Above-ground woody biomass and organic soil layers
Fire intensity	Low	Low	Moderate	High	High	High
Fire frequency	Very low	High	Medium	Low	Low	Low
Climatic conditions	Severe drought	Drought with antecedent wet period	Dry period with moderate fire weather	Drought, extreme dry weather and strong winds	Extreme drought and strong winds	Drought, extreme dry weather and strong winds
Fire stimulated recruitment (establishes seedlings immediately post-fire)	Nil	Low	Moderate	High	High	High
Crown sprouting (replaces photosynthetic area following defoliation)	Low	Low	Low	High	High	Low
Root suckering (replaces fire-killed stems from existing root system)	Moderate	Low	Low	Low	Moderate	Low
Self-pruning of dead lower branches (removes fuel ladders to inhibit canopy fire spread)	Low	High	High	Nil	Nil	High
Thick bark (protects cambium)	Low	High	High	Nil	Low-Moderate	Low
Fire severity	High	Low	Low	Moderate	High	High
Characteristic vegetation type	Tropical rainforest	Dry ponderosa pine forest, western USA	Dry eucalypt forest, Australia	Fire-dependent Mediterranean shrublands	High-elevation conifer forests, western USA	Wet eucalypt. forests, Australia

so, in the case of invasive grasses burning Hawaiian forests. The complexity of fire, vegetation and climate interactions has frustrated the development of a general model for fire effects on vegetation and has impeded unified global accounts of fire activity. Indeed, until recently, the geographic remit of fire ecology was primarily regional and local description of fire effects on various ecosystems, as there was no simple way to create a global synthesis of landscape fire activity.

6.6   Fire ecology

It is clear from the above that fire has numerous direct and indirect effects on vegetation patterns, situated within a web of ecological interactions (Figure 6.16).

This realization is important, as it argues against ecologists searching for simple and singular 'causal' relationships in controlling vegetation patterns. Furthermore, the interactive nature of fire highlights the fundamental difference between fire and other large scale disturbances, such as tropical storms or tectonic activity, that are independent of vegetation. Rather, fire is more akin to a generalist herbivore, as the effects are modulated by direct and indirect feedbacks; hence, fire genuinely has an 'ecology'.

Such complexity has made studying fire a difficult and vexatious subject, demanding integrative and synthetic thinking, rather than simpler linear and reductionist approaches. Experiments involving manipulation of fire regimes beyond simple fire exclusion are notoriously difficult to execute, given the need to operate at landscape scales and over long time periods (in some cases, given the lifespan of many trees, over centuries). Even over short time spans, fire treatments can become confounded with climate variation and herbivory. For this reason, fire ecologists have relied on historical records to understand how fire regimes have varied in the past.

It is important to acknowledge that it is difficult to identify causal factors conclusively in historical techniques. For instance, there is often robust debate about the interpretation of variation in past fire activity, such as the relative importance of human and climatic influences. Obviously, the complexity of fire ecology demands both mechanistic and conceptual models that can capture multiple feedbacks that operate at different spatial and temporal scales. To work effectively, such models must combine a diversity of data on plant responses to fire, ecosystems effects of fire and the geographic and spatial pattern of fire activity in different environments.

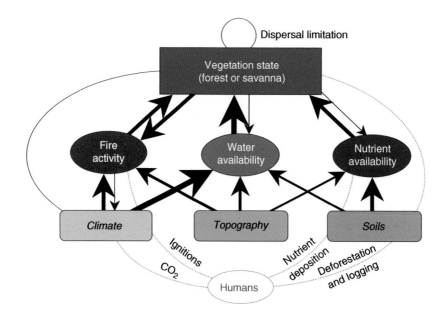

Figure 6.16   Web of interactions that influence the control of savanna and rainforest boundaries. The width of the solid black lines indicates the strength of the interaction. The dashed lines the potential effect of humans. (After Murphy and Bowman, 2012).

However, there remain very large gaps in knowledge about most of these subjects, demanding more fundamental discoveries about the 'biology of fire'. Filling these gaps demands more field research, remote sensing analyses, experimentation and numerical model development. All these research programs require conceptual frameworks about how fire shapes biological and environmental processes. These issues are explored in the following chapters.

6.7 Conclusion

Understanding the geography of fire – pyrogeography – is a rapidly developing field that is illuminating how fire activity has changed and is showing distinct patterns at local, regional and global scales. The control of fire activity is not simply a function of climate or vegetation. Furthermore, fire activity, derived by both natural and anthropogenic ignitions, has had a major effect in shaping vegetation patterns from global and local scales.

A key organizing principle in pyrogeography is the fire regime, yet the complexity of interaction between fire, plants and the environment frustrates the ability to map or predict fire regimes beyond broad generalizations. Fire ecology is the science that grapples with the interaction of fire with life, using an array of approaches including field observations, experiments and modelling. This science is also rapidly developing, and it is increasingly bringing to bear ecological and evolutionary principles to understand the diversification of flammable ecosystems on Earth.

Further reading

Bond, W.J., Woodward, F.I., Midgley, G.F. (2005). The global distribution of ecosystems in a world without fire. *New Phytologist* **165**, 525–538.

Bradstock, R.A. (2010). A biogeographic model of fire regimes in Australia: current and future implications. *Global Ecology and Biogeography* **19**, 145–158.

Holdridge, L.R. (1967). *Life zone ecology.* Tropical Science Center. San Jose, Costa Rica.

Marlon, J.R., Bartlein, P.J., Carcaillet, C., Gavin, D.G., Harrison, S.P., Higuera, P.E., Joos, F., Power, M.J., Prentice, I.C. (2008) Climate and human influences on global biomass burning over the past two millennia. *Nature Geoscience* **1**, 697–702.

Odion, D.C., Moritz, M.A., DellaSala, D.A. (2010). Alternative community states maintained by fire in the Klamath Mountains, USA. *Journal of Ecology* **98**, 96–105.

Power, M.J., Marlon, J., Ortiz, N., *et al.* (2008). Changes in fire regimes since the Last Glacial Maximum: an assessment based on a global synthesis and analysis of charcoal data. *Climate Dynamics* **30**, 887–907.

Shugart, H.H., Woodward, F.I. (2011). *Global Change and the Terrestrial Biosphere: Achievements and Challenges.* Wiley-Blackwell.

Chapter 7

Plants and fire

7.1 Introduction

As described in the previous chapter, fires are very extensive across the globe and fire is a major factor in the distribution of world vegetation. However, if fire is so influential, why do flammable ecosystems vary so much in their structure, composition and dynamics? One answer would be that climate is the major determinant of the mix of growth forms in any region, and that fire is merely a consequence of these growth forms. An alternative explanation is that although climate selects which growth forms could occur, fire acts a major filter, sorting which of these dominate in flammable ecosystems. The relative importance of fire as cause or consequence of ecosystem structure may also vary in different climate regions of the world. In this chapter, we explore the fire biology of plants in order to help understand why flammable ecosystems vary so much in their structure and functioning. We first consider the diverse ways in which plants resist, recover and recruit after a single fire. We then consider how the pattern of repeated fires – the fire regime – selects for different suites of plant traits. The fire regime is, itself, partly determined by the vegetation and partly by extrinsic factors, particularly climate and topography. We use examples of different fire regimes to explore these synergies.

7.2 Fire and plant traits

How does fire damage or kill plants, and how do plants and plant populations respond? The main impact of fire on the vegetative parts of plants has traditionally been considered to be in causing heat damage to regenerative tissue, such as the cambium and buds. The location of perennating buds is the basis of one of the first growth form classifications, that of Raunkiaer. In his system, phanerophytes are woody plants with buds in the canopy, hemicryptophytes are plants such as grasses with buds at ground level, and geophytes have buds below the ground, while therophytes survive the dormant season only as seeds. Raunkiaer's system was built around how plants survive climatic stress, such as winter cold, but bud location is also critical for fire survival. Phanerophytes (trees and shrubs with bud tissues above the ground) are most vulnerable to fire damage, because these regenerative tissues are directly exposed to flame.

Fire survival modes are particularly diverse in woody plants and, since even the largest tree starts as a seed, these modes often change through the different life stages of the tree. Bark thickness strongly influences stem survival. Thick bark provides an insulating layer, reducing heat transfer to the stem and protecting living tissue in the cambium. Recent studies suggest that heat damage to cambial tissue is less

Fire on Earth: An Introduction, First Edition. Andrew C. Scott, David M.J.S. Bowman, William J. Bond, Stephen J. Pyne and Martin E. Alexander.
© 2014 John Wiley & Sons, Ltd. Published 2014 by John Wiley & Sons, Ltd.

Figure 7.1 Modes of survival and post-burn recruitment in crown-fire regimes. (Photos from W.J. Bond). (A) Re-sprouting from lignotuber, *Arctostaphylos* in chaparral. (B) Stand-replacing crown fire in a stand of *Pinus coulteri*, a Californian serotinous pine. (C) Rapid post-burn flowering of a geophyte, *Cyrtanthus purpurea*, Cape fynbos. (D) Serotiny in Leucadendron, *Proteaceae*, Cape fynbos. (E) Serotiny in a Cape endemic family, *Bruniaceae*. (F) Epicormic sprouting in a conifer, *Pseudotsuga macrocarpa*, after a low intensity chaparral burn.

(G) Serotinous daisy, *Phaenocoma prolifera*, fynbos Asteraceae. (H) Non-sprouting ground *Protea* whose seeds survived the fire in the insulated cones with re-sprouting sedge behind. (I) Deep-seated epicormic bud characteristic of eucalypts, allowing rapid canopy recovery after severe fires. (J) Stand of jarrah (*Eucalyptus marginata*) with a heathland understorey, showing rapid recovery of the eucalypt by epicormic sprouting.

important than hydraulic failure in causing stem death (Midgley *et al.*, 2010). Heating of water under tension in the xylem causes cavitation and embolisms, stopping water supply to the leaves. Canopy die-back after a fire is similar to 'drought' – a much more rapid cause of death then ring-barking (cambial injury).

Plants have numerous methods of recovery following fire, each of which has different costs and benefits for the functioning of plants. The distribution of different recovery modes within an ecosystem is shaped by competitive interactions and the fire regime. Some trees are able to recover most of their canopy after fire by activating epicormic buds insulated by bark. Eucalypts are particularly remarkable in this respect, with buds deeply embedded in the bark and even the wood. Thus, buds are able to re-sprout even after very severe fires where the bark has been burnt off (Figure 7.1I).

Many shrubs and juvenile trees have thin bark and no bud protection. Stems are killed by fire, but plants re-sprout from buds at the base. Some woody plants sprout from large plate-like structures called lignotubers, containing numerous buds and storage tissue (Figure 7.1A).

Basally sprouting woody plants have to regenerate their entire canopy from post-burn regrowth. Some woody plants spread vegetatively by clonal expansion from the mother plant from buds sprouting from rhizomes or roots. Clonal spread is often stimulated by removal of above-ground stems, so that fires can promote vegetative expansion of a population of genetically related shoots.

Aspen is a well-known example of clonal spread, with groves of genetically identical trees spreading over hundreds of hectares. The phenomenon has also been reported for Californian redwoods, which not only sprout from the base after the main stem is damaged, but also spread vegetatively by root suckers (Douhovnikoff *et al.*, 2004). Finally, many woody plants have thin bark or poorly protected buds and are killed by fire. Fire-killed trees are common in extensive rainforests, where fires are rare and catastrophic. However, many woody species living in flammable ecosystems are also routinely killed by fires (Figure 7.1B). The next generation emerges as germinants from seeds stored in the soil or the canopy of the pre-burn generation.

Among herbaceous plants, grasses are seldom killed by fire and re-sprout rapidly from buds situated at, or just below, the soil surface, where they are insulated by old leaf sheaths. Because leaves grow from basal meristems, grasses can recover leaf area very rapidly after a burn. Geophytes are particularly common amongst monocots (Amaryllidaceae, Asphodelaceae, Hyacinthaceae, Haemadoraceae, Iridaceae, Orchidaceae, etc.) and occur in many flammable shrublands and grasslands. They are capable of rapid re-sprouting from buds insulated by the surface soil layers, with growth subsidized by stored reserves in bulbs, corms and tubers (Figure 7.1C).

Seeds in the soil are also insulated from fire injury, though the degree of protection varies with the depth of burial in relation to fire severity. In some flammable ecosystems, seeds are stored above ground in insulated woody structures. This phenomenon, serotiny, occurs in conifers in the northern hemisphere (e.g. *Pinus, Picea, Cupressus, Sequoiadendron*), and in conifer genera and angiosperm families in the southern hemisphere (e.g. conifers: *Callitris, Widdringtonia*; angiosperms: Asteraceae, Bruniaceae, Casuarinaceae, Ericaceae, Myrtaceae and Proteaceae) (Figure 7.1D–H).

7.2.1 Post-burn recovery: vegetative re-growth and storage tissue

Plants that re-sprout after fire depend on the survival of buds to initiate regrowth and on stored reserves for rapid recovery. The reserves are commonly stored in underground storage organs (USOs) containing large amounts of carbohydrate (starch), and these are mobilized to subsidize re-sprouting after fire. They may also be used to supply seasonal resource demands.

As might be expected, comparisons of closely related re-sprouters and non-sprouters in flammable Mediterranean-type climate shrublands have shown much larger carbohydrate reserves in the roots of re-sprouters, but similar concentrations in stems which are top-killed in crown fires (see Pate *et al.*, 1990; Bell and Ojeda, 1999).

Mature re-sprouters recover much more rapidly than non-sprouters when burnt, and they are able to flower and fruit within one or two growing seasons. However, there is a cost to sprouting, in that reserves have to be allocated to future use rather than to current needs. This cost is expressed as much slower growth rates of seedlings of re-sprouters compared to related non-sprouters, with increased risks of mortality, and much longer delays until first reproduction. This trade-off between reserves for sprouting and growth rate helps explain why non-sprouters often dominate in flammable shrublands such as Cape fynbos and south-west Australian heathlands (Bond and Midgley, 2001).

However the growth/storage trade-off is not universal amongst re-sprouters and non-sprouters. In chaparral, for example, members of the Rhamnaceae do not show clear differences in seedling growth rates related to whether they are re-sprouters or not. In this case, it has been suggested that resources are allocated to tissues in re-sprouters that contribute to tolerance of stresses such as shading and drought, rather than to underground storage organs. In other words, the costs of allocating resources to re-sprouting may be expressed in ways other than reduced biomass production in seedlings (Pratt *et al.*, 2012).

There are also genetic costs to sprouting. Deleterious mutations accumulate over the much longer lifespan of re-sprouters relative to non-sprouters. These mutations are not purged after each successive generation, as happens with seedlings of non-sprouters after each fire. The genetic load hypothesis helps explain the low seed set, high abortion rates and low production of viable seeds characteristic of many strongly sprouting species (Lamont and Wiens, 2003; Lamont *et al.*, 2011).

Sprouting behaviour often changes during the different life histories stages of a plant. Savanna trees provide good examples. Very frequent grass-fuelled surface fires select for tree seedlings with reserves stored in swollen roots. Seedlings acquire sprouting ability

Figure 7.2 Modes of survival in frequently burnt savannas and grasslands. (Photos A-C from W.J. Bond, D,E from N. Zaloumis). (A) Canopy recovery from epicormic sprouting is common in many savanna trees, Cerrado, Brazil. (B) Thick bark is characteristic of many savanna trees, Cerrado, Brazil. (C) The combination of underground storage organs to facilitate rapid post-burn re-sprouting, and a pole-like stem to enable rapid bolting above flame height, are characteristic of savanna trees exposed to frequent fires. *Terminalia sericea*, South Africa. (D) Some frequently burnt, tropical and subtropical grasslands are rich in herbaceous species with large root systems and (E) underground storage organs which re-sprout rapidly after fire. These examples from South African upland grasslands.

very rapidly (less than a year), in contrast to the seedlings of sister taxa growing in fire-intolerant closed forests that are killed by fire. Seedlings develop into multi-stemmed juveniles and then 'pole-like' juveniles, with large swollen burls (lignotubers) containing large carbohydrate reserves which, after fire, subsidize stem growth and help maintain the root system as the canopy re-builds (Schutz *et al.*, 2009; Figure 7.2C).

If stems grow to fireproof size (stems thick enough and tall enough to escape fire damage) between successive fires, then they escape the flame zone and continue to grow to full maturity. At this stage,

root starch reserves are mobilized and used for tree growth, so that mature trees may have low starch content in their roots. In frequently burnt mesic savannas, stems that fail to reach fireproof size before being topkilled by fire have to start regrowth all over again. Saplings can persist for decades, re-sprouting after each successive fire, either from the base of the plant or higher up the stem after less intense burns. Non-sprouters are rare or non-existent in these frequently burnt savannas. The trade-off is between allocation to current growth (facilitating the probability of reaching escape size and maturity) and setting aside root reserves to ensure that the plant survives the next fire if the stem fails to reach escape size. Savanna trees may spend decades juggling carbon between these competing goals before they die or, at last, succeed in escaping the flame zone and growing into mature reproductive trees.

The C4 grasses of tropical and subtropical grassy ecosystems can recover very rapidly after a burn. The grasses are strong competitors and restrict recruitment opportunities for non-grassy woody or herbaceous species. Tropical grasslands in Africa and South America contain many forb species with large underground storage organs, including dicots from diverse families, but especially Fabaceae and Asteraceae (Figure 7.2D, E). In the savannas of both these regions, there are also species that look superficially like forbs but are, on closer inspection, woody plants with extensive underground stems and short-lived above-ground shoots. These 'dwarf' trees have close relatives that are tall trees in savannas or closed forests. The underground trees seem to have evolved early maturation (neoteny) where fires are too frequent and growth rates too slow for saplings ever to escape the fire trap.

Those tropical and sub-tropical grasslands, which are rich in species with underground storage organs,

are extremely resilient to frequent burning. However, these plants cannot tolerate shading and can be eliminated by fire suppression. The very persistent perennial forbs of tropical grasslands contrast strongly with short-lived annuals and biennials that are common in flammable grasslands of temperate regions. These differences may lead to divergent responses to burning. In North American prairies, for example, forb diversity in grasslands increases where fires are suppressed. Thus, the response of grassland diversity to different fire frequencies can be quite different, depending on the plant functional types that are present.

7.2.2 Post-burn recovery: seedling recruitment

In some flammable ecosystems, seedlings are very common in the first couple of growing seasons after a fire (Figure 7.3), particularly where the post-burn environment has high light, a flush of soil nutrients, low competition and low risk of herbivory. Such conditions are typical of pyrophilic shrublands of Mediterranean-climate regions in the Mediterranean

Figure 7.3 A post-burn flush of seedlings is a clear indication of a fire-adapted ecosystem, such as this example from burnt fynbos in the Cape region of South Africa. (Photo from W.J. Bond).

basin, in Californian chaparral and in the heathlands of the Cape region of South Africa and southern Australia. Many plants in these systems have soil-stored seeds with fire-stimulated germination (Keeley and Fotheringham, 2000).

The heat shock associated with fire is a common cue in the hard-seeded members of the Anacardiaceae (*Rhus*), Apiaceae, Cistaceae, Ericaceae (*Arctostaphylos*), Fabaceae, Malvaceae, Rhamnaceae (*Ceanothus*), Sterculiaceae, etc. Smoke is also a common germination cue, stimulating germination in many soft-seeded species in these flammable shrublands. Hundreds of species require smoke as a germination cue, and the discovery of the phenomenon has released many species to horticulture that were previously very difficult to cultivate. Smoke can stimulate not only germination but also seedling growth, including those of grasses and tree seedlings in savannas. There is still much to be discovered about the wider implications of the phenomenon.

Instead of germinating from soil-stored seed banks, many herbaceous species have fire-stimulated flowering, which is also effective in synchronizing recruitment to the post-burn environment (Figure 7.1C). This is a common trait not only in many monocots (grasses, orchids, members of Iridaceae, Liliaceae *sensu lato*), but also in some re-sprouting woody plants. Fire-stimulated seed release from serotinous woody structures has the same effect of timing recruitment to post-burn conditions. Plants with fire-stimulated flowering or fire-stimulated seed release are already cued to exploit post-burn conditions, and usually lack other specialized germination cues.

7.3 Fire regimes and the characteristic suite of fire plant traits

Why do flammable ecosystems vary so much in their structure and dynamics? The answer seems to lie partly in the pool of growth forms physiologically compatible for a given climate and the sorting of these potential growth forms by fire. The effects of fire in selecting among growth forms depend on the fire regime. Fire regimes are the recurrent patterns of fire frequency, severity (amount of biomass burnt) and area burnt (see Chapter 6). Different fire regimes select for different suites of fire response traits.

Crown fire regimes burn with sufficient intensity to consume the canopies of all woody plants (Figure 7.4A). In these high-severity fire regimes, woody stems are typically thin-barked (as investment in bark is a waste of resources) and vegetative sprouting is basal rather than epicormic. Non-sprouting plants that are killed by fire and regenerate from aerial or soil seed banks are often present.

Surface fire regimes, in contrast, are fuelled by litter or grass and are of low severity and do not burn leaves in the canopy of tall trees (Figure 7.4B). Trees in these systems typically have thick bark insulating the stem, coupled with self-pruning so that there are no fuel ladders carrying surface flames into the canopy. Many trees in surface fire regimes re-sprout from epicormic buds and are able to restore the canopy rapidly after a burn.

These two examples indicate that, despite the diversity of fire resistance, recovery and recruitment traits, only a subset of these traits is typical for any given fire regime. The implication is that the fire regime acts as a filter, selecting potential growth forms so that only a small subset of the flora is able to grow in any given locality.

7.3.1 Determinants of fire regime

Fire regimes depend on the synergy between external physical factors and the properties of vegetation. Prior to hominid use of fire, key physical preconditions for wildfires were climatic conditions conducive to lightning strikes, periods dry enough for ignition to occur and enough oxygen in the atmosphere to sustain burning (see Chapter 4). To influence the geography of plant communities, another physical precondition is landscapes free of major barriers to fire spread. Such landscape barriers include rivers, lakes, ice and snow, gravel beds and other areas hostile to plant growth. Within these physical constraints, the vegetation itself is the major contributor to the fire regime.

The way in which plants construct the fire regime is most apparent where entirely different fire regimes occur in the same landscape, such as mosaics of fire-resistant tropical forests and flammable grasslands, or sclerophyll shrublands and conifer forests (see Chapter 6). Changes in the fire regime can be brought about by changes in external conditions, such as climate change, but also by changes in the growth form mix, producing changes in flammability or productivity. The shortest interval between successive fires is constrained by the time taken to build sufficient,

Figure 7.4 Fire regimes. (A) A crown fire in Australian *Banksia* shrubland (image from B. Lamont). (B) A surface fire in a mesic savanna in northern Australia (image from A. Andersen, reproduced with permission of CSIRO).

continuous flammable biomass for fires to spread. This depends on the plants present, their intrinsic productivity and their growing conditions.

7.3.2 Flammability

In the previous chapter, we introduced the fire triangle concept, where the necessary conditions for fire to occur are often expressed as a triangle, with biomass representing the biological contribution to fire. However, it is not only the amount of biomass, but also the type that is critical. For example, tropical grasslands have much lower biomass than tropical forests with which they often co-occur. However, the grasses are highly flammable when they dry out during the dry season and support very frequent fires – in some instances burning twice in a year – whereas the forests, with high biomass, seldom burn.

There are also striking differences in flammability between woody plant communities with similar above-ground biomass. For example, Cape fynbos shrublands, like other Mediterranean-climate shrublands, burn readily, and their fire responses have been well studied. However, a dense shrubland ('thicket') occurs in the same landscapes as fynbos, in regions

Figure 7.5 Biomass alone is a poor predictor of fire occurrence. The distribution of biomass in a plant community strongly influences flammability and the probability of burning. (A) thicket vegetation from the winter rainfall regions of south-western Africa. Fires are unknown in this ecosystem and large vertebrates are the main consumers. (B) Cape fynbos, a flammable shrubland with regular fires and many fire-adapted species, which occurs under similar climate conditions. (Photos from W.J. Bond).

with winter rainfall, and does not burn at all (Figure 7.5A). Fynbos shrubs differ from thicket shrubs in leaf size, leaf texture, shrub architecture and retention of dead branches (Figure 7.5B). These differences in the arrangement of biomass account for striking differences in flammability, regardless of similarities in total above-ground biomass.

Where the spread of fires is determined by litter fuels, the characteristics of litter also differ, helping to promote or retard fire spread. For example, in conifer woodlands of the south-eastern United States, pines produce a highly flammable litter layer that promotes fire spread. In contrast, co-occurring oaks produce litter that packs densely, tending to reduce fire spread.

In the mixed conifer forests of the Sierra Nevada, flammability properties of leaf litter vary with leaf length. Pines with longer leaves produce lightly packed litter which burns readily promoting rapid fire spread. Trees with short leaves, such as white fir, sequoia and incense cedar, produce densely packed litter which slows fire spread. Many years of fire suppression have led to replacement of pyrophilic pines by the more shade-tolerant but fire-sensitive cedars and firs, whose litter dampens fire spread.

Changes in the fuel properties of litter may have many repercussions for fire spread and the nature of plant assemblages. In the deep past, differences in the flammability of leaf litter amongst different conifers are thought to have altered fire regimes hundreds of

millions of years ago, causing a change in species dominance (Belcher *et al.*, 2010).

An important implication of the biological contribution to characteristics of the fire regime is that changes in the growth form mix, or species composition, can alter fire regimes without any change in the fire weather conditions. Biologically determined differences in fire regimes have been documented in natural ecosystems, that match changes in fire regime observed as a consequence of the invasion of exotic species. The alternative stable states discussed in the previous chapter are examples of how intrinsic flammability can alter fire regimes in vegetation exposed to the same climates and within the same landscapes.

Examples include chaparral adjacent to conifer forests in the Cascade Mountains of western North America; while conifers burn with low severity litter-fuelled surface fires, the chaparral burns with high severity crown fires (Odion *et al.*, 2009). Another example of alternative flammable ecosystems built around growth forms with different flammable properties comes from central and northern Australia. Here, shrub-like tussock grasses *Triodia*, known as 'spinifex', burn every few decades when rainfall is sufficient to produce continuous fuels. Woody species within the spinifex grasslands have attributes typical of crown fire regimes such as serotiny, fire-stimulated seed germination and basal re-sprouting. Spinifex communities often form abrupt boundaries with eucalypt savannas, associated with changes from sand to sandy loam soils. These eucalypt savannas have an understorey of soft-leaved bunch grasses that recover very rapidly after burning and can support several fires in a decade given sufficient precipitation. The savannas support different woody species, with a very different suite of woody plant traits. Fire-stimulated recruitment is rare or absent. Instead, juvenile trees typically have underground storage organs and erect, pole-like stems, facilitating rapid growth above the low flames characteristic of low-severity fire regimes.

7.4 Evolution of fire traits

Plant traits that facilitate resistance, recovery and recruitment from fire have often been called fire adaptations. The assumption of 'adaptation' has also been common in interpreting other features of plants growing in particular environments. For example, in chaparral and other Mediterranean-type climates, shrubs with leathery (sclerophyllous) evergreen leaves have been considered adaptive to the winter-wet/summer-dry climates. However, closer analysis of the chaparral flora has shown that few of the plants evolved their characteristics within the Mediterranean climate. Rather, the onset of a Mediterranean climate caused ecological sorting of those species with compatible traits from the existing species pool. Sclerophyllous leaves existed long before the onset of winter rainfall climates (Ackerly, 2004).

There are similar problems in interpreting fire response traits, such as re-sprouting, as adaptations rather than the outcome of species sorting by the fire regime from a larger species pool. As noted previously, a fire survival trait such as re-sprouting may not only be of benefit in surviving fire but also in surviving frost, wind storms or herbivory. Adaptation implies a change from the ancestral condition. Differences in flammability between the litter of pines and oaks could be interpreted as an adaptation of pines to promote fires which help exclude the more competitive oaks. However, the litter properties are much more likely to be ancestral features of pines versus oaks, regardless of whether they are exposed to fires.

For reasons such as these, one should be cautious in labelling traits as 'adaptive' when they seem to help fit an organism to its environment. However, it is also unreasonable to reject the idea that plants have features that confer fitness under specific fire regimes.

Phylogenetic methods have helped to resolve these difficulties by revealing the age and ancestry of putative fire traits. For example, Simon et al. (2009) analyzed sprouting and other traits of woody plants living in highly flammable Brazilian savannas (Cerrado). They developed a molecular phylogeny of legume genera to help trace the origin of the savanna flora and shifts in traits related to their flammable habitat (Figure 7.6).

All the savanna trees were derived from forest ancestors. Molecular dating methods showed that savanna lineages evolved from forest ancestors in the last ten million years, with most appearing less than 5 ma. Independent palaeoecological evidence has shown that the C4 grasses dominating these savannas first assembled into the tropical grassy biomes from about nine million years ago. The savanna tree lineages are strikingly different from their forest ancestors in a range of traits compatible with their new flammable habitat. Many have swollen underground roots, enabling vigorous post-burn re-sprouting, along with thick bark and a general reduction in plant size. Thus, for the savanna tree lineages, re-sprouting is a novel trait, an adaptation associated with open habitats with frequent surface fires, divergent from forest ancestors. The forest lineages lack the necessary attributes for surviving frequent grass-fuelled fires.

7.4.1 Fire survival traits

Phylogenetic methods have also been useful for exploring the origins of fire-survival traits in pines. Life history evolution in pines has followed different paths in the two sub-genera, Strobus (Haploxylon) and Pinus (Diploxylon) (Keeley and Zedler, 1998; Keeley, 2012). Nearly all species in the Strobus lineage occur in low-productivity alpine or desert environments, where fuel is too discontinuous to burn in most years. Alpine species of Strobus, such as *Pinus cembra* in Europe, *P. sibirica* in Asia and the bristlecone pine, *P. aristata*, in North America live in fire-free habitats. Desert-dwelling species, such as the pinyon pines of western North America, are not regularly challenged by fire. Collectively, these species have the thinnest bark in the genus.

Members of the Pinus sub-genus commonly occur in environments where fire is a regular occurrence. Traits diverge, depending on the fire regime. Pines growing in surface fire regimes, such as *Pinus pinea*,

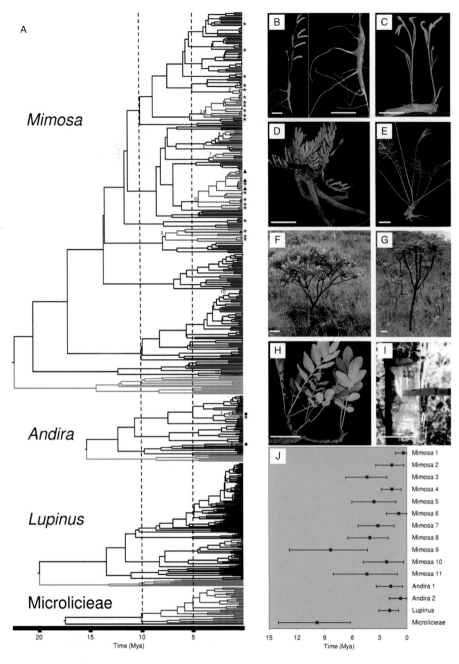

Figure 7.6 Evolution of the fire-adapted Cerrado flora of Brazil. (A) Chronograms for 15 Cerrado lineages (red) with outgroups (forest species) depicted in gray. (B–E) Examples of underground storage organs facilitating rapid post-burn sprouting. (F) rosulate shrub. (G) Pachycaul treelet with few thick branches. (H) A branch of the geoxylic suffrutex or "underground tree". (I) Thick corky bark. (J) Divergence-time estimates for 15 Cerrado lineages (crown nodes). (Figure from Simon *et al.*, 2009).

P. pinaster, *P. sylvestris* in Eurasia and *P. ponderosa*, *P. jeffreyi* and related pines in North America, share thick bark, self-pruning of dead branches (reducing fire-laddering from surface to canopy fuels) and long needles producing flammable litter. Pines subjected to crown fires (*P. halepensis* and *P. brutia* in Europe, *P. contorta* and *P. banksiana* in high altitude and boreal North American forests, and *P. attenuata* and others in Californian chaparral) have traits that 'embrace'

fires. These pines retain dead branches, creating ladder fuels and thereby ensuring crown fires. Consistent with the stand-replacing crown fires, these pines have thin bark and recruit from seeds stored in serotinous cones.

Recent phylogenetic studies indicate that the ancestral condition of the Pinus lineage was thin-bark, branch-retaining and with seed release at maturity (He *et al.*, 2012; Figure 7.7). According

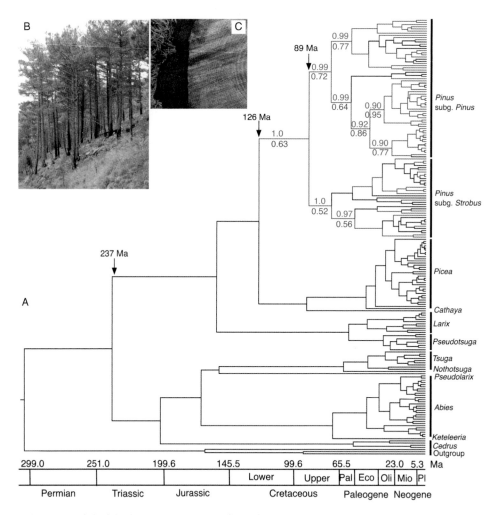

Figure 7.7   The origin of thick bark in pines estimated from phylogenetic methods. (A) A phylogeny of conifers with red lines indicating evolutionary pathways for modern species with thick bark (> 15 mm). The scale at the bottom indicates time, estimated from molecular dating of the phylogenetic tree. This analysis points to a mid-Cretaceous (89 ma) origin of conifers with thick bark adapted to surface fires. Serotiny, retention of seeds in closed cones with massed seed release after stand-replacing crown fires, also first evolved at this time. (B) Pine (*Pinus nigra*) forest following surface fire, with trees about 20 m tall, showing old branch shedding. (C) Cross-section of trunk of *Pinus* sp., showing bark 45 mm thick at its maximum. (Figure from He *et al.*, 2012).

to their analysis, thick bark ($>$ 15 mm) originated in the early Cretaceous ($\approx$ 126 ma). Fossil evidence indicates the existence of flammable fern savannas as this time. Very thick bark ($>$ 30 mm), such as that of *P. ponderosa* in America and *P. pinaster* in Europe, first appeared in the mid-Cretaceous ($\approx$ 89 ma), when fossil evidence indicates that fires were frequent. From about this time, fossil flowers become common, beautifully preserved as charcoal.

The charcoalified flowers and other plant parts indicate that flowering plants had now become major contributors to fuel, launching novel fire regimes that may have contributed to the spread of flowering plants in these ancient landscapes (Bond and Scott, 2010). The pine phylogeny also reveals a mid-Cretaceous origin (89 ma) for serotiny, indicating the appearance of crown fire regimes also associated with fires fuelled, for the first time, by biomass created by flowering plants. Today, pines thrive in diverse fire regimes, including crown fires and surface fire regimes of boreal forests, litter- and grass-fuelled surface fires in ponderosa pine-type forests and even frequent tropical savanna fires. The phylogenetic evidence indicates that fire-adaptive traits in *Pinus* are, indeed, fire adaptations which first evolved from fire-sensitive ancestors in flammable ecosystems of the Cretaceous, when dinosaurs still roamed the earth.

Dated molecular phylogenies have also been used to explore the origins of epicormic sprouting in Australian eucalypts. Epicormic sprouting is common in many savanna trees after surface fires have scorched their canopies. However, no trees other than eucalypts survive and re-sprout after the severe fires characteristic of sclerophyll forests in wetter parts of Australia.

Epicormic sprouting from meristematic strands buried deep in wood tissue is thought to be the functional basis for the remarkable ability of eucalypts to re-sprout after these fires (Figure 7.8). Crisp *et al.* (2011) used dated molecular phylogenies to show that early ancestors of eucalypts lacked epicormic buds, which first evolved about 60 ma. These results suggest fires were sufficiently common to select for epicormic sprouting 50 million years earlier than dates suggested by fossil charcoal records (Late Miocene, from 10 ma).

Dormant bud protection

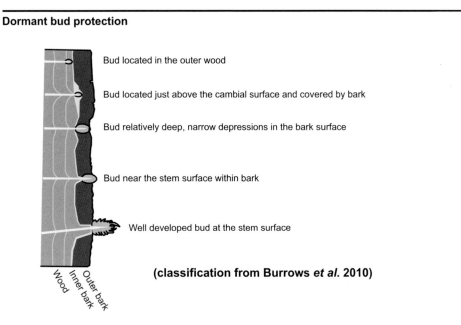

Bud located in the outer wood

Bud located just above the cambial surface and covered by bark

Bud relatively deep, narrow depressions in the bark surface

Bud near the stem surface within bark

Well developed bud at the stem surface

(classification from Burrows *et al.* 2010)

Outer bark
Inner bark
Outer Wood

Figure 7.8 The location of buds in relation to insulating bark influences the recovery of crowns burnt in a fire. Resistance to canopy fires decreases from top to bottom. The most fire-resistant buds are those of eucalypts, since they are located below the bark and in the outer wood. (Drawn by Tristan Charles-Dominique from information in Burrows *et al.*, 2010). Courtesy Tristan Charles-Dominique.

7.4.2   Fire-stimulated reproduction

Phylogenetic studies are also being used to explore the origins of fire-stimulated recruitment traits. Fire-stimulated flowering is common in many herbaceous plants in flammable shrublands and grasslands. Though this is clearly adaptive in that recruitment is timed to periods of high resources, low competition and low predation, it is less clear whether the fire cue is an adaptation. Fire is seldom an obligate requirement for flowering, and other cues, such as increased resources when competitors are removed, will also promote flowering. However, a recent phylogenetic study of fire-stimulated flowering in orchids has shown that the trait is restricted to particular lineages, is derived (the ancestral state is not fire-stimulated), and evolved in lineages in Cape fynbos from about 13 ma. The nature of the cue – facultative or obligate – is irrelevant in establishing that fire-stimulated flowering is an adaptation linked to increased fire activity in this system.

The origin of another fire recruitment trait, serotiny, has also been explored with phylogenies for the genus *Banksia*, Proteaceae, in Australia. Serotiny is a trait found in crown fire systems where seeds are retained in insulated woody structures until burnt by a fire, when seeds are released into the post-burn seed bed. Using a dated molecular phylogeny, He *et al.* (2011) estimated the origin of serotiny as $\approx$ 60 ma, implying that crown fire regimes existed in Australia from the earliest Cenozoic. This ancient origin of fire-adapted vegetation is consistent with phylogenetic dating of the origins of epicormic sprouting in eucalypts, but it is at variance with the much later origin estimated from fossil charcoal.

It is important to note that one of the difficulties with fossil data is that conditions most suitable for fossil preservation, including lakes, wetlands and river banks, are those where fires are least likely to burn. Upland vegetation, where fires are most likely to be burning, would be 'invisible'. However, development of new methods, such as the coupling of phylogenetic and fossil data, are beginning to reveal that fire was a prominent process in organizing ecosystems in the deep past, just as it is today (see Chapters 4 and 5).

7.4.3   The evolution of flammability

Plants from different biomes show great differences in flammability, with more open, low-biomass biomes generally showing higher flammability than closed forest vegetation. Flammability is influenced by many plant traits such as leaf size, plant architecture and the presence of flammable secondary chemicals. Fuel particle size and moisture content is crucial, with small leaves, small diameter shoots and retention of many dead branches promoting flammability.

That flammability is an emergent property of some plant communities has long been recognized. The problem has been in determining whether flammability can evolve in any of the constituent species and whether this would feed back on the fire regime. Bond & Midgley (1995) used a kin selection argument to show that flammability could evolve if a more flammable morph, by burning fiercely, could kill its less flammable neighbours, and if its offspring were better able to colonize the gaps. Since one of the quickest ways of colonizing a post-burn gap is fire-stimulated recruitment, they predicted that plants with flammable morphologies should be associated with fire-stimulated recruitment (Figure 7.9). This prediction has been tested for North American pines by Schwilk & Ackerly (2001).

Relative flammability can be enhanced in pines by the retention of dead branches. Surface fires climb the fuel ladder, promoting stand-replacing crown fires that kill neighbouring trees. Species which retain dead branches are more likely to burn in crown fires and should therefore be associated with traits such as serotiny, which promote rapid colonization of the gaps created by burning (Figure 7.9).

In contrast, trees that self-prune their lower branches, such as ponderosa pine, lower the risk of crown fires and should release their seeds in older stands without a fire cue. Schwilk and Ackerly (2001) found strong support for these predictions. They identified a suite of 'fire embracing' traits, including branch retention, thin bark and serotiny, contrasting strongly with 'fire-resistant' traits such as branch pruning, thick bark and no serotiny. These and other studies in flammable shrublands have found support for the predicted correlation between flammable vegetative traits, especially dead branch retention and post-burn recruitment consistent with selection for flammability as an 'adaptation'. Where flammability does evolve, there is a possibility for 'niche construction', meaning that selection for a trait alters the selective environment to further promote organisms with that trait. Fire and evolution of flammability is an excellent candidate for further studies of niche construction.

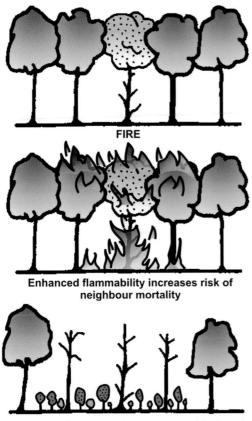

FIRE

Enhanced flammability increases risk of neighbour mortality

Flammable plants exploit gaps created by burning, especially if they have additional growth advantages

Figure 7.9 A kin-selection hypothesis for the evolution of flammability (ex Bond and Midgley, 1995). A more flammable genotype (shaded) can create gaps if it causes fire to spread to less flammable neighbours. In this model, flammability traits are most likely to evolve if the more flammable genotype has additional traits that increase the chances of its seedlings taking the gap. These could include fire-stimulated recruitment, such as serotiny, and faster seedling growth relative to less flammable forms. (Redrawn from Bond and van Wilgen, 1996). Reproduced by permission of Elsevier.

7.5 Summary and implications

There have been major advances in our understanding of the functional and evolutionary relationships between plants and fire over the last few decades with, no doubt, more discoveries yet to be made. The fire regime concept has been of central importance in making sense of the plant-fire nexus. The suite of traits in surface fire versus crown fire regimes, for example, is quite different, so that changes in the fire regime – and not only the presence or absence of fire – will select for quite different kinds of species. The recent use of phylogenetic methods to trace the evolutionary history of particular fire-adaptive traits is, along with fossil evidence, revealing the long history of fire as a major factor shaping the biota and biomes of the world. An understanding of the biology of fire responses is key to recognizing fire-dependent versus fire-sensitive species, and pyrophilic versus pyrophobic ecosystems, and therefore in devising appropriate fire management systems.

Further reading

Ackerly, D.D. (2004). Adaptation, niche conservatism, and convergence: comparative studies of leaf evolution in the California chaparral. *The American Naturalist* **163**, 654–671.

Belcher, C.M., Mander, L., Rein, G., Jervis, F.X., Haworth, M., Hesselbo, S.P., Glasspool, I.J., McElwain, J.C. (2010). Increased fire activity at the Triassic/Jurassic boundary in Greenland due to climate-driven floral change. *Nature Geoscience* **3**, 426–429.

Bell, T. L., Ojeda, F. (1999). Underground starch storage in *Erica* species of the Cape Floristic Region – differences between seeders and resprouters. *New Phytologist* **144**, 143–152.

Bond, W.J., Midgley, J.J. (1995). Kill thy neighbour: an individualistic argument for the evolution of flammability. *Oikos* **73**, 79–85.

Bond, W.J., Midgley, J.J. (2001). Ecology of sprouting in woody plants: the persistence niche. *Trends in Ecology & Evolution* **16**, 45–51.

Bond, W.J., & Scott, A.C. (2010). Fire and the spread of flowering plants in the Cretaceous. *New Phytologist* **188**, 1137–1150.

Crisp, M.D., Burrows, G.E., Cook, L.G., Thornhill, A.H., & Bowman, D.M. J.S. (2011). Flammable biomes dominated by eucalypts originated at the Cretaceous-Palaeogene boundary. *Nature Communications* **2**, 193.

Douhovnikoff, V., Cheng, A.M., Dodd, R.S. (2004). Incidence, size and spatial structure of clones in second-growth stands of coast redwood, *Sequoia sempervirens* (Cupressaceae). *American Journal of Botany* **91**, 1140–1146.

He, T., Lamont, B.B., Downes, K.S. (2011). *Banksia* born to burn. *New Phytologist* **191**, 184–196.

He, T., Pausas, J.G., Belcher, C.M., Schwilk, D.W., Lamont, B.B. (2012). Fire-adapted traits of *Pinus* arose in the fiery Cretaceous. *New Phytologist* **194**, 751–759.

Keeley, J.E. (2012). Ecology and evolution of pine life histories. *Annals of Forest Science* **69**, 445–453.

Keeley, J.E., Fotheringham, C.J. (2000). Role of fire in regeneration from seed. In: Fenner, M. (ed). *Seeds: The Ecology of Regeneration in Plant Communities*. 2nd ed, pp. 311–330. CAB International.

Keeley, J.E., Zedler, P.H. (1998). Evolution of life histories in Pinus. In: Richardson, D.M. (ed). *Ecology and Biogeography of Pinus*, pp 219–250. Cambridge University Press, Cambridge, UK.

Lamont, B.B., Enright, N.J., He, T. (2011). Fitness and evolution of resprouters in relation to fire. *Plant Ecology* **212**, 1945–1957.

Lamont, B.B., Wiens, D. (2003). Are seed set and speciation rates always low among species that resprout after fire, and why? *Evolutionary Ecology* **17**, 277–292.

Midgley, J. J., Lawes, M.J., Chamaillé-Jammes, S. (2010). Savanna woody plant dynamics: the role of fire and herbivory, separately and synergistically. *Australian Journal of Botany* **58**, 1–11.

Odion, D.C., Moritz, M.A., DellaSala, D.A. (2009). Alternative community states maintained by fire in the Klamath Mountains, *USA. Journal of Ecology* **98**, 96–105.

Pate, J.S., Froend, R.H., Bowen, B.J., Hansen, A., Kuo, J. (1990). Seedling growth and storage characteristics of seeder and resprouter species of Mediterranean-type ecosystems of SW Australia. *Annals of Botany* **65**, 585–601.

Pratt, R.B., Jacobsen, A.L., Hernandez, J., Ewers, F.W., North, G.B., Davis, S.D. (2012). Allocation tradeoffs among chaparral shrub seedlings with different life history types (Rhamnaceae). *American Journal of Botany* **99**, 1464–1476.

Schutz, A.E.N., Bond, W.J., Cramer, M.D. (2009). Juggling carbon: allocation patterns of a dominant tree in a fire-prone savanna. *Oecologia* **160**, 235–246.

Schwilk, D.W., Ackerly, D.D. (2001). Flammability and serotiny as strategies: correlated evolution in pines. *Oikos* **94**, 326–336.

Simon, M.F., Grether, R., De Queiroz, L.P., Skema, C., Pennington, R.T., Hughes, C.E. (2009). Recent assembly of the Cerrado, a neotropical plant diversity hotspot, by *in situ* evolution of adaptations to fire. *Proceedings of the National Academy of Sciences* **106**, 20359–20364.

General reading

Brown, J.K., Smith, J.K. (eds) (2000). *Wildland fire in ecosystems: effects of fire on flora*. General Technical Report RMRS-GTR-42-vol. 2. Ogden, UT: U.S. Department of Agriculture, Forest Service, Rocky Mountain Research Station.

Bond, W.J., Keeley, J.E. (2005). Fire as a global 'herbivore': the ecology and evolution of flammable ecosystems. *Trends in Ecology & Evolution* **20**, 387–394.

Bond, W.J., Midgley, J.J. (2001). Ecology of sprouting in woody plants: the persistence niche. *Trends in Ecology & Evolution* **16**, 45–51.

Bradstock, R.A., Gill, A.M., Williams, R.J. (eds) (2012). *Fire Regimes, Biodiversity and Ecosystems in a Changing World*. CSIRO Publishing, Canberra, Australia.

Clarke, P.J., Lawes, M.J., Midgley, J.J., Lamont, B.B., Ojeda, F., Burrows, G.E., Enright, N.J., Knox, K.J.E. (2012). Resprouting as a key functional trait: how buds, protection and resources drive persistence after fire. *New Phytologist* **197**, 19–35.

Keeley, J.E., Bond, W.J., Bradstock, R.A., Pausas, J.G., Rundel, P.W. (2012). *Fire in Mediterranean ecosystems: ecology, evolution and management*. Cambridge University Press, Cambridge, UK.

Lamont, B.B., Enright, N.J., He, T. (2011). Fitness and evolution of resprouters in relation to fire. *Plant Ecology* **212**, 1945–1957.

Pausas, J.G., Bradstock, R.A., Keith, D.A., Keeley, J.E. (2004). Plant functional traits in relation to fire in crown-fire ecosystems. *Ecology* **85**, 1085–1100.

Chapter 8
Fire and fauna

8.1 Direct effects of fire on fauna

Human fear of fire has shaped public perceptions on its impact on fauna. Images such as Bambi and Smokey the Bear have influenced generations to believe that fire is a disaster disrupting nature's Eden. Unfortunately, fire suppression policies stemming from antipathy to fire have led to the decline of many animal species living in pyrophilic ecosystems. But does fire kill animals – and how do animal populations respond to a burn?

The direct effects of fire on animals, like plants, depend on the location of the animals with respect to flame height. Unlike plants, animals can move, so the direct effects of fire also vary with the mobility of the animals. Among vertebrates, adult birds and large mammals are seldom killed by fire, but their young can suffer high mortality, especially if fires are burnt in seasons when these young animals are not yet mobile enough to take refuge.

Ground-dwelling ants largely escape fire damage, while the fate of tree-living invertebrates varies with fire severity and flame height. They are more likely to be killed in crown fires but survive in taller trees in surface fire regimes. Animals may also die of asphyxiation; for example, most of the large mammals that died in the Yellowstone fires of the late 1980s are thought to have died from smoke inhalation.

However, there is also a positive side to the balance sheet: fires, smoke and recently burnt areas attract many animals. Insect-feeding birds are a common sight along an actively burning fire front and in the charred world behind it. Some insects actively seek fires. In conifer forests of the western USA, Buprestid wood-boring beetles, *Melanophila* spp., use infra-red sensors to find burning trees where they mate, lay eggs and hatch larvae that live in the dead wood.

The direct effects of fire on animal populations are generally minor, relative to the indirect effects on the post-burn environment. A crown fire in a woody ecosystem transforms the habitat for vertebrates within minutes, stripping away taller trees and the resources they provide. The results of this can be catastrophic, especially in tropical rainforests with no history of repeated burning.

Although there is little data on the longer term effects of fire, Barlow and Peres (2004) reported that fires in Amazon forests led to the decline of arboreal fruit-eating mammals, especially primates and birds such as large toucans, pigeons, and oropendulas. A repeat fire in the recovering forest was particularly disastrous for such species. The disruption of habitat also made animals vulnerable to other threats so that, for example, large-bodied diurnal primates forced to forage in post-burn habitats were killed in large numbers by human hunters.

Where fires occur more predictably in flammable ecosystems, they initiate a post-burn succession, with different species colonizing the post-burn habitats at different post-burn ages. In shrublands of Mediterranean-climate regions, for example, a succession of rodent species takes place as the habitat changes after

Fire on Earth: An Introduction, First Edition. Andrew C. Scott, David M.J.S. Bowman, William J. Bond, Stephen J. Pyne and Martin E. Alexander.
© 2014 John Wiley & Sons, Ltd. Published 2014 by John Wiley & Sons, Ltd.

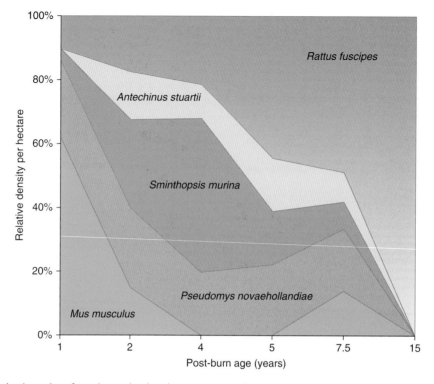

Figure 8.1 Animal species often change in abundance as vegetation re-grows after a fire. To accommodate the differing requirements of different species, conservation managers aim for greater heterogeneity of post-burn age to promote biodiversity. This example shows changes in relative abundance (%) of small mammal species in a eucalypt woodland recovering from fire. Note that each species reaches maximum relative abundance at different post-burn ages. (Figure drawn from data in Fox, 1990).

a burn. An Australian example is shown in Figure 8.1. Fires alter the structure and, therefore, the availability and quality of food. Basally sprouting woody plants are more accessible to ungulate browsers and the coppicing shoots are generally rich in nitrogen. In grasslands and savannas, moribund dead leaves and stems burn off in the fire, so that the newly sprouting leaves are more accessible to grazers and are of higher quality, being also rich in nitrogen. Consequently, burnt areas of grassland act as a magnet for drawing off grazer herds from unburnt patches.

Changes in vegetation structure also affect visibility, opening up the habitat so that larger herbivores can graze with less vigilance for predators. In contrast, smaller herbivores that hide from predators in dense cover are more vulnerable in the post-burn environment. By altering food availability, quality and the 'landscape of fear', fires create post-burn environments that are attractive to many large mammal

herbivores – a behavioural response long known to human hunters.

8.2 The effect of fire regimes on fauna

Different animal species have different habitat requirements. Among those most influenced by fire are vegetation structure, food distribution, abundance and quality, risk of predation, the availability of nesting sites and the landscape mosaic. As a useful generalization, fires change habitats for animals in proportion to the extent of post-burn changes in vegetation structure. The vertical and horizontal density of vegetation is a key niche dimension for many birds, mammals and flying insects. Thus, some bird species prefer open vegetation with low structural

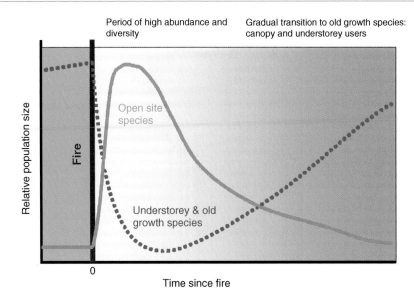

Figure 8.2 Post-burn changes in populations of animal species are driven by changes in vegetation structure as the vegetation recovers from the burn. Species prefer different vegetation structure, driving compositional changes in faunal communities. The extent of structural change caused by a fire determines the magnitude of compositional shifts, while the rates of vegetative recovery determine the rate of faunal change. The figure represents the most extreme case where a forest experiences a severe stand-replacing crown fire. Following surface fires, such as those in woodlands and open pine or eucalypt forests, there are small compositional shifts and fast recovery rates. (Figure re-drawn after Smith, 2000, USDA).

diversity, others shrubby vegetation with more vertical layering, while some species prefer closed tree canopies. In addition, some animals require a diversity of vegetation structures, often occurring at habitat edges.

Fire causes changes in animal communities, depending on the magnitude of fire effects on vegetation structure. The most extreme changes follow stand-replacing crown fires in forest systems. The least changes occur after surface fire regimes burning through woodland or forest understoreys. Fires initiate a successional sequence of animal species (Figure 8.2), with the magnitude of compositional turnover being proportional to these fire-induced changes in vertical structure. The rate of turnover increases with the rate of post-burn recovery.

As an example of extreme structural change, Raphael *et al* (1987) reported the response of bird species in the first 20 years following a large stand-replacing fire in mixed conifer forest in the Sierra Nevada. During the survey period, 32 bird species were recorded, of which 28% were restricted to the burnt area, compared to 19% restricted to unburnt forest. The species that thrived in the burnt area included those characteristic of low brush and open

ground. Woodpeckers, which exploited dead standing trees, also showed a marked increase in early post-burn vegetation. As shrubs began to increase in the recovering vegetation, birds that utilize shrubs increased by 500%, while woodpeckers steadily declined. After 25 years, vegetation structure had still not recovered to unburnt forest structure, and the bird community was still changing, albeit at a slower rate.

In contrast to the large impacts of crown fires, surface fires causing minor change in vegetation structure have much smaller effects on bird communities. In North America, bird communities return to pre-fire composition as soon as three years post-burn in grasslands, and even sooner than that after surface fires in conifer woodlands.

While changes in faunal composition are linked to changes in vegetation structure, fires can also cause striking changes in faunal abundance. The causes include an increase in food resources, such as dead trees for insects and hence for the birds that feed on them; seeds of herbaceous plants that flourish in the open post-burn habitat; and browse and graze quality and quantity. In chaparral, for example, mule deer densities increased 2–4 fold after fire and fawn production doubled. The increase is comparatively short-lived, with populations decreasing to pre-burn levels within a decade of chaparral fires.

In African savannas, the post-burn flush of tender grass shoots attracts grazers from unburnt areas, but only for a few months until grass quality declines to pre-burn levels. In low productivity ecosystems, fires have the reverse effect, causing reduced large mammal abundance. Mule deer avoid recently burnt pinyon-juniper vegetation, preferring stands that have been unburnt for several decades, presumably because of slow vegetation recovery in these arid ecosystems. In cold, unproductive northern regions, caribou avoid boreal forest burnt by crown fires for 50 years or more until lichens, a major food source, become re-established.

Changes in abundance, both positive and negative, may also depend on direct and indirect effects of predators. More open post-burn habitats may be preferred by large mammal herbivores because of the decrease in perceived, or actual, risk of predation. Elk at Yellowstone, for example, avoid unburnt aspen stands since wolves have been re-introduced. Smaller animals, such as rabbits, that use concealment rather than flight to avoid predators, avoid large burnt areas but may increase their foraging activity near the burn margins, where they can hide in adjacent unburnt vegetation. The relative importance of changes in food quality and quantity versus changes in risk of predation is an active area of research of considerable relevance to understanding faunal responses to burns.

8.3   The landscape mosaic and pyrodiversity

Since fires alter habitats and fire regimes alter landscape configuration of habitats, land managers are increasingly manipulating fire regimes to preserve biodiversity. In the south-eastern United States, for example, some of the most threatened species occur in fire-maintained pine savanna habitats. Fire suppression has led to extensive replacement of these savannas by oak woodlands. Land managers have reintroduced fire to maintain the savanna habitat and species dependent on it, such as the red-cockaded woodpecker (*Picoides borealis*), Bachman's sparrow (*Aimophila aestivalis*), the gopher tortoise (*Gopherus polyphemus*) and savanna invertebrates.

Where fires are widespread and frequent, such as in the savannas of northern Australia, the most threatened species are those that prefer infrequently burnt patches. Many animals require patchy habitats with a mix of different post-burn successional stages. For example, the rufous hare-wallaby (*Lagorchestes hirsutus*) population collapsed following the cessation of patch burning by Aborigines in central Australia. In northern Australia, the partridge pigeon (*Geophaps smithii*) has become extremely rare, because this ground-nesting bird has insufficient unburnt habitat.

In Scotland, red grouse (*Lagopus lagopus scoticus*) feed mostly on young post-burn re-sprouting shoots of *Calluna*, an ericaceous shrub. However, older vegetation is used for nesting sites and as cover from predators. The birds are territorial, so that the optimal habitat is a finely divided mosaic of young and old heathland. To boost grouse numbers, land managers burn narrow strips of heathland to provide the preferred mix of young and old shrubland (Figure 8.3).

The diverse habitat requirements of fauna have led to the concept of pyrodiversity – that mosaics of vegetation of different post-burn ages and different fire histories within a landscape are necessary to preserve faunal diversity. The key assumption is that pyrodiversity begets habitat diversity, which promotes biodiversity. The pyrodiversity concept has driven the development of explicit fire management regimes, such as the patch mosaic burning system applied in southern African savannas (Figure 8.4).

Appreciation of the effects of fire on the landscape mosaic, and that, in turn, on different faunal elements, has led to explicit attempts to design fire regimes for promoting pyrodiversity. However, practical implementation can be difficult. First, faunal responses to pyrodiversity vary with different species and different animal groups, which may have different requirements for the optimal spacing and patch area of preferred habitats. For example, birds that prefer open grassy habitats in African savannas require much larger habitat patches than do birds preferring densely wooded vegetation.

Second, designing fire regimes for pyrodiversity is a complex scientific problem of combining metapopulation dynamics of the target species with the shifting mosaic created by fires burning under different conditions. Different species will require different habitat mixtures, and one solution may not fit all species.

Managing for pyrodiversity is also constrained by the resources available for fire management and the

Figure 8.3 Burning can promote habitat diversity which, in turn, may promote biodiversity. (Reproduced by permission of CNPA).

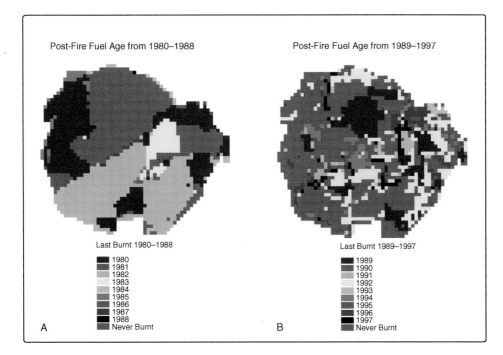

Figure 8.4 The effects of contrasting fire management systems on pyrodiversity as measured by the mosaic of post-burn age. The (A) panel indicates the post-fire fuel age in 1988 after applying block burns from 1980 to 1988. The (B) panel indicates the same in 1997 after applying a patch mosaic burning system explicitly designed to promote pyrodiversity. (From Brockett *et al.*, 2001). Reproduced by permission of CSIRO.

scale of the area being managed. In northern Australia, extensive fires introduced after European settlement are thought to be implicated in the catastrophic decline of small mammals in the region. The reintroduction of aboriginal burning practices, which produced much smaller and more intricate fire mosaics, is being attempted on a large scale, partly motivated by reduction of fire-related greenhouse gas emissions and partly to support biodiversity by promoting pyrodiversity. Innovative thinking on how to manage for more heterogeneous fire regimes, such as this example, can contribute to greater pyrodiversity. However, we still need a far better understanding of the links between pyrodiversity and biodiversity.

8.4 The effect of fauna on fire regimes

While fires alter animal habitats and fire regimes generate habitat mosaics in flammable ecosystems, animals also have impacts on the fire regime. Some of the best examples come from arid grasslands and savannas, where grass productivity, and therefore fuel accumulation, is relatively low in most years. Heavy grazing by animals reduces fuel for fire, so that grazers can have major impacts on the fire regime. In the ponderosa pine

forests of the south-western USA for example, fire frequencies were closely associated with changing livestock densities. Light grazing by Navajo and Hispanic sheep flocks was replaced by heavy grazing by large flocks of sheep and herds of cattle in the late 19th century, causing a sharp reduction in fire activity several decades before fire suppression. Subsequent reduction in grazing, coupled with fire suppression by land management agencies in the 20th century, allowed dense recruitment of the pines, which changed the entire structure of the ecosystem from an open woodland to a densely wooded forest with dense 'dog-hair' thickets of young pines.

Another example of fire and grazers as competing 'herbivores' comes from Dubinin *et al.* (2011), who studied a 19 000 km^2 area of steppe grasslands in southern European Russia. Since the collapse of the USSR in the early 1990s, there has been a steep decline in livestock (mostly sheep). The resultant reduction in grazing led to the recovery of the grasslands to burnable fuel loads and huge increases in burnt areas, especially after high rainfall years (Figure 8.5). The large scale of the switch from grazer to fire-dominated herbivory in this study suggests that the effect of grazing on fire is not merely a local patch phenomenon, but can have regional impacts on fire regimes.

In a South African savanna, the white rhino, a 2.5 ton megaherbivore, creates and maintains short

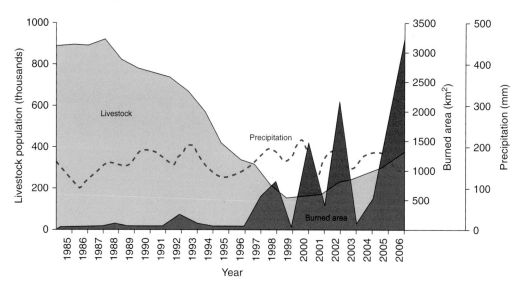

Figure 8.5 Sheep population, annual precipitation and burnt area in steppe grassland of southern Russia. The livestock numbers declined steeply from 1991, after the collapse of the Soviet Union, causing grassland fuels to increase, which then precipitated an increase in burnt area. Note that rainfall effects on burnt area only appeared after the collapse of the grazers. (From Dubinin *et al.*, 2011). Reproduced with permission from Springer Science+Business Media B.V.

Figure 8.6 Large mammal grazers, such as these white rhinoceros from Africa, can graze grasses very short, producing grazing lawns. The lawns act as firebreaks, reducing fire size and thereby altering the fire regime. Over the last few thousand years, many large mammal grazers have become extinct, allowing more frequent and extensive grass-fuelled fires. The cage-like architecture of the trees here are characteristic of heavily grazed areas in Africa and contrast with pole-like tree architectures where fires are common.

grass grazing lawns. The lawns do not burn and act as grazer-maintained firebreaks (Figure 8.6). They are utilized by a suite of short grass specialist grazers and a distinctive bird and invertebrate fauna. The removal of rhino in controlled capture operations has led to the loss of lawns and their associated fauna. This removal has also led to much larger fires. Larger burnt areas provide nutritious post-burn grazing which, in turn, attracts grazers away from the lawns and onto burnt taller bunch grasslands. Since concentrated grazing is needed to create and maintain the lawns, areas with large, frequent, management burns have few lawns. The poaching of rhinos for their horns not only threatens the species but also the lawn grass fauna that is indirectly dependent on the rhino.

Prairie dogs in the western USA (*Cynomys ludovicianus*) may seem unlikely substitutes for rhinos but they, too, create extensive short grass areas which do not burn. Animals that graze grasses too short to burn, such as white rhinos and prairie dogs, are ecosystem engineers, having profound effects on ecosystems through their ability to alter fire regimes.

Observations of contemporary reduction of fire due to localized heavy grazing has led to the suggestion that the extinction of the Pleistocene megafauna in many parts of the world was followed by the loss of grazer-maintained short grass habitats and the many species dependent on them. The megafaunal extinctions are thus hypothesized to have led to a switch from grazer dominance to fire dominance in diverse regions of the world (see Chapter 10). Contemporary observations suggest that release of fire following mammal extinction would have been most striking in the more arid grasslands. In higher rainfall regions, grass productivity exceeds the consumption rates of grazers in the growing season so that, when grasses lose their palatability in the dry season, enough surplus fuel persists to support extensive fires.

Forest/grassland mosaics are common in higher rainfall savannas of the world. Animal activity may also play a role in preserving forest patches within a matrix of highly flammable grasses. The forests in these landscapes often have a sparse understorey, in part because of forest browsers, so that savanna fires are less likely to spread in the forest understorey.

In Australia, orange-footed scrubfowl (*Megapodius reinwardt*, a bird the size of a chicken) produce enormous ground nests of leaf litter, resembling a large compost heap, in which they incubate their eggs. Forest patches with scrub fowl present may be less likely to burn because leaf litter loads are reduced by the nesting behaviour of the bird.

Figure 8.7 *Atta* leafcutter ant nests in a Brazilian Atlantic forest (A) and an ant trail (B). Litter removal in forests creates fire belts, which help to protect forest margins from burning in low-intensity fires entering the forest from adjacent savannas. (A: photo from Rainer Wirth and Inara Leal B: photo from Karine Carvalho).

In Brazil, leaf harvesting ants (*Atta* spp.) may play a similar role (Carvalho *et al.*, 2012). The ants strip the forest understory of leaves near their large nests and along their trails (Figure 8.7). The bare patches created then reduce the probability of fires spreading from neighbouring savannas. Since the ant nests are concentrated near the forest margins, *Atta* leaf harvesting may contribute to the stability of the forest/savanna boundaries in these landscape mosaics.

8.5   Fire and the evolution of fauna

In contrast to plants, there are few, if any, studies of the influence of fire on the evolution of animals (or animals on the evolutionary history of fire and flammable vegetation). The fossil record of Bovid radiations (the dominant herbivores of African savannas) shows major diversification from the Pliocene, a few million years after the initial expansion of savannas as recorded by carbon isotope records of the spread of C4 grasses. Thus, the large grazing herds seem to have exploited a new resource rather than helping to create the biome in the first place. Evolutionary links between fauna and fire could provide a rich resource for exploring the origins and spread of flammable

ecosystems, in a manner similar to recent studies of plant adaptations (Chapter 7).

The large volume of literature on faunal responses to vegetation fires provides many examples of behavioural responses to vegetation fires that warrant further study as fire adaptations. For instance, careful research has shown that populations of large herbivores such as moose differ in their response to large carnivores, depending on the previous history of exposure to predators. Analogous divergence in behavioural responses to fire might be expected in large herbivore populations exposed to frequent fires and those where fires are absent, such as on islands. The manner and rates of evolution of fire avoidance behaviour would be of particular interest where flammable grasses are transforming ecosystems where fires have been rare or non-existent. For example, can animals adapt to the novel fire regimes in Sonoran desert vegetation invaded by Buffel grass?

Faunal attraction to fire, such as that of Melanophora beetles to burnt trees, also offers scope for evolutionary studies. Birds of prey often patrol active fire fronts in an orgy of feeding on fleeing animals. Some, such as the black kite (*Milvus migrans*), are very widespread species, with populations occurring in regions of strongly contrasting fire activity (e.g. temperate Europe versus Australia). In Australia, black kites are reputed to carry burning twigs that ignite new fires. If this is true, it would seem to be the only example of an animal other than humans igniting landscape fires.

Critical studies of behavioural responses to burning, comparing populations or closely related species in ecosystems with contrasting fire regimes, could help provide a framework for predicting which faunal components might be at risk following biome shifts or alien plant invasion.

8.6 Summary

Growing appreciation of the effects of fire on fauna has helped change attitudes to fire management, including the risks of fire suppression. Many conservation agencies are attempting to design and maintain fire regimes that provide the full array of habitat types for key species. This is a difficult endeavour, especially where fire-maintained habitats are also small, fragmented and adjacent to urbanized areas. Complete biome switches further complicate management for fauna, such as in wooded vegetation invaded by flammable grasses, or in grassy habitats where trees are thickening up and fragmenting the flammable vegetation. In some ecosystems, it is useful to consider fire and vertebrate herbivores as competing consumers for food/fuel. Manipulation of one consumer will inevitably affect the other. Mammal herbivores and fire both operate over large spatial scales, requiring new levels of analysis and integration for a holistic understanding of their interdependent ecologies.

Further reading

Barlow, J., Peres, C.A. (2004). Ecological responses to El Niño-induced surface fires in central Brazilian Amazonia: management implications for flammable tropical forests. *Philosophical Transactions of the Royal Society of London. Series B: Biological Sciences* **359**, 367–380.

Brockett, B.H., Biggs, H.C., Van Wilgen, B.W. (2001). A patch mosaic burning system for conservation areas in southern African savannas. *International Journal of Wildland Fire* **10**, 169–183.

Carvalho, K.S., Alencar, A., Balch, J., Moutinho, P. (2012). Leafcutter ant nests inhibit low-intensity fire spread in the understory of transitional forests at the Amazon's forest-savanna boundary. *Psyche: A Journal of Entomology* **2012**, Article ID 780713.

Dubinin, M., Luschekina, A., Radeloff, V.C. (2011). Climate, livestock, and vegetation: what drives fire increase in the arid ecosystems of Southern Russia? *Ecosystems* **14**, 547–562.

Fox, B.J. (1990). Changes in the structure of mammal communities over successional time scales. *Oikos* **59**, 321–329.

Raphael, M.G., Morrison, M.L., Yoder-Williams, M.P. (1987). Breeding bird populations during twenty-five years of postfire succession in the Sierra Nevada. *Condor* **89**, 614–626.

Smith, J.K. (2000). Wildland fire in ecosystems: effects of fire on fauna. General Technical Report RMRS-GTR-42-vol. 1. Ogden, UT: US Department of Agriculture, Forest Service, Rocky Mountain Research Station.

Chapter 9

Fire as an ecosystem process

9.1 Introduction

According to classical ancient Greek natural philosophy, the existence of all things is the result of the interactions between the 'elements' fire, air, water and earth. While this does not make any scientific sense, it is true that many critical ecosystems processes are driven by the interactions of fire with climate (air), hydrology (water) and geomorphology and soils (earth). These interactions can operate near instantaneously on the local environment, or over geological time scales with global effects. As a consequence, fire can affect ecosystem services upon which human societies depend, including the provision of clean air, the maintenance of soil fertility, the storage of carbon and climate stability.

The past and current use of fire by humans influences these processes, sometimes with unanticipated, but not always detrimental, knock-on effects. Given rapid global environmental changes, understanding fire as an ecosystem process and how it affects the Earth System, is pivotal for achieving sustainable fire management. In this chapter, we explore how fire affects soils, hydrology and climate.

9.2 Fire and erosion

Most fires have little effect on geomorphological process. However, severe fires can be powerful erosive agents that shape landscape and soils (see Chapter 1,

Part One). The heat from severe fires can cause widespread fracturing and shattering of rocks and boulders, thereby vastly accelerating background physical weathering rates. In some soil types, high temperatures can create friable aggregates of soil particles that can repel water (hydrophobic), thereby reducing infiltration of rainfall and rendering soils highly erodible, yet, in other soil types, high temperatures can remove water repellency, increasing rainfall infiltration (see Chapter 1, Figure 1.14). The cause of these differences remains unknown, including the relative importance of biological processes and soil parent materials.

The destruction of vegetation cover, ground surface litter layer and organic soil horizon exposes mineral soils to desiccation, and the full impact of raindrops thereby increasing the risk of post-fire erosion (Figure 9.1).

The review by Parise and Cannon (2012) provides a concise summary of the impact of fire on soils and landforms (see also Doerr and Shakesby, 2013). They show that there is very high variability in the effect of fire on erosional processes, reflecting differences in fire intensity, soil types, vegetation and climate (Figure 9.2). For example, studies of burnt hillsides from different continents show a range of increases in post fire erosion between 50 and 870 times pre-fire rates. Severe fires in steep terrain can trigger mass wasting, including landslips and gully erosion (Figure 1.19 (Chapter 1), Figure 9.3).

Increased erosion following severe fires can result in heavy loads of suspended organic and inorganic sediment in streams and rivers (Figure 9.4). Heavy sediment loads can reduce aquatic primary productivity because of reduced light penetration through the water

Fire on Earth: An Introduction, First Edition. Andrew C. Scott, David M.J.S. Bowman, William J. Bond, Stephen J. Pyne and Martin E. Alexander.
© 2014 John Wiley & Sons, Ltd. Published 2014 by John Wiley & Sons, Ltd.

Figure 9.1 A denuded land surface after a high-severity fire in the Coastal Ranges, California. The limited recovery of the ground layer due to drought conditions, and the steep slopes, increase the risk of severe erosion (photo: David Bowman).

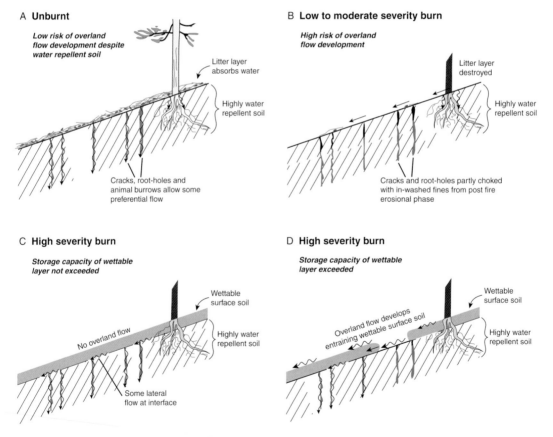

Figure 9.2 Hydrological response scenarios of forested terrain with high natural background levels of soil water repellency for (A) unburnt conditions and following fire of (B) low-moderate (C) and high burn severity (D). (From Doerr *et al.*, 2006). Reproduced by permission of Elsevier.

Figure 9.3 Large erosive channel caused by post-fire erosion following the Missionary Ridge Fire, Colorado, USA, in 2002 (photo: J. Moody, USGS).

column. Sediment input can change water chemistry (especially pH and nitrogen availability in the form of nitrate). If stream bank vegetation has been destroyed, water temperatures may increase. The combination of increased nutrients and higher water temperatures can lead to greater biological activity (such as algal blooms), thereby reducing dissolved oxygen that triggers localized dying off of fish populations.

Figure 9.4 Charcoal-rich flows after rainstorm, from a forest fire by overland flow from the Rodeo-Chediski Fire, Apache-Sitgreaves National Forest, Arizona, USA, 2002 (photo: D. Neary, USDA Forest Services).

Figure 9.5 Large alluvial fan caused by post-fire erosion following fire, Wren Creek/North Fork Boise River, Idaho, USA (photo: J. Moody, USGS).

Post-fire erosion can have dramatic impacts on watersheds, polluting water supplies (Moody and Martin, 2009). This is well illustrated by a severe bushfire that burnt almost all of the catchment of the main reservoir for the city of Canberra, Australia's capital, in 2003. A series of subsequent intense rainfall events flushed massive quantities of sediments and ash into the reservoir. Worthy and Wasson (2004) estimated that the sediment yield in the catchment increased from 2.3 tonnes/km^2/year to 9.2 tonnes/km^2/year after the fire with an overall input of 1663 tonnes of inorganic and 137 tonnes of organic sediments compared to the annual pre-fire input of 316 tonnes inorganic and 23 tonnes organic sediment. This massive input made the water unusable for human consumption and required the construction of a water treatment plant.

Similar erosion events have been recorded in sediments, allowing scientists to reconstruct extreme fire events that date back over tens of thousands of years (Figure 9.5). For example, Meyer *et al.* (1995) studied sedimentary deposits by rivers (alluvial fans) following the 1988 Yellowstone National Park wildfires to identify other similar older deposits. They concluded that over the last 10 000 years, severe fires that caused large-scale catchment erosion have occurred approximately every 300–450 years.

These dramatic soil losses are by no means general to all fires and all ecosystems. It is interesting to contrast Californian chaparral with Cape fynbos, both Mediterranean-type climate pyrophilic shrublands occurring in steep mountainous terrain. Chaparral is notorious for soil slumps and major sediment losses post-burn. Soils losses of as much as 197 tonnes/ha have been recorded in small bounded plots. In contrast, soil losses from burnt fynbos, measured in similar ways, were less than 1 tonne/ha. The highest losses were recorded when pine plantations, an artificial ecosystem replacing fynbos, burnt under severe conditions when sediment loads increased to 16 tonnes/ha.

Striking differences such as these mean that studies on soil and geomorphological effects of fire cannot be easily generalized across geographical regions, even those sharing similar climates and vegetation. Even after the most extreme fires, recovery of vegetation following fire reduces post-fire erosion, and catchment hydrologies can return to their pre-fire condition relatively quickly.

9.3 Fire and nutrient cycling

Fire influences the chemical composition of soil by abruptly changing nutrient cycling between plants and soils. Different chemical elements have different

Figure 9.6   Thick ash on the forest floor following the Overland fire, Colorado, USA, 2003 (photo: D. Martin, USGS).

temperatures from which they are transformed from solid to gas (volatilization or vaporization). Broadly speaking, nitrogen and sulphur vaporize at >300 °C, phosphorus and potassium at >750 °C, and calcium and magnesium at >1500 °C (albeit the precise temperatures can vary between inorganic and organic forms of some elements). The temperature of the fire therefore determines the proportion of plant nutrients that are volatilized.

The type of fire also influences nutrient loss from volatilization. The greatest losses are from crown fires killing green leaves before they could reach senescence, when the whole plant reabsorbs nutrients. Surface fires (see Fig 1.8, Chapter 1, Part One) burning leaf litter, or dry grass fuels in savannas, cause much lower nutrient losses because the plants have already reabsorbed nutrients from senescing leaves.

The solid residue of combustion is ash, containing charcoal and highly soluble oxides of plant nutrients that raise soil pH (Figure 9.6). A small fraction can also be exported from a site as 'fly ash' that becomes entrained in the smoke column, and the remainder forms a mantle on the burnt soil surface. Plants are very effective at capturing nutrients released into the soil by fire. For example, Kellman *et al.* (1985) showed that in a neotropical savanna, the nutrients leached into the soil were captured by plant roots within one week following the fire.

Nutrient losses following fire are controlled by vegetation type, fire intensity and the nutrient content in the above ground biomass. This is well illustrated by the study of Kauffman *et al.* (1994), who compared an adjacent grassland (Campo Limpo) and open evergreen woodland (Cerrado) in the Brazil. They showed that the grassland had lower fuel loads and smaller total above-ground nutrient stocks than the woodland. Yet, because of the high abundance of fine fuels, the fire-line intensity was greater in the grassland than in the forest. The higher combustion temperatures in the grassland resulted in a greater loss, relative to the woodland, of above-ground nitrogen, carbon and sulphur pools due to higher levels of volatilization. Although the mass of phosphorus, calcium and potassium losses from the grassland and the woodland were similar, the proportional loss was greater in the grassland, reflecting the small mass of these elements in this system.

Most of the nutrients lost from fires can be replaced through local re-deposition of ash and inputs from rainfall. For example, Cook (1994) showed that, excluding the fallout of entrained ash, rainfall and background atmospheric deposition would replace sulphur, phosphorus, potassium, calcium and magnesium lost from a fire in a north Australian eucalypt savanna within one year (Figure 9.7).

Figure 9.7   A smoke plume from a savanna fire in northern Australia. Most of the nutrients are replaced by atmospheric inputs, with the exception of nitrogen, for which biological fixation is a critical source (photo: David Bowman).

It is often assumed that nutrient losses from repeated fires cause long-term decline in soil nutrient and carbon content following repeated fires. However, in savannas, the cumulative direct effects in volatilizing nitrogen and carbon are relatively small. For example, Coetsee *et al.* (2010) compared soil nitrogen and soil organic carbon in an African savanna exposed to 50 years of burning or complete fire exclusion. The effects of burning were small relative to the effects of the presence or absence of a tree on nutrient stocks. Both nitrogen and carbon content of the topsoil were much greater under tree canopies than in the open (1.5–2 times) – a common pattern in savannas. While there were no significant direct fire treatment effects on total soil nitrogen and carbon, fire history did alter tree cover and then indirectly influenced soil nitrogen and carbon. Thus, in savannas, and possibly other systems, the indirect effect of fire on soil nutrients caused by changes in vegetation structure can be at least as important as the direct effects of volatilization losses.

Soil is a very effective insulator, and heat from fires goes upward. However, extremely intense fires can cause major changes to soil nutrient pools. For instance, heating tall eucalypt forest soil over 400 °C caused a substantial release of phosphorous and nitrogen, largely due to combustion of the soil microflora, with some minor release of phosphorus from the inorganic soil fraction. The availability of nutrients for plants is also enhanced, because high temperatures reduce the clay fraction that binds nutrients in soil. Chambers and Attiwill (1994) suggested that the pulse of nutrients following intense heating of soils, particularly phosphorus and nitrogen, largely explains the rapid growth of eucalypts on soils exposed to intense heat – a phenomenon known in Australia as the 'ashbed effect' (Figure 9.8).

The combined effects of heat causing changes to soil chemistry (see Chapter 1.7, Part One) have substantial effects on the soil biota, which can influence nutrient availability, soil development and plant growth. Severe fire may kill both the beneficial and detrimental components of the soil biota. For example, symbiotic root fungi (mycorrhiza), which enhance plant acquisition of nutrients, can be killed along with plant root pathogens Soil bacteria are involved in the formation of ammonium (NH_4^+) by fixing nitrogen gas (N_2), the breakdown of organic matter and the transforming of ammonium to nitrate (NO_3^-). Fire substantially impacts on the bacteria that

Figure 9.8 Aftermath of a high intensity regeneration fire in Tasmania designed to provide a nutrient pulse for eucalypt seedlings. Such fires typically consume all of the organic horizon and sterilize the soil (photo: David Bowman).

drive the nitrogen cycle, although these do recover quickly after fire, thereby restoring the nitrogen cycle.

9.4 Fire and pedogenesis

Over thousands of years, fire regimes can shape the development of soils, a process known as pedogenesis. McIntosh *et al.* (2005), for example, attribute the strongly contrasting soils that develop under similar climates and on similar geologies in New Zealand and Tasmania as reflecting the effects of contrasting fire regimes. They suggest that a long history of frequent burning in Tasmania has caused the development of infertile soils due to nutrient depletion, loss of organic matter and migration of clays to the subsoils (a process known as 'eluviation'). Whether these differences are attributable to human fire use, as suggested by McIntosh *et al.* (2005), is debatable, given the differences in vegetation types between Tasmania (dominated by fire-adapted eucalypts) and New Zealand (which lacks well-developed fire-adapted tree flora).

Many different indigenous cultures that live in tropical rainforests used fire to increase nutrient availability temporarily in an agricultural system known as 'slash and burn' or swidden (see Part Three). This agrosystem involves cutting and burning trees to release nutrients to support crops (Figure 9.9). Once the fertility drops, the fields are abandoned and nutrient stocks are able to re-accumulate. Over thousands of years, this practice has affected small areas, resulting in soils with high concentrations of charcoal and clay minerals altered by fire.

A study in Central Amazonia by Lehmann *et al.* (2003) showed that relict soil from pre-Columbian settlements (anthrosols) were more fertile compared with similar non-modified soils (ferralsols), because of higher concentrations of charcoal, which boosts nutrients retention by reducing leaching. This example shows that, with care, fire can be incorporated into an agroecological system without degrading the soil nutrient capital – indeed, increasing it. Of course, the critical issue is the spatial scale of the fields and the rotation times. The discovery of the use of charcoal in the Amazonian farming system has led to attempts to simulate the practice by creating 'bio-char' (charcoal for soil improvement), both to promote soil productivity and as a means of sequestering carbon. Bio-char is a stable form of carbon that can persist in the soil for thousands of years (see Chapters 1 and 3, Part One).

Figure 9.9   Slash and burning in the Papua New Guinea (PNG). Clearing and burning rain forest in preparation for a Taro and Sweet Potato garden, Tari Basin, PNG (A). Sweet potato gardens with emergent Pandanus sp. in the Levani Valley, PNG, are maintained by slash and burning agriculture (B) (photos: Simon Haberle).

In summary, fire affects the physical, biological and chemical dimensions of soil (Table 9.1). There is very large variation in these effects in response to different fire intensities, soil and vegetation types and landscape settings, and such variability limits the ability to reach global generalizations.

9.5   Fire and atmospheric chemistry

Fires inject heat and matter into the atmosphere, including large quantities of water vapour, inorganic carbon particles, carbon monoxide (CO), carbon dioxide (CO_2), methane (CH_4), nitrogen oxides (NO_x), sulphur oxides (SO_x) and very small quantities of a very wide diversity of organic and inorganic compounds in liquid, solid and gaseous forms, some of which are highly toxic to humans (see Chapter 1, Part One). The majority of the emissions are water vapour and CO_2, with the remaining proportion of the chemical components depending upon the intensity of the fire, the availability of oxygen (whether the fire is flaming or smouldering) and the chemical make-up of the fuel.

Smouldering combustion (Rein, 2013) produces higher emissions of carbon monoxide and particulates formed from condensation and aggregation of unburnt

Table 9.1   Soil properties modifiable by fires. (From Certini, 2005). With kind permission from Springer Science+Business Media B.V.

Physical properties	Chemical properties	Biological properties
Water repellence	pH	Quantity and type of organic matter
Structure stability	Availability of nutrients	Composition and abundance of soil microbiota
Bulk density	Exchangeable capacity	Composition and abundance soil invertebrates
Particle-size distribution	Base saturation	Rate of biological fixation of nitrogen and conversion to nitrate
Mineralogical assemblage		
Colour		
Temperature regime		

Table 9.2 Turnover time of some key chemical and physical emissions from combustion and the geographic effect on climate (From Chapin *et al.*, 2008). Reproduced with permission of Ecological Society of America.

Emission	Turnover time	Spatial scale
CO_2	3 years	Global
CH_4	8.4 years	Global
N_2O	120 years	Global
H_2O	10 days	Sub-continental
Aerosols	Days to weeks	Regional to continental
NO_x	<1 day	Regional
SO_2	<1 day	Regional

organic compounds. The greenhouse gas effect (or radiative forcing, measured in watts per m^2) varies among gases and aerosols, so that a small concentration of one gas can have the same radiative forcing as a large quantity of another gas. For example, methane has around 25 times more radiative forcing capacity as CO_2.

Gases and particulates from fires are eventually reabsorbed by the biosphere or oceans, resulting in global equilibrium, although the residence times of gases in the atmosphere varies from days to hundreds of years. The contrasting turnover of gases and particulates means that some gases and aerosols have long-term global effects, while others only have short-term regional effects (Table 9.2). Historically, the effect of fire on the atmospheric chemistry of the Earth was negligible, because emissions were balanced by sequestration. However, combustion of fossil fuels and deforestation has resulted in substantial disequilibrium, and this is affecting the entire Earth System.

9.6   Fire and climate

The gases and particulates in smoke plumes can affect regional and global climates. Some effects are immediate, while others are longer term. Aerosols in smoke have a warming effect on the lower atmosphere (troposphere), thereby inhibiting convection of moisture and reducing precipitation regionally (Figure 9.10). Further, aerosols released by fire can lead to a reduction in precipitation as they act as cloud condensation nuclei. This means that the available water vapour aggregates around an increased number of particles in the atmosphere and thus slows the aggregation of these particles to form cloud droplets (spheres of

Figure 9.10   A smoke haze over the Australian tropical city Darwin. During the dry season, near continuous burning results in a thick smoke layer in the lower atmosphere (photo: David Bowman).

Figure 9.11   Smoke from plantation fire, generating a pyrocumulous cloud in Mato Grosso, Brazil, the southern Amazon, 2006 (photo: Jennifer K. Balch).

water around 10 microns in diameter), thereby reducing precipitation.

However, severe fires may trigger intense convection storms with pyrocumulous clouds (Figure 9.11) that can generate strong winds, intense rainstorms and lightning activity. Under the most extreme fires, 'pyrocumulonimbus' storms can form, which eject smoke through the troposphere into the lower stratosphere in the same way as do volcanic eruptions.

Gases and aerosols from fires can spread globally through the atmosphere (Figures 1.38-1.42, Chapter 1, Part One), thereby affecting the global energy balance. Some aerosols, such as sulphates, reflect sunlight back into space and have a cooling effect, while other aerosols, like black carbon, absorb heat and warm the atmosphere. Blackened land surfaces after fire also reduce the reflectance of solar energy (reducing albedo, the proportion of solar irradiation absorbed by the land surface) and result in warming the lower atmosphere (albeit this is strongly influenced by the spatial size and severity of the fire and consumption of vegetation cover). If grass cover replaces tree cover, the albedo of a site may increase following fire. Thus, fire results in a number of positive and negative climate forcing effects, many of which cancel each other out, with the exception of CO_2 (Figure 9.12).

Bowman *et al.* (2009) suggest that, since the Industrial Revolution, CO_2 emissions from deforestation fires alone have contributed about 20% of the total increased radiative forcing. The concern is that global warming will increase the recurrence of severe fires, reducing the capacity of vegetation – and particularly forests – to absorb CO_2 and resulting in a positive feedback between climate change and landscape burning.

It is important to note that such global assessments generalize the effects across biomes. For example, the effect of fire in causing deforestation in the tropics is different from that in the boreal zone. In the boreal forests, wildfire releases carbon from stores in both biomass and soils, and this represents a significant flux of greenhouse gas that consequently contributes to global warming. However, this effect may be counteracted by increased albedo during winter months because of increased snow cover, although the deposition of black carbon from fires over sea ice sheets reduces albedo and also contributes to global warming and to the loss of ice. The severity of fire and the time since the fire are thought to strongly influence the release of carbon and the changes in albedo in the boreal zone (Figure 9.13).

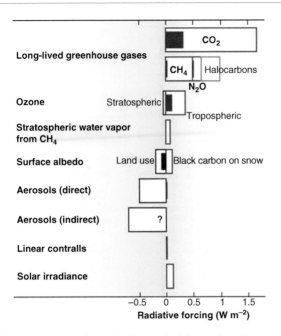

Figure 9.12 Estimated effects of deforestation fires on radiative forcing to the global climate (inner bars), relative to all radiative forcing identified by the IPCC (open slid bars) since the start of the Industrial Revolution (c. 1750). Red indicates warming effect and blue a cooling effect. All of the components are assumed to cancel out, with the exception of CO_2, which is thought to have contributed around 20% of the total warming. All other fire emissions are assumed to be in a steady state, although this assumption may be wrong, as the world increasingly warms. (From Bowman *et al.*, 2009). Reproduced by permission of the American Association for the Advancement of Science.

Evapotranspiration from tropical forests transfers large quantities of water from the soil to the atmosphere, recycling precipitation. Burning of tropical forests results in a drier climate that is more conducive to burning because of reducing rainfall and increasing lightning. Indeed, Beerling and Osborne (2006) outline how fire and climate may have interacted to increase the extent of savanna vegetation at the expense of forest over the last eight million years (Figure 9.14).

The spread of agriculture through European and Asia was associated with clearing of forests by burning. There is an ongoing controversy as to whether or not this has resulted in a significant release of greenhouse gas that may have changed the Earth's climate. Ice cores record a mid-Holocene (ca. 6000 years ago) reversal of the downward trend of CO_2 and CH_4 greenhouse gas concentrations in ice cores that had begun during the start of the Holocene. While most scientists believe this is a natural process, Ruddiman (2003) has argued that it is the result of anthropogenic burning of forests for agriculture. Controversially, he has posited that the release of these greenhouse gases avoided the return to ice age conditions.

There is serious concern that current combustion of fossil fuels and destruction of forests may trigger a feedback between climate change and fire activity. For instance, global warming may increase dangerous fire weather and increase the flammability and abundance of fuel, thereby promoting more frequent fires that will result in increased emissions

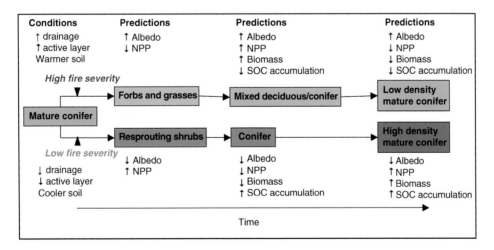

Figure 9.13 Predicted effects of high and low severity fires on Alaskan boreal forests over a hundred years of recovery. (From Goetz *et al.*, 2007). Reproduced with permission © IOP Publishing Ltd.

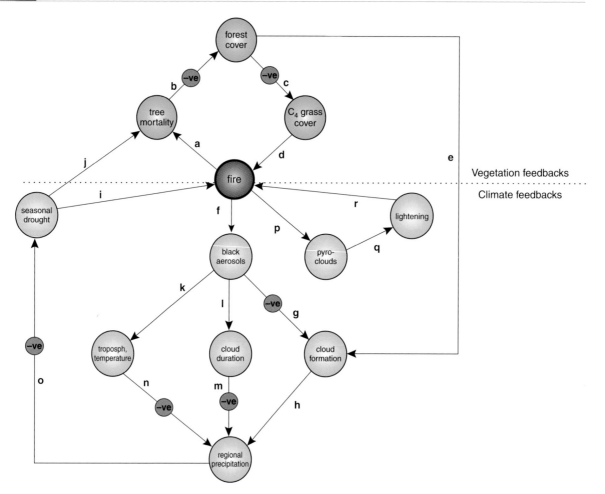

Figure 9.14 Diagrammatic representation of fire-climate-vegetation feedbacks in the savanna biome. Frequent fires increase grass dominance and reduce forest cover by killing trees. Pathway a-b-c-d: Loss of forest cover reduces evapotranspiration and cloud formation, causing a drier regional (pathway e-h) with a longer dry season that favours more frequent fire (pathway e-h-o-i-a-b) further favouring the loss of tree cover (pathway e-h-o-j-b). Aerosols in smoke plumes amplify the trend for a drier climate by affecting clouds (pathways e-f-g-l) and the temperature of the lower atmosphere (pathway k). (From Beerling and Osborne, 2006). (See Chapter 3.2 and 3.3, (Part One) for a discussion of fire feedbacks and systems diagrams).

and reduced carbon storage capacity of many vegetation types.

9.7　Summary

Fire instantaneously links landscape to the atmosphere through the release of heat, gases and organic and inorganic compounds. The effects of fire ramify through ecosystems and the Earth System, influencing vegetation, soils, landforms, stream flows and regional and global climate, resulting in devilishly complicated interactions. While there has been considerable progress in understanding fire as an ecosystem process, considerable gaps remain in our knowledge. Furthermore, the enormous variability in the effects of fire make it difficult to generalize findings geographically and through time.

Unravelling the role of biology, climate and geology in creating this variability in ecosystem processes is an enormous research challenge. This challenge can only tackled by research teams made up of scientists with a range of disciplinary experience and with the capacity to make measurements that use the same methodologies and, therefore, create data sets that are truly comparable. A major limitation to re-analyses of existing published data on various ecosystem processes (so-called 'meta-analyses') is that many of the observations are not directly comparable, being based on strongly contrasting sampling strategies and different analytical procedures. For example, there is no universally agreed method to measure 'soil phosphorus', the choice of method depending on the specific question being asked.

There is no doubt that the fire regime concept is pivotal to any comparative analyses of fire's ecological effects, because it provides a common currency for describing and classifying landscape fire. Nonetheless, care is required to ensure that the components of fire regimes are measured using consistent methodologies.

International research teams, aided by the internet and air travel, are increasingly undertaking global comparative studies of the effect of fire on ecosystem processes. One motivation for such research is the recognition that fire regimes are changing in response to global environmental change associated with human activities, with knock-on effects to ecosystems services. The quest for the sustainable use of fire demands high-quality evidence to balance the often opposing effects of combustion on ecosystem services, biodiversity and society. The subject of the next chapter is anthropogenic influence on landscape fire in the past, present and future.

Further reading

Bowman, D.M.J.S., Balch, J.K., Artaxo, P., Bond, W.J., Carlson, J.M., Cochrane, M.A., D'Antonio, C.M., DeFries, R.S., Doyle, J.C., Harrison, S.P., Johnston, F.H., Keeley, J.E., Krawchuk, M.A., Kull, C.A., Marston, J.B., Moritz, M.A., Prentice, I.C., Roos, C.I., Scott, A.C., Swetnam, T.W., van der Werf, G.R., Pyne, S.J. (2009). Fire in the Earth System. *Science* **324**, 481–484.

Chambers, D.P., Attiwill, P.M. (1994). The ash-bed effect in *Eucalyptus regnans* forest – chemical, physical and micro-biological changes in soil after heating or partial sterilization. *Australian Journal of Botany* **42**, 739–749.

Coetsee, C., Bond, W.J., February, E.C. (2010). Frequent fire affects soil nitrogen and carbon in an African savanna by changing woody cover. *Oecologia* **162**, 1027–1034.

Cook, G.D. (1994) The fate of nutrients during fires in a tropical savanna. *Australian Journal of Ecology* **19**, 359–365.

Kauffman, J.B., Ward, D.E., Cummings, D.L. (1994). Relationships of fire, biomass and nutrient dynamics along a vegetation gradient in the Brazilian Cerrado. *Journal of Ecology* **82**, 519–531.

Kellman, M., Miyanishi, K., Hiebert, P. (1985). Nutrient retention by savanna ecosystems: II. Retention after fire. *Journal of Ecology* **73**, 953–962.

Lehmann, J., da Silva, J.P., Steiner, C., Nehls, T., Zech, W., Glaser, B. (2003). Nutrient availability and leaching in an archaeological Anthrosol and a Ferralsol of the Central Amazon basin: fertilizer, manure and charcoal amendments. *Plant and Soil* **249**, 343–357.

McIntosh, P.D., Laffan, M.D., Hewitt, A.E. (2005). The role of fire and nutrient loss in the genesis of the forest soils of Tasmania and southern New Zealand. *Forest Ecology and Management* **220**, 185–215.

Meyer, G.A., Wells, S.G., Jull, A.J.T. (1995). Fire and alluvial chronology in Yellowstone National Park – climatic and intrinsic controls on Holocene geomorphic processes. *Geological Society of America Bulletin* **107**, 1211–1230.

Parise, M., Cannon, S.H. (2012). Wildfire impacts on the processes that generate debris flows in burned watersheds. *Natural Hazards* **61**, 217–227.

Ruddiman, W.F. (2003). The anthropogenic greenhouse era began thousands of years ago. *Climatic Change* **61**, 261–293.

Worthy, M., Wasson, R.J. (2004). Fire as an agent of geomorphic change in southeastern Australia: implications for water quality in the Australian Capital Territory. In: Roach, I.C. (ed). *Regolith 2004*, pp. 417–418. CRC LEME, Canberra, Australia (http://crcleme.org.au/Pubs/Monographs/regolith 2004/Worthy%26Wasson.pdf).

Chapter 10

Fire and anthropogenic environmental change

10.1 Introduction

Fire scientists must grapple with the fact that biomass combustion and humans are deeply entwined (Bowman *et al.*, 2011). Although landscape fire has an ancient history on Earth (Chapter 4, Part One; Belcher *et al.*, 2013), humans and our ancestors (hominins) have changed the geographic extent, severity and timing of landscape fires since at least the late Pleistocene (about 50 000 years ago; see Chapter 5, Part One). In some landscapes, humans have altered fire regimes so much that they have changed vegetation distributions.

It is now impossible to precisely define 'natural' background fire activity, given the ubiquity of human landscape burning for almost every environment on Earth. Indeed, there remains considerable academic debate about whether the use of fire by colonizing people into uninhabited flammable landscapes had a transformative or a minor role in driving environmental change, compared with climatic variation.

Why does understanding the nexus between nature-human fire matter? Because resolution of this debate is important to understanding our species ecology, and indeed our identity. Furthermore, it is crucial to answering a number of pressing questions that are at the heart of contemporary fire management, including:

- Are economically destructive wildfires true natural disasters, over which humans have little control, or do they reflect a failure in management?
- Could anthropogenic climate change outstrip the capacity of humans to manage landscape fire?
- Could uncontrolled fires become a major accelerant to global warming via the release of large stocks of carbon?

The purpose of the chapter is to explore the interplay between humans and fire in shaping biotas, fire regimes and ecosystem processes in the past, present and future. This chapter cannot provide definitive answers to the above questions, but it provides an ecological framework for thinking about them.

10.2 Prehistoric impacts

There are competing views concerning human prehistoric impacts that are relevant to debates about anthropogenic landscape burning. They can be described broadly as the 'noble savage' and 'human

Fire on Earth: An Introduction, First Edition. Andrew C. Scott, David M.J.S. Bowman, William J. Bond, Stephen J. Pyne and Martin E. Alexander.
© 2014 John Wiley & Sons, Ltd. Published 2014 by John Wiley & Sons, Ltd.

Figure 10.1 'Aborigines using fire to hunt kangaroos', a watercolour by Joseph Lycett (c. 1775–1828), set in New South Wales c. 1817. Many plant ecologists have overlooked the fundamental importance of fire to Aboriginal economies throughout Australia, being primarily used to manage wildlife habitat and as a hunting tool (National Library of Australia).

the destroyer' narratives. The noble savage narrative stems from observations by 18th and 19th century explorers recording the biodiversity-rich landscapes that supported healthy hunter-gatherer and subsistence agriculturalist societies (Figure 10.1). Fire was routinely noted as a tool used by diverse cultures globally to manipulate their local environment.

However, the significance of these historical records remains contested. Some researchers have asserted that prehistoric human fire use was negligible, being geographically restricted, and that climate is the real driver of fire activity. This view is particularly prominent in North America. Alternatively, it has been claimed that humans skilfully used fire at a broad scale to shape landscapes and, thereby, increase the abundance of resources they valued. This perspective is increasingly accepted in Australia; for instance, the historian Gammage (2011) recently claimed that the Australian continent, at the time of European discovery, was the 'biggest estate on Earth' because of the stewardship of indigenous people!

The 'man the destroyer' narrative has developed with the discovery of dramatic environmental changes that followed the human colonization of pristine environments, particularly islands. The most important environmental change was the extinction of large animals (megafauna). Flannery (1994), for example,

popularized the view that Australia had been completely transformed by human colonization in the late Pleistocene (Figure 10.2). The divergent narratives reflect contrasting temporal perspectives, and are therefore not necessarily mutually exclusive. This is illustrated below by reviewing contrasting interpretations of the role human colonization played in the extinction of the Australian and New Zealand megafauna.

New Zealand was colonized around 750 years ago by the Māori (a people descended from Polynesians), whose economy was a combination of hunter-gather and subsistence agriculture. There is no question that the arrival of humans caused the extinction of many large flightless bird species (e.g. moa). There is also evidence, from lake sediments, of sustained burning immediately following colonization, which caused widespread forest loss.

Prior to human colonization, the incidence of fire in the forests was low, and correspondingly the tree flora were not highly fire adapted, rendering this environment particularly vulnerable to human-set fires. There is no evidence of any anomalous climates at the time of human colonization that could provide an alternative explanation for this spike in fire activity. Recently, mathematical modelling by Perry et al. (2012) suggests that colonizing humans may have intentionally set fires in areas of flammable non-forest vegetation, setting off a feedback that created larger areas of flammable

Figure 10.2 A skeleton of an extinct leaf-eating kangaroo. A wide diversity of reptiles, birds and marsupial became extinct in late-Pleistocene Australia. The cause remains contested, but human agency is almost certainly involved, including over-hunting and habitat transformation associated with changed fire regimes (photo: David Bowman).

vegetation. In this instance, there is a strong case for a transformative effect of anthropogenic landscape burning on the New Zealand environment.

The Australian situation is more ambiguous, reflecting the sparse and much more ancient palaeoecological record. Disentangling human influence in Australia is complicated by the fact that it is far harder to detect the signature of anthropogenic burning in highly flammable Australian biomes, compared with the infrequently burnt New Zealand biomes. Indeed, this is one reason why such uncertainty remains as to when hominins first started to use fire: early hominins coexisted with fire in flammable savanna landscapes long before they commenced domesticating it.

A very diverse assemblage of Australian animals became extinct in the late-Pleistocene. The timing of the extinctions are generally accepted as being coincident with the colonization of Australia by the ancestors of the Australian Aborigines around 45 000 years ago, although a minority of scientists believe that megafauna and humans coexisted until the height of the last ice age (known as the late glacial maximum, or LGM), around 20 000 years ago, when climate change contributed to the extinctions. Because there is no evidence of climate change at the time of colonization, it is generally accepted that humans were responsible for the extinctions.

Some scientists assert that the cause of the extinctions was sustained burning of closed canopy vegetation, resulting in the increased extent of grasslands and flammable eucalypt forests. The loss of broadleaf vegetation is thought to have had far-reaching effects, including disruption of food webs by eliminating larger browsers, causing a 'trophic cascade' of extinctions (Figure 10.2). It has been claimed that the loss of tree cover caused aridification of the continent by reducing evapotranspiration and increasing albedo. Miller *et al.* (2005) described these combined fire induced effects as triggering 'ecosystem collapse' across the Australian continent. Other researchers have claimed that fire regimes changed in response to the loss of large herbivores, which resulted in increased fuel loads that then caused increased fire activity, leading to the dominance of fire-adapted vegetation.

Because of the antiquity of human colonization of Australia, the evidentiary basis for these claims relies on ingenious use of sparse environmental archives, including fossilized pollen and dung spores, charcoal trapped in lake sediments and the stable isotope content of fossils, to broadly reconstruct vegetation, herbivore abundance and fire regimes. Mathematical modelling has also been used to support these different arguments, although the results of these modelling exercises remain contested.

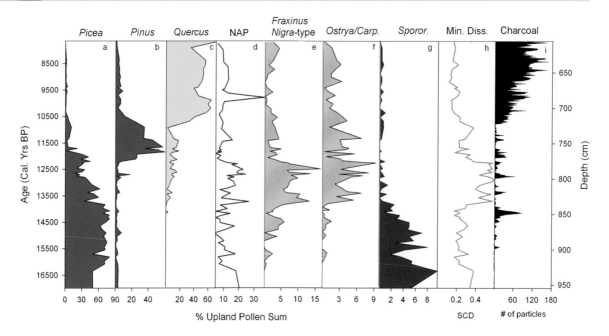

Figure 10.3 Changes in the pollen of selected tree taxa (a to f), Dung spores (g) and charcoal (i) at Appleton Lake in Indiana from 17 000 to 7000 years ago. The increase in charcoal lags the decline of the dung spores by 2000 years, possibly reflecting the loss of megafauna that released some broadleaf trees, resulting in more severe fires following increased fuel loads. (From Gill *et al.*, 2009). Reproduced with permission of the American Association for the Advancement of Science.

Parallels with Australian prehistory have been drawn with the extinction of megafauna and changes in vegetation types and fire activity that occurred when humans colonized North America some 14 000 years ago. Gill *et al.* (2009) used the dung fungus *Sporormiella* as a proxy for megafauna abundance around lake sites in Indiana and New York. Their analysis showed that the decline of megafauna around 14 000 years ago preceded increased fire activity by 1000 years (Figure 10.3). They interpreted this increased fire activity as reflecting increased fuel loads following the release of palatable hardwoods from herbivory pressure. These authors suggested that these ecological changes were caused by human hunting pressure (and possibly associated increased burning), but acknowledged that this did not exclude the role of climate.

10.3 Prehistoric fire management

The catastrophic effect of human activity needs to be contextualized with observations of indigenous landscape burning that point to an interdependence of biodiversity with managed fires. The majority of direct observations come from Australia, where some Aboriginal hunter-gatherer cultures have persisted into the 21st century. Aboriginal hunter-gatherers used fire extensively to clear campsites, to facilitate travel, to rid areas of insects and snakes, to encircle game while hunting and to create areas of fresh grass to attract kangaroos (Figure 10.4).

A feature of Aboriginal fire use was the creation of a patchwork burnt and unburnt areas (see Figure 10.1). There is evidence from northern Australia that the breakdown of Aboriginal burning practices following European colonization in the late 19th and early 20th century has resulted in the decline of a range of plant and animal species, and that fuel loads have built up, resulting in more destructive fires. Thus, it is possible that fire use by people initially had a dramatic impact on the Australian environment but, over a 40 000-year period, extensive and frequent Aboriginal fire use created habitat 'pyrodiversity' upon which biota depended.

Records of indigenous burning from around the world reveal a common pattern of numerous small fires set before the onset of the fire season, thereby creating a mosaic of burnt and unburnt areas. The

Figure 10.4 Aboriginal people in Arnhem Land use fire to burning landscape to achieve a range of objectives. They have a relaxed and familiar attitude to landscape fires that is often confronting for people from other cultures, where landscape fires are feared. (photos: A. Clay Trauernicht; B. Glenn Hunt).

motives for such burning practices are similar to those recorded in Australia. This patchwork of burnt and unburnt areas would have limited the spread of any naturally ignited fires.

Contemporary fire managers employ a similar strategy by burning flammable landscape under controlled conditions to reduce fuel loads, thus reducing the severity of any subsequent wildfires. This system works well in frequently (e.g. biennial) burnt biomes like savannas, where mild early season fires consume fuel that would otherwise be burnt in more intense late season fires. However, in less frequently burnt forest ecosystems, frequent management burning can be ecologically destructive, given the need of some plant and animal life cycles to be free from fire disturbance for many years. Indeed, very frequent burning can promote the abundance of flammable plants, increasing fire risk rather than decreasing it. Furthermore, unless extensive areas are treated, the planned burning is ineffective in reducing the extent of episodic wildfires. For example, analysis of the effectiveness of planned burning in Australian forests suggests that avoiding one hectare of wildfires requires 3–4 hectares of planned burning.

Landscape burning was probably an emergent property of hunter-gatherer fire management, given that people were moving to different habitats, tracking seasonal variation of resources. Over thousands of years, indigenous patch burning may have created areas of treeless vegetation in some areas. The rapid colonization of trees into some grasslands patches (sometimes called 'balds', pockets or meadows) is consistent with this view (Figure 10.5).

Figure 10.5 Grassland 'balds' on the Bunya Mountains in southeast Queensland, Australia. Aboriginal fire management maintained these treeless areas, which are infilling with forests following the cessation of burning in the 20th century. (photo: R.J. Fensham).

It is important to acknowledge that the idea that indigenous people practiced skilful fire management that substantially modified the occurrence of widespread fires is not universally supported by all scientists, with many arguing that climate is the primary driver of fire activity, especially in forests with long fire return intervals. For example, on the basis of charcoal records, Mooney *et al.* (2011) claimed that Aborigines had negligible influence on fire activity in Australia. Likewise, compilation of charcoal records from around the world point to the predominance of climate in driving fire activity globally (Power, 2013). Uncertainty about past fire regimes and the that role indigenous people or climate has played has led to philosophical friction amongst fire managers, who have argued against attempts to 'restore' prehistorical fire regimes or to adapt such regimes to suit modern circumstances.

10.4   Contemporary fire management

Until recently, a widespread approach to fire management was suppression, using an ever-increasing array of technologies and techniques (see Chapter 16, Part Four; Figure 10.6). However, this approach has led to ballooning budgets, increased fuel loads and changes in forest structure, rendering many forest types susceptible to severe fires.

The ponderosa pine (*Pinus ponderosa*) forests of the western USA are a prime example where there has been a switch from a surface fire regime that caused limited tree mortality to severe crown fires that can kill entire forest stands (Roos and Swetnam, 2012). There is increasing recognition that suppression policies have proved unworkable, and that the exclusion of fire is detrimental for ecosystem health and biodiversity conservation. This has led some fire managers to instigate a policy where 'natural' fires set by lightning are allowed to burn freely, so long as they do not threaten human communities or other values.

This approach is, of course, constrained by the intensity of human settlement and the type and condition of the vegetation. Allowing naturally set fires to burn through forests that have been subject to decades of fire suppression can result in extreme fires, because of the build-up of fuels and the structure of the vegetation that results in massive crown fires. For

Figure 10.6   Fire-fighters directly attacking a fire front in Turkey. Such strategies to control fire are costly and dangerous and in the long term unsustainable (photo: Aykut Ince).

this reason, some managers have advocated mechanical thinning of the forests to reduce fuel loads and reduce the risk of destructive crown fires.

In many contemporary landscapes, it is impossible to adopt broad scale burning because of habitat fragmentation, land tenure boundaries and the presence of vulnerable infrastructure. Some fire managers have argued that fire management is better focused on the perimeter of human settlements (the 'wildland-urban interface') because of the difficulty in applying enough fuel reduction burning to influence effectively the extent and severity of wildfire (Figure 10.7A–C). In addition to planned burning and thinning of vegetation, an obvious approach on the wildland-urban interface is to use land use planning to limit the coexistence of housing with flammable vegetation (Figure 10.7D). Planning rarely enjoys any political support, because of the impact on property values and perceived constraints on the lifestyle choices of ex-urbanites attracted to residing in the green, leafy wildland-urban interface.

It is now generally accepted that fire management is critical for biodiversity conservation within natural landscapes. However, there are rarely simple solutions, because any particular combination of fire severity and fire frequency will advantage some organisms and disadvantage others, raising the vexed question of 'what is the prime objective of fire

management for biodiversity conservation?' (Keeley *et al.*, 2012). In the case of indigenous burning, thought to have created pyrodiversity, it is almost certain that the motive for burning was not to conserve biodiversity but, rather, to increase the abundance of valued resources. The creation of biodiverse landscapes was a side-effect.

In contrast, contemporary conservation biologists are rarely able to design and implement fire management with a narrow set of objectives; rather, they need to balance fire regimes that suit species threatened with extinction against the requirements of other species, some of which are valued for other reasons. Furthermore, managers must also ensure the protection of human life and property, and of ecosystem services, especially minimizing smoke pollution that is known to affect human health. Climate change is making these difficult decisions even more complicated.

10.5   Climate change

Anthropogenic climate change has the capacity to alter fire regimes fundamentally via both direct and indirect effects. The direct effects are through changing seasonal patterns of temperature, wind and precipitation, especially the occurrence of extreme fire weather events (see Chapter 1, Part One). In Chapter 6, we discussed how

Figure 10.7   The wildland-urban interface (WUI) in (A) Santa Barbara, California, (B) Hobart, Australia and (C) Cape Town, South Africa. Urban sprawl into flammable landscapes is a driving force in wildfire disasters and is as much a social issue as an environmental one. Attempts to create safe area on the WUI are one strategy (D), but there is a high risk of catastrophic failure (E). A typical response after a fire disaster is to immediately rebuild perpetuating the problem (F). (Photos: A, B, C, David Bowman; D, Deborah Martin, USGS; E, Courtesy Channel 7 Denver; F, Chris Dicus).

indirect effects of climate influence fuel availability. In arid areas fuel is limited by available moisture for plant growth whereas, in mesic areas, fuel availability is controlled by periods of moisture deficit that dry out biomass.

Increased CO_2 in the atmosphere may increase biomass production by increasing whole-plant water use efficiency and photosynthetic rates. Increased CO_2 may also substantially change the competitive balance among plants, especially grasses and trees in tropical savannas. Given the complexity of the climate-plant interactions, and limited ability to predict future regional climates (especially rainfall and wind), it remains impossible to predict confidently how future fire regimes and vegetation types will change from global to local scales. Some of these changes may be ecologically transformative, amplifying climate change effects, while others may have negligible effects because the changes cancel out. The following examples provides a taste of these uncertainties and complexities.

- In some settings, such as deserts, it is possible that climate change may result in sharply reduced fire activity, because climate drying will further reduce available fuel loads.
- In productive environments, such as tropical rainforests, extreme drought may increase fuel availability.
- In some flammable forest environments, increased CO_2 may accelerate the production of fuels by increasing tree growth rates and reducing their life spans, thereby creating more intense and more frequent fires.
- Mild winter temperatures in some temperate regions, such as the western USA, may increase fuel loads by allowing irruptions of defoliating insects that kill trees.
- Increased frequencies of fire may disadvantage organisms that have life cycles that demand specific periods without fire, such as obligate seeding plants that dominant Mediterranean ecosystems (Keeley et al., 2012).
- Increased CO_2 could reduce fire activity in tropical savannas by changing the competitive balance of trees and grasses, causing a decline in the abundance of flammable grass fuels and a decrease in fire activity.

All of these scenarios may be wrong, because they ignore a major unknown: how ignitions will vary due to changes in lightning activity and human activities.

Resolving the unknowns associated with the effect of climate change on fire activity has become a research focus, with a range of different approaches, including analysis of meteorological data, statistical analysis of global climate model projections and analysis of dynamic global vegetation models (DGVMs) (see Chapter 1, Part One). Trends from meteorological data suggest that, for some regions, there is evidence of changes to fire season length and increase in fire weather severity. For instance, an analysis by Westerling et al. (2006) of fire and climate records showed an abrupt increase in large forest fires in the western United States since the 1980s, was associated with the earlier onset of the spring melt and higher spring and summer temperatures (see Chapter 1, Figure 1.27). Such analyses are constrained by the limited historical data on fire extent and high quality meteorological records.

Analysis of the outputs of Global Climate Models (GCM) are also increasingly being used to project statistically how current patterns of fire activity may change in response to projected future climates at global to regional scales (Chapter 1). Using this approach, Westerling et al. (2011) developed a predictive model of the relationship between climate and past large fires for the Greater Yellowstone region. They then investigated the consequences of future climates, based on regional projections (downscaled) of three global climate models. This analysis predicted a dramatic increase in fire frequency that may convert conifer forests to woodlands and other non-forest vegetation. Working at the global scale, Moritz et al. (2012) used a similar approach and predicted that the most significant increase in fire activity will be in the boreal zone (Figure 1.43, Chapter 1, Part One).

It is important to note that this statistical approach cannot disentangle change in human ignitions, nor can it incorporate the effect of climate change and CO_2 enrichment on fuel production. Dynamic global vegetation models (DGVM) have the potential to incorporate change human ignitions, the CO_2 fertilizer effect and changes in climate to explore how fire activity in the future may change. However, such models are hampered by inadequate data and the daunting complexity of the interactions, resulting in the need to explore a wide range of plausible scenarios. DGVMs do, though, provide a powerful means to explore the potential of feedbacks between fire and other components of the Earth system. For example, they can explore interaction of climate, CO_2, fire and major plant growth forms, including new mixtures from invasive plants.

10.6 Fire and carbon management

Despite numerous uncertainties there is no question that climate change will compound the already fraught problems of fire management. This is especially so, given increasing societal pressure to manage fire regimes to optimize carbon benefits through reduced emissions of greenhouse gases and enhanced carbon sequestration. There are, currently, only a few examples of adapting fire management to serve the 'carbon economy', but these are expected to grow in step with concerns about global climate change. In northern Australian savannas, managers are setting fires early in the dry season, when mild fire-weather conditions prevail, to reduce fuel consumption and fire severity as part of a voluntary carbon offset program supported by a major oil and gas company. Because the Kyoto Protocol is based on the assumption that all CO_2 emissions from landscape fires are reabsorbed by vegetation, the northern Australian scheme is based around reducing non-CO_2 greenhouse gas emissions (e.g. methane and NO_x).

There remains debate about whether using fire to manage greenhouse gases can work in non-savanna flammable biomes, where fire is much less frequent. For example, the modelling by Campbell *et al.* (2012) shows that fuel reduction by thinning and burning would be ineffective in reducing greenhouse gas emissions from ponderosa pine forests, because of the low ratio of avoided wildfire to prescribed fire. However, such interventions may be very effective if they can avoid biome switching of high-biomass forests to low-biomass, non-forest states, they cannot be applied to all fire dependent forests (Figure 10.8).

Reducing fire activity in fire sensitive tropical rain-forests is a central tenet of many carbon sequestration schemes such as the Reduced Deforestation and Degradation scheme (REDD+) (Figure 10.9). Tropical forest deforestation is an important contributor to greenhouse gas pollution, contributing an estimated 20% of the CO_2 released since the industrial revolution. Tropical forests are being cleared and burnt to create agricultural land (Figure 10.9).

Logging tropical forests changes the microclimate, making them more vulnerable to fire during dry seasons. Reducing the clearing and burning of tropical forests can have substantial benefits, not only by reducing carbon emission but also by conserving biodiversity and limiting air pollution. Severe air pollution from the destruction of tropical rain forests

is contributing to the premature deaths of around 400 000 people per year (Figure 10.10).

It is important to acknowledge that carbon sequestration schemes have the potential for perverse outcomes. For example, suppressing fire in flammable environments to increase carbon stores is an unsustainable strategy that could result in greater emissions of carbon than if regular burning were applied (Figure 10.11). Furthermore, converting savannas to forests, while increasing carbon storage and reducing fire activity, may have negative effects on biodiversity and could actually contribute to regional warming by increasing albedo.

10.7 Fire regime switches: a major challenge for fire ecology

Fire regime switches occur when there are major changes in the composition of vegetation which, in turn, triggers changes in land cover (biome type). Such changes in vegetation type and fire activity have cascading consequences for land use, biodiversity, delivery of ecosystem services and, where the changes occur over large scales, Earth-atmosphere feedbacks.

Understanding, managing and predicting future fire regime switches is a major challenge for fire science in general, and fire ecology in particular. A climate-centred view of the distribution of vegetation is that dominant growth forms and the biomes they represent will change their distribution as their preferred climate changes. There is no doubt that climate is a key driver of the kinds of plants and the kinds of biomes that occur in a given region but, as we have seen in earlier chapters, the same climate can support strikingly different biomes, depending on whether they promote or exclude fire.

The implication is that fire activity is not simply a consequence of climate selecting particular growth forms, which then generate fire regimes; unlike other disturbances, such as cyclones, floods and other extreme meteorological phenomena, fire is both a product and a cause of the vegetation. The important point is that changes in vegetation can alter the probability of fires occurring and, in so doing, mould the ecological landscape. Below, we provide some examples of biome switches.

In contrast to human caused deforestation, many tropical areas are experiencing forest expansion at the expense of grasslands and savannas. Over the last century, a general trend of increasing woody plants

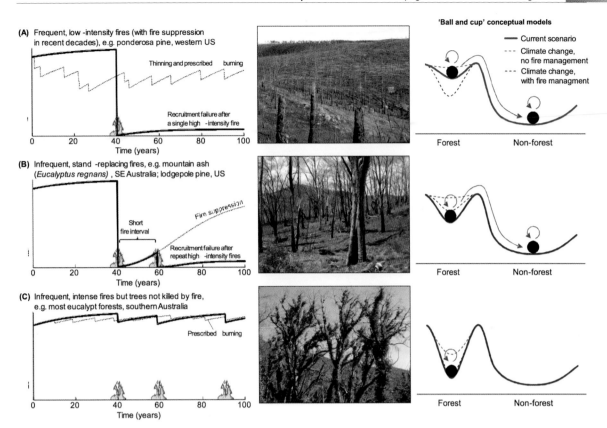

Figure 10.8 Contrasting responses of forest biomass to wildfire and climate change, and possible effects of alternative fire management scenarios. Possible management response to *Pinus ponderosa* forest adapted to frequent, low-severity fire, where historical fire suppression has increased the density of small trees and the risk of stand-replacing fire (A). Because seeds do not survive the fires, these forests generally have limited regenerative capacity after stand-replacing fires. Climate change is thought to increase the risk of such extreme fires, and potentially thinning and prescribed burning can decrease the risk of stand-replacement, preventing long-term shifts to low-biomass states following regeneration failure. This system vulnerability is depicted in the cup-and-ball diagram that shows climate change increases the likelihood of a switch to a non-forest state. Management intervention may reduce the risk of these changes. Possible response of forest adapted to infrequent, stand-replacing fire, such as *Eucalyptus regnans* (B). Although seeds survive high-intensity fires (e.g. stored in canopy-borne serotinous fruits), climate-driven reductions in intervals between stand-replacing fires can kill off immature regrowth, leading to subsequent regeneration failure. Under climate-change scenarios, the most appropriate management option for minimizing the risk of regeneration failure may be total fire suppression. For forests experiencing infrequent, high-severity fires, such as most eucalypt forests, the trees are highly resistant to fire because of their ability to resprout making such forests resilient to changes in fire frequency (C). Management to prevent fire-driven state shifts are probably not required (eg most eucalypt forests). (From Bowman *et al.*, 2013).

has been observed in many savannas worldwide. In some instances, there is not only a change in tree densities, but also in tree species composition and their functional attributes. Forest trees, unlike savanna trees, exclude grasses by shading. Since grasses fuel savanna fires, the loss of grasses from the system causes a biome switch by changing ecosystem function.

The causes of the phenomenon are disputed, involving changes in land use such as heavy grazing and/or reduced fire activity, but also global drivers, of which increasing CO_2 is probably a significant factor. The importance of global change drivers implies that the kind of fire management which was effective in controlling trees in the past will no longer be so, either today or in the future. This is the experience of many rangeland managers, where standard management methods have failed to stop the process.

The consequences of scrub forest invasion are loss of grazing capacity, loss of endemic plants and animals in conservation areas, and reduced economic value of

Figure 10.9 A deforestation fire in Mato Grosso, Brazil, the southern Amazon, 2006. The destruction of tropical rainforest is a biodiversity disaster and a major contributor to the greenhouse effect (photo: Guido van der Werf).

the land. Mechanical and chemical methods of reducing woody cover are expensive and impractical at the scale at which the phenomenon is occurring. However, in some settings, radical new approaches are being explored to disrupt biome switches. The setting of 'firestorms' is being explored in Texas by ranchers and in some African savanna parks, to open up heavily bushed areas in an attempt to return open grassy habitats. Instead of the traditional practice of using prescribed fires under mild weather conditions, fires are burnt under extreme conditions to generate extreme fires, which penetrate dense bush regardless of the presence or absence of grasses as fuel. These new approaches to fire management to prevent biome switches have the additional benefit of training fire crews for a future in which increasingly severe fires are predicted as a result of climate change.

Other radical approaches to managing biomes switch include using targeted grazing and browsing to consume fuel. Controversial ideas include

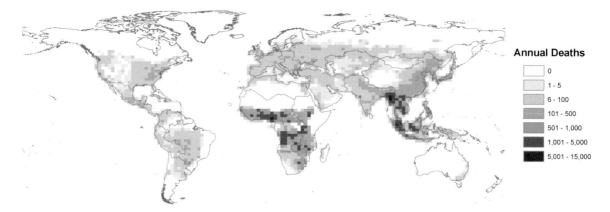

Figure 10.10 Global pattern of mortality caused by exposure to wildfire smoke. This map was created by combining multiple global datasets and following protocols developed by the World Health Organization. (From Johnston *et al.*, 2012).

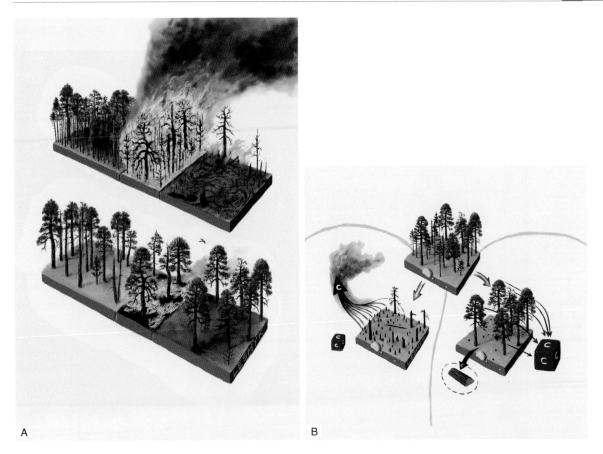

Figure 10.11 Contrasting scenarios of the effect of fire management on carbon storage. (A) High intensity crown fires that develop because of an overstocked *Pinus ponderosa* forest that has been long unburnt, or a surface fire in a thinned forests. (B) The effect of carbon storage is greater following the crown fire, as represented by the black cubes of carbon and the off-take of timber. (From Hurteau *et al.*, 2008). Reproduced with permission of Ecological Society of America.

introducing megaherbivores to ecosystems that lost them in the Pleistocene, in order to change fire regimes (Figure 10.12).

In the boreal zone, shrubs and trees are also expanding into tundra in vast areas of North America and Eurasia. The biome shift from tundra to boreal forest (taiga) is thought to be triggering significant changes in Arctic climates by altering albedo. Snow, which packs deeply on tundra vegetation, is a highly reflective surface, whereas emergent conifers have dark canopies which absorb irradiation.

Fire is a natural feature of boreal ecosystems occurring over a vast scale. Future global warming is predicted to nearly double the annual burnt area in Canada within the next 50 years. Increased fire activity might cause loss of forest cover and increased carbon

emissions to the atmosphere, but these will be counterbalanced, to some degree, by increased reflectance if changing fire activity reduces tree and shrub cover in tundra.

The vast scale of boreal vegetation has inhibited significant human intervention in the past but, given the global imperative for managing our planetary futures, prescribed fire for geo-engineering albedo and radiative forcing may become a plausible new land use objective in boreal regions in the future. Indeed, pioneering work on geo-engineering using prescribed fire to reduce greenhouse gas emissions is already being explored in another vast but thinly populated area, the Northern Australian savannas. A key concern in the boreal zone is, of course, to preserve the soil carbon stores; thus, any

Figure 10.12   Replacing extinct megafauna with non-native species is a potential management approach to control fire regimes disrupted by invasive grasses. These ideas are very confronting to the general public, even though they make scientific sense. (© News Limited). Reproduced with permission.

use of fire must not result in the combustion of organic soils.

10.8   Invasive plants and altered fire regimes

The spread of invasive plants is a major contributor to 'global change'. The 19th century biogeographers who first recognized the importance of climate in determining patterns of world vegetation saw biomes constructed from the indigenous pool of species.

However, over the last century, thousands of plant and animal species have been transported to new lands.

This vast, unplanned experiment is revealing unexpected insights into just how important the organisms, the species and their attributes are in constructing vegetation independently of the physical environment. Nowhere is the effect of the organism more apparent than in the impacts of non-native species on fire regimes. The classical interpretation has been that organisms are optimally suited to their environment through natural selection. Thus, large stem succulents dominate in warm deserts because of their ability to store water in arid environments, while specialized root systems are common in heathlands because they can access nutrients in low nutrient soils.

However, the invasion of succulent deserts and nutrient-poor heathlands by novel growth forms, including highly flammable grasses, shows that these functional traits are not the only 'solution' to the physiological problems posed by their distinctive environments. The magnificent succulent communities of the Sonoran desert are being threatened by invasive grasses, particularly buffel grass (*Pennisetum* (= *Cenchrus*) *ciliaris*), which competes with native species and carries fire into semi-arid systems where they did not previously occur (Figure 10.13).

Buffel grass is also transforming the unique spinifex tussock grasslands (*Triodia* spp.) of central Australia. Spinifex grasses are C4 grasses, which have evolved a shrub-like habit and evergreen leaves. They are thought to be adapted to the nutrient-poor soils and erratic rainfall of the region. Spinifex communities burn at multi-decadal intervals and contain plants with fire life histories characteristic of crown fire regimes. In both the Sonoran desert and central Australian arid lands, the familiar hypothesis that the dominant growth forms are those optimally suited to the soils and climate can be rejected. The disastrous spread of a fire-promoting grass shows us that completely different growth forms could dominate the system, and that the dominant growth forms in the native ecosystems are optimal not for the environment, but only for the species pool in which they occur.

These and other examples of grass invasion, cheat grass (*Bromus tectorum*) in sagebrush country, *Ehrharta calycina* in Western Australian heathlands, Gamba grass (*Andropogon gayanus*) in Australian savannas (Figure 10.14A), are ominous signs of how the addition of new growth forms can trigger major biome shifts through their ability to alter fire regimes (see Chapter 1, Part One and Balch *et al.*, 2013) (Figure 10.14B).

These examples of invasives altering biomes by altering fire regimes also expose the lack of a sound conceptual framework for predicting fire regimes from first principles. The impacts of climate change on future biome distributions are increasingly likely to be influenced by the presence of novel, non-native growth forms. Where these novel growth forms also alter fire regimes, plenty of surprises can be anticipated. Eucalypts, the ubiquitous Australian tree clade, are remarkably fire-promoting and fire-adapted. There seem to be no analogues to this plant functional type anywhere else in the world. Eucalypts have been transported around the world as forestry plantation

Figure 10.13 Invasive grasses spread between the Suguaro cacti in the Sonoran desert, near Tucson, Arizona, USA. These grasses spread fire between the cacti, thus threatening the whole ecosystem. Note the wildland-urban interface in the background, which is a source of ignitions into this system (Courtesy Aaryn Olsson, University of Arizona).

Figure 10.14   Fire in Gamba grass that has invaded northern Australian savannas, following its introduction as a pasture grass for cattle (A) (photo http://worldblog.eu/2012/02/06/page/3/). Because of its height and biomass, fuel load is up to eight times greater than that of native grasses, meaning it burns with greater intensity that the comparatively fire-tolerant vegetation can tolerate. Such intense fires are driving a collapse of the eucalypt-dominated ecosystem. Grass fire-cycles are transforming ecosystem worldwide, such as the destruction of this pine forest near Flagstaff, Arizona (B), USA (photo: Grant Martin, Cronkite News Service). Reproduced with permission of Cronkite News.

trees and ornamentals. Will they be a disaster for native ecosystems in the future, just as introduced grasses have been in the present?

The current spread of non-native species provides abundant empirical evidence that the pool of potential growth forms in an area is a major contributor to the kinds of fire regimes and biomes that result. Observations of the effects of non-native species on fire regimes have also helped in interpreting evolutionary changes in the geological past. The evolution of novel growth forms can now be seen to have triggered major ecological revolutions by promoting novel fire regimes in the case of ferns, Cretaceous angiosperms (e.g. Bond and Scott, 2010) and C4 grasses, or by suppressing fire activity in the case of broadleaved angiosperm trees forming closed tropical and temperate forests (see Chapter 4, Part One).

10.9 Conclusion

Fire management is undergoing rapid changes. The old paradigm of attempting suppression of all forest fires has ended as it is now recognized as being neither economically nor ecologically sustainable. Yet, there remains great uncertainty over how to achieve sustainable fire management, especially during a period of extremely rapid global environmental changes. This demands a fundamental rethink of the aims and limitations of fire management.

This quest for sustainable fire management demands holistic thinking and an awareness that a solution to one problem may create other problems. Thus, solutions will necessarily vary according to environment, and will involve more trial and error. It is important to recognize that ecological and economic trade-offs are inevitable, as it is most unlikely that fire interventions can simultaneously benefit biodiversity, maximize ecosystem services and reduce occurrence of economically destructive fires.

A vital step requires learning important lessons from indigenous cultures that have coexisted successfully with flammable and fire sensitive vegetation for millennia. This demands careful historical and palaeoecological research to chronicle how fire regimes have evolved over time. There is no doubt that this research will reveal substantial ecological effects of indigenous fire management, such as the extinction of the megafauna.

The reassortment of biotas, particularly the liberal spreading of highly flammable herbaceous and woody plants, and drastic changes to both land cover and climate associated with rapidly growing human populations, means that fire managers cannot look to the past and present for solutions to emerging problems. They must be imaginative and apply the logic of evolutionary biology to help discern an uncertain future. In developing future fire management approaches, novel and challenging ideas need to be carefully considered, such

as the role of megaherbivores in controlling fuel loads, the breaking of out-of-control grass fire cycles or geo-engineering landscape carbon stocks and albedo. New management interventions must be carefully evaluated and monitored, not only against the stated aims but also on the impacts of ecosystem services, biodiversity and human security.

Satellite surveillance has revolutionized our understanding of the geographic patterns of fire activity, but on-ground measurement of the ecological effects of different fire regimes lags far behind. To manage fire meaningfully, we need much better understanding of, and the capacity to measure, the variability of landscape fire and how this influences biodiversity and ecosystem services, especially carbon greenhouse gas emission and carbon storage.

We are conducting a truly grand global fire ecology experiment, replete with feedbacks, tipping points and inevitable surprises. Understanding what is happening is a prerequisite for a sustainable relationship with fire – a strange phenomenon that is part biology, part physics and part chemistry, and which is both our ally and our competitor.

Further reading

Campbell, J.L., Harmon, M.E., Mitchell, S.R. (2012). Can fuel-reduction treatments really increase forest carbon storage in the western US by reducing future fire emissions? *Frontiers in Ecology and the Environment* **10**, 83–90.

Flannery, T. (1994). *The Future Eaters*. Reed Books, Chatswood, NSW, Australia.

Gammage, B. (2011). *The biggest estate on Earth: How Aborigines made Australia*. Allen and Unwin, Melbourne, Australia.

Gill, J.L., Williams, J.W., Jackson, S.T., Lininger, K.B., Robinson, G.S. (2009). Pleistocene megafaunal collapse, novel plant communities, and enhanced fire regimes in North America. *Science* **326**, 1100–1103.

Miller, G.H., Fogel, M.L., Magee, J.W., Gagan, M.K., Clarke, S.J., Johnson, B.J. (2005). Ecosystem collapse in Pleistocene Australia and a human role in megafaunal extinction. *Science* **309**, 287–290.

Mooney, S.D., Harrison, S.P., Bartlein, P.J., Daniau, A.-L., Stevenson, J., Brownlie, K.C., Buckman, S., Cupper, M., Luly, J., Black, M., Colhoun, E., D'Costa, D., Dodson, J., Haberle, S., Hope, G.S., Kershaw, P., Kenyon, C., McKenzie, M., Williams, N. (2012). Late Quaternary fire regimes of Australasia. *Quaternary Science Reviews* **30**, 28–46.

Moritz, M.A., Parisien, M-A., Batllori, E., Krawchuk, M.A., Van Dorn, J., Ganz, D.J., Hayhoe, K. (2012). Climate change and disruptions to global fire activity. *Ecosphere* **3** A49, 1–22.

Perry, G.L.W., Wilmshurst, J.M., McGlone, M.S., Napier, A. (2012). Reconstructing spatial vulnerability to forest loss by fire in pre-historic New Zealand. *Global Ecology and Biogeography* **18**, 1609–1621.

Westerling, A.L., Hidalgo, H.G., Cayan, D.R., Swetnam, T.W. (2006). Warming and earlier spring increase western U.S. forest wildfire activity. *Science* **313**, 940–943.

Westerling, A.L., Turner, M.G., Smithwick, E.A.H., Romme, W.H., Ryan, M.G. (2011). Continued warming could transform Greater Yellowstone fire regimes by mid-21st century. *Proceedings of the National Academy of Sciences of the United States of America* **108**, 13165–13170.

References for part two

Ackerly, D.D. (2004). Adaptation, niche conservatism, and convergence: comparative studies of leaf evolution in the California chaparral. *The American Naturalist* **163**, 654–671.

Balch, J.K., Bradley, B.A., D'Antonio, C.M. & Gomez-Dans, J. (2013). Introduced annual grass increases regional fire activity across the arid western USA (1980–2009). *Global Change Biology* **19**, 173–183.

Barlow, J., Peres, C.A. (2004). Ecological responses to El Niño-induced surface fires in central Brazilian Amazonia: management implications for flammable tropical forests. *Philosophical Transactions of the Royal Society of London. Series B: Biological Sciences* **359**, 367–380.

Beerling, D.J., Osborne, C.P. (2006). The origin of the savanna biome. *Global Change Biology* **12**, 2023–2031.

Belcher, C.M., Collinson, M.E., Scott, A.C. (2013). A 450 million year record of fire. In: Belcher, C.M. (ed). *Fire phenomena and the Earth System: An interdisciplinary guide to fire science.* 1st edition, pp. 229–249. J. Wiley & Sons, Ltd, Chichester.

Belcher, C.M., Mander, L., Rein, G., Jervis, F.X., Haworth, M., Hesselbo, S.P., Glasspool, I.J., McElwain, J.C. (2010). Increased fire activity at the Triassic/Jurassic boundary in Greenland due to climate-driven floral change. *Nature Geoscience* **3**, 426–429.

Bell, T. L., Ojeda, F. (1999). Underground starch storage in *Erica* species of the Cape Floristic Region – differences between seeders and resprouters. *New Phytologist* **144**, 143–152.

Bond, W.J., Scott, A.C. (2010). Fire and the spread of flowering plants in the Cretaceous. *New Phytologist* **188**, 1137–1150.

Bond, W.J., Midgley, J.J. (1995). Kill thy neighbour: an individualistic argument for the evolution of flammability. *Oikos* **73**, 79–85.

Bond, W.J., Midgley, J.J. (2001). Ecology of sprouting in woody plants: the persistence niche. *Trends in Ecology & Evolution* **16**, 45–51.

Bond, W.J., van Wilgin, B.W. (1996). *Fire and Plants.* London, Chapman & Hall.

Bond, W.J., Woodward F.I., Midgley G.F. (2005). The global distribution of ecosystems in a world without fire. *New Phytologist* **165**, 525–538.

Bowman, D.M.J.S. (1992). Monsoon forests in north-western Australia. II. Forest-savanna transitions. *Australian Journal of Botany* **40**, 89–102.

Bowman, D.M.J.S., Balch, J.K., Artaxo, P., Bond, W.J., Carlson, J.M., Cochrane, M.A., D'Antonio, C.M., DeFries, R.S., Doyle, J.C., Harrison, S.P., Johnston, F.H., Keeley, J. E., Krawchuk, M.A., Kull, C.A., Marston, J.B., Moritz, M. A., Prentice, I.C., Roos, C.I., Scott, A.C., Swetnam, T.W., van der Werf, G.R., Pyne, S.J. (2009). Fire in the Earth System. *Science* **324**, 481–484.

Bowman, D.J.M.S., Balch, J., Artaxo, P., Bond, W.J., Cochrane, M.A., D'Antonio, C.M., DeFries, R., Johnston, F.H. Keeley, J.E., Krawchuk, M.A., Kull, C.A., Mack, M., Moritz, M.A., Pyne, S.J., Roos, C.I., Scott, A.C., Sodhi, N. S., Swetnam, T.W. (2011). The human dimension of fire regimes on Earth. *Journal of Biogeography* **38**, 2223–2236.

Bowman, D.M.J.S., Murphy, B.P., Boer, M.M., Bradstock, R. A., Cary, G.J., Cochrane, M.A., Fensham, R.J., Krawchuk, M.A., Price, O.F., Williams, R.J. (2013). Forest fire management, climate change and the risk of catastrophic carbon losses. *Frontiers in Ecology and Evolution* **11**, 66–67.

Bradstock, R.A. (2010). A biogeographic model of fire regimes in Australia: current and future implications. *Global Ecology and Biogeography* **19**, 145–158.

Brockett, B.H., Biggs, H.C., Van Wilgen, B.W. (2001). A patch mosaic burning system for conservation areas in

Fire on Earth: An Introduction, First Edition. Andrew C. Scott, David M.J.S. Bowman, William J. Bond, Stephen J. Pyne and Martin E. Alexander.
© 2014 John Wiley & Sons, Ltd. Published 2014 by John Wiley & Sons, Ltd.

southern African savannas. *International Journal of Wildland Fire* **10**, 169–183.

Burrows, G.E., Hornby, S.K., Waters, D.A., Bellairs, S.M., Prior, L.D., Bowman, D.M.J.S. (2010). A wide diversity of epicormic structures is present in Myrtaceae species in the northern Australian savanna biome – implications for adaptation to fire. *Australian Journal of Botany* **58**, 493–507.

Butler, D.W., Fensham. R.J., Murphy, B.P., Haberle, S.G. Bury, S.J., Bowman, D.M.J.S., (in press). Aborigines managed forest, savanna and grassland – biome switching in montane eastern Australia. *Journal of Biogeography.*

Campbell, J.L., Harmon, M.E., Mitchell, S.R. (2012). Can fuel-reduction treatments really increase forest carbon storage in the western US by reducing future fire emissions? *Frontiers in Ecology and the Environment* **10**, 83–90.

Carvalho, K.S., Alencar, A., Balch, J., Moutinho, P. (2012). Leafcutter ant nests inhibit low-intensity fire spread in the understory of transitional forests at the Amazon's forest-savanna boundary. *Psyche: A Journal of Entomology* **2012**, Article ID 780713.

Certini, G. (2005). Effects of fire on properties of forest soils: a review. *Oecologia* **143**, 1–10.

Chambers, D.P., Attiwill, P.M. (1994). The ash-bed effect in *Eucalyptus regnans* forest – chemical, physical and microbiological changes in soil after heating or partial sterilization. *Australian Journal of Botany* **42**, 739–749.

Chapin, F.S., Randerson, J.T., McGuire, A.D., Foley, J.A., Field, C.B. (2008). Changing feedbacks in the climate-biosphere system. *Frontiers in Ecology and the Environment* **6**, 313–320.

Coetsee, C., Bond, W.J., February, E.C. (2010). Frequent fire affects soil nitrogen and carbon in an African savanna by changing woody cover. *Oecologia* **162**, 1027–1034.

Cook, G.D. (1994) The fate of nutrients during fires in a tropical savanna. *Australian Journal of Ecology* **19**, 359–365.

Crisp, M.D., Burrows, G.E., Cook, L.G., Thornhill, A.H., & Bowman, D.M. (2011). Flammable biomes dominated by eucalypts originated at the Cretaceous-Palaeogene boundary. *Nature Communications* **2**, 193.

Doerr, S.H., Shakesby R.H. (2013). Fire and the Land Surface. In: Belcher, C.M. (ed). *Fire phenomena and the Earth System: An interdisciplinary guide to fire science.* 1st edition, pp 135–155. J. Wiley & Sons, Ltd, Chichester.

Doerr, S.H., Shakesby, R.A., Blake, W.H., Chafer, C.J., Humphreys, G.S., Wallbrink, P.J. (2006). Effects of differing wildfire severities on soil wettability and implications for hydrological response. *Journal of Hydrology* **319**, 295–311.

Douhovnikoff, V., Cheng, A.M., Dodd, R.S. (2004). Incidence, size and spatial structure of clones in second-growth stands of coast redwood, *Sequoia sempervirens* (Cupressaceae). *American Journal of Botany* **91**, 1140–1146.

Dubinin, M., Luschekina, A., Radeloff, V.C. (2011). Climate, livestock, and vegetation: what drives fire increase in the arid ecosystems of Southern Russia? *Ecosystems* **14**, 547–562.

Flannery, T. (1994). *The Future Eaters.* Reed Books, Chatswood, NSW, Australia.

Food and Agriculture Organization (2001). *Global Ecological Zoning for the Global Forest Resources Assessment 2000 Final Report.* Food and Agriculture Organization of the United Nations, Rome. Available at: http://www.fao.org/geonetwork/srv/en/metadata.show?id=1255.

Fox, B.J. (1990). Changes in the structure of mammal communities over successional time scales. *Oikos* **59**, 321–329.

Gammage, B. (2011). *The Biggest Estate in the World: How Aborigines Made Australia.* Allen and Unwin, Melbourne, Australia.

Gill, J.L., Williams, J.W., Jackson, S.T., Lininger, K.B., Robinson, G.S. (2009). Pleistocene megafaunal collapse, novel plant communities, and enhanced fire regimes in North America. *Science* **326**, 1100–1103.

Glasspool, I.J., Scott, A.C. (2013). Identifying past fire events. In: Belcher, C.M. (ed). *Fire phenomena and the Earth System: An interdisciplinary guide to fire science.* 1st edition, pp. 179–205. J. Wiley & Sons, Ltd, Chichester, UK.

Goetz, S.J., Mack, M.C., Gurney, K.R., Randerson, J.T., Houghton, R.A. (2007). Ecosystem responses to recent climate change and fire disturbance at northern high latitudes: observations and model results contrasting northern Eurasia and North America. *Environmental Research Letters* **2**. doi: 10.1088/1748-9326/2/4/045031.

He, T., Lamont, B.B., Downes, K.S. (2011). *Banksia* born to burn. *New Phytologist* **191**, 184–196.

He, T., Pausas, J.G., Belcher, C.M., Schwilk, D.W., Lamont, B.B. (2012). Fire-adapted traits of *Pinus* arose in the fiery Cretaceous. *New Phytologist* **194**, 751–759.

Hoffmann, W.A., Geiger, E.L., Gotsch, S.G., Rossatto, D.R., Silva, L.C.R., Lau, O.L., Haridasan, M., Franco, A.C. (2012). Ecological thresholds at the savanna-forest boundary: how plant traits, resources and fire govern the distribution of tropical biomes. *Ecology Letters* **15**, 759–768.

Holdridge, L.R. (1967). *Life zone ecology.* Tropical Science Center, San Jose, Costa Rica.

Hurteau, M.D., Koch, G.W., Hungate, B.A. (2008). Carbon protection and fire risk reduction: toward a full accounting of forest carbon offsets. *Frontiers in Ecology and the Environment* **6**, 493–498.

Johnston, F. H. Henderson, S.B., Chen, Y., Randerson, J.T., Marlier, M., DeFries, R.S., Kinney, P., Bowman, D.M.S., Brauer, M. (2012). Estimated global mortality attributable to smoke from landscape fires. *Environmental Health Perspectives* **120**, 695–701.

Kauffman, J.B., Ward, D.E., Cummings, D.L. (1994). Relationships of fire, biomass and nutrient dynamics along a vegetation gradient in the Brazilian Cerrado. *Journal of Ecology* **82**, 519–531.

Keeley, J.E. (2012). Ecology and evolution of pine life histories. *Annals of Forest Science* **69**, 445–453.

Keeley, J.E., Bond, W.J., Bradstock, R.A., Pausas, J.G., Rundel, P.W. (2012). *Fire in Mediterranean Climate Ecosystems: Ecology, evolution and management*. Cambridge University Press, Cambridge, UK.

Keeley, J.E., Fotheringham, C.J. (2000). Role of fire in regeneration from seed. In: Fenner, M. (ed). *Seeds: The Ecology of Regeneration in Plant Communities*. 2nd ed, pp. 311–330. CAB International.

Keeley, J.E., Pausas, J.G., Rundel, P.W., Bond, W.J., Bradstock, R.A. (2011). Fire as an evolutionary pressure shaping plant traits. *Trends in Plant Science* **16**, 406–411.

Keeley, J.E., Zedler, P.H. (1998). Evolution of life histories in *Pinus*. In: Richardson, D.M. (ed). *Ecology and Biogeography of Pinus*, pp 219– 250. Cambridge University Press, Cambridge, UK.

Kellman, M., Miyanishi, K., Hiebert, P. (1985). Nutrient retention by savanna ecosystems: II. Retention after fire. *Journal of Ecology* **73**, 953–962.

Lamont, B.B., Enright, N.J., He, T. (2011). Fitness and evolution of resprouters in relation to fire. *Plant Ecology* **212**, 1945–1957.

Lamont, B.B., Wiens, D. (2003). Are seed set and speciation rates always low among species that resprout after fire, and why? *Evolutionary Ecology* **17**, 277–292.

Lehmann, J., da Silva, J.P., Steiner, C., Nehls, T., Zech, W., Glaser, B. (2003). Nutrient availability and leaching in an archaeological Anthrosol and a Ferralsol of the Central Amazon basin: fertilizer, manure and charcoal amendments. *Plant and Soil* **249**, 343–357.

Marlon, J.R., Bartlein, P.J., Walsh, M.K., Harrison, S.P., Brown, K.J., Edwards, M.E., Higuera, P.E., Power, M.J., Anderson, R.S., Briles, C., Brunelle, A., Carcaillet, C., Daniels, M., Hu, F.S., Lavoie, M., Long, C., Minckley, T., Richard, P.J.H., Scott, A.C. Shafer, D.S., Tinner, W., Umbanhowar, C.E., Whitlock, C. (2009). Wildfire responses to abrupt climate change in North America. *Proceedings of the National Academy of Sciences, USA* **106**, 2519–2524.

McIntosh, P.D., Laffan, M.D., Hewitt, A.E. (2005). The role of fire and nutrient loss in the genesis of the forest soils of Tasmania and southern New Zealand. *Forest Ecology and Management* **220**, 185–215.

McWethy, D.B., Whitlock, C., Wilmshurst, J.M., McGlone, M.S., Li, X. (2009). Rapid deforestation of South Island, New Zealand, by early Polynesian fires. *The Holocene* **19**, 883–897.

Meyer, G.A., Wells, S.G., Jull, A.J.T. (1995). Fire and alluvial chronology in Yellowstone National Park – climatic and intrinsic controls on Holocene geomorphic processes. *Geological Society of America Bulletin* **107**, 1211–1230.

Midgley, J. J., Lawes, M.J., Chamaillé-Jammes, S. (2010). Savanna woody plant dynamics: the role of fire and herbivory, separately and synergistically. *Australian Journal of Botany* **58**, 1–11.

Miller, G.H., Fogel, M.L., Magee, J.W., Gagan, M.K., Clarke, S.J., Johnson, B.J. (2005). Ecosystem collapse in pleistocene Australia and a human role in megafaunal extinction. *Science* **309**, 287–290.

Moody, J.A., Martin, D.A. (2009). Forest fire effects on geomorphic processes. In: Cerdá, A., Robichaud, P. (Eds.), *Fire Effects on Soils and Restoration Strategies*, pp. 41–79. Science Publishers, Inc, Enfield, NH.

Mooney, S.D., Harrison, S.P., Bartlein, P.J., Daniau, A.-L., Stevenson, J., Brownlie, K.C., Buckman, S., Cupper, M., Luly, J., Black, M., Colhoun, E., D'Costa, D., Dodson, J., Haberle, S., Hope, G.S., Kershaw, P., Kenyon, C., McKenzie, M., Williams, N. (2011). Late Quaternary fire regimes of Australasia. *Quaternary Science Reviews* **30**, 28–46.

Moritz, M.A., Parisien, M-A., Batllori, E., Krawchuk, M.A., Van Dorn, J., Ganz, D.J., Hayhoe, K. (2012). Climate change and disruptions to global fire activity. *Ecosphere* **3** A49, 1–22.

Murphy, B.P. and Bowman, D.M.J.S. (2012). What controls the distribution of tropical forest and savanna? *Ecology Letters* **15**, 748–758.

Murphy, B.P., Williamson, G.J., Bowman, D.M.J.S. (2011). Fire regimes: moving from a fuzzy concept to geographic entity. *New Phytologist* **192**, 316–318.

Odion, D.C., Moritz, M.A., DellaSala, D.A. (2010). Alternative community states maintained by fire in the Klamath Mountains, USA. *Journal of Ecology* **98**, 96–105.

Parise, M., Cannon, S.H. (2012). Wildfire impacts on the processes that generate debris flows in burned watersheds. *Natural Hazards* **61**, 217–227.

Pate, J.S., Froend, R.H., Bowen, B.J., Hansen, A., Kuo, J. (1990). Seedling growth and storage characteristics of seeder and resprouter species of Mediterranean-type ecosystems of SW Australia. *Annals of Botany* **65**, 585–601.

Perry, G.L.W., Wilmshurst, J.M., McGlone, M.S., Napier, A. (2012). Reconstructing spatial vulnerability to forest loss by fire in pre-historic New Zealand. *Global Ecology and Biogeography* **18**, 1609–1621.

Power, M.J. (2013). A 21,000-year history of fire. In: Belcher, C.M. (ed). *Fire phenomena and the Earth System: An interdisciplinary guide to fire science*. 1st edition, pp. 207–227. J. Wiley & Sons, Ltd, Chichester, UK.

Power, M.J., Marlon, J., Ortiz, N., *et al.* (2008). Changes in fire regimes since the Last Glacial Maximum: an assessment based on a global synthesis and analysis of charcoal data. *Climate Dynamics* **30**, 887–907.

Pratt, R.B., Jacobsen, A.L., Hernandez, J., Ewers, F.W., North, G.B., Davis, S.D. (2012). Allocation tradeoffs among chaparral shrub seedlings with different life history types (Rhamnaceae). *American Journal of Botany* **99**, 1464–1476.

Raphael, M.G., Morrison, M.L., Yoder-Williams, M.P. (1987). Breeding bird populations during twenty-five years of postfire succession in the Sierra Nevada. *Condor* **89**, 614–626.

Rein, G. (2013). Smouldering fires and natural fuels. In: Belcher, C.M. (ed.). *Fire phenomena and the Earth System: An interdisciplinary guide to fire science.* 1st edition, pp. 15–33. J. Wiley & Sons, Ltd, Chichester.

Roos, C.I., Swetnam, T.W. (2012). A 1414-year reconstruction of annual, multidecadal, and centennial variability in area burned for ponderosa pine forests of southern Colorado Plateau region, Southwest USA. *The Holocene* **22**, 281–290.

Ruddiman, W.F. (2003). The anthropogenic greenhouse era began thousands of years ago. *Climatic Change* **61**, 261–293.

Schutz, A.E.N., Bond, W.J., Cramer, M.D. (2009). Juggling carbon: allocation patterns of a dominant tree in a fire-prone savanna. *Oecologia* **160**, 235–246.

Schwilk, D.W., Ackerly, D.D. (2001). Flammability and serotiny as strategies: correlated evolution in pines. *Oikos* **94**, 326–336.

Shugart, H.H., Woodward, F.I. (2011). *Global Change and the Terrestrial Biosphere: Achievements and Challenges*, Wiley-Blackwell, Hoboken, NJ.

Simon, M.F., Grether, R., De Queiroz, L.P., Skema, C., Pennington, R.T., Hughes, C.E. (2009). Recent assembly of the Cerrado, a neotropical plant diversity hotspot, by *in situ* evolution of adaptations to fire. *Proceedings of the National Academy of Sciences* **106**, 20359–20364.

Smith, J.K. (2000). Wildland fire in ecosystems: effects of fire on fauna. General Technical Report RMRS-GTR-42-vol. 1. Ogden, UT: US Department of Agriculture, Forest Service, Rocky Mountain Research Station.

Staver, A. C., Archibald, S., Levin, S. A. (2011). The global extent and determinants of savanna and forest as alternative biome states. *Science* **334**, 230–232.

Westerling, A.L., Hidalgo, H.G., Cayan, D.R., Swetnam, T.W. (2006). Warming and earlier spring increase western U.S. forest wildfire activity. *Science* **313**, 940–943.

Westerling, A.L., Turner, M.G., Smithwick, E.A.H., Romme, W.H., Ryan, M.G. (2011). Continued warming could transform Greater Yellowstone fire regimes by mid-21st century. *Proceedings of the National Academy of Sciences of the United States of America* **108**, 13165–13170.

Wood, S.W., Murphy, B.P., Bowman, D.M.J.S. (2011). Fire-scape ecology: how topography determines the contrasting distribution of fire and rain forest in the south-west of the Tasmanian Wilderness World Heritage Area. *Journal of Biogeography* **38**, 1807–1820.

Worthy, M., Wasson, R.J. (2004). Fire as an agent of geomorphic change in southeastern Australia: implications for water quality in the Australian Capital Territory. In: Roach, I.C. (ed). *Regolith 2004*, pp. 417–418. CRC LEME, Canberra, Australia (http://crcleme.org.au/Pubs/Monographs/regolith2004/Worthy%26Wasson.pdf).

PART THREE
Anthropogenic fire

Photo

Field burning in the Black Forest of Germany in the 19th century. The painting reminds us that humans are active agents of fire in the landscape, that this unique role applies to agricultural and cultural settings as much as to wildlands, and that much of agriculture relies on applied fire ecology. (Image from J.G. Goldammer, Global Fire Monitoring Centre).

Fire on Earth: An Introduction, First Edition. Andrew C. Scott, David M.J.S. Bowman, William J. Bond, Stephen J. Pyne and Martin E. Alexander.
© 2014 John Wiley & Sons, Ltd. Published 2014 by John Wiley & Sons, Ltd.

Preface to part three

Sometime during the Pleistocene, hominins acquired the capacity to manipulate fire. The living world could thus begin to control ignition – could compete with lightning – in ways to complement its control over oxygen and fuel. By the end of the Pleistocene, only one creature, *Homo sapiens*, possessed that capacity.

Part Three explores what this extraordinary power has meant for Earth and for humanity. It views fire not simply as a phenomenon of nature for which the natural sciences are the typical source of analysis, but as a human activity for which other disciplines are relevant, including history, anthropology, economics and even the humanities. Because people choose their behaviour, and choose on the basis of how they understand and value the world and organize societies, the natural world must align with cultural, political and moral worlds. The story remains based on evidence – but evidence and experiences that are not restricted to the purview of science.

We first review the ways humanity and fire have co-existed, if not co-evolved, and what patterns might be evident in that record. We then consider the acceleration of the human presence over the past 200 years, during what many observers have come to call the Anthropocene, an epoch triggered by a shift in humanity's combustion habits. We end by considering how today humanity attempts to manage fire in landscapes.

Chapter 11
Fire creature

Sometime during the Pleistocene, perhaps as much as two million years ago (almost certainly by half a million years ago), the grand narrative of earthly fire underwent a fundamental change – probably the greatest since the Devonian. The living world acquired the means to control ignition. An organism – a hominin, to put a name on the genus – learned how to maintain and use fire, and then to start it. Over the last 200 000 years, one species, *Homo sapiens*, transformed that know-how into a species monopoly. Life could now regulate all the ingredients for combustion. A unique fire planet acquired a unique fire species to inhabit it.

The effect was profoundly mutual. The possession of fire granted humanity a presence – literally a fire-power – far beyond its genetic assets. Equally, humanity remade the pyrogeography of the planet, reconfiguring regimes where fire already existed and taking fire to places it did not exist naturally. If humanity could never have come to dominate Earth without fire, it is also true that fire would never have assumed the informing role it has claimed without an agent to act on its behalf. In each other, humanity and fire found pyric doubles (Pyne, 2001).

11.1 Early hominins: spark of creation

Just when that evolutionary spark occurred is unknown, and likely unknowable. After all, no one invented fire: it was simply there. In fact, fire's ideal regimen fits almost exactly the putative hearth of hominins, the African savanna and the Rift Valley. Early hominins would have grown up amid fires as they did among antelopes and acacias. They would have scavenged among its burns and occasional kills. At some point, they would have picked up a brand or gathered embers as they might a bone or stone flakes.[1]

The early hominins made tools. The earliest, *Homo habilis*, is in fact named for its nominal ability to make implements that could extend its physical capabilities. However, such technologies were replicas and compensations for other meagre faculties. Hard-edged stone substituted for talons, teeth and claws. Axes added heft to limbs and power to muscle. Heavy choppers could smash what lighter mandibles could not. Throwing objects offset poor leaping. Such tools imitated anatomical features. They endowed hominins with abilities that, for other creatures, natural selection knapped out of genetic cores.

Yet fire represented a change in kind. It was not a material object, but a chemical reaction; to survive, it had to spread into fresh fuels or else, if contained at some place, had to be fed combustibles. Fire differed from other technologies in that it could be not merely picked up but perpetuated and then rekindled, and it is hard to conceive any other model for such activity than tending children.

Fire could only be scavenged from the landscape seasonally; otherwise, it had to be preserved, an eternal flame. This was complicated, not only because it

---

[1] This sub-section quotes or paraphrases a chapter in Pyne (2012).

Fire on Earth: An Introduction, First Edition. Andrew C. Scott, David M.J.S. Bowman, William J. Bond, Stephen J. Pyne and Martin E. Alexander.

required relentless foraging for fuel, but because it demanded a set of social relationships and because the coals would inevitably expire from time to time. The solution was to make fire on demand.

Fortunately, the simplest techniques involve casting sparks or briskly rubbing, eventually by drills, and these were technologies fundamental to other tool-making. Even *Homo erectus* must have witnessed sparks thrown from struck stones. If the sparks landed on tinder, then the tool-maker became a fire-maker. By the time of *Homo sapiens*, the ability to make fire at will was commonplace. Interestingly, however, most aboriginal societies recorded well into modern times – Australian Aborigines and Andaman Islanders, for example – preferred to carry fire or coals with them continually. Early on, it seems, people identified themselves as a fire-carrier.

They also identified fire with life. That intuition had reason behind it because fire, though not alive, is a creation of the living world and displays many of its outward properties. Like living creatures, it breathes and eats; it warms, it moves, it sounds; it must be tended; it must be bred and trained. It must be sheltered – given a *domus* – and may well be the prototype for domestication. When left or expired, it is buried.

11.1.1   *Cooking as pyrotechnology*

The paradigm for all the pyrotechnologies that followed is likely a usage so common as to be overlooked – cooking. Roasting, broiling, and sautéing foodstuffs led to cooking stone, sand, metal, liquids, wood, whatever might be converted into usable forms by controlled heating. A common flame transmuted them all.

That is literally true, for to cook is to begin the digestive process early. Heating adds value to raw biomass: it makes eating easier and more efficient and amplifies nutritional value. It converts lumps of hydrocarbons into physiological fuel. It denatures protein, gelatinizes starches, and otherwise renders foodstuffs more digestible. It remakes barely edible starches into higher-caloric carbohydrates. It detoxifies foods of many harmful chemicals and kills off worms, bacteria and other disease-bearers and parasites. It leverages a given harvest of biomass, perhaps marginal, into foodstuffs capable of sustaining an organism. More items can be eaten, with greater payoff. Cooked tubers have a higher caloric content than uncooked meat (Wrangham, 2009).

So radical a change in diet encouraged a restructuring of physiology and morphology. The outcome saw the most dramatic anatomical changes ever recorded among the hominins. With less need to break down biomass, mechanically and chemically, people have (compared to cognate primates) a down-sized mouth, stomach and guts. We developed smaller teeth and jaws, since fire has already begun preliminary mastication. Our stomachs and intestines were also reduced in size, since fire had started the biochemical breakdown of plants and meat. We no longer needed a massively muscular skull or a gargantuan digestive tract. Our head could become big and our gut small. We could process ideas rather than herbage. Such reforms cannot be deduced from environmental changes alone but, once they occurred, they became encoded in our genome. We became physiologically dependent on cooking.

In recent times, various food cultists have experimented with diets dedicated solely to raw fare. Since cooking, as Claude Levi-Strauss observed in *The Raw and the Cooked*, is identified with civilization, it is also bound with civilization's discontents, among them a legitimate wariness toward the excessive processing of foodstuffs that has rendered so much of modern food unhealthy. Yet Levi-Strauss' fundamental insight was correct: humans cannot survive on raw offerings alone, even when they have access to fresh foods all year round and add nutritional supplements. People without fire cease to be people, not just symbolically but physiologically. They cannot find sufficient energy to survive; they cannot reproduce. In eerie ways, the trials echo old myths about the origin of fire, which universally consider uncooked food as the most desperate of deprivations.

11.1.2   *Pyric paradigm, from hearth to habitat*

In thus fashion, the process by which pyrotechnology has been able to remake the world began with humanity first remaking itself. Control over fire changed human anatomy and physiology and became encoded in the evolving genome. Equally, fire altered human behaviour, since people now had to commit to its ceaseless tending, and around the fire people gathered to eat, socialize, instruct and tell stories. Not least, fire affected humanity's sense of itself. Other tools had cognates among fellow creatures, but only humanity had fire. Its possession was a Faustian bargain that self-consciously set humanity apart from the rest of creation.

The cooking fire became a pyrotechnic paradigm for nearly all fire's uses, as hearth morphed into furnace, forge and Bunsen burner. One could cook wood to harden it into spear points or convert it to charcoal; cook, broil, or roast stone to soften it for easier flaking, melt sand into glass, smelt ore into metals or harden clay into ceramics. With flame, one could fell and hollow out trees. With fire one could crack rock to tunnel mines. Whatever the task, fire could quicken and strengthen the core reactions. It was the *organum* of alchemy, the methodology of modern chemistry, and a broad-spectrum catalyst for practically everything people did. Through fire, they made Earth habitable, and through fire they have left the planet for other worlds.

Humanity's fires no more stayed by the hearth than humanity did. They foraged, they hunted, they wandered. Instead of bringing objects to the fire, early humans took fire to a wider world. They used flame not just as a physico-chemical reaction but as a biochemical one. They burnt landscapes. They began to cook the Earth itself.

Modern thought, like modern science, tends to be reductionist. It imagines the simple, then builds into the complex by adding new factors. The hominin experience with fire, however, was likely the reverse. They experienced an Earth that burnt routinely; fire was something in the ambient world, inextricably a part of that whole. They would have seen how fire worked on landscapes, because that was the world in which they hunted and foraged. They would have noted, as even casual tourists to African savannas can today, that wildlife gravitate to newly greening areas that succeed burns, since those grasses are the most succulent and protein-rich. They saw how nature had made foodstuffs edible by cooking, and how such sites became uninhabitable if long unburnt. An extraordinary pantry of game and plants clustered on the burnt patches; that was where you went to find food, and then you re-burnt – cooked – the fire-attracted grazer or tuber in more isolated circumstances to extract further nutrition. Early hominins were among those fire-drawn species.

Fire spreads, fire interacts, fire catalyzes – these make fire unique among ecological processes. To control it, even if only to determine where and when to kindle and then let natural circumstances dictate the outcome, is an extraordinary capacity. So,

although control was often tenuous – the nominally tame could go feral, and loosing fire could resemble training a grizzly bear to dance – its firepower allowed humans to shape their habitat in ways far beyond those available to other creatures. Until the arrival of *Homo*, no creature ever had the capacity to restructure landscapes wholesale. It is as though African wildebeests or North American bison had the power to determine when and where the greening patches on which they grazed would appear. Yet humanity's firepower granted it a biotic leveraging far beyond stone choppers or atlatls. It allowed them to engineer whole ecosystems. Increasingly, the relevant fire regime was the one that humans directly and indirectly established.

For most of humanity's existence, fire had a dual identity as both a feature of the natural world and an implement that humans fashioned. It was something in nature, independent of people, an ecological presence like rain and sunlight. At the same time, it was a technology that could help them make their world more habitable. But it was a peculiar, shape-shifting technology, very unlike implements made from stone and wood. In some respects, fire resembled a mechanical tool; a candle held flame as a handle did a claw head. In other respects, it more resembled a domesticated species; it was a tamed artefact of the living world, like a sheep dog or milk cow, residing within a cultivated landscape. In still other forms it more resembled a captured ecological process that could be used toward human ends, not unlike a working elephant or a cheetah trained to hunt. To most minds, it was simply fire – a universal companion to humanity's varied journeys, something ever with us and something on which we have always depended.

It was as a biological pyrotechnology that fire use expanded. It became the basis for fire hunting, fire foraging, fire fishing, fire farming, fire herding, even fire fishing (the lights would draw fish where they could be speared). In effect, the hominin diet was twice cooked, once in the field, and again in the hearth. Almost everything people did on the land, somewhere in its great chain of technic being, had fire. If, as Cicero observed, humans had made a second nature out of the first nature they inherited, they did so with fire. The eternal fire in humanity's temples had its analogue in the inextinguishable fires they distributed over the lands they worked.

And if humans became physiologically adapted to cooked food, so the landscapes they occupied became ecologically adapted to anthropogenic burning. Over the millennia, as anthropogenic fire trickled, probed and colonized more and more lands, anthropogenic fire became the new norm for fire on Earth. Culture after culture acknowledged in myth and ceremony that fire's possession is what segregated humanity from the rest of creation – that fire is the kindler and enabler of the technology behind culture.

The Earth, as a fire planet, had found its keystone species.

11.2 Aboriginal fire: control over ignition

Tending fire was a model – perhaps the exemplar – for domestication, and cooking was a paradigm for pyro-technologies generally. But this was only a point of cultural kindling, for fire was a universal catalyst, never far distant from whatever people did – and, by changing its settings, people in turn changed fire. Over time, wild fire assumed new identities. It was tamed, hunted, foraged, fished, cultivated, urbanized and machined. Almost everything people did they did with fire and fire with them. The gamut of fire practices is as diverse as the planet's combustible surfaces. Anthropogenic fire's expressions vary as much as people and places do.

11.2.1 Principles and patterns

Still, some principles exist and a few patterns do emerge. A first-order distinction depends on the degree of control that people exercise over burning, which is to say, their ability to start and spread flame. In the simplest expression, they control ignition. They can decide time, place, frequency and size. Fire regimes result from the interaction of torch (or fire-stick) and landscape. To give these practices a name, call them collectively *aboriginal fire*.

The ability to cast sparks, however, does not guarantee that a fire will start or that it will spread. Only a tiny fraction of lightning strikes, after all, actually kindle fires. In brief, humanity's firepower does not depend solely on its capacity to create a point of ignition, but equally on the capacity of the setting to receive and carry that spark. A first principle of aboriginal fire, then, is that it works best in places that

already burn or that have the ingredients for burning but lack a dependable ignition source – a deficiency that people can satisfy. So it is that a surprising number of fire-origin myths relate how fire, once liberated from some hoarding potentate, penetrates into its surroundings and has to be coaxed out again by people.

The power of landscape fire derives from its power to propagate – a capacity that lies not with the torch itself but with its context. Accordingly, societies that rely on fire to extract their means of living will gravitate to places that are intrinsically combustible. They will flourish where fire is possible and struggle where it is not. A map of aboriginal populations will be largely a map of fire-prone landscapes, with the exception of economies that depend on fishing or maritime resources. Fish, shellfish and marine mammals all thrive quite apart from fire. Even so, fishing has its fire component, as people typically carry it in boats and shine it into waters to attract fish, then cook or smoke the catch; and they supplement that haul with fire-catalyzed gardens or foraging. Without fire, the harvest would be smaller and inedible. But these are examples of fire as a direct cooking technology, not of indirectly cooking the landscape.

Even economies based on digging sticks, spears and torches can influence a fire environment. A legacy of past anthropogenic burning, of course, shapes the capacity for future burning, but people can leverage their firepower – ever an interactive technology – by hunting, foraging and otherwise altering their surroundings, which is, to say, fire's context. They promote those species they want and weed out those they do not. Ethnographic (particularly ethnobotanical) studies often show an astonishing number of species used by long-resident peoples. One interpretation is that they are clever and adaptable and have learned to co-inhabit landscapes with the natural world. Another is that, over time, they have found ways to promote species that serve useful purposes and remove those that do not. Such selective rearranging affects fire regimes (Balée, 1994).

The more obvious interaction, however, involves hunting. Grazers and browsers compete with fire for the small combustibles that serve as fodder for the one and fuel for the other. Animals, moreover, reorganize their habitats – 'niche construction' is the preferred term – in ways that can help or hinder burning. Large herbivores like elephants open up landscapes; they can help keep fire in a place that otherwise might close tree canopies shut and retard fire's spread. Depending on whether the landscape is naturally disposed to fire, there are two general outcomes possible as a result of hunting such creatures. In fire-limited settings, reducing the population of megafauna through hunting may lead to a lessened capacity for fire, since the firestick alone cannot beat back encroaching bush or closing canopy. In fire-prone settings, though, driving down the numbers of herbivores liberates more grass and shrubs for fire to feed on. In this way, a common practice – hunting – can yield opposite consequences for fire. The first example leaves a firestick-wielding creature more feeble; the second, more powerful.

Similarly, how people actually apply fire to the land, even when they burn with the same techniques, can express itself in dramatically different outcomes, depending on the landscape's disposition for burning. Identical fire practices can, in one setting, spread over a landscape, while in another they barely sustain themselves. As with so much of anthropogenic fire, common principles can lead to very different outcomes, depending on the particulars of the circumstances in which they occur.

11.2.2   Principles and practices

What are these shared principles? The simplest is to note that people seek to replace nature's fires with their own. This competition is both temporal and spatial. People burn outside the natural fire season and they inscribe a different pattern of burning on the land. A creature whose power depends on fire seeks, as Henry Lewis once put it, to substitute fires of choice for fires of chance.

That said, not all anthropogenic fires are deliberate, rational, or utilitarian. Fire follows people like vermin do; people leave fire as they do other debris; they can burn accidentally, wantonly and maliciously. A goodly fraction of anthropogenic burning may be characterized as fire littering. However, just as other untoward human behaviours – the distribution of accidents or crimes, for example – reveal patterns, then so do fires. They contribute toward that statistical composite of burning called a fire regime.

11.2.2.1   Patterns in time

The usual temporal arrangement is for burning to begin as soon as the land can take it. Parts may still be under water from the rainy season, or may be dappled with snow banks, but patches that can accept fire get burnt, and then the patches between them. Some patches may be shielded from fire altogether. They may be left unburnt or prepared to resist burning by later, hotter fires by having their peripheries early-burnt into buffers. As the season matures, the burning increases accordingly, quickening as the dry season prepares to segue into the wet (Figure 11.1).

The first dry thunderstorms will pack the greatest punch for natural fire-starting. By then, however, those areas that people most care about will have been protected from wildfire, either because they have greened up, recently burnt or are surrounded by an incombustible perimeter. Lightning fire will have to pick among the residual patches. Where anthropogenic fire is robust, it can overwhelm lightning fire so completely that it renders lightning trivial as a source. The prevailing fire regime becomes anthropogenic.

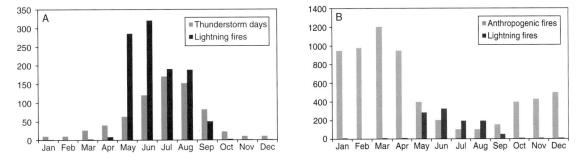

Figure 11.1 Two graphs for protected forests in Florida, USA. The left graph (A) shows the seasonal evolution of lightning-caused fires, which cluster at the start of the wet season. The right graph (B) puts those same starts within the larger context of anthropogenic fires, which are far more numerous and occur at a different season. (Data from Komarek, 1964).

From this common template, variations are endless. Where the landscape is rough, where the patches are broken and the scene dappled with niches, where society is lightly settled (or unsettled) or where anthropogenic fire is legally outlawed, lightning will boost its relative contribution and may dominate, since its pre-emptive competitors have been removed; such is certainly the case where parks or other reserves promote 'natural' processes. However, where prescribed burning is allowed, the old pattern reasserts itself: people burn early and often and lightning takes what people are unable, too indifferent towards or are unwilling to burn.

11.2.2.2 Patterns in space

The other competition is spatial and begins with the commonplace observation that anthropogenic fire goes where people go. It follows their routes of travel, it accompanies them to the sites where they hunt, forage, fish and trap and it resides where they do. Even where the fires are accidental or malicious, as with arson or warfare, they accompany people. To give this amorphous observation some structure, we might characterize the resulting patterns as lines of fire and fields of fire (Figure 11.2).

Lines of fire refers to corridors along which people routinely move. It is customary to keep such thoroughfares open for unimpeded transit and to make ambush difficult. Outside of deserts or sparse grasslands, this means burning. Historical accounts describe such corridor burning in Siberia, Canada's Northwest Territory, Tasmania, India's Central Provinces – almost anywhere that a rank biota will seal off openings if not forcibly beaten back by axe or flame. Typically, the burning occurs very early in the season, so the fires burn out from the pathway to wet understoreys along the flanks and expire.

As with much of anthropogenic fire, however, practice is rarely perfect and unintended consequences abound. There are few normal years; most are too wet or too dry, too rich in fuels or too lean. In a conifer forest, a line of fire may barely burn out the corridor or creep to the incombustible edge in 19 years out of 20. A droughty 20th year, however, may let those flames bolt wildly into the backcountry. Such is hardly a controlled burn or, strictly speaking, a utilitarian one, but it certainly qualifies as anthropogenic and it will shape the resulting fire regime as surely as the others.

Fields of fire are patches that people burn in order to accomplish some fire-catalyzed purpose. They may be places of encampment, hunting grounds or sites to

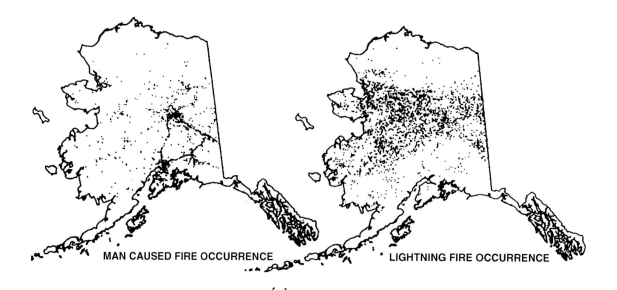

Figure 11.2 The two geographies of fire, natural and anthropogenic. Because of its political geography and boreal environment, Alaska shows the two regimes with particular starkness. Natural fire concentrates in the Yukon Valley. Anthropogenic fire follows routes of travel and clusters of settlement (From Gabriel and Tande, 1983).

harvest fruiting shrubs and trees, collect honey or to otherwise gather whatever game or edible plants hold interest.

Foraging by fire can prune berry patches and trim shrubs harvested for fresh twigs used for baskets. It can stimulate select tubers such as camas and geotrophs. It can assist the harvesting of piñon nuts, chestnuts and *mowia* flowers. Burning can clean out, for a while, nasty pests such as ticks and chiggers. It can create defensible space around combustible habitations such as wattle wickiups and thatched lean-tos; it can expose to view possible threats from tigers, snakes or enemy raiders. It can communicate through smoke or, by means of controlled smoke, alert local communities that a party of outsiders is passing through with peaceful intent. More malevolently, it can be a weapon of war.

By means of selective burning, habitats can become more hospitable to favoured fauna. A good fire can stall tendencies toward closed forests and, instead, refresh the sun-thirsty browse and grasses that game animals feed upon. In the short term, burning can effectively 'bait' the landscape in ways that draw animals to the sumptuous green pick poking through the ash (in fly- and mosquito-infested areas, smudge pots can serve the same purpose). Over the long term, repeated burning sustains a habitat against pressures to overgrow into scrub or closed forest and, thus, promote a higher population of hunted species than nature would otherwise allow. Elk, springbok, chital – none thrive in old-growth forest. Seasonal burning, with places blackened or greening up, can direct the seasonal movement of ungulates. Trappers in the boreal forest burn along strips that will become routes of travel for hunted mammals over the winter – ideal places in which to set snares.

Most spectacularly, fire hunting can take the form of active drives in which flames serve as beaters to force the animals through a gauntlet, or into places where they can be killed more easily. In this way, flames accompany spear and arrow; animals so hunted include whatever the primary grazer might be in places amenable to surface burns. In North America, deer were fire-herded into tidal peninsulas and into lakes, bison driven over cliffs, rabbits and moose surrounded, and even grasshoppers gathered by rings of flame and smoke. The genius of the system is, of course, that it promotes and renews the habitat that makes the valued game possible. Far from being an act of vandalism, the practice is a model of sustainability.

Some practices transcend the simple geometries of both lines and fields. One is ease of access. Tall grass or scrubby bush makes travel onerous and perhaps hazardous; it restricts the ability to see dangers; and it can threaten villages and encampments with wildfire. The burning is protective in the broadest sense. Such values merge with aesthetic ones; people generally prefer open vistas or mixed landscapes in which some woody covert mingles with far-seeing panoramas. And properly burnt landscapes – openly visible to inhabitants – are typically valued for their testimony to good husbandry.

'Cleaning up the country' is widely seen among aboriginal economies as an index of both ownership and good biotic citizenship. People lay claim to land by their burning and, by the skill and care with which they do the burning, they invite judgment about the character of their claim. For many pre-industrial societies, a land unburnt is a land unkept, as shameful as a house full of hoarded junk.

Their fire practices may be taken as an index of their stewardship generally. If people have burnt suitably, they have probably also managed the countryside well in their other interactions with it, since it is impossible to decouple fire from everything else that shapes fire's environment. Such sentiments express cultural norms; outsiders, or others with a different economy, will view the burnt land with eyes trained to judge by other standards, and may see burnt land as a mark of ignorance or vandalism. Both parties can agree, however, that a land burnt badly bears witness to a slovenly people.

11.2.3   *By the numbers*

Intuitively, it would seem that more people mean more fire, and that some direct formula should join numbers of people to numbers of fires. This isn't so, of course, any more than it holds that more lightning means more ignitions. Not only does the character of the countryside determine what sparks take, the source of anthropogenic ignition varies. People move, fire propagates; the most extensive burning will accompany groups (usually small) that trek seasonally across fire-prone landscapes. In brief, the relationship depends on how a society organizes its fire practices.

It depends, too, whether a people are colonizing a landscape or maintaining one. In colonizing, the populations of people and fire obey a rude curve in which both rise with increasing numbers, then fall as population thickens and fires become more and more restrained. The number of fires may continue to rise, but they are housed in hearths or appliances, not on landscapes. Burnt area declines as other pyrotechnologies substitute for, or redirect, free-burning flame. Open fire is left to migrate primarily to nature preserves or the frontier. In a roughly settled core, fire is tamed into cultivated fields or confined to hearths and furnaces, or absorbed into more sophisticated technologies, replaced by devices that disaggregate heat and light from flame.

However, this formula assumes a regular progression of fire evolution that is, in truth, not universal. It also assumes that, once a riotous frontier phase has passed, aboriginal fire burns with rational and predictable rhythms – which it does not. The landscape is always unsettled, societies are always in turmoil, and their interaction, while broadly regular, varies year by year. Some years are wet, some dry; some are lush with combustibles, some drained of them. Societies must interact with other groups, undergo outbreaks of disease and bouts of political instability, gain and lose experience, acquire and shed knowledge, relocate, go to war, change norms. Control over members is no more total than over flame on the land. The interaction of firestick and landscape is ever negotiated and tentative. In some years, fires will barely spread along lines of ignition. In others, they will bolt into the bush. The one constant is fire in ever-changing forms. In aboriginal economies, fire becomes a ceaseless presence, always ready to spread as conditions allow.

Yet, just as storms – their yearly appearances full of variability – can be organized into climates, so fires can assume patterns as regimes. Over hundreds or thousands of years, even unstable rhythms of burning can begin to operate within statistical borders. Anthropogenic fire and land will, like predator-prey relationships with all their fluctuations, display something like regularities. The longer the human presence, and the more complete the human control over the landscape, the more predictable are the patterns that emerge and the more disruptive any interruption of those regularities can be for the resident biota.

The firestick is a powerful tool, its use so extensive that it can emulate a kind of pyric horticulture, or what Rhys Jones felicitously termed 'firestick farming.' Suitably placed with the proper fulcrum, it can, like an Archimedean lever, move whole landscapes. However,

its power comes with sharp constraints; it may be within reach but beyond grasp. It cannot burn where the land will not let it. A million sparks will do nothing in snow or sandy desert, or on a rainforest floor immersed in deep shadow and stripped bare of organic matter. People can carry fire everywhere they go, but they cannot make the land they visit carry the fires they bring.

11.2.4   The firestick in action: selected examples

In the end, principles are abstractions; the reality is what fire etches on the land, and few of its manifestations are pure. They are alloys of mixed practices, legacy landscapes and social minglings. How could it be otherwise when fire integrates weather, terrain and vegetation, with all its evolutionary history, and people, with all their historical contingencies and cultural quirks? Aboriginal burning cannot exist apart from its setting, and it happens typically as part of a composite of fire regimes.

By way of illustration, consider the following suite of examples.

11.2.4.1   Kalapuya in the Willamette Valley, USA

The Willamette Valley runs north-south between the volcanic Cascade Mountains and the Coast Range of Oregon in the north-western USA. Prior to European contact, it hosted a medley of environmental niches, with wetlands to the north and drier, hummocky landscapes to the south. Most of it was prairie-dappled, with splashes of woods and riparian forest of one kind or another. Nearly all of it was burnt by the resident Kalapuya (Boyd, 1999; Johannessen *et al.*, 1971).

A reconstructed almanac has the Kalapuya gathered in the 'wet prairies' in late spring and summer, where they collected tubers such as camas and harvested waterfowl. The winter wet season still lingered. In July and August, the weather became dry and the people moved out of the lowlands and, as conditions permitted, they did some patch burning after harvesting grass seeds, sunflower seeds, hazelnuts and berries. At this time, too, the camas plots had dried, were harvested and were then burnt.

The scope of burning widened as the seasonal dryness permitted – part of a 'cleaning up' of debris and reopening of corridors. By late summer, the higher prairies were being burnt after tarweed had been gathered, and for collecting insects like grasshoppers. In October, following an acorn harvest, the oak savannas – the most extensive of Willamette landscapes – were fired. The last phase was to burn along the valley edges as part of communal drives for deer.

These specific tasks were supplemented, as it were, with fire littering and pushed into both mountain flanks through well-worn routes of travel. Though not mentioned in historic accounts, it is likely that parts of the marshlands suitable for waterfowl were also fired outside of the rhythms of nesting. The entire cycle, allowing for peculiarities of setting, is typical of aboriginal fire regimes. Without routine burning, trees quickly swarm over the landscape.

There is little lightning fire. Certainly, there is not enough to have kept the valley in the grasses and shrubs that defined the landscape which greeted early European explorers. Settlement, beginning in the 1840s soon destroyed or redefined that mottled landscape and its sustaining fire regimes. Diseases (including malaria) became endemic and caused an expected population crash among the Kalapuya. In some places, pastoralists replaced them and kept the land in grass by burning for forage (and, in later years, for the production of commercial grass seed). In others, they fenced the patches into farms, whose stubble they burnt, or placed into field rotation. Elsewhere, they drained wetlands and, by preventing fire, so allowed the peripheral Douglas-fir forest to overrun the land, so that a major logging industry appeared to fell and mill the timber in the early 20th century. While selected pieces of the landscape ensemble still burned, this was done so for different purposes, and their interaction as a system crumbled.

11.2.4.2 The Krahô, Tocantins, Brazil

The *cerrado* is another biota in a broadly wet-dry climate that is capable of supporting woods but persists as a grassland complex, one that is often loaded with woody shrubs, through routine burning. Modern anthropologists have recorded the fire practices of many indigenous tribes. Consider the Krahô as representative in offering a common smorgasbord of aboriginal burning practices that also integrate with elements of swidden (Figure 11.3).

The reasons given (and techniques used) are the usual ones, and the seasonal calendar unfolds accordingly. The opening of landscape with the advent of drier conditions begins with lines and fields of fire. Corridors are burnt as encountered. Patches used for hunting are kindled – many small pieces over the course of several days for each trip. Fires early in the dry season (April/May), which burn weakly, are set to shield swidden plots (*roças*) and patches (*carrascos*)

Figure 11.3 Ka'apor swidden in Amazonia (Photo courtesy of William Balée).

rich with fruiting trees, game in need of covert or flora that needs protection from later, hotter fires. Some fruiting flora (e.g., *mangaba*) require selective burning on a less than annual basis; as with berries, they produce best when pruned with flame, and some may require smoke to germinate.

Hunting fires, too, begin early and then continue throughout the middle of the dry season until the flames become too intense. The preferred method is a variant of a surround. In one version, an escape route from the flames is left open to deer and other game, but is lined with hunters. In another, the fire may simply encase the field and burn all those creatures unable to flee into burrows. Throughout May, if some livestock is part of the economic mix, pastures are fired, piece building on piece. Other patches are burnt in ways that fill in and stabilize the larger expanses of *cerrado* against more ferocious fires that can come toward the end of the dry season.

The swidden plots require more intense drying than does grass, so they are burnt later, in August or September. Honey – a valued resource (honey bees were introduced) – is collected in September or October, for which smoke is vital to stun or remove the bees from their hives. Once used, torches may be dropped to the ground, or the ground around the tree may be fired. Since the season is late, fires can spread if surrounding sites have not been burnt earlier. Throughout, opportunistic cleaning fires reduce pockets of unruly grasses and scrub, places perceived as hoarding such unwelcome creatures as snakes, scorpions, ticks and wasps. These pockets can also be difficult to walk through (or see into) and can explode into damaging fire if left unburnt for too long. Meanwhile, littering fires consume combustibles. With much of the landscape already torched under controlled conditions, late season cleansing fires are possible without damaging more than they help (Mistry *et al.*, 2005; Pivello, 2011).

The ethnographic studies have yielded some other interesting insights not possible from historic accounts. One concerns governance, or a distinction between fire practices that belong to individuals and those that remain under communal jurisdiction. Fire usage under the purview of individuals includes not only the burning of swidden and livestock plots, since both relate to personal sites, but also honey-gathering fires and fires set for cleaning.

A gender divide is also apparent. Men deal with hunting and large-area fires; women with domestic fires (among other tribes, women assist with some swidden and fire-catalyzed gathering). Children grow up playing with fire (as one observer noted, 'A favourite pastime for children during the night was to run around with burning palm leaves' – palm leaves being a common source of ignition for landscape fires).

There is a general sense, too, that outside of dedicated fire-free sites (what in many cultures are granted special standing as sacred groves), long-unburnt patches are a hazard, because they will burn and, when they do, they burn with damaging effects – hence, the urgency for constant clean-up.

11.2.4.3 Nyungar, jarrah and tuart forests, Western Australia

Although there are abundant first-contact reports by Europeans of fire use among Australian Aborigines, few accounts outside the tropical north provide the fine-grained details that modern fire science seeks. However, the forests of Western Australia suggest another perspective on the cumulative impact of Aboriginal burning by tracking what happened when the natives and their firesticks departed. The primary text in this case is the local grasstree, or as it is known in the Nyungar language, the *balga* (Ward and Sneeuwjagt, 1999).

The balga records fire scars, along with annual growth rings, and it lives long enough to hold a chronicle over 200 years. The pre-contact era exhibits three (or more) fires every ten years, which is, to say, a fire about as frequently as the land will carry it. This is a biota prone to burn: a broadly mediterranean climate, abundant lightning, a flora tempered by long evolutionary exposure to fire. In such settings, it is often tricky to isolate the anthropogenic contribution, which may substitute one kind of fire for another, or shift the seasonality of burning, or simply prune and sculpt existing patterns, rather than impose a wholly new order. The dry season burns of the Nyungar segue seamlessly into those caused by lightning. What makes the balga scene especially valuable, however, is that the record of burning closely tracks the indigenous population (Figure 11.4).

One study area holds mostly tuart (a eucalypt). For nearly a century following contact, from the 1860s to the 1960s, the frequency of fire dropped and fluctuated between two to three burns a decade. Cattle compete with flame, but the Nyungar were absorbed into herding and adapted their fire practices

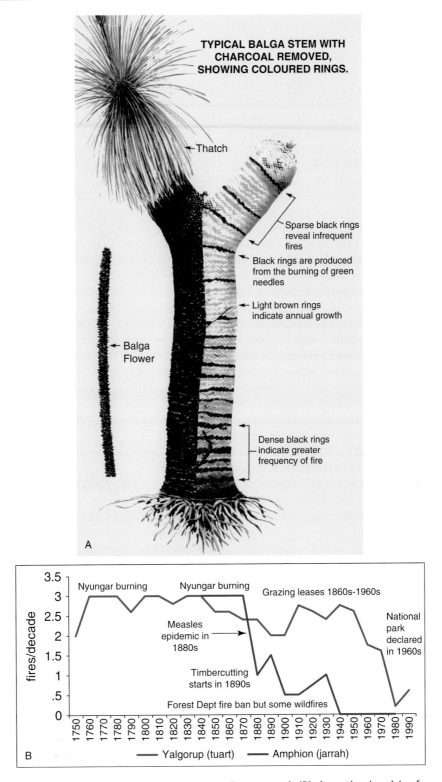

Figure 11.4   Top image (A) shows fire scarring on grasstrees. Bottom graph (B) shows the chronicle of scars for two sites (From Ward and Sneeuwjagt, 1999).

accordingly. They burnt as before, leaving fire refugia along riverine wetlands and rocky ridges. Then, in 1960, the forest became Yalgorup National Park, the existing economies were evicted and fire frequency plummeted.

A second study area in jarrah forest (another eucalypt) traces the impact of a 1880s measles epidemic on the local community. As the population of Nyungars crashed, so did the population of fires. A spat of logging in the 1890s brought a small but temporary boost before the site was gazetted as the Amphion Block under the administration of the Forests Department, which (like colonial forestry everywhere) attempted a fire ban. As the agency's capacity improved to the point that it was able not only to halt anthropogenic fire but to suppress lightning fire, fires vanished from the record.

There is no enduring way to abolish fire, and the mere attempt to do so may be destabilizing. Fire must re-enter, if not deliberately then by force and happenstance. Today, for a public estate aspiring to promote ecological goods and services, the ideal fire regime in jarrah looks toward a mix of burning on a calendar that resembles the historic regimen does not restore the *ancien regime*, but it keeps fire on the land in mixed ways, perpetuates a culture of burning and advances the goals of an industrial society rather than one that must live on what it can extract, through the firestick, from the land. These are important points because fire, once removed, is difficult to restore safely and an active fire culture, once extinguished, is not easily rekindled.

11.3 Cultivated fire: control over combustibles

The environmental constraints on aboriginal fire mean that, if humanity wants to increase its firepower, it has to control other aspects of the fire environment as fully as it does ignition. In practical terms, this means tinkering with fuels. It means slashing woods, moving combustibles, draining wetlands, digging peat, sending flocks to chew and trample, or otherwise converting the landscape into something that can burn more readily and according to rhythms more useful to humans. It means that the pulses and patches of burning – the regimen of fire's appearance on landscapes – will push beyond the natural limits imposed by climate and geography and more and more fall under human control. Hand and head will compete more vigorously with drought and wind.

Mostly, such tinkering takes the form of what, in vernacular terms, is regarded as agriculture. For fire history, 'agriculture' embraces those actions that modify a landscape's vegetation, i.e. its fuel structure. Instead of relying on seasonal rhythms to ready fuels, lightning to ignite them and indigenous flora to reclaim the site over the course of several years, farmers artificially create combustible patches, kindle them according to their own calendar and then plant exotic cultigens in the ash until the local biota crowds them out. Instead of depending opportunistically on pasture and browse, freshened by lightning's lottery, pastoralists can burn in places and at times that enhance forage and align fire's schedule better with their own.

Call the first an example of a *fire-fallow system*, and the second a *fire-forage* one. Both depend on manipulating the landscape to create conditions – to fashion fuels – suitable for the kinds of fires people desire. Since fire assumes the character of its setting, both are varieties of what might be considered cultivated fire and both obey principles of fire ecology.

11.3.1 Fire-fallow farming

In fire-dependent farming, burning is the critical catalyst; it makes possible the conditions that allow for planting. However, fire requires fuel, so the rhythm of burning is set by the rhythm of combustibles. The classic agronomic term for such fuels is *fallow*. The plot is abandoned, for years or maybe decades, during which it returns to scrub or woods, or it is untreated (or perhaps lightly grazed) for a year, during which time it accumulates unharvested straw and weeds of various sorts. Some of the intruders are part of a package of anthropogenic invasives (of which crops are one), while some come from a resurgent native biota reclaiming its sites. The fallowed land provides the fuels to support the next round of burning.

European agronomy has long condemned the practice. It saw untended fields as simple waste and an expression of slovenly stewardship – as land out of production, as messy places susceptible to disease, social disorder or wildfire. The fallow was worthless and, worse, it was set afire. Critics saw burning fields in the same light they did burning trash – it was landscape rubbish, removed lazily and carelessly by fire.

The perspective of a fire partisan, however, inverts that perception. The fallow was not burnt to get rid of it. It was grown in order to be burnt. The system needed fire, and fallowing was the means to create the combustibles to allow the fire to do its ecological work. Where the local conditions did not allow sufficient fuels to accumulate, farmers would gather pine needles and branches, perhaps dung, and even dried seaweed, to supplement it. Their fields required the ecological jolt provided by fire, and fallow was the means to get it. The cycle of burning followed the cycle of fallowing. Outside of floodplains, almost all agriculture required fire somewhere in its almanac of events.

In a general way, fire-fallowing takes two forms. In one, the farm moves around the landscape. In the others, the landscape, in effect, passes through the farm. Both are exercises in applied fire ecology.

11.3.1.1 Swidden

The first form is classic swidden agriculture, more colloquially known as slash-and-burn. A suitable patch is identified, its woods and shrubs felled or its large trees girdled, and the debris allowed to dry and then burnt. Over the first year, crops thrive. During the second year, they struggle, perhaps helped with a second (much lighter) burn or through active hoeing. By the third year, the native species, dismissed as weeds, reclaim the site. The cycle of swiddening, in brief, obeys the normal pattern of post-burn recovery.

Swiddening is a capacious, amorphous practice, almost infinitely variable. Farmers have applied it to old-growth and second-growth forests, to shrubs and orchards, to sod and peat – in fact, to anything that can be cut, dried and burnt. Since not all sites are amendable for crops, swidden patches dapple landscapes with patterns peculiar to local circumstances. Moisture may make bottomlands better than uplands. Frost may argue for mid-slope plots on mountains rather than those on crests or valleys. Solar warming may favour south-facing slopes (in the northern hemisphere) over north-facing ones.

Once used, a site will be revisited according to some rhythm set by demographics and the time required to re-grow suitable fallow. The revisited sites are easier to clear: the woody vegetation is smaller and more readily chopped and scattered. What farmers want is a hot, uniform fire, which depends on small-diameter fuels spread uniformly over the field. Moreover, particularly in tropical and subtropical settings, some large trees may be left to provide shade and lessen the ferocity of sunlight on exotic cultigens. Eventually, over centuries, the landscape becomes dappled with patches in various states of cultivation and fallow – an ideal arrangement not only for shifting cultivation but for biodiversity. The fallow furnishes rough pasture, hunting and trapping grounds, foraging opportunities and perhaps decorative or medicinal plants (Figure 11.5).

11.3.1.2 Field rotation

In the second variant of fire-fallowing, what might be called field rotation, the plot remains fixed and the farm cycles through it by a series of calculated cultivations. Following a burn, a deliberate sequence of crops is planted in a kind of landscape succession, with the choice of crops determined by long experience. It might, for example, begin with cereals and end with tubers as soil fertility changes. The sequence may be brief (a year or two, much as with swidden), or it might extend for six or eight years with care and weeding. At some point, however, the plot will be left to fallow and then burnt, which will allow the sequence to be renewed. A given farm may have two or three fields, one of which will be in fallow (Figure 11.6).

Fire practices thus reflect respective economies, but the farm economies themselves express not only environmental circumstances but social ones. Some landscapes will not support continuous cropping without constant tilling, fertilizing and other intensive treatments, but it is equally the case that social preferences will shape the choice of cultivation. Some societies distrust footloose groups that move through the countryside; they want people as fixed to their assigned places as fields are. Thus, cultures both reflect and shape their geographic surroundings. Social norms will express themselves in choices about land use and, hence, in fire practices. The resulting fire regimes are as much the outcome of laws, aspirations and cultural mores as of terrain, lightning and the density of pyrophytes.

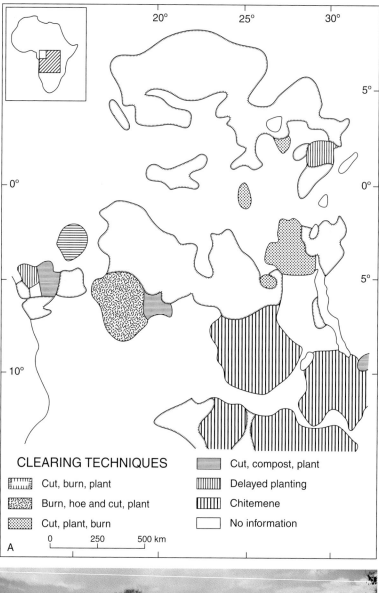

Figure 11.5 Swidden cultivation (chitemene) in miombo woodlands, central Africa.

A. Mosaic effect of cutting, burning, and planting (Map from Batchelder and Hirt, 1966).
B. Ground-level perspective from Zambia (Photo from Kkibumba/CC-public domain).

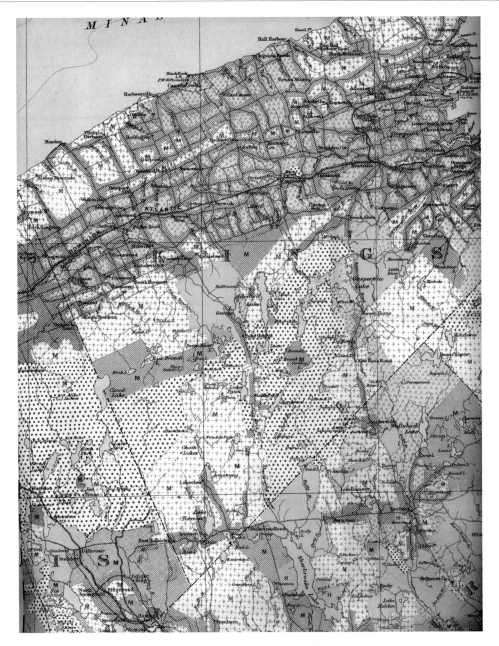

Figure 11.6 Burning patterns (pink, both solid and stippled) west of Halifax, Nova Scotia, showing clear lines and fields in an agricultural setting. (From Fernow *et al.*, 1912).

11.3.2 Fire-fallow farming: selected examples

11.3.2.1 Swidden, northern Thailand

As an example of classic swidden, consider this description from the mountain fringe of northern Thailand in the late 1960s. A monsoonal climate combines with hills and valleys to yield two subsistence economies – swidden for the dry and irrigation cropping for the wet. The region's tribes use similar technologies, but with differing degrees of social control. The Lua' encode the annual rhythms with ritual and firm strictures (even religious); the Karen allow more individual choice and tolerate a degree of social

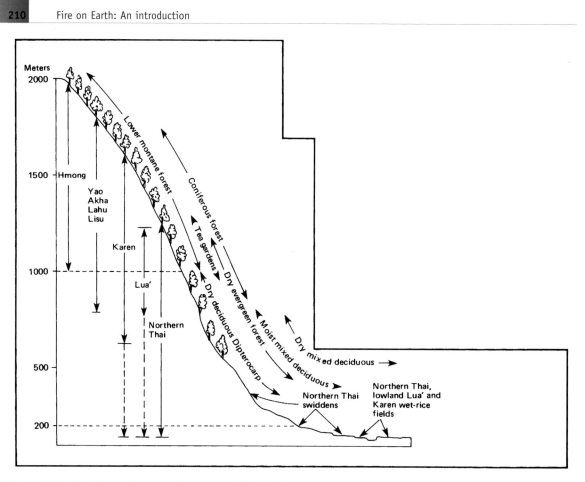

Figure 11.7 Landforms, forests, and patterns of farming in northern Thailand (From Kunstadter, 1978). Reproduced by permission of University Press of Hawaii.

and environmental slopover. The period of fallowing is roughly 10 years (Figure 11.7).

The annual cycle begins in January, a period of cool and wet, in which the village elders identify the sites to be cleared. As the climate warms into February, the plots are cut and left to dry. Some large trees will be spared or will have their branches trimmed. The Lua' leave woods along ridgetops and streams untouched to lessen erosion; they are also inclined to clear firebreaks around the felled plots.

By March, the dry season is maturing, the fields are kindled and what does not burn (the larger particles) is stacked and re-burnt, again and again. The firing commences from top to bottom, and the men stand with guns to wait for game that the fire might flush from the site. If flames spill over the borders of the plot, Lua' villagers have an obligation to fight it (Karen appear less concerned). The fate of unburnt logs varies: the smaller pieces are hauled to the village for use as domestic firewood; some remain on site

for fencing; and the largest and longest are arranged as check dams to blunt hillside erosion.

Successful plots are planted to root crops, and sometimes to maize and cotton, along with sorghum, peppers, beans, soya, mustard greens, marigolds and other flowers, viney vegetables and even tobacco. Besides relying on fences, swiddeners set traps for rats and begin weeding, a process that is as relentless as the weeds themselves (farmers believe that crop yield increases proportionally to weeding). By June, the rains have begun. By September, the rains ease, the final weeding occurs and the earliest crops begin to ripen. Attention shifts to irrigated fields. Harvest continues into December.

The sites persist, for a while, as pasture for domestic animals, as a source for firewood and as a reservoir of useful plants (several hundred species are identified by the Lua'). After centuries, the landscape has become dappled with swidden plots at various stages of fallowing. Most trees re-sprout from the root collar. The old biota reclaims the burn. The fallowing lasts ten years, when the site is re-cleared and burnt. The landscape is a model fire mosaic, although the pieces and dynamics reside with farmers (Kunstadter, 1978).

11.3.2.2   Brandwirtschaft, Black Forest, Germany

Temperate Europe has no routine fire season; its seasons vary by temperature, not precipitation. Controlled fire of many varieties, however, has made agriculture possible. Each practice has its own term, and the collective expression for them is *Brandwirtschaft* (Figure 11.8).

In the Black Forest, the system that evolved over millennia. In the 19th century, the pattern looked like this. As much as 30% of the woodland was high forest (*Hochwald*), 50–70 years of age; it was typically logged and the slash then burned, out of which arose a new conifer woods. The remainder of the landscape was low forest (*Niederwald*) that was the scene for repeated cutting, grazing, and burning on a rotation of 12–20 years. The rhythm was calculated to maximize the assorted products sought. There was coppice for ready

conversion to fuelwood or charcoal for hearth or furnace; woody slash for swidden farming; broom and thatch for housing and for byre bedding; oak bark for tanning (ideally, 10–11 years old, but not over 17). On swiddened plots, depending on what trees were most valued, oak or pine seeds were sown into the ash along with rye or potatoes. If the seeding failed, then hazel, birch, and hawthorn sprang up, although these, too, had their uses (Pyne, 1997).

In 1871, V. Vogelmann noted that most of the population of the Schwarzwald survived through a mixed regime of swidden, known locally as *Reutberge* ('hacked mountains'), a practice that was 'many centuries old.' The forested mountains, he continued, were 'stocked with coppice forests which are hacked after clearcut, fertilized by burning of litter (through a flaming fire) or by smouldering of grass and utilized for growing rye, potatoes or oats.'

Soon afterwards, a British forester visited the region and, astonished, likened it to what he had seen in Africa: 'All down the valley is poor, open, or devastated private forest with patches of Kumri [swidden], now burning

Figure 11.8 Forest swidden in Finland, an example of temperate European techniques extended into the boreal forest. Note how larger timber may be split and spared for use as fences or construction, and how the burning requires continual attention to ensure that the remaining large pieces are consumed. For an example of field reburning, see the image on Part 3 title page. (By Berndt Lindholm (1877), from J.G. Goldammer).

about every quarter of a mile. The burning is done down hill with ten men on a width of about 30 yards . . . The scene is like Kaffirland in spring.' Worse, to his mind, the practice was 'recognized by law and the burning of the forest not punished.' The state even furnished conifer seedlings for replanting.

The burning was regulated, as it had to be. Swiddeners fired fields in strips separated by firebreaks, not *en masse*; they smoothed out debris to ensure an even burn; they laid heavier logs along contours to prevent rolling brands from escaping and to retard soil erosion; they left some mature trees for seeding; they followed a rigorous, if local, prescription for crop sequencing; they restricted grazing, often denying access altogether for two years, then regulating access by sheep and goats in particular; and they notified neighbours, obtained permits, staffed fire lines and burnt with free-burning flame or in piles as conditions warranted.

This agro-forestry landscape unravelled as woods became more valued for timber than ash, as peasants left the country for the city, as rapid transport made local tanning bark uncompetitive with South American imports and industrial chemicals as, in brief, foresters pushed out farmers, silviculture replaced agriculture and fossil fuels supplanted the fallows and firewood. Fire vanished, along with those who used it.

11.3.3 Fire-forage pastoralism

Fallowing is a system for managing fire and flora. There is no equivalent term for fire and fauna, but the system again hinges on fine particles of biomass – what fire scientists call fuel and agronomists term forage (or fodder). The simple formulation is that burning renews many landscapes favoured by grazers and browsers; it freshens grasses, especially subtropical grasses (like C_4), that are unpalatable unless newly growing and it promotes grasses in places that would otherwise become woody shrublands or forests unless beaten back by flame. Fire can rejuvenate pasture as it does arable land.

With farming, the fire-fuel relationship is fairly straightforward. With pastoralism, however, it is trickier because livestock and fire compete for the same potential fuels. One slow-combusts that biomass by eating it, while the other, fast-combusts it through flame. The resulting dynamic is the ecological equivalent of a three-body problem, for which there is no exact solution. Free-ranging livestock and free-

burning fire are always out of sync, and that slight misalignment can worsen into outright instabilities amid the compounding effects of climate and economic cycles. What fire-catalyzed pastoralism shares with fire-catalyzed farming is that the land must be taken out of active production in order to accumulate fuels.

What people bring to the system is a degree of regularity to both burning and herding. They ensure a linkage of grazers, forage and fire, even amid drought and other disturbances inherent in the larger geography. However, they also add to those environmental instabilities by introducing others grounded in a social order. Often, shepherds reside outside normal village life, and not only do animals and arable crops clash, so do the demands for common fuels. In brief, pastoral fires do not always integrate with those of farmers and horticulturalists.

Moreover, the economy of large-scale herding relies on markets that obey their own rhythms, for which grazing seems a half-step out of kilter (there always seem to be too many or too few animals). Pastoralism succeeds when animals (not too many) can be moved over usable landscapes (not too remote), or if markets are distant, where the capacity exists to shift animals to fresh forage (or forage to animals). The social capacity to burn and relocate herds between pastures can only synchronize the various factors so far, yet such factors shape the resulting fire regimes.

Grazing and burning together form a constellation of points that can be variously configured. In a way loosely comparable to fire-catalyzed farming, pastoralism can mean either moving the flock through the landscape or moving the landscape through the flock. In the former, the animals are brought to pasture as forage becomes available by the turning of seasons and the ecological spark of fire. In the latter arrangement, forage is essentially brought to flock and herd. In extreme forms, it can result in large feedlots to which both forage and livestock are relocated.

However, this simple analogue fails to capture the astonishing variety of adaptations and local expressions. Perhaps the easiest formulation is to arrange those options according to the degree of manipulation people exercise over them, not only their control over those combustibles available on a patch of ground but over the ability to move animals and forage. Europe is particularly rich in variations (as well as records by which to analyze them). Consider two classic expressions: one in which flocks move seasonally between mountain and valley and one in which the flocks move between fixed patches (infield and outfield) on a farmstead.

11.3.3.1 Transhumance

Transhumance is an ancient practice, common where mountains permit forage in different seasons. It was widespread, in various forms, throughout Europe, but especially defined pastoralism within the Mediterranean basin. Shepherds drive flocks into the mountains for the summer, then return them to valleys for winter. The mountain pastures, along with corridors of transit, are burnt – sometimes in the spring ahead of the migrating flocks, but most often in the fall, behind the departing flocks. The winter pasture may be rough pasture, perhaps in shrublands or even woods, and so burnt with some regularity, or it may be fallowing fields (Evans, 1940; Figure 11.9).

Unsurprisingly, considering its diversity of lands and societies, Mediterranean transhumance displays many variations. Some expressions are strictly local, as flocks move seasonally from village to mountain pasture, but some are organized regionally. The most famous – Spain's, which had been granted a royal charter (and monopoly) known as the Mesta – trekked across Iberia's central plateau as well as from *meseta* to *monte*. Colonization exported both variations to North America. Shepherds, notably Basques, relocated their style of transhumance from the Pyrenees to the Sierra Nevada and Cascades, while authorities sought to recreate the Mesta amid Mexico's *altiplano*. Later, stripped of medieval guilds and state control, the pattern of vast cattle and sheep drives migrated into Texas and then spilled throughout the Great Plains (Figure 11.10).

11.3.3.2 Infield-outfield

At the other end of the gamut lies a temperate European pattern that strives to integrate animals with crops. The flocks and herds move between infield and outfield, or seasonally, between rough pasture during the summer and a barn for the winter. Fire can freshen those outfields and may help to turn the crank of rotating plots in the infield, but the shift from extensive (or fire-allied) herding of this sort to a more intensive husbandry is possible by feeding the livestock from cultivated fields and returning manure from barns and corrals to those acres. The flow of materials – fodder, manure – is more closely managed, and so is the flow of energy manifest by fire, which becomes more indirect. The two

economies of ploughed field and grazed paddock begin to merge (Figure 11.11).

This pattern has become, for European agronomy, an ideal and an exemplar of how to boost productivity and how to lessen the need for fallow and fire. It has served, with reasonable success, as a model for colonization in other temperate climes. Elsewhere, its vision is a chimera and it has failed in practice but, by establishing a niche geography as normative and by bonding with European forestry, it has stimulated a belief that agricultural burning of all kinds can be removed by right thinking and diligent labour. Outside its peculiar conditions of origin, however, the model works only with vast external inputs of fertilizer, agrochemicals and energy, plus a stable society that typically requires substantial subsidies from the state or sanction from a shared religion.

11.3.3.3 Pastoral-fire tourism

The Flint Hills puncture the formidable steppes of North America like a low-mounded massif of chert. Critically, they prevented the grasslands from being ploughed. Instead, sustained by tallgrass prairie, they became ranchlands; soon, they created an industrial version of transhumance that has since segued into tourism as well.

In the 19th century, the Flint Hills were a waystation for cattle drives across the Great Plains to railheads and markets. Contracts even specified that the grass should be burnt at particular times in the early spring, so that the herds shuffling from the south would find fresh forage (and be protected from wildfire). As railways and paved highways replaced dusty trails, the cattle were shipped by railcar and truck. The lands would be burnt in April, the cattle would arrive as the prairie re-greened and the herds would be fattened before being shipped to market. The prairie pastures served as a kind of open-air, free-range feed lot.

Unlike subsistence pastoralism, which is buffered from major markets by distance, Flint Hills ranching is closely aligned with national economies and must adapt constantly. It must adjust its imported cattle not only according to the amount of forage (set by rainfall) but by social demand. Thanks to modern rangeland management, its pastoralism is strongly influenced by a reductionist model that mimics science and mimes the market. The burning has one purpose: to maximize beef production. This emphasis is changing (slowly), however, as ranchers are

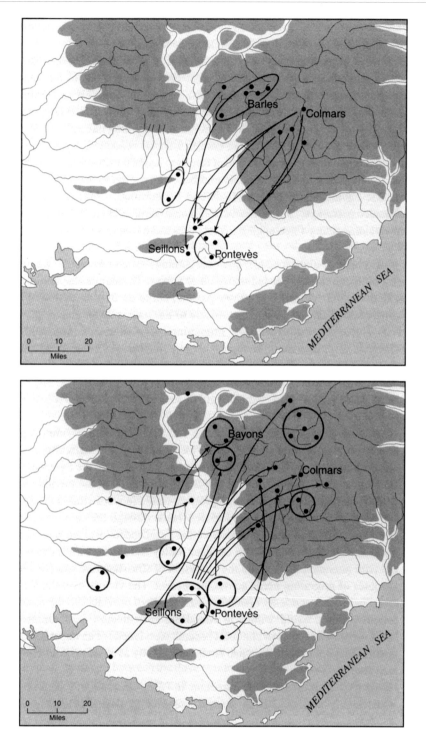

Figure 11.9   Transhumance in southern France. Top map shows the autumn movement of flocks from mountains to valleys, and the bottom map the spring reversal back to the mountains. (From Pyne, 1997). Reproduced with permission of the University of Washington Press.

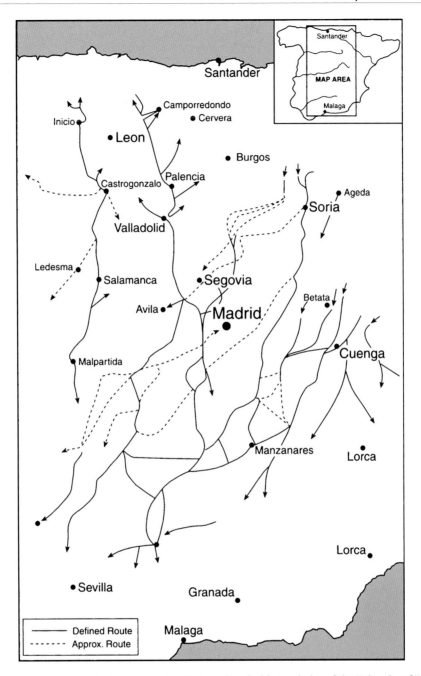

Figure 11.10 Transhumance in Spain (From Pyne, 1997). Reproduced with permission of the University of Washington Press.

rediscovering the ancient value of patch burning, and as the public rekindles interest in the nation's historic grasslands, its indigenous fauna and their cultural heritage (Fuhlendorf *et al.*, 2009).

Today, the Flint Hills encompass 96% of the residual tallgrass prairie. Four nature preserves sprawl over the hills, one administered by the National Park Service, two by the Nature Conservancy and one by Kansas State University as a biological research station. The Tallgrass Prairie Preserve is now stocked with free-range bison and, because of a three-year burning rotation, with such threatened

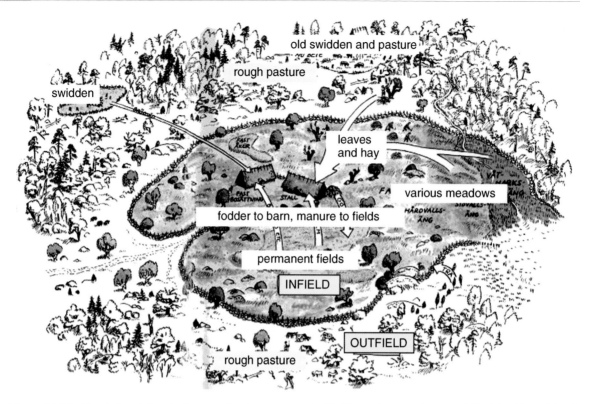

Figure 11.11 Classic forest farm in Sweden, showing movement of fodder and manure between infield (arable and barn) and outfield. The diagram shows, once again, the extension of a temperature European model into northern forests (Adapted from Ekstam *et al.*, 1988).

native species as the greater prairie chicken. With this as a working model, some private ranchers may similarly begin to vary their regimes. On the Kansas prairie, a form of pyro-tourism, or recreational burning, has emerged in which visitors pay for the opportunity to participate. By such means, a fire-based pastoralism may survive into modern times (Figure 11.12).

Figure 11.12 Tallgrass Preserve, Flint Hills, USA. Bison grazing on recently burnt grass (Photo courtesy of S.J. Pyne).

11.3.4   Agro-economies of mixed fire: selected examples

In fact, the world abounds in other patterns that depend on anthropogenic burning, nearly all involving some mixture of fire practices. Some have vanished – lost echoes from the past. Some endure, more or less in a traditional form. Yet others have metamorphosed into modern avatars better suited for an industrial society.

11.3.4.1   Northern Sweden

The temperate European model faltered when pushed northward into the boreal forest. The soil was poor and acidic, the climate harsh, the human population small relative to the landscape. What evolved was a mixed economy of swidden farming and migratory husbandry. The livestock consisted of cattle and goats that fed on rough pasture. Where soils and close settlement were possible, a variation of infields and outfields were used. However, as agrarian colonization pushed northward, the links between arable and pasture became more tenuous. A pattern of seasonal herding called *säter* (or *saeter*) evolved, in which young women and children relocated to remote pastures, where they tended a cottage industry of dairy products, particularly cheeses, that could be stored for winter. The rough pastures were renewed by burning. The more closely tended outfields near the homestead could be harvested for winter fodder when the herds returned (Figure 11.13).

Over time, as settlement penetrated inland and northward, two fire frontiers evolved. One brought the fire-requiring *säter* economy of Swedes into conflict with the fire-avoiding economy of reindeer herding by the Saami. For their winter range, reindeer fed on lichens, which can burn fiercely and widely, but which recover slowly – a lichens 'prairie' may need 40–60 years to restore itself. Although it seems that reindeer welcome (and may need) some alternative fodder, of the sort created by patch burning in the spring, the pattern of burning determined the pattern of transhumance between mountains and lowlands. The border between fire regimes was thus incommensurable, and vast burns are reported as a means of driving off Saami settlements and of maintaining a barrier between farms and herds of reindeer. The conflicts reached a climax in the late 18th century. The most visible expression of an eventual reconciliation was a truce in their mutually exclusive fire practices.

A century later, that newly established fire economy was, itself, threatened by an advancing frontier of logging, the vanguard of a spreading industrialization. Foresters anxious for timber and mining companies keen on fuelwood for smelting did not want the woods burnt for rye and rough forbs, and they saw heathlands as wastelands best converted to conifer plantations. A fire-catalyzed agriculture was gradually replaced by a fire-phobic silviculture, and farmstead and *säter* yielded to tree farms. By the mid-20th century, fire had become, in the words of Anders Granström, of 'archaeological interest'. The Swedish economy had shifted, and society's sense of what constituted appropriate fire regimes had shifted with it (Zackrisson, 1976).

11.3.4.2   Madagascar

Most of Madagascar – its central savannas – hosts an ancient form of herding, supplemented by small gardens, in which fire is an undeniable, necessary and even overriding presence. The landscape constantly feeds flame or beast or, where done with skill and luck, both in synchronization.

Burning of some sort has gone on for millennia. But the historic era commenced between 400–1000 CE, when waves of Malagasies arrived, along with cattle. Much of the indigenous megafauna soon went extinct. Range fires broadened synergistically with the swelling prairies of the interior, such that herding and burning came to define the bulk of the island, or what was effectively a micro-continent. What did not burn as grasslands burnt in swidden (*tavy*) plots around its mountain periphery. French colonial officials, in the person of foresters, soon attempted (unsuccessfully) to suppress the burning, but banning burning only criminalized it without removing it and created the ecological equivalent of a black market for fire. 'Red Madagascar' became the island's enduring sobriquet (Kull, 2002).

The reasons for burning are the usual ones and the pattern is typical. The herders burn to prevent brush from encroaching, to rejuvenate forage, to contain pests (primarily ticks), to prevent against wildfires and to help move herds – to attract cattle to fresh fodder and to ease movement. The biological reality is that old pantropical grasses have little nutritional value. Unless they are burnt well, woody shrubs invade, dead grasses crowd out and inhibit new growth, and cattle have less to eat and less that is palatable. As one observer noted, 'A pasture without fire is not a pasture' (Figure 11.14A, B).

Reasons vary with the purpose of burning, which changes over the year as the island turns from its wet to its dry season. Most burning is opportunistic; or, as

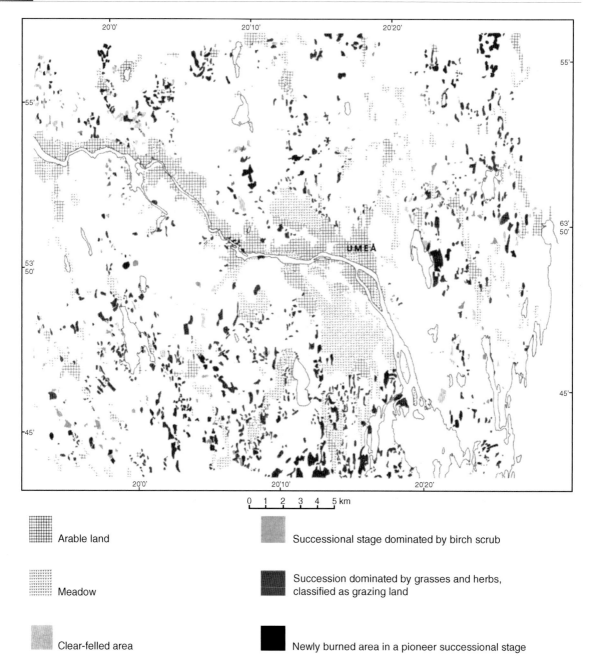

Figure 11.13 Land use around Umeå, a dense mosaic of sites cut, burnt, grazed, planted and re-growing (From Zackrisson, 1976). Courtesy Kungl. Vitterhetsakademien.

Christian Kull observes, 'People burn pastures in patches, and they burn these patches at different times, taking advantage of the best moments – both ecological and political – to do so.' Patches vary from 0.5 ha to more than 200 ha.

A detailed study at Afotsara found fires in April and June associated with a locust invasion (which also allowed the burning to proceed with fewer problems from the authorities). June and July commenced the 'normal pattern' of winter burning, which readied

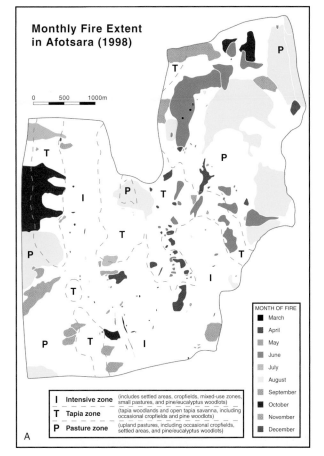

Figure 11.14 Red Madagascar.

A. Monthly calendar of fire in Afotsara.
B. Patch burns.
Reproduced by permission of Taylor & Francis.

forage for herds that move out of small farmlands and rice fields at that time. As August became drier and warmer, the fires increased in number. By September and October, burning was limited to those patches not yet fired; then began fires attached to croplands. The burning faded with the onset of the rainy season in November and December.

The upshot of such a pattern is a complex patchiness – patches of differing size, of burning severity, of biological effects. Burning is not a singular event but a seasonal process. For an extensive pastoral economy of much land and relatively few herds and herders, the arrangement is ideal, even if it is denounced by the authorities, who seek to regulate it out of existence.

Independence replaced foresters with environmentalists as critics of pastoral burning, and the colonial bureaucracy with more modern successors. The awkwardness is that the indigenous economy cannot exist without folk burning. The only way to eliminate fire is to overturn and instantly industrialize the island, and such a move, utopian in its expectations, will still require burning to maintain the character of most landscapes. The torch will pass from peasants to officials and its purpose will mutate from stimulating forage to sustaining creatures that ecotourism finds enchanting.

11.4 Ideas and institutions: lore and ritual

The essence of anthropogenic fire is that humans intervene into the workings of fire on Earth. They move points of ignition, tweak the environment of combustibles and reshape regimes. They redefine the logic of fire, i.e. its dynamics. The fire landscape becomes a negotiation between culture and nature. A society's technological sophistication thus matters as much as a biota's primary productivity. Patterns of fire will respond to social mores and political organization as fully as they do to terrain and wind. A kind of cultural teleconnection comes into being, such that a distant event may, by being transmitted through ideas and institutions, influence how fires appear a continent away. Such factors guide fire behaviour as fully as physical ones do.

In some settings, humans' influence is meagre. There is not much to work with, either because the geography is deeply hostile to fire or because humanity's capabilities are feeble. In others, people can commandeer the fire regime and redirect it to serve their own interests

better. In a few, the built environment dominates and, in principle at least, the resulting fire regime is almost wholly the outcome of human choices and deeds.

But not quite. Houses burn, and villages suffer conflagrations when dry winds blow sparks across combustible roofs and drive flames through shambles as congested as slash piles. Until modern times, most built landscapes were reconstituted out of natural landscapes; buildings were reconstructed woods and prairies (with thatch roofing), and both arrangements burnt in the same ways. In the geography of planetary burning, urban fires have been trivial in area, yet they have shaped fire regimes widely by affecting how people have thought about fire generally, and consequently how they have applied or withheld fire beyond the city walls.

How have people perceived fire? It was, as the folk sayings go, a good servant but a bad master, the best of friends and the worst of enemies. People understood in their genetic bones that they could not survive without controlled fire. They knew, too, that wildfire was among their greatest threats. They needed ways to distinguish among fires, to code their collective experiences acquired over many years and to regulate the fire behaviour of the community. Fire was too essential to ban, and potentially too dangerous not to constrain without some system of governance.

Throughout most of human history, however, both knowledge and governance have been local. While some laws were written, most fire mores were traditional, passed by one generation to another through spoken word and practice. Children learned by doing; they played with fire as they did with toys that emulated tools and other simulacra of adult duties, and they did so within a social setting. They learned fire's environmental and cultural limits. Collective fire practices were institutionalized, if only through ritual and the counsel of elders.

Not until an alliance of imperialism and industrialization broke this rough arrangement was traditional lore systematically replaced by putatively scientific knowledge, ancient habits superseded by constructed institutions (including national bureaucracies), fire skills removed from childhood experience and granted only to certified groups taught by the methods of formal education, and customary experience wiped away in favour of written fire codes enforced by the police powers of the state. That such reforms still struggle to match the tempo of the pyric transition to industrial burning helps to account for the unsettled state of fire on Earth today, and they are why a description of planetary fire based solely on physical and biological parameters cannot account for how and why fire looks and behaves as it does. That requires a new understanding of fire ecology, anthropogenic fire and the necessary hybrids of fire management.

11.5 Narrative arcs (and equants)

Among fire regimes, hybrids are the rule, mosaics of fire practices the norm and firescapes syncretic. People do not – cannot – use fire in isolation; it is the original interactive technology and, on landscapes, the various manifestations of fire influence one another. In the Pre-Columbian Tennessee Valley, people used one set of agricultural burning in bottomlands and another, of a more aboriginal character, on uplands. Both evolved over time and, in their interplay as societies migrated, cultigens arrived and experience improved.

Another way of expressing this arrangement is that fire practices are not simple stages in some evolutionary progression, such that one must follow another in strict order. Only in the grossest sense is such an evolutionary order apparent: people had to know how to burn before they had a reason to slash, and understanding how to combust biomass in machines had to precede burning fossil biomass. Otherwise, anthropogenic practices are typically mixed, like the fires that make up a regime. Some old habits linger, new ones only partially suppress or substitute, and fire regimes may cascade one to another without any inherent order. The upshot is a long, relatively lean narrative arc that relates the saga of anthropogenic fire. However, it is one made of innumerable equants and fragments that behave like pieces in a kaleidoscope, rather than a progression of one inevitable step after another. At fine scales, fire history looks less like a grand staircase and more like a random walk. At larger scales, patterns emerge.

Consider the following examples as illustrations of histories that, in a loose way, might furnish narrative templates. One organizes around the notion of a frontier; the other, of long-wave narrative.

11.5.1 Frontier narratives

One common organizing device is the concept of a fire frontier, in which regimes change in a rudely wave-like pattern. The conversion of Amazonia is a contemporary expression of a frontier of fire advancing by active slashing and burning. The scene in Iberia, particularly Portugal, demonstrates the reverse process in which land abandonment, a frontier of settlement retreat, feeds a fire plague. The recorded narrative in the American Ozarks captures both the rise and fall – the full cycle – of a fire frontier over 400 years. These examples are especially useful, because they involve regions that have little or no natural fire. The record they hold testifies to the character of anthropogenic fire practices and the kinds of regimes that can result.

11.5.1.1 A frontier, advancing

Amazonia – a term with both legal and biotic referents – occupies most of northern Brazil. Rainforest claims the bulk of the regional biota, although it is purest in the northwest, where it thrives under constant tropic rains and seasonal flooding It is more mixed toward the south and east, which show annual shifts in wetting and drying and experience occasional droughts while grading into *cerrado*. There is almost no natural ignition (Davidson *et al.*, 2012; Figure 11.15A, B).

In the early 1960s, when a military junta ruled Brazil, geopolitical and socio-political concerns combined to promote a program of colonization. The geopolitics concerned lightly administered regions that the junta feared might be susceptible to insurgencies and even secessional movements. The socio-politics focused on a 'surplus' population in the impoverished north-east that might be encouraged to colonize Amazonia, along with a belief that opening up the region to modern industry could spur economic growth generally. It was, in brief, a nationalistic project not unlike schemes undertaken at the same time in the Soviet Union and Australia, and later adopted by Indonesia and Malaysia. The architects pointed to American colonization as a rough model. A system of roads and incentives thus set into motion a transmigration of peoples who made the Amazonian fringe habitable by slashing and burning on what, collectively, amounted to a huge scale.

An 'arc of deforestation' emerged along that fire-prone perimeter, from a combination of small and large landholders. Eventually, much of the land, following repeated burnings, each of which gnawed away at the felled logs, ended up as pasture. The burnt areas became, in turn, points of entry for further fires to leak into neighbouring unburnt land. Viewed over several

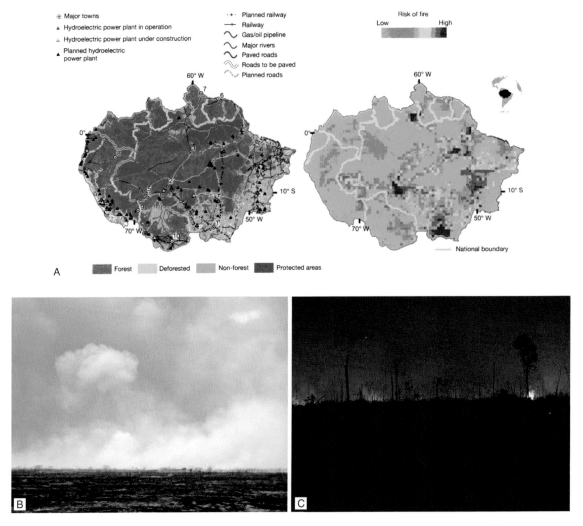

Figure 11.15 (A) The Amazonian frontier, showing land categories and development corridors (left) and fire consequences (right; from Davidson *et al.*, 2012). Photos from Mato Grosso: Reproduced by permission of Nature Publications. (B) shows extensive burning for land conversion and pasture. (C) evening image is how the world has interpreted the global consequences (Photos courtesy of Jennifer Balch).

decades, the burning proceeded in a large wave composed of smaller waves, which aligned with both the cycles of climate and those of the Brazilian economy which, by the late 1980s, suffered hyperinflation.

The ugly burns and sub-continental smoke palls attracted global attention and condemnation. The twisting flames and smoking rainforest made for graphic images that focused attention on the fires, rather than the underlying dynamics behind the land clearing. This was a distraction, or a case of misdirection. 'Without fire... there is no life', as the Brazilian proverb says. Until alternative livelihoods or techniques appear for settlers, the transmuting frontier will continue.

The likelihood is that, as with American examples, the process will continue until a combination of climate and economics causes the master wave to arc downward. Where the climate does not allow for any seasonal drying, fire will be difficult to sustain; and, when the economy matures, rural colonization will not be viewed as necessary. As the zone of colonization matures, fires will fade away, or will cease to be viewed as the vanguard of rainforest ruination. Already, agricultural conversion is moving into the *cerrado*. This will reduce overall burning, since it will transform frequently burnt prairie into soya fields, but it will not reduce ecological losses or solve Brazil's socio-economic difficulties, since the farms are mechanized. If the American example holds, the wave of forest-converting fires will pass, and Brazil will be left with some relic parks, a lot of abandoned land and an environmental cleanup program.

11.5.1.2   A frontier, retreating

The Mediterranean has an ideal fire climate – an annual rhythm of winter wetting and summer drying, broken by frequent droughts. Locally, this combines with favourable terrain and winds (a few famous, like the mistral). Along its mountainous perimeter, there are some lightning ignitions. However, for thousands of years, the region has routinely burnt within an agricultural landscape, rich with fire but contained by cultivation. From time to time, under the blows of invasion or plague, the land has gone feral and the fires with it. In recent times, however, the unravelling has come from political and economic reform. Few areas have been as hard hit as Portugal (Peireira *et al.*, 2006).

A century ago, Portugal had a pre-industrial, mostly pastoral economy, based on small landholdings. The ancient division of the landscape into fields (*ager*), woods (*silva*) and pasture (*saltus*) prevailed, organized around villages. Burning was common, not only in fields and on pastures, but for general cleaning and trash disposal. Large patches of land were communal. During the Salazar regime, however, the state embarked on a wholesale programme of afforestation in an attempt to use forestry as a means of industrializing. Pine plantations (later, eucalypt) were planted on communal lands, which were then closed to grazing. The populace began to seep away into Oporto and Lisbon. The programme accelerated during the 1950s and 1960s, before the regime collapsed in 1974.

The Portuguese landscape, no less than Portuguese society, began unravelling. A new political order stabilized and Portugal gained entry to the European Union in 1986, but the rapid modernization only exposed the countryside to a global market and EU development funds, and it quickened the depopulation of the landscape, whose old economy could not compete. Millennia of settlement began to retreat, leaving a disordered even feral landscape prone to an outbreak of disorganized, feral fires. The formerly tended countryside overgrew with combustibles, and the old small burns and cleaning fires gorged on them. The villagers, who had once fought those fires that did break free, were now out of the hills and into the cities. In the 1930s Portugal

recorded only 5000 ha burnt from wildfire. From 2000–2010, some 160 000 ha burnt annually. In 2003, the fires swept over 425 000 ha, or 5% of the national estate. No country in the world experienced a higher proportional rate of wildfire (Figure 11.16A, B).

The immediate crisis, of course, was wildfire. The state responded accordingly with research, efforts to build capacity for a formal fire suppression organization and relief projects. Wildfire, however, was a manifestation, not a cause; controlling fires beyond reflex reactions would require a recolonized countryside. Some form of the old order had to be restored or recreated in more modern forms. The sprawling woods might be constrained by cooperatives of small landholders, perhaps as "sustainable forest management systems". Other landscapes might be converted into parks or nature preserves and managed by a combination of tending and prescribed burning. The ideal would be to place the countryside within the EU's agricultural system and subsidize cultivation and herding, as the EU does with other products. In this case, such a subsidy would underwrite fire management rather than grain or butter.

The financial crises of the EU make such an expansion unlikely. As with Brazil's advancing fire frontier, Portugal's retreating one will probably have to pass through to a natural terminus, at which point the fever will break, fire will have drained away, and the countryside can rebuild around a more favourable regimen of controlled burning.

11.5.1.3   A frontier, full cycle

The Ozark Mountains are an erosional feature – a plateau dissected by streams into hills and hollows – that lies close to the center of the United States. The region has a fascinating fire history that researchers have recreated through fire-scarred trees and written records. The population of fires tracks closely with the population of migrating peoples, rising when settlement spreads, cresting as active settlement ceases, then falling as settlement matures and then shifts its economic foundations (Guyette *et al.*, 2002; Figure 11.17).

When records began, in the early 17th century, fires were few. This may, however, reflect a long wave of depopulation by diseases and dislocations that spread along trade routes prior to actual European colonization. The introduction of the horse in the early 18th century resulted in a small spike, possibly caused by increased mobility, new hunting strategies and burning

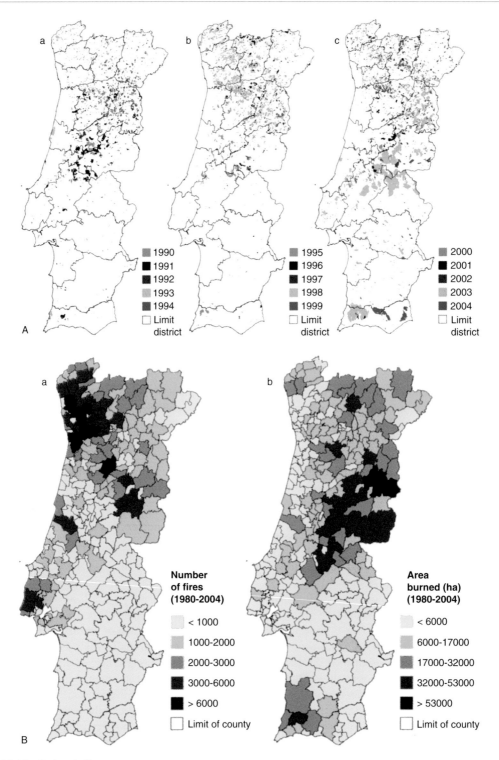

Figure 11.16 Portugal aflame.

A. Burnt area annually from 1990–2004 (From Pereira *et al.*, 2006).

B. Numbers of fires and composite burnt area from 1980–2004 (From Pereira *et al.*, 2006). Reproduced with permission.

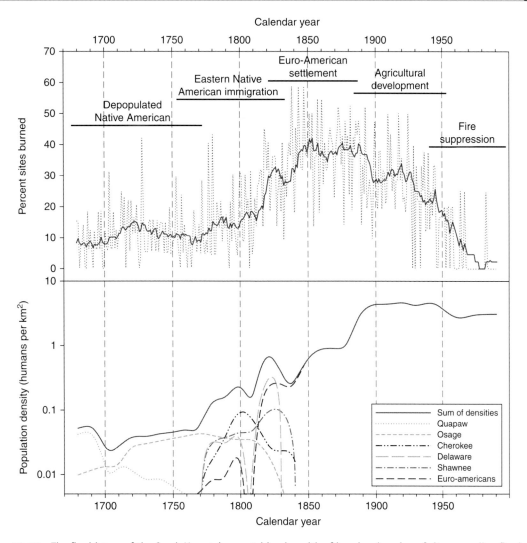

Figure 11.17 The fire history of the Ozark Mountains, matching breadth of burning (number of sites recording fires) with changes in population and settlement (From Guyette *et al.*, 2002). Reproduced with permission from Springer Science+ Business Media B.V.

for bison and other ungulates. A slow surge began in the late 18th century as the Native Americans were forced out of the eastern US and moved into the relative vacuum of the Ozarks. European-descended colonists soon followed and, between 1820 and 1850, the wave of burning steepened into a crest as settlers filled one valley after another.

As settlement ripened, the land became dissected again, not by ridges and stream-cut valleys but by roads, ploughed fields, pastures, grazed woodlands and forest clearing. The lush fuels were stripped, like the ploughed soils. Some social norms regulated burning. By the end of the 19th century the crest had passed and the wave of

fires began a descent. In the 20th century, both private and state-sponsored forestry attacked fires and campaigned to abolish traditional burning. Some rural lands were transferred to state forests, and some to private reserves such as the Pioneer Forest. All sought to root fire out of the landscape.

The primary mechanisms, however, were the exhaustion of the fuels that powered the firescapes of settlement and, though it came fitfully, what can be termed the pyric transition that accompanies industrialization. Active fire suppression mopped up what remained. By the latter 20th century, fire loads were lower than at any time over the past 400 years. As the conversion of rural landscapes to those typical of industrial societies continued, fire began to return through deliberate reintroduction in the name of nature protection.

The likely outcome will be the renewal of burnt sites, although at a level comparable to that at the onset of the narrative. Most anthropogenic combustion will burn fossil fuels.

11.5.2 Grand or long-wave narratives

A second grouping of narrative types might take a very long-term view, organized around a coherent theme. The following two examples show contrasting realms. In one, fire arrives at an incombustible site and then changes character through human use. In the other, fire seems an enduring feature but, while human use alters its regime, fire influences anthropogenic usage as much as people do fire.

11.5.2.1   Finland: from the Mesolithic to Modernism

Finland is a recent landscape: it only emerged from beneath ice and sea some 6000 years ago and it continues to rise, converting tidelands to marsh to dry land. From records in the southwest, the story is one in which fire comes to a place incapable of burning, co-evolves with human history and then morphs in more recent times to an industrial fire regime. Charcoal records this chronicle with unusual, nearly unique, thoroughness (Bradshaw et al., 1997; Figure 11.18).

When the first record of fire appears, it is minor and embedded within the Mesolithic era. There is little basis for burning. Seasonality is weak, the forest consists of hardwoods dominated by lime. The human presence is limited to openings around lakes and coasts. That changes abruptly with the advent of the Neolithic, i.e. with the entry of agriculture accompanied by axe, livestock and swidden. Probably assisted by warming and the emergence of some seasonality, a mixed forest of coniferous and deciduous trees is both an aid and an outcome to small-scale slashing and burning. The process accelerates during the Iron Age, and then through the Medieval warm period. Paradoxically, it rises through the Little Ice Age, before reaching a maximum in the mid-19th century. Because of the intensity of *burn-beating* (as Finns translate the practice into English), annual charcoal declines somewhat, due to a shortening

of the forest-fallow cycle and a shift from conifers to birch. There is less fuel to burn. The record then pauses.

When combustion revives, it continues to deposit charcoal at roughly the same volume as before, but in the form of soot from the industrial burning of coal and oil, not from biomass. Charcoal from wood drops nearly to levels not seen since the Mesolithic. There are several reasons. The economy, with state support, valorizes forests for timber, not tar, ash and charcoal, and enacts laws to regulate folk burning. The shift to an early industrial economy replaces biomass combustion with fossil fuels. Open fire disappears from the landscape, with the exception in recent years of interest in restoring some flame.

In brief, anthropogenic fire came, saw(ed), conquered and then yielded to a more modern variant. Finland is a cameo of Earth, and is extraordinary largely for having the means, with its labyrinth of lakes, to preserve a record.

11.5.2.2   Florida: from lightning to drip torch

Florida, too, is a relatively recent landscape, having also emerged from the sea, but in this case, its origins were rife with fire. Lightning is abundant – the densest in North America. Seasonality is pronounced, with a lush rainy season (152 cm per year) recreating rank fuel arrays annually. Humans have used fire from their initial colonizations. Each newcomer has relearned and adapted to landscape fire. All in all, Florida is probably the most intrinsically fire-prone part of the United States (Myers and Ewell, 1990; Pyne, 2011a).

The chronicle of fire is continuous, although varied according local biotas and human activities, from the northern pineries of the Panhandle to the swampy sawgrass of the Everglades. The variation in ecosystems depends, as one researcher noted, on how deep is the sub-surface water and how frequent are the surface fires.

The geography of the Pre-Columbian population is unknown. It was almost certainly denser in the north than the south, and it is likely that it was all but obliterated with European contact. Most European settlements remained on the coast. The interior was reoccupied by tribes driven from the north. American colonization came late, in the mid-19th century and sparsely, mostly by the free-range herding of cattle. Burning was relentless. In the early 20th century, the saying was, 'Florida burned twice a year'. Ranchers burnt late in the dry season, just before the rains, and often they burnt again at the end of the dry season to

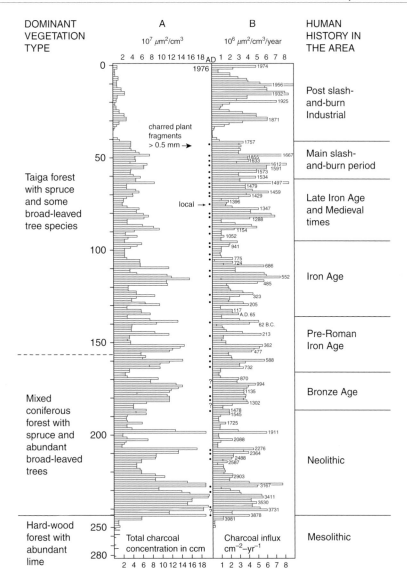

Figure 11.18 A charcoal record of Finland's long chronicle of fire, from the onset of the Neolithic to the pyric transition to industrial combustion (From Bradshaw *et al.*, 1997). Reproduced with permission from Springer Science+Business Media B.V.

encourage a second growth of forage. This is no doubt an exaggeration, but one with a hefty kernel of truth. In the early 20th century, the state forester announced that 115% of the state had burnt over the past year.

A detailed description of persistence in fire culture is available from what today is Eglin Air Force Base in Florida's panhandle. Here, the aboriginal torch changed hands without really altering the landscape. As an illustration of fire practice persistence, consider this account of burning by white American settlers in 1911, in and around a reserved national forest. Land use included low-grade herding and horticulture,

although the loose pastoralism was little distinguished from hunting, and cultivation consisted of crudely extracting sap from native longleaf pines. The description by the administrating forester nicely describes the multiplicity of purposes for which residents used fire and the long duration of burning, well outside the natural season:

The popular sentiment of the residents within the forests, in common with nearly all of the people of the South, is unqualifiedly in favour of the annual burning over of the pineries. The homesteader and the cattleman burn the woods to keep down the blackjack [conifer reproduction], undergrowth and to better the cattle range. The turpentine operator burns over his woods annually, after raking around his boxed trees, and at the time when the burning will do least harm in order to protect his timber from the later burnings that are sure to occur. He burns also with the idea of keeping the turpentine orchards clear of undergrowth and free from snakes, in order that the Negro labourers may gather the gum with ease and safety. The camp hunters, of whom there is a large number during the fall and winter months, set out fires in order to drive out game from the thickets. All of these different classes of people have for a great number of years

Figure 11.19   The Florida fire scene in profile.

A. Wildfire, control and the reintroduction of fire through prescribed (Rx) burning (Data from Florida Forest Service).
B. Florida pine and palmetto burning (Photo courtesy of Ray Lovett).

been accustomed to burning the woods freely and without hindrance of any kind and it is done without the knowledge or the feeling that they are breaking the laws or in any way doing damage. On the contrary, they all have the most positive belief that burning is necessary and best in the long run... The turpentine operator burns his woods and all other neighbouring woods during the winter months, generally in December, January, or February. The cattleman sets fire during March, April and May to such areas as the turpentine operator has left unburnt. During the summer there are almost daily severe thunder-storms, and many forest fires are started by lightning. In the dry fall months hunters set fire to such 'rough' places as may harbour game. It is only by chance that any area of unenclosed land escapes burning at least once in two years (Eldredge, 1911).

Open-range ranching (and its fires) continued until 1948, when immigration into the state began its astonishing run. However, unless the land was converted into incombustible forms, such as cities or citrus orchards, the burning continued. Alarmed at the conversion of rural land to suburb, the state commenced a major campaign to purchase and protect prime landscapes. All of these lands held some kind of burning regimen. If the fires did not come on human terms, they would come on nature's. The imperative to burn transcends all categories of land use. It is practised on national forests, parks and wildlife refuges; on state forests, parks and wildlife reserves; on recreational lands under the jurisdiction of local authorities; on military reserves; and on private lands, particularly where ranching dominates the economy. Nowhere in the United States is prescribed fire practised on a scale comparable to that in Florida (Figure 11.19).

To reconsider the Panhandle in particular, in 1940, the Choctahawchee National Forest was transferred to the War Department in what evolved into Eglin Air Force Base. Today, most of Eglin AFB is open to the public and managed much like a national forest. The fire programme prescribe-burns roughly 40 000 ha a year. The endangered red-cockaded woodpecker has taken the place of tapped pines and free-ranging cattle. As it matures, the programme is moving toward a more seasonally mixed batch of burns, much like what people had maintained before the land assumed its modern institutional form. In that transformation, Eglin might

stand for many modern parks and nature preserves as they seek to reintroduce fire with as little collateral damage as possible.

All this has made Florida the hearth of prescribed burning in the US and, in some respects, the world. The state led resistance to federal policies and forestry doctrines that tried to abolish burning. So, too, it has pioneered ways to ensure that prescribed fire will flourish. Through both governmental and private institutions (such as the Tall Timbers Research Station), it has promoted the research and technology necessary to validate this practice.

Everglades and Big Cypress account for 80% of the prescribed burning done nationally by the National Park Service. Its very uniqueness, however, makes it difficult to export the Florida experience, as it results from a peculiar fusion of fire-prone settings and a persistent fire culture – or, to paraphrase the old expression, of blood, soil and fire. Unsurprisingly, it works best in the neighbouring south-eastern US and the Caribbean.

All in all, Florida's storied experience makes a fitting coda for a long narrative of fire's persistence and the role of humans in writing that environmental text.

Further reading

Balée, W. (1994). *Footprints of the Forest: Ka'apor Ethnobotany – the Historical Ecology of Plant Utilization by an Amazonian People.* Columbia University Press, New York, NY.

Boyd, R. (1999). Strategies of Indian Burning in the Willamette Valley. In: Boyd, R. (ed.) *Indians, Fire, and Land in the Pacific Northwest*, pp. 94–139. Oregon State University Press.

Bradshaw, R.H.W., Tolonen, K., Tolonen, M. (1997). Holocene Records of Fire from the Boreal and Temperate Zones of Europe. In: Clark, J.S. et al. *Sediment Records of Biomass burning and Global Change*, pp. 347–365. *NATO ASI Series I*, Vol. **51**. Springer-Verlag.

Davidson, E.A., de Araújo, A.C., Artaxo, P., Balch, J.K., Foster Brown, I., Bustamante, M.M.C., Coe, M.T., DeFries, R.S., Keller, M., Longo, M., Munger, J. W., Schroeder, W., Soares-Filho, B.S., Souza, C.M. Jr., Wofsy, S.C. (2012). The Amazon basin in transition. *Nature* **481**, 321–327.

Eldredge, I. (1911). Fire Problems of the Florida National Forest. *Proceedings of the Society of American Foresters* 164–171.

Evans, E.E. (1940) Transhumance in Europe. *Geography* **25**, 172–180.

Fuhlendorf, S.D., Engle, D.M., Kerby, J., Hamilton, R.G. (2009). Pyric Herbivory: Rewilding Landscapes through the Recoupling of Fire and Grazing. *Conservation Biology* **23**(3), 588–598.

Guyette, R.P., Mjuzik, R.M., Dey, D.C. (2002). Dynamics of an Anthropogenic Fire Regime. *Ecosystems* **5**, 472–486.

Johannessen, Carl L. Davenport, W.A., Millet, A., McWilliams, S. (1971). The Vegetation of the Willamette Valley. *Annals of the Association of American Geographers* **61**, 286–302.

Kull, C.A. (2002). Madagascar's burning issue: the persistent conflict over fire. *Environment* **44**(3), 8–19.

Kunstadter, P. (1978). Subsistence Agricultural Economies of Lua' and Karen Hill Farmers, Mae Sariang District, Northwestern Thailand. In: Kunstadter, P., Chapman, E.C., Sabhasri, S. (eds.) *Farmers in the Forest. Economic Development and Marginal Agriculture in Northern Thailand*, pp. 74–133. University Press of Hawaii for the East-West Center: Honolulu, HI.

Mistry, J., Berardi, A., Andrade, V., Krahô, T., Krahô, P., Leonardos, O. (2005). Indigenous fire management in the cerrado of Brazil: The case of the Krahô of Tocantins. *Human Ecology* **33**, 365–386.

Pereira, J.M.C., Carreiras, J.M.B., Silva, J.M.N., Vasconcelos, M.J.P. (2006) Alguns conceitos básicos sobre os fogos rurais em Portugal. In: Pereira, J.S., Pereira, J.M.C., Rego, F.C., Silva, J.N., Silva, T.P. (eds.). *Incêndios Florestais em Portugal*. ISA Press, Lisbon, Portugal.

Pivello, V.R. (2011). The use of fire in the cerrado and Amazonian rainforests of Brazil: past and present. *Fire Ecology* **7**(1), 24–39.

Pyne, S. (2001). *Fire: A Brief History*. University of Washington Press.

Pyne, S. (2012) *Fire: Nature and Culture*. Reaktion Books.

Ward, D., Sneeuwjagt, R. (1999) Believing the Balga. *LANDSCOPE* **14**(3), 10–16.

Wrangham, R. (2009). *Catching Fire: How Cooking Made Us Human*. Basic Books.

Zackrisson, O. (1976). Vegetation dynamics and land use in the lower reaches of the river Umeälven. *Early Norrland* **9**, 7–74.

Chapter 12

A new epoch of fire: the anthropocene

12.1 The Great Disruption

Over time people found new ways to expand their firepower, but the more fire they possessed, the greater their need for fuels. The living world could only supply so much. If people harvested further fuel wood for hearth and furnace, if they shortened the cycle of fallowing, if they exhausted new lands available for conversion by fire, if they continued to extract more than the natural world could restore in a timely way, then the landscape degraded. Burning became a kind of biotic strip mining. As the great naturalist Linnaeus observed in the 18th century, the world would have rich parents and poor children. If humanity's power derived from its control over combustion, then it could only expand that firepower by finding another cache of fuel.

It discovered that source in fossil biomass. These new landscapes were actually the landscapes of the past, long buried and now excavated, but they were different in that they could not be strewn over fields, like weeds and slashed wood or pine needles raked out of forests. They had to burn in special chambers, and their energy – heat, light, power – often had to be transmitted to humanity's various habitations indirectly. That was the pattern overall; fuels were extracted from the geologic past, burnt in the present,

and their effluent dispatched to the geologic future. Apart from noxious air pollution (which was no less true for burning peat or wood), the ill effects of the new combustion were displaced from the present into another time and place.

So pervasive is fire on Earth, and so fundamental is it to humanity, that a change in how the planet's species monopolist uses combustion will cascade through almost every aspect of the biosphere, as well as the atmosphere and even the geosphere. The perception has grown that humanity has gone beyond the usual dynamics of niche construction and has become a global geologic force, which deserves to have its own geologic epoch, the Anthropocene. Advocates further identify a second phase that has occurred since World War II, which has been termed the Great Acceleration (Steffen *et al.*, 2011).

The Anthropocene maps onto what is loosely known as industrialization. For fire history, 'industrialization' is shorthand for that shift in fuels from surface biomass to fossil biomass, with all that means for how humanity applies and withholds fire on the land. Usefully, the general culture agrees, since popular imagination has long identified the Industrial Revolution with William Blake's 'dark satanic mills' belching soot from combusted coal. In fact, all the potential environmental maladjustments – the onset of global warming, the explosion

Fire on Earth: An Introduction, First Edition. Andrew C. Scott, David M.J.S. Bowman, William J. Bond, Stephen J. Pyne and Martin E. Alexander.
© 2014 John Wiley & Sons, Ltd. Published 2014 by John Wiley & Sons, Ltd.

of human population, modern planetary pollution, the triggering of mass extinctions – align with a relatively simple index of industrial fire. It seems that 200 years is accomplishing what had taken the Pleistocene over two million and the Mesozoic over 200 million. The Anthropocene is the epoch in which anthropogenic fire mutated and metastasized.

12.2   The pyric transition

12.2.1   The mechanisms of regime change

The mechanics of the transition involve two processes. One is technological substitution; the other, outright suppression. The first occurs with industrial maturity and proceeds according to a logic of economic efficiency and legal requirements. The second depends also on a degree of sophistication, in that active firefighting may require the apparatus of industrial fire, in the form of fire engines and pumps and electrical-powered communication systems. The upshot is a landscape underwritten by drastically different fire regimes.

The quick story is that the new combustion supplants the old. Electric lights substitute for candles, gas stoves replace wood-burners, diesel-guzzling tractors supersede hay-fuelled oxen. Instead of open flame restructuring woods, there are chainsaws and wood-chippers. Instead of fires that liberate nutrients into soil, air and stream, fossil-fuelled factories break down fossil biomass to produce nitrogen, phosphorus and other fertilizers, while trucks haul them to distant sites and tractors-drawn spreaders sow them. Instead of temporarily fumigating a place by smoke and heat, or hoeing by selective fire, farmers spread artificial pesticides and herbicides, and they scarify with mechanical harrows. Blueberry fields are 'burned' under propane jets drawn by tractors. Flaming fallow recedes before a fossil fallow combusted in special chambers. Free-burning fire disappears and closed combustion takes its place.

Where open fire occurs, its internal-combustion competitor quickly extinguishes it. In the past, the best form of fire control was to cultivate the landscape in such ways that it could not burn, or had built-in firewalls of incombustible barriers or fire-resistant plants, or was already deliberately burnt. Such pre-emptive protective burning was a way to substitute

controlled fire for wildfire. It had the advantage of keeping fire on the land. However, as societies industrialize, they counter open fire by directly attacking it. The only means available is through the machinery of internal combustion and other industrial inventions; they pump water or chemicals, or drop it from aircraft, or transport crews by vehicles to the flanks of free-burning fire, where they typically run power machinery. They rely less on presuppression burning and backfiring. The area that is burnt shrinks and, in built landscapes, it can all but vanish. In wildlands, it will shrivel until, interestingly, the trend may eventually reverse under considerations of cost, firefighter safety and ecological integrity.

The realm of industrial fire, too, has its lines of fire and fields of fire. Internal combustion runs on a matrix of roads (or flight paths) and burns within the patches of the built landscape. These are the energy pathways that carry and organize nutrients and species for an industrial ecology. They connect primary producers to consumers, pump water for irrigation and distribute the derived fertilizers and fumigants. They sustain the agents (people) that run this top-down ecosystem. They constitute a refashioned matrix of fire effects. Remove that combustion source and the system will collapse. If the outcome seemingly echoes what preceded it, that is because internal combustion often substitutes for open burning, much as silica substitutes for lignin in petrified wood. So, too, they remake the conceptual landscape, standing to the traditional notions of fire ecology as their burning does to traditional firescapes (Figure 12.1A, B).

Pause for a moment to ponder what the reformed transport means, since it hugely influences the human structuring of geography and the determination of what constitutes resources, commodities and markets, that is, how people interact with their surroundings. Terrestrial transport by steam (and later petroleum distillates) has replaced draft animals. The transition is symptomatic of a shift from surface biomass and slow combustion (metabolism) to fossil biomass and fast combustion. In fact, the train of combustion-catalyzed reactions goes further, since much of agriculture was devoted to growing pasture or grain to stoke the domesticated fauna. Those fields are now put to other purposes (e.g. oats for breakfast cereals to feed humans), their fires have been sublimated into engines, and draft livestock have become – like flame – largely ceremonial.

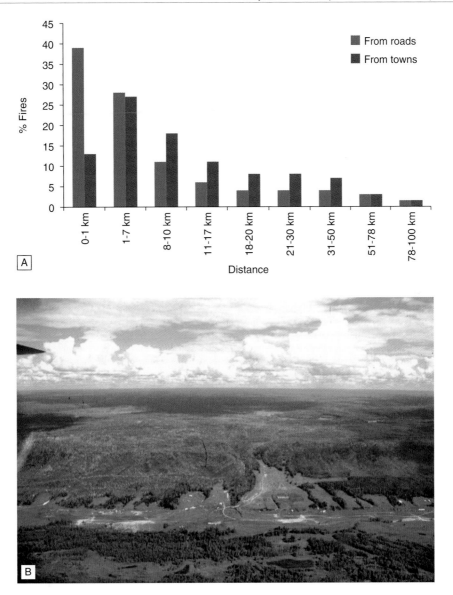

Figure 12.1 Modern lines and fields of fire, Siberia.

A. Russian fires by proximity to roads and villages generally. The decrease in fires adjacent is due to fields and intensive cultivation (Data from Avialesokhrana).

B. Aerial photo over Lena River and parallel road, showing pattern of burning along corridors. The distinctions shows particularly well in boreal and temperate forests, but less well in grasslands where fires can bleed across the landscape (Photo courtesy of S.J. Pyne).

12.2.2 Combustion's new fire regimes

Most pre-industrial landscapes were agricultural, which is to say, they were shaped by anthropogenic fire in the service of farming and herding. The pyric transition rebuilt those environments. No longer did farmers need to cultivate fuels through fallowing in order to power draft animals, fertilize fields or kill noxious weeds and pests. They could run tractors, reapers and tillers on gasoline and diesel; they could lavish nitrogenous fertilizers on fields; they could spray herbicides and pesticides at will, typically from tractors, pumps or aircraft. Those chemicals derived largely from fossil biomass, distilled through processes of industrial combustion (as Howard Odum famously remarked, today the potato consists partly of petroleum). So likewise does the fire regime in which it resides). Moreover, what crops are grown (and when) depends on transport, which feeds on more petroleum. The entire economy of agricultural production morphs from open flame to internal combustion, and from living fallow to fossil fallow. With the loss of fallow, and as cultivation intensifies, ecological services shrink and

ecological goods such as biodiversity retreat to fence lines, right-of-ways and parks.

After the transition to industrial combustion, a fire mosaic still exists, but it is often no longer visible. In much the same way that modern buildings are shaped by the threat of fire but fire itself rarely occurs, so modern landscapes result from the application of anthropogenic fire, but it is a combustion rarely expressed in flame. It occurs internally or invisibly, distilled into petroleum fuels or sublimated into electricity. Of course, not all modern energy derives from fire, any more than all organic decomposition results from burning fossil fuels, but most does, as does

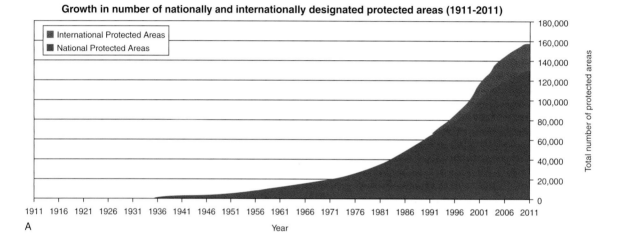

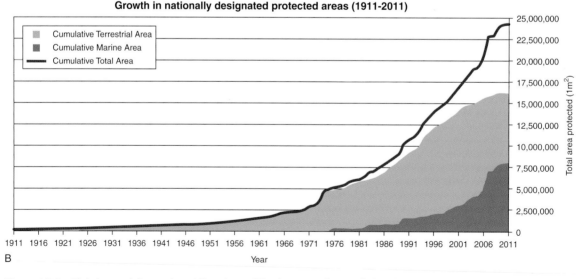

Figure 12.2 Global growth in numbers (A) and area (B) of protected natural sites (Data from UNEP – World Conservation Monitoring Centre).

almost all transport, which structures the flows of energy and goods. Industrial fire ecology has its regimes and mosaics. It just displays them covertly.

There is one grand exception – nature reserves, which industrial societies like to establish. In a sense, these are the fallow of modern economies and, like fallowed lands of old, they hold most of the native biodiversity. Not all such sites are naturally fire-prone but, if they can burn, then they need to. In such circumstances, removing fire can be as ecologically disruptive as inserting it into places that do not have it on their own. Moreover, not all of those reserves are natural; they are often cultural landscapes, for which anthropogenic fire is a vital catalyst. In either case, removing fire may be ecologically disruptive, even catastrophic. If fire has been present for a long time, then it may have to remain in order to sustain ecological integrity. Paradoxically, even as industrializing societies strive to eliminate open fire generally, they struggle to retain or restore it on nature preserves (Figure 12.2).

This much is easily observed. The pyric transition, however, has not been systematically studied, despite its significance as the entry into the modern world and as prime catalyst for global change. How any particular society manages the transition will vary by environment, history and cultural peculiarities, in ways that can yield, superficially, diametrically opposite effects. In linking hinterlands to urban markets, industrialization will stimulate burning in tropical landscapes by encouraging conversion of woods to pastures, but the same linkages will dampen fire in arid landscapes, since overgrazing, which strips away fine fuels, is a likely outcome.

Countries that have a legacy of colonial fire management may well retain institutions and inherited ideas from that era which will affect how they approach fire management; they will display different patterns of land use (as, for example, with the creation of forest reserves); and they may have agencies for fire research that will not be found in places outside that imperial network. And, of course, there will be cultural differences. Canadians and Americans, for example, despite enormous similarities, have distinctive ways of managing wildland fire that derive from their different political structures and sense of national identity.

Such considerations may seem far removed from a genuine ecology of fire, but the pyric transition creates an ecology that operates on different principles from the past. Previously, people had to interact with their natural environments, and each could check or amplify the other. People could only burn so much in defiance of climate and indigenous biota, yet climate and biota could only retard human changes so much. Remove people from such contexts and fires would rearrange themselves but could flourish regardless.

Industrial fire, however, cannot exist without people. Remove the people and combustion ceases. In such settings, how people see the world and how they organize themselves through institutions will have enormous consequences for how fire appears on the land.

12.2.3   Fire's industrial revolution

The industrial revolution – emerging during the age of political revolutions – fundamentally rewired the dynamic of fire on Earth. Humanity's ability to burn was no longer limited by what it could extract from the living landscape; it could excavate landscapes from the geologic past and export their effluent into a geologic future.

Fossil biomass promised an almost unbounded source of combustibles not constrained by season, place or biota. Humanity's fires could burn winter and summer, night and day, through drought and deluge, amid desert, tundra, tropical rainforest, boreal woodland or temperate grassland. The conditions that had shaped fire's behaviour and effects for hundreds of millions of years had no effect on combustion conducted in special engines and furnaces and distributed covertly. Fire became dissociated from its ecological foundations. Humanity's firepower became more diffuse, sublimated, potent and unchecked (Figure 12.3).

The reason is that the new fuels did not burn openly on the land – they combusted in special chambers. They created power that had to be applied indirectly through machines, transmitted by electrical lines or used to create chemicals and transport them outside traditional biogeochemical cycles. They energized an artificial ecosystem of mechanical producers and consumers. What is important to recognize is that this acquisition was not only something applied by people from the outside – a biotic innovation that ecosystems had to accommodate – but that it also replaced processes that humanity had nurtured during its tenure on Earth.

Industrial combustion was more than a new fire: it forced a reconstitution of that symbiotic relationship between people and planet that anthropogenic fire had

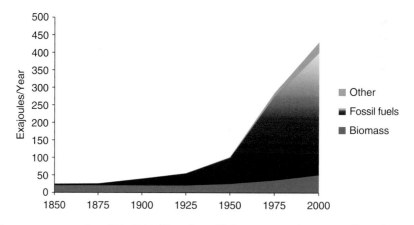

Figure 12.3 Primary energy supply, 1850–2000 (Data from US Department of Energy, Office of Energy Efficiency and Renewable Energy).

long mediated. Its effects cascaded wherever people lived and then beyond, to wherever the output of humanity's new fire practices could penetrate, be these electricity from coal-fired dynamos carried hundreds of kilometres from their point of origin, or greenhouse gases loosed into the atmosphere. This is a new kind of fire ecology, for which the traditional descriptors of fire effects hardly apply and for which inherited concepts seem hollow.

The transformation – still unfolding after 200 years, but instantaneous in biotic and geologic time – is the pyric equivalent of bow shock or a hydraulic jump. In short order, industrializing societies have restructured fire on their landscapes. In habitat after human habitat, the new fire has replaced the old. Flames and biofuelled burning disappeared from homes, from factories, from cities, and then from fields and forests. They lingered longest in lightly used landscapes and protected natural areas such as parks and reserved woods although, even here, societies charged with administering such areas sought to abolish open fires – not only those set by people, but those set by nature as well.

This reformation might aptly be termed the *pyric transition*, after the well-known demographic transition that accompanies industrialization. In both cases, the immediate effect is a population explosion – in the one of people, in the other of fires – as old traditions persist and new practices appear. Then, over decades, the new replaces the old.

Certainly, something like this has characterized fire history. The shock of the transition begins with an overpopulation of fires, often abusive and promiscuous. It ends with the removal of open burning, so much so that the numbers plunge below replacement values. In this way, early industrializing societies experience a wave of destructive conflagrations, while fully industrialized ones undergo a fire famine as fire's abrupt removal stuns biotas.

12.3 Enlightenment and empire

The Anthropocene coincides with two other historic developments of fundamental significance for contemporary pyrogeography. One centres in the realm of ideas and the other in that of institutions. Over the course of 150 years, both have profoundly restructured earthly fire. Humanity's ideas and institutions are as critical to fire's presence on Earth, as are wind, fine-particled combustibles and drought. They are as much the drivers of contemporary fire as climate and evolutionary ecology.

12.3.1 Enlightenment fire

The big idea was the penetration of the scientific revolution, which matured during the 17th century, throughout general culture into what became known as the Enlightenment. At its core was a commitment to Reason, as this seemed to be expressed in such scientific theories as Newton's model of the solar system. The belief spread that scientific knowledge was superior to other species of knowing, and that the model could be extended indefinitely to all realms of inquiry. The upshot was to align the pyric transition with an intellectual, as well as a technological, revolution.

Specifically, the Enlightenment altered the status of fire in both high culture and folk culture. Among elites, fire as an idea – as an organizing concept for understanding Earth and humanity – waned at the same time as industrial combustion waxed. Enlightenment agronomists famously denounced fire (and fallow). Modern chemistry made fire an epiphenomenon of oxygen; physics, of thermodynamics; forestry, an act of human vandalism. The integrative role that fire had long enjoyed broke apart under the picking of reductionist science. It became possible to isolate the pieces of fire and then to devise technologies to exploit each – a trend that suited the pyric transition as it moved burning from free-burning landscapes to machines.

Such are the common dynamics of modern science as it redefines common experience into entities that it can interrogate in their simplest form. But fire was different; alone of the four ancient elements, fire did not claim a discipline or academic standing in itself (the only fire department at a university is the one that sends an emergency vehicle when an alarm sounds). Instead, it was something derived from other, now more elemental subjects.

At the start of the 18th century, the celebrated naturalist Herman Boerhaave could declare fire essential to any universal understanding of natural philosophy. By the end of that century, fire as a subject, much less as an informing principle, yielded primacy, just as it was receding from the practical experience of intellectuals. No less significantly, fire's grounding in the living world was replaced by one lodged in a built and mechanical one.

Equally, fire began to lose its rooting in popular culture and practice, the basis for its use over millennia, as traditional knowledge became suspect. Instead, European intellectuals, and those who looked to Europe for inspiration, condemned fire as wasteful and dangerous. They sought to replace indigenous fire practices with a suite informed by science – or, better, they sought to abolish them altogether as unsubstantiated superstition. They treated traditional European fire lore with the same disdain as they did indigenous knowledge elsewhere throughout their imperium.

This process has resulted in another of the pyric transition's dramatic divisions of Earth. The places with scientific capacity tend to lack fire, while the places with lots of fires lack a robust scientific establishment. In other words, the experience of fire for industrial societies is remote and hypothetical or confined to nature preserves; it is absent from vernacular life. Out of sight tends to mean out of mind; residents know fire indirectly or virtually through transmitted images, TV or remote-sensing satellites. It is often hard for such observers to appreciate the ecological value of fire in working landscapes.

Ideas, however, need institutions to gain traction if they are to install reforms on the ground. Change often came slowly because fire, and the lore behind its use, remained indispensable to practitioners, who would only yield if they had a working substitute. A revolution in fire required not only laws but social organs and, ideally, a group committed to the project.

12.3.2 Imperial fire

The undertaking found that agent in forestry, a self-identified guild that hated and feared fire, claimed academic standing and was willing to fight fire both politically and in the field. By the 18th century, forestry had established valences to the state in Prussia and France. As they evolved, forestry bureaus became, in a sense, the social equivalent to putting fire into mechanical combustion chambers.

What made the guild powerful, however, was the peculiar convergence of industrialization and imperialism to create a new category of land use – or, more properly, a reformed and expanded category of reserved lands. Unoccupied or lightly inhabited landscapes, places vacated of indigenous peoples through colonization, formerly rural scenes – all could be redefined as public estates and committed to common purposes in the name of environmental health. State-sponsored conservation appeared on a vast scale, notably through systems of forest reserves and, less extravagantly, as archipelagos of nature parks. Revealingly, the process emerges in lock-step with the Anthropocene.

Foresters claimed the administration of those reserves and the creation of reserves made forestry powerful; together, they justified a creed of fire protection. Ideally, foresters would have been happy with fire's exclusion; practically, they were content to prevent fire where possible and to extinguish it where it broke out. The reserves, however, were not uniform. They varied by their intrinsic susceptibility to fire: some could burn annually, some only episodically. No less important, some were effectively empty of people, while others were either inhabited or subject to seasonal occupation by villagers and pastoralists. In the former case, fire control depended on climate and the biota's capacity to burn. In the latter, it also depended on the ability to control people. Unsurprisingly, fire control was more effective where the authorities had to deal only with nature than where they had to deal with permutations of nature and people interacting.

The imperial outburst that accompanied the Renaissance was led by Iberians and, coming before the scientific revolution and state-sponsored forestry, it did not establish reserves or cede its rule to foresters. However, a second outburst that swelled during the 18th century, principally powered by northern Europeans, did so. Great Britain, France, Holland, Russia, and those latter imperialists such as the United States, Canada and Australia, all accepted that the state had responsibilities for conserving natural resources for the common good. Interestingly, these focused not so much on forests as a source of wood, but instead on broader 'influences' that affected climate, particularly with respect to droughts and floods. The concept expanded to include parks and strict nature preserves, not just managed forests. All such sites – but particularly those that were forested – were a public good. Where they existed, they needed to be protected, and where they had been stripped away, they needed to be restored.

Fires were deemed an obvious threat, certainly to foresters and to like-minded theorists of state-sponsored conservation. Uncontained fires appeared to be everywhere, and everywhere they seemed to be overwhelming. That perception was partly due to the simple shock of people who took temperate Europe's fire scene as a norm, rather than a niche, and then discovered that the rest of the world burnt freely. However, magnifying the shock was the fact that the theory of state conservation emerged during the time of the pyric transition, when open burning typically experiences a population explosion and sprawls uncontrollably across landscapes. Even lands that had been long tended by people became unhinged. Until fire was contained, forestry and any other 'rational' enterprises declared that their missions were impossible. Moreover, that forestry claimed the same standing as engineering, a branch of applied science, argued for a programme of research as well as field treatments. One could remake folk knowledge as the other would folk landscapes.

With enormous commitment, the project appeared to succeed. Such institutions were instruments of the pyric transition as fully as steam engines and dynamos, for they directed how the new combustion would replace the old. In truth, the outcome was ironic, in that the more state forestry removed fire from fire-prone landscapes, the more those places deteriorated. The ecological tapestry unravelled; critical species failed to regenerate or if present, to thrive; and, most tellingly, fuels built up such that fires became more virulent and damaging. The pyric transition could remove fire from cities and factories without destroying their capacity to remain habitable and productive; it could, in fact, improve conditions by promoting industrial fire's technological benefits without its proximate side-effects. In reserves, however, the transition sapped the properties that were the reason for protecting those lands. Eventually, authorities watched fire return, either as feral outbreaks or through deliberate reintroductions.

The upshot is a paradox prominent in contemporary pyrogeography. Within industrialized societies, where open burning is gone, or going, or ceremonial, there are landscapes that are burning as much, or even more, than in the past. One of the arguments for reserves was that they would make it possible to control fire. They could, in principle, better regulate the people who started so many fires, and they would have at their disposal an agency ready and able to fight fires from any source. Instead, where fire was a natural occurrence, the forest and park reserves became permanent abodes for free-burning fire. More encouragingly, they have become the fallow of industrial societies – places outside commodity production that hold most of the biotic variety of their landscapes.

12.4   Scaling the transition

Places that have made the pyric transition exhibit a very different pastiche of fire regimes than those – even those with identical natural circumstances – that have not done so. The environmental effects that people had formerly achieved from open burning they now get from surrogates distilled out of fossil biomass or transcended into machines. Since most industrial inhabitants reside in urban areas, they no longer have a need for fire or wish to have it around, except in very constrained ceremonial settings.

Strikingly, too, fire science shows a similar segregation into places that study fire and places that have it. The parting between traditional lore and formal learning continues a very old story about the distinctions between those who use fire and those who study it, and it stands as a reminder that the transition wiped out not only landscapes but the knowledge by which long-resident peoples had sustained them. The transition may be slow or fast but, at any particular site, either open burning or industrial combustion predominates, and this phenomenon seems to operate at scales that range from the local to the global.

12.4.1 Global

Satellite surveillance testifies vividly to the ways the planet is rapidly segregating into two grand realms of combustion. The night lights of industrialized regions contrast strikingly with the surface fires of those starting the transition (or who are still outside it) (Figure 12.4). Europe and Africa show the division clearly, especially since the Mediterranean Sea and Sahara Desert are almost wholly empty of burning, save for islands and petroleum drilling sites.

Of course, not all European lights result from combustion-powered dynamos; hydropower is important in Scandinavia, and France derives much of its energy from nuclear fission. Similarly, not all the lights of sub-Saharan Africa indicate open fires. The bright glares in the Sahara and off-shore in the Gulf of Benin result from the flaring of natural gas associated with oil drilling, and patches of bright lights cluster around a few cities. However, in a gross sense, the pattern is clear, along with the trend.

12.4.2 Subcontinental

At a finer-grained scale, consider two situations in which countries that have not made the transition sit within a matrix of those that have. Perhaps the most dramatic is the divide between the two Koreas. The two nations are identical in environment, ethnicity and traditional culture, but the Korean War (1950–1953) split their recent human histories. The resulting

demilitarized zone (DMZ) along the 38th parallel divides two regimens of fire.

After the war, South Korea began to industrialize. The record of fuels consumed shows clearly an abrupt shift from biofuels to fossil fuels and thus the pyric transition in the built environment. What is fascinating is that the same transformation occurs in the surrounding landscape. North Korea failed to make that transition in a meaningful way. The two images in Figure 12.5 demonstrate the consequences.

The upper photo from satellite imagery shows evening lights. North Korea is a blank in what is otherwise a region aglow. The lower photo is a MODIS satellite snapshot of the Korean Peninsula, in which hot spots map open fires (the cluster of fires along the DMZ reflects the practice of burning off the vegetation in the spring, when dry cold fronts will push the fires south).

The pyrogeography mocks the oft-heard canard that 'fire does not respect borders'. It certainly will, if those borders reflect differences in land use, which will, in turn, reflect distinctions in human behaviour – or, in this instance, the maturation of the pyric transition.

Slightly less dramatic, perhaps, is the situation between Myanmar (Burma) and Thailand. Again, apart from Rangoon, the capital city, evening lights are sparse in Myanmar, leaving the country almost a void within the region. By contrast, open fires are rife and, although these are largely confined within the country, there is some slop-over in the less developed hill regions along the borders, which are also scenes of insurgencies, refugee camps and other expressions of unrest that typically result in fires.

12.4.3 National

Now consider the situation within a single nation. To simplify the matter, two general circumstances apply. In one, a country is undergoing the pyric transition and displays uneven outcomes. The landscape of industrial combustion is patchy. The other circumstance involves a developed nation with ample back country, not amenable to cultivation and left open for rude grazing or logging or gazetted into nature reserves. In such places, a deep competition evolves between combustion wild, prescribed and feral. Interestingly, these are among Earth's primary firepowers.

Figure 12.4 DMSP night-time lights for three years (which accounts for the different colors in Africa) processed by the NOAA National Geophysical Data Center (Image from Chris Elvidge).

12.4.3.1 The USA

Here is a country that has undergone the pyric transition but which, as a result of its history, set aside a large fraction of its national estate as forest reserves or for other nature-conserving purposes. The two trends, fire prevention and fire promotion, persist. The US continues to extirpate fire from its urban and suburban landscapes, and from its agricultural fields and pastures. More and more, economics and law (informed by public health and safety considerations) are squeezing fire out of the countryside. At the same time, nearly a third of the national estate is public land (although a third of that is in Alaska), and there is a small but growing accumulation of private land dedicated to public purposes, such as biodiversity.

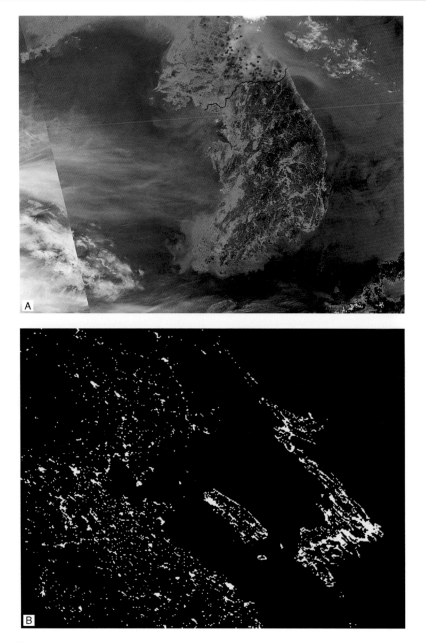

Figure 12.5 Two Koreas.

A. MODIS view of open fires on Korean Peninsula, May 2005. (Image from US National Oceanographic and Atmospheric Agency).
B. Evening satellite picture of lights, an index of electricity, with North Korea conspicuously dark (Image from US Defense
    Meteorological Satellite Program).

These uncultivated lands are, today, the scene for most major fires, for natural fires and for a considerable fraction of the nation's prescribed fires (Wildland Fire Leadership Council, 2011; Figure 12.6A, B).

There are a few exceptions, all of them where fire remains integral to working landscapes and their associated cultures. Mostly, these find regional expression, notably in parts of the Great Plains and the south-east. There is also an organized movement to protect burning (and a landowner's right to burn) on private lands, primarily for agricultural purposes, for commercial forestry or for brush removal.

The national Coalition of Prescribed Fire Councils, which has grown with extraordinary vigour, helps facilitate with advocacy and legal advice. The national trend toward fire's removal from urban, suburban and exurban landscapes, however, is well advanced and, while NCPFC and organizations of similar mind can help stabilize the geography of rural burning, they are unlikely to reverse it (Melvin, 2012).

This split pattern is undeniable, but detecting it requires a finer-grained mapping than coarse satellite reconnaissance alone can provide. Viewed from space, the United States has abundant fires in its western half and south-east. The western fires are almost all on public lands, and most are wildfires or naturally kindled fires deliberately allowed to burn more or less freely. The south-eastern fires are the result of anthropogenic burning, mostly prescribed fire. Neither that distinction among fires, nor the actual association with public land, shows without recourse to closer scaling.

12.4.3.2 Australia

Then there is the one country that is also a continent. Largely for environmental reasons, Australia has industrialized unequally, with large swathes of its tropical north and central Outback lightly inhabited (if at all). Australia thus displays a schizophrenic settlement pattern: it is both the least densely settled of inhabited continents and the most urbanized, all amidst one of the most fire-prone environments on the planet. The texture of its fire-mosaicked land ranges from very coarse to minute (Figure 12.7).

While Australia contains the various paradoxes of the pyric transition, it has woven into it several strands of its own devising. Those regions that have become

densely settled (and industrialized) show the usual symptoms, but most of the landscape is so flammable that even city parks within Sydney can burn. The most striking feature of Australia's history is its anomalous acceptance, for a long time, of rural burning. Alone among forestry's colonial outposts, it adopted controlled fire as a basis for bushfire protection and forest management overall.

Yet this tolerance is proving difficult to sustain in the face of political and popular pressures to reduce the pervasiveness of fire, even in nature reserves. What the Australian scene highlights is the need to include scale with any discussion about suitable fire levels, practices and intensity. A national policy may not mean much, and may even prove damaging, if applied uniformly across such a variety of settings. What makes sense in the monsoonal forest of the Cape York Peninsula may be meaningless amid the Mediterranean woodlands of the Adelaide Hills. How settlement and fire reconcile themselves in the jarrah and karri forests around Manjimup may have little to say regarding the savage outbreaks in the Victorian mountains. Rural towns in Queensland will differ from exurbs on the perimeter national parks in the Blue Mountains of New South Wales. Like many fire-rich nations, Australia has many kinds of fires and many kinds of fire problems (Ellis *et al.*, 2004).

12.4.3.3 Canada

To complete the sweep of former colonies that now constitute major firepowers, consider Canada, the nation with the second largest national estate and a country of enormous combustibility. Except for patches along its fringes, its interior grasslands and boreal forest are as fire-prone as any biota on the planet. Equally, Canada has vigorously industrialized. As few places can, it combines an extensive, sparsely settled outback with a high-intensity modern society committed to industrial combustion (Pyne, 2007).

Like Australia, Canada's cities align along its border – in this case, the southern boundary with the USA (the exception, Edmonton, lies 480 km north). As it industrialized, it sought to project industrial fire along roads, rails and flight paths, and to replace open flame with closed flame where possible. Settlement itself converted burning prairie and the mixed forests around the Great Lakes to arable fields and pasture.

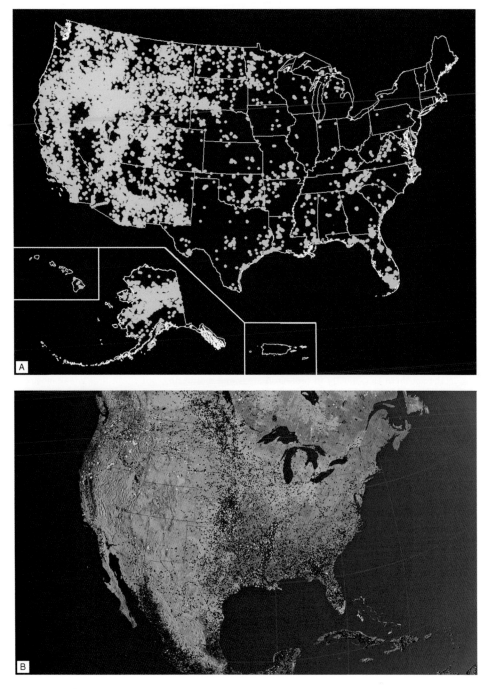

Figure 12.6   Two Americas.

A. The geography of large (>100 acre [40 ha]) fires in the USA from 1980–2005. The distribution aligns well with public lands or privately owned nature preserves (Image from US Geological Survey).

B. The geography of all burning in 2012 (Image from NASA).

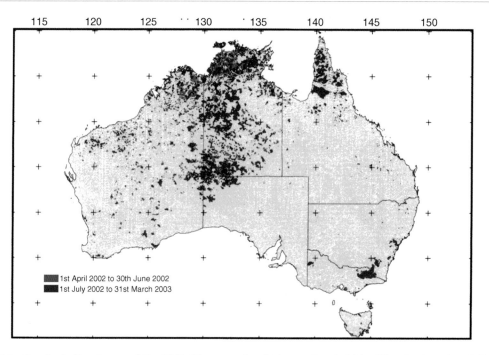

Figure 12.7 Two Australias. A map of the 2002–03 season, in which an estimated 54 million hectares burnt. Typically, 65–85% of annual burning occurs in the tropical north and, during exceptional years, in the desert interior. Much smaller burning occurs in the southern tier, mostly in protected areas and mountains, but these fires wreak most of the damage (From Ellis *et al.*, 2004).

Elsewhere (and spectacularly), the national government sought to extend fire protection over most of the western lands it inherited after Confederation. This proved quixotic. In 1930, the federal government ceded its lands to the western provinces, effectively dissolving its national forest service. Today, control over natural resources (and hence fire) belongs with the provinces. The dominion manages fire on national parks and supports research through the Canadian Forest Service.

Canada is a hot spot for both realms of Earthly combustion. Its boreal forests ripple with fire, somewhere in the country, on a roughly decadal scale. These fires appear to be increasing in size and savagery and have begun to slam into communities as they once did in settlement times. Meanwhile, Canada consumes energy at a higher rate than the US. Fire protection has relied on industrial equipment, and Canadians have excelled at pumps, aircraft and fire control technology. Canada has developed fossil fuels and continues to do so, most controversially the tar-sands of Alberta, which demand large investments of energy to extract and

process. By the 1990s, the carbon output from industrial combustion exceeded that from landscape fire. In sum, Canada is a high-end site – a global hotspot – for combustion of both varieties (Figure 12.8).

What matters is how this competition is expressed on the land and how society reconciles the rhythms of a boreal environment (and burning regime) best characterized by extreme values, rather than norms, with the cadences of institutions that seek a middle, that try to dampen those outbursts, and must do so within the context of a political confederation that cedes primary control to the provinces. Mostly, the large provinces drew an effective line of control, beyond which they did not attempt fire control.

In 1983, Canada resolved the question of how to cooperate on wildfire emergencies by establishing a public corporation, the Canadian Interagency Forest Fire Centre. It continues to rely on the Canadian Forest Service, a national institution, to supply scientific research of value to all the agencies. It has not, however, found an equivalent means to coordinate the resources needed to conduct prescribed burning or

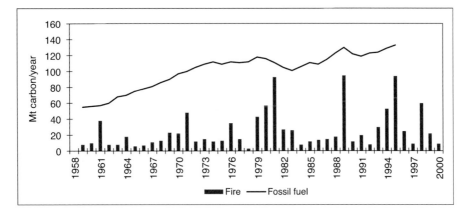

Figure 12.8   Canadian direct carbon emissions, 1958–2000 (Data from Amiro *et al.*, 2001).

otherwise to manage wildland fire. Even the largest provinces struggle to muster the critical mass to manage fire on the scale required.

Put differently, there are two competitions at work. One is between industrial and landscape fire; the other, between the provinces and the national government. This is a contest that lies at the political core of confederation. The future of the first will likely find its narrative in climate change – boreal Canada is projected to experience major shifts. The future of the second will depend on how Canadians negotiate their diverse national and regional needs and desires.

And so it goes. The pyric transition is not an abstract principle imposing itself on Earth. It has emerged out of human practices, and even the global partitioning of combustion is the cumulative outcome of choices made by individuals, communities and agencies.

12.5   After the revolution

Over time, the combustion competition for inhabited landscapes ends with the triumph of industrial over free-burning fire, but how rapidly (and completely) this occurs can vary widely. The conversion is quickest and fullest in built environments, stutters over rural landscapes and passes over wildlands patchily. The fact is that the pyric transition has simply not been studied sufficiently to establish patterns and estimate rates with much statistical confidence. Some places make the transition briskly and others tediously, while a few seem to have stalled. With respect to fire, scale applies to history as well as to geography.

A small historical sample suggests that it requires the removal of the local, fire-wielding population, a process that can take many years, depending on relevant politics and cultural readiness. In the United States, it appears to have taken 50–60 years on average. The breakneck industrialization of China will provide an interesting test on the scenario and the rates at which it can occur, as will India, whose democratic politics prohibits the kind of authoritarian changes in land use apparent in China.

Equally, in a few places, the process seems to have arrested or split, such that the two economies of fire do not engage, even when fossil fuels are abundant. Nigeria and Angola, for example, are oil-rich states which have not applied that fuel (and the revenues earned from it) to the reformation of their societies.

As part of its revolution, Mexico enshrined local authority, codified in the cession of land to local communities, known as *ejidos*. From a fire history perspective, these cessions had a similar effect as establishing nature reserves, because they spared the protected places from the full impact of the pyric transition. The landscape is fragmented, with some places remade and others not. The removal of that legislation in 1996 offers a test on the model.

Perhaps even more unclear is what happens after the transition. What kind of fire ecology characterizes the newly remade landscapes? What kind of institutions are suitable? One answer is that open fire will disappear from built landscapes and is likely to return to natural ones. However, those strands will braid with the trend toward technological substitution and suppression that characterize the transition overall. The first will work to remove combustion as a source of power, while the second will work to remove flame from landscapes.

12.5.1   Trends in pyrotechnology

Industrialization marked a phase change in humanity's long history of energy production. Viewed as technology, however, the pyric transition has its own trends and phase changes. Two currently dominate. One works to increase the production of fossil fuels, while the other labours both to increase the energy intensity of those fuels and to decrease their carbon content. The paradox is that, even as the world reduces its carbon economy on a *per capita* or *per machina* basis, its overall combustion will likely rise for the foreseeable future. Humanity will continue to run on a combustion economy for many decades, if not centuries.

The amount of fossil fuel seems, in practical terms, boundless. New reserves are constantly being discovered offshore (e.g. Angola and Brazil). Drilling plunges ever deeper for oil, and the entire Arctic Ocean basin will soon open to exploitation. Technology has brought Canadian tar sands into production. It has allowed for reworking of former reservoirs. With hydraulic fracturing (fracking), it is liberating unimaginable amounts of natural gas. What will limit use are constraints imposed by pollution, climate change, economic costs, politics and global population.

Equally, pyrotechnologies are encouraging a progressive decarbonization of the combustion economy.

As fossil fuels have displaced biofuels, so more distilled fossil biomass, with lesser carbon content, will replace those with higher ratios of carbon. Combustion will rely less and less on bulk hydrocarbons, rich with carbon, and will burn more refined fuels that boast higher energy and emit fewer greenhouse gases and pollutants. Eventually, humanity will shrink its dependence on combustion as a power source further, in favour of renewables, fuel cells or nuclear plants. Technology may undergo a second pyric transition, in which 'fire' itself disappears. Prime movers will run on power sources that transcend combustion. What Plato once imagined as a world lit only by fire will become a world lit only by virtual flames (Figure 12.9).

Such technologies will bolster the drive to 'suppress' – really, to exclude – fire on vernacular landscapes. A decarbonized world will evolve first in the built landscape, then likely it will happen on agricultural fields and pastures, and then there will be pressures to extend its reach into wildlands. An easy segue is the argument, all too prevalent, that prescribed fire is no more than another implement in the managerial toolkit, one especially useful for reducing fuels. If open burning is a tool, then the pyric transition will seek to replace it with a better tool not based on flame, and eventually to dispose of it altogether in favour of tools that do not escape, leak smoke or belch pollutants, and that can do their work with more precision than inattentive flames, which can have all the discipline of a snuffling hound. If reducing or rearranging fuels is the point, then there are lots of technologies, many powered by internal combustion, capable of doing that task.

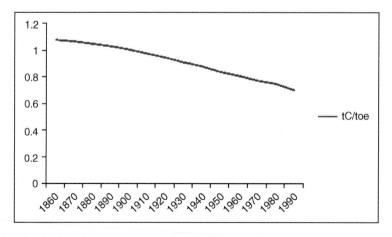

Figure 12.9   The decarbonization of energy expressed as a ratio of tons of carbon per ton of oil equivalent energy (Data from Nakićenović, 1996). Reproduced by permisson of MIT Press.

The argument that prescribed burning is merely a tool misses the point. The justification for holding, or reinstating, fire in nature reserves is that it does ecological work that nothing else can. Fire is not simply a flaming wood-chipper, a mechanical slasher of hydrocarbon fuels, but a biochemical catalyst for biotas. Where humanity wants patches of the natural world, it will have to tolerate or encourage free-burning fire as an essential feature of many protected reserves.

The two faces of fire – as a technology and as an ecological process – will have different outcomes. Despite intentions, removing fire from human society will be slow and incomplete, yet gradually technology will replace the power that combustion has traditionally supplied by power from other sources. The decarbonization of energy is already under way.

Paradoxically, the need for reintroducing fire to maintain ecological integrity is likely to increase, as will the opportunities for wild or feral fire. The nature preserves of most industrial societies suffer not from too much fire, but from too little or from too much of the wrong kind of fire. Over time, the solar cell will replace the Bunsen burner, but lightning and anthropogenic fire will continue to interact in wildlands.

These are trends made by human choice, not by laws of nature. The future has not been written. What the future promises is that humanity will remain the keeper of the planetary flame.

12.5.2 Trends in pyrogeography: selected examples

Each phase of anthropogenic fire has its characteristic pathologies or outcomes that appear problematical to its times. The Anthropocene is no exception. The pyric transition has solved some earlier fire problems, notably the noxious effects of smoke and open flame in cities and the interminable fuel crisis prompted by the insatiable appetite of furnaces and prime movers for wood, charcoal and oil. However, by reorganizing landscapes and changing the character of anthropogenic fire, it has introduced opportunities and problems of its own.

Like fire matters generally, these do not fall into tidy categories, for it is worth repeating that fire is an expression of all its circumstances. They may, though, be grouped according to the intentions behind a change in land use. Call them the fires of new lands, the fires of abandoned lands, the fires of recolonized lands, the fires of reclassified lands and the fires of restored lands. Each expresses it own dynamic, but behind them all lies the Big Burn of industrial combustion.

12.5.2.1 New lands

In the post-WWII, era many countries with large hinterlands sought to develop them for export commodities or internal settlement. In effect, they renewed colonizing, this time with the power of industrialization behind them. Some, such as the Soviet Union, Australia, Canada and the USA, were already well industrialized, but others were not. For them, large-scale land clearing became an issue not only for an emerging environmentalism but for Earth's pyrogeography. Those lands were cleared by fire.

These nations experienced, in the 20th century, the pyric transition that developed nations had undergone in the 19th. The colonizing resulted from pushes and pulls. The pushes were often political – a desire to convert 'unproductive' land into export crops, to establish an administrative presence in places perhaps subject to political unrest, or to satisfy internal pressures. The pulls were the attraction of relatively free land and a generally stable form of wealth that was not so readily eroded by inflation. In the end, the 'new lands' were colonized out of economic incentives and with resettlement schemes that intended to move people from overpopulated areas to what were deemed underpopulated areas.

The two most notorious episodes have occurred in the tropics, as Brazil began to develop Amazonia and as Indonesia and Malaysia turned on Borneo. Between 1964 and 2010, the population of Amazonia increased from six million to 25 million and the forest cover shrank by 20%, most of it transformed into pasture. In Indonesia the wholesale draining of organic soils led to stubborn fires that immersed the entire region seasonally in a noxious haze that closed airports and undermined public health. Fossil-fuel powered transportation allowed, respectively, for commercial pastures and plantations of palm oil. Massive industrial slash-and-burn replaced small-scale swidden. Roads introduced weeds, many of them pyrophytic, and fires from the cleared sites crept into the surrounding forests, gnawing at their frontiers and, where further penetrating, subverting their ecological integrity.

During major ENSO droughts, the burns could be colossal. In 1982–83, fires burnt an estimated 3.5 million ha, of which 0.8 million ha were primary rainforest. In 1991, some 200,000 ha burnt in forest (other land categories were not included). In 1994, some 5.1 million ha burnt across the medley of landscapes. The 1997–98 season claimed a whopping 9.655 million ha. With each outbreak, a dense, cloying smog, euphemistically termed 'haze', smothered the region, prompting calls for trans-border agreements and fire control (Figure 12.10).

Still, like pre-industrial colonization, these were mostly one-off burns. They fed on fuels that, once burnt, would not re-burn in the same way. At any site, the magnitude of the burning would decline. Once completed, the wave of clearing would give way, on pastures, to milder annual burns, and on palm oil plantations to an export of biofuels. Compared to past settlements, the difference was the rapidity and thoroughness of conversion, as what had traditionally taken centuries occurred in decades, and what had previously re-grown as complex fields and fallows was now simplified by the throughput of industrial energy and chemicals. Land colonization had passed through the pyric transition (Hecht and Cockburn, 2011; Goldammer, 2002).

12.5.2.2 Abandoned land

An eerily inverted process – abandonment – is occurring on many long-settled landscapes that are rapidly subjected to an industrial economy. The best expressions are found along Mediterranean Europe. The upshot is the replacement in the built environment not only of open by closed combustion, but of tamed fire with feral fire on the land (Figure 12.11).

The political economy behind the process involves the rapid industrialization of nations with established agricultural countrysides, catalyzed by accession into the European Union and often by the collapse of dictatorships. The old fields can no longer compete with a global market; the population, particularly the young, flee the countryside for the cities; the land, no longer closely cultivated, overgrows with vegetation; and fires, accidental and deliberate, scale up from light seasonal washings to deep scourings. In many places, the landscape combustibles are further stoked by conscious attempts to convert pastoral commons

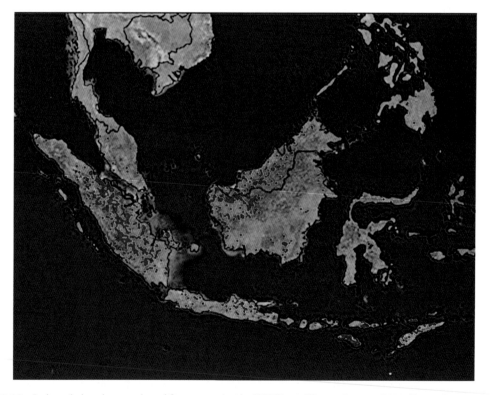

Figure 12.10 Indonesia burning, as viewed from space by the MODIS satellite on August, 2008 (Image from NASA).

Figure 12.11   Europe's fires concentrate into the Mediterranean Basin (and developing Balkans), 2000–2009. Note the particular density associated with north-western Iberia (Product of MODIS satellite, as mapped by European Forest Fire Information System).

into commercial forest plantations, often with flammable exotics such as eucalypts.

The outcome is to boost the burning, while dampening the means to control the fires. The more radical the removals, the more unchecked the wildfires. The process is worse where winter rains are normally sufficient to produce lush growth that the droughty summers burn off. In short, natural and human geography interact with almost fissionable explosiveness. The phenomenon is common throughout Mediterranean Europe, but most damaging in northwest Iberia and Greece. Galicia and northern Portugal have become the scenes for deadly conflagrations.

But characterizing the scene as one of abandonment can apply to the response as much as to the cause. The Mediterranean biota is intrinsically fire-prone, but what has held it in thrall is a close cultivation that is an intimate expression of the prevailing social order. In European reckoning, wildfire is something that follows from the breakdown of that order, often associated with unrest and arson, and control will be re-imposed with the restoration of order. Clearly, the solution is to find an industrial-age equivalent to cultivation and light burning. Instead, however, the political response has too often been to treat the fires as simple insurrection, or a threat to the social model, and to attempt to fight back with crews, engines and aircraft. Since the originating irritant is land use, a firefight, however telegenic, does nothing to correct the cause and by itself will only aggravate the

scene. Where the land has been abandoned, it spawns abandoned fires.

12.5.2.3   Recolonized lands

Industrialization reduces pressure on rural lands for commodities and allows them to be put to recreational uses or dedicated to nature protection. In many countries, however, they have become subject to a new frontier of colonization by urbanites; they fill not with pastures or plantations, but with houses. They display a patchy mosaic of exurbs, protected wildlands and disused agricultural lands. If the geography is naturally fire-prone, then this frontier, like its land-clearing predecessors, can erupt in a wave of flame. In the United States, some 10% of the national estate and a third of housing units lie within this zone of recolonization (Radeloff *et al.*, 2005).

Such communities are not living off the land – theirs is not an agricultural economy. They are living on it. Their income comes from elsewhere. These are often second or retirement homes, and they exist for amenity values. Because the land is not being utilized to produce food, fibre or wood, it suffers an economical abandonment and can overgrow, and this tendency is often reinforced by a desire for a 'natural' look and the screening properties of dense vegetation. Worse, aesthetic preferences can lead to wood-structure houses that behave, with respect to fire, like slash piles. Because the settlement reaches beyond municipalities, it often lacks an infrastructure for firefighting, much less fire management within a context of land management. The outcome, paradoxically, is to create a wave of combustibles not unlike that created by earlier land-clearing frontiers that, in turn, stokes a new wave of deadly fires.

The dynamics can be found throughout the developed world, but they have been studied most fully in Australia and the United States. Research consistently points to the house itself as the site of failure. Most houses succumb not to a wall of flame but to surface fires that can move from nearby vegetation to vulnerable points, or to ember attacks that send swarms of sparks, like a hive of angry bees, to search out spots of weakness. Combustible roofing, flammable vegetation adjacent to wood construction, vents and screens that can carry embers into the structure – these are where houses catch fire. So, while some landscaping is useful, the structure and the density of other outbuildings, garages and nearby houses are the critical determinants of damage.

In brief, these are built environments that need to be treated as such: they are patches of urban fire and they can be contained by the same strategies used successfully in cities, by land planning, zoning and fire codes for construction. Such measures are easier to identify than to enact but even in the United States, although sprawl is universal and interbreeds with whatever natural hazards are around, the losses due to fire are relatively minor compared to those from wind and water, and they are specific to certain sites. From 1990–2010, for example, some 85% of the houses burnt (and even more of damages) in the US have occurred in California. Such fires are to that state as hurricanes are to Florida. Elsewhere, the losses are comparable to those due to tornadoes (Figure 12.12).

Once they become large, such fires exceed the capacity of fire agencies to control or to provide structure protection or public security. One solution is to evacuate residents ahead of approaching fires, a strategy that merges nicely with the general thinking of emergency services, who want civilians out of the line of fire and away from their operations. Another is to train able-bodied residents to protect themselves and their homes, on the reasonable theory that the formal fire services can never protect all the structures as risk, that last-minute evacuations are a primary cause of fatalities and that, knowing they will be called upon to defend their property, residents will be more inclined to prepare.

The controversy pivots on matters of timing and preparation. Evacuations, to be done well, must be done early. The choice to stay and defend or shelter in place is worse than ineffective if not accompanied by suitable preparations.

12.5.2.4 Reclassified lands

The reservation of lands for nature protection, generously defined, is a feature of industrial societies. Where fire is possible (or even natural), these areas can become sites for problem burning and, where they are large, they can kindle massive fires – or what has come to be termed 'megafires'.

The surprise is that such fires have emerged on lands that have been 'protected' for over a century, and where public fire agencies seem to have brought fire under control. They have broken out in precisely those countries that, outfitted with public lands and agencies to administer them since colonial times, had evolved the most complex apparatus for fire protection. What upset that assumption, and why now? The reason is that such lands can be dedicated to different uses, each of which supports distinctive fire regimes, and that, over the past 30–40 years, the reserved lands have been reoriented to new purposes, with a new suite of appropriate practices. Reclassifying them has created new categories of fire (Figure 12.13).

The tributary causes are several. One is the simple ecological aftershock of fire's abrupt removal – or just the attempt at removal. Available biomass that fire had long 'cultivated' turned weedy and effulgent and grew in patterns that made it more accessible. The issue was not large chunks of live or dead boles, but reproduction, shrubs, tangles of scrub and close-packed stands of conifers, all ready to burn. These not only ceased to burn, but were no longer subject to logging, thinning, slash burning, grazing or other practices that might have held them in check and were made less accessible by limitations on access by machines. They went, in brief, from being state forests to being parks, preserves and wilderness.

That legal designation constrained what people might do or not do. Of special concern were old-growth stands – always rare and, as logging had scalped many, targets for strict protection. These expanded, as advocates had hoped. The blowback, however, was that these biotas were often prone to stand-replacing fires and had reproduced over aeons from such events. Now, eruptions of this sort might be

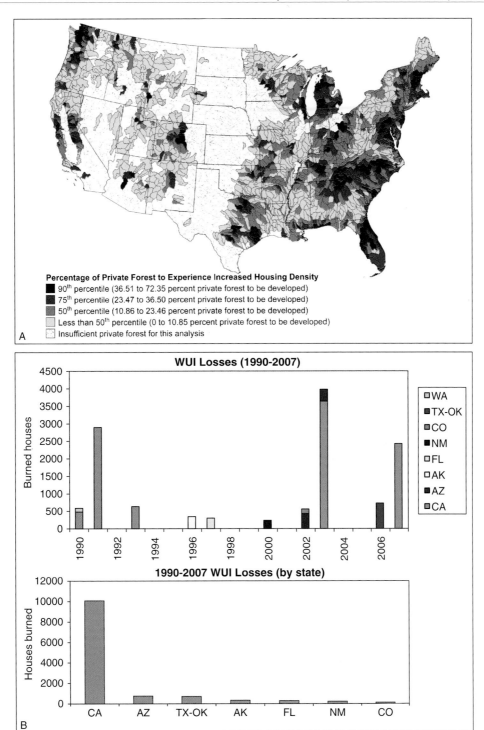

Figure 12.12   The wildland-urban interface in the United States, 1990–2007.

A. Map of forest areas at risk (From U.S. Forest Service).

B. Graphs track the growth of burnt houses, and the concentration in a handful of states, overwhelmingly California (Data from Cohen 2008).

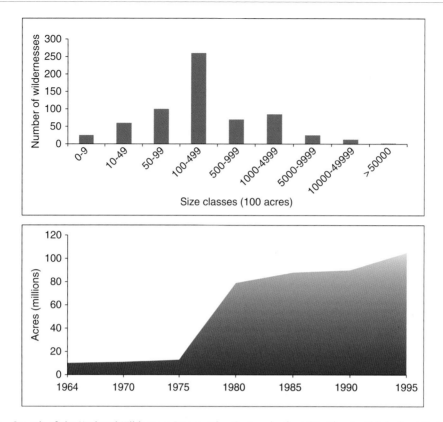

Figure 12.13 Growth of the National Wilderness Preservation System in the USA. About a third of public land lies in the west, and about a third of that in Alaska; the distribution of wilderness follows accordingly. The big increase in acreage in the late 1970s comes from Alaska lands (Data from Gorte, 2005).

likened to clear-cutting, in that they stripped the precious woods away. It mattered little whether the trees were lost to saws or flames; they were gone and were unlikely to be replaced in less than centuries. The fires they needed biologically would be denied them by their protectors.

Thus, the accepted value of some intense burns, the likelihood of big fires and the inability to intervene except through an aggressive fire suppression (which was doomed to fail while authorities became increasingly reluctant to risk firefighters against remote fires) all grew and converged. Crews backed off and burned out routinely, while fires blew up more frequently and from smaller sparks. More hectares burned. Fires merged. Out of the flames, the spectre of the megafire appeared (Figure 12.14).

Critics pointed to climate change as the underwriter of the increase in big burns (a fiery counterpart to increased hurricanes), and favourable weather was certainly essential. However, the fires were the product of humans interacting with nature in complex ways; they appeared when and how they did because people decided to live on the land and to relate to fire differently. Large wildland fires erupted because there were large wildlands set aside for them to inhabit.

12.5.2.5 Restored fire: public lands

If megafires were the unwanted or accidental by-product of changes in public land use, deliberately restored fires of various kinds were their benign counterpart. Lands that had witnessed less and less fire over decades of protection began to experience fire again as a result of conscious policy and practice. Restored fire began to challenge wildfire and, in some locales, to replace it with haste and deliberation.

This quest confronts two grand challenges – one philosophical, the other practical. Interestingly, the two are intertwined. Principles only have meaning to the extent that they can be expressed on the ground.

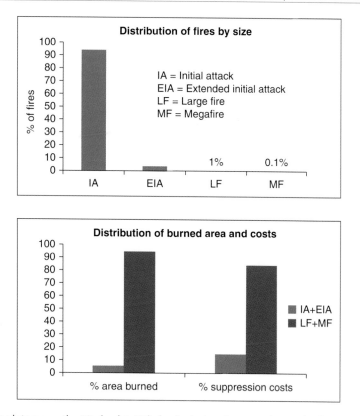

Figure 12.14 The fire plutocracy: the 1% (and 0.1%) dominate burnt area and costs in the US (Data from Williams and Hyde, 2009).

Equally, practice has no direction without bonding to some larger purpose that can easily be called 'philosophy'.

The principle behind restoration is simply stated. Natural areas deserve – require – natural processes to function, and banishing those processes can cause upheavals that lead to problems and even to ecological insurgencies, of which wildfire is an excellent example. In fire-prone or historically burnt areas, fire belongs as much as wolves – it is a matter of principle, but also of ecological soundness. Without the right regimen of burning, the system does not work properly. It becomes unstable. It fails to yield the ecological goods and services desired.

The philosophical issue is what exactly is being restored – a former fire regime, or just fire, or a fire-resilient landscape capable of adapting to altered and, perhaps, unforeseeable conditions? By what methods ought that restoration be accomplished? Should the land be more or less opened to fire and let the ash fall where it may? Or should people consciously set targets, apply fire and monitor the outcomes against announced goals? Are people reinstating a natural regime by natural means, or is it enough to simply put fire back, even if that means slashing and burning? Is any human intervention justified in places specifically reserved to be untrammelled by the hand of humanity? Having upset a quasi-natural order, should people be allowed to create suitable conditions so fire can perform its appropriate role? How should administrators accommodate the fact that many 'natural' landscapes are, in truth, 'cultural' landscapes of a peculiar kind? In brief, how, in principle, should fire be reinstated? To what ends?

Principles, however, only have meaning to the extent that they can be applied, which leads to the flip-side challenge – practice. How, exactly, might fire be restored? By what means? At what costs? Under what social compact? In reality, restoring fire is akin to reinstating a lost species. Its success requires a suitable habitat. Fire synthesizes its surroundings; it will take its character – do its biological work – from its context. Simply dumping fire onto the land may or may not restore the old regime; it might depend on how long fire had been withheld, and what changes had occurred by way of enhanced fuels and species vanishing and invading.

Again, two strategies are common, both of which recognize the need to reconcile fire and land. One emphasizes land, and other fire, as its primary focus. The land strategy wants to get land into a proper condition so that it can promote the fires people want and help contain those they do not want. In this way, choices about land use determine the character of fire. The second strategy reverses the emphasis. The choice of fire practices will determine what kind of land use results. In remote areas, typically vast or self-contained by mountains or deserts, amid biotas that can take a wide range of burning, it may be possible to allow fire to ramble on a landscape scale. Such distinctions, however, tend to clarify what is, in reality, muddled. Most fire management programs, if they succeed, mix and match from among whatever practices can achieve the agency's goals (Figure 12.15).

While there are many ways and examples by which to characterize fire's restoration, perhaps the most useful is between wild and working landscapes. In wildlands, the preference is to adopt a natural-fire strategy that essentially returns (or, by another perspective, outsources) the task to nature. It grants as much space as possible for lightning fire to hew its way, and allows the embers to fall where they might.

The advantages of this approach are that managers need not understand everything about the system (nature will provide). They do not need to commit staff and equipment (nature will oversee) and they are not culpable in the same way as someone setting a fire would be (the burn is an act of God or nature). The disadvantages are that the process may not, in fact, restore the landscape to its former state; that some fraction of fires (and their smoke) will inevitably escape and require expensive efforts to contain them; and that the absence of human agency is a legal fiction that may not survive challenges. By a variety of titles – prescribed natural fires, wildland fire uses, resource benefit fire – the practice has succeeded in getting hectares burnt. In most settings, despite considerable resolve and expense, prescribed fire remains a niche enterprise.

In working landscapes, the preferred strategy is to rely on prescribed fire. Managers set fires at times and places and under circumstances that permit a tamed, or at least captured, fire to do the work asked by humans. Since people are already fussing with the landscape, there is less concern with other manipulations, such as tinkering with the fuel array to encourage the kinds of fires sought. Controlled fire is another means to work the landscape. If a tradition of burning has survived, such fires may be accepted by local communities.

The downside to prescribed fire is that it can be expensive, its fires can also escape and it (or its instigators) can be brought to court. Generally, success requires that the practice be part of a system in which there is cultural acceptance, enough space to accommodate slop-overs and escapes and legal tolerance for the unexpected.

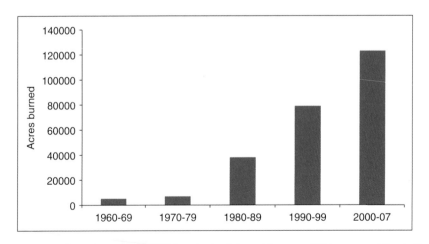

Figure 12.15   The growth of area burnt by natural fires in the Selway-Bitterroot Wilderness and Frank Church River of No Return Wilderness in the Northern Rockies, USA (Data from US Forest Service).

The United States offers good illustrations of both approaches on its public domain. Carolina Sandhills National Wildlife Refuge provides an especially sharp demonstration of a program that began with a surfeit of wildfire and virtually eliminated the problem by substituting prescribed fire. Sequoia-Kings Canyon National Parks, with a larger and more mixed landscape, succeeded in reinstating fire to the Big Trees by prescribed burning, while allowing naturally ignited fires to flourish in the backcountry (Figure 12.16).

Both drew on fire cultures. For Carolina Sandhills, there was a resident culture of rural burning that never fully vanished before it was resuscitated and redirected to species protection. For Sequoia-Kings, an invented culture grew out of convictions that fire had to return for the vaunted sequoias and wilderness woods to flourish.

However, both also had administrative approval and even legal heft behind their efforts. Carolina Sandhills hosted the endangered red-cockaded woodpecker, so had a legislative imperative to intervene; since the bird depended on the fire-thirsty longleaf pine, refuge managers needed to burn and take whatever steps were needed to make the burning possible. Sequoia-Kings had the 1963 Leopold Report that had identified the Sierra parks as places profoundly out of whack because of fire exclusion. That became, in 1968, the foundation for a new manual of park administration that encouraged fire's restoration. Both the Endangered Species Act and the National Park Service's Administrative Guidelines for Natural Areas removed (or at least suspended) the philosophical arguments. Both sites could concentrate on technique.

The differences between them matter: they are not interchangeable. Prescribed fire is considered an unwarranted intrusion in Sierra wilderness, while legal wilderness in Florida will not only accept prescribed burning but make it a foundational practice. In such ways, philosophy translates in practice and vice versa. What is hard to accept is the charge that prescribed burning is somehow alien, as though it were the operational equivalent of an invasive species. It is fundamental to humanity's ecological agency and, through us, to many landscapes. To remove it may be as great a shock to ecosystems as to remove naturally ignited burns.

12.5.2.6 Restored fire: private lands

Left to its own ends, the pyric transition would abolish open fire so far as it is possible. No open fires would exist in built landscapes and only wild settings, where fire did indispensable ecological work, would retain 'friendly flame'. The middle landscape between the urban and the wild would slowly trend toward fire's exclusion. However, such predictions depend on the transition – so new a phenomenon that its ultimate end points are unknown – proceeding unchecked toward that pyric peneplain. They ignore the stubborn resistance offered by local societies and settings.

It appears that the transition slows as it ages and, in many places, it may stabilize after overshooting. It may, like the imperialism that guided its global reach, be forced to rely on indirect rule; it may hybridize with indigenous practices; and it may adapt to local circumstances. For decades, it appeared that the torch had passed into the hands of governmental authorities, and that it had become something average citizens could not be trusted to use. It was an intrinsic threat to public safety and must, therefore, be entrusted only to sanctioned agents of the state. Instead, a counter-reformation has emerged to create a civil society for fire.

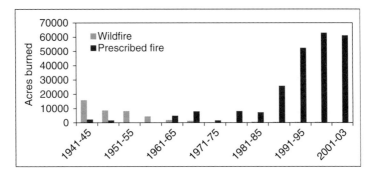

Figure 12.16 Wildfire replaced by controlled fire, Carolina Sandhills National Wildlife Refuge, USA (Data from Ingram and Robinson, 1998).

Since it has progressed so far through the pyric transition, the United States offers, again, a fascinating profile of what a hybridized fire community of local and nation, private and state institutions might look like. The effort to reinstate fire into the country's landscapes, to ensure a landowner's right to burn, to accommodate a mixed realm of combustion and purposes, to encourage a variety of intellectual understandings and research traditions, all resulted from protest by citizens in the 1960s against what they regarded as an unwarranted state hegemony over fire. To be effective, though, dissatisfaction had be organized. Prescribed burning, in particular, had to operate on a landscape scale among like-minded and similarly skilled practitioners. This trend differs from simple privatization, in that the latter only outsources to the private sector what the public sector wants done. The emergence of a genuine civil society challenges the ends and means of the public interest in fire.

The upshot is the evolution of private-sector associations for fire research and management, the appearance of local fire associations (and a national Coalition of Prescribed Fire Councils) and a renegotiation of rights and responsibilities among all the responsible parties to the American fire scene (a National Cohesive Strategy for Wildland Fire Management). The rural fire scene has firefighting capabilities, notably with volunteer fire departments. It does not have equivalent capacity to apply fire. Unless a landowner's estate is very large, it is hard to muster the necessary resources, which means individual landowners must aggregate their lands and interests – and even then, they will likely require a supplement in the form of equipment, funding or research. Very few operatives have the social and political space and economic reserve to manoeuvre on their own, or to counter the compelling processes set in motion by the pyric transition. Only by 2012 did the US have a rough atlas of what kind of burning occurs, and by whom (Figure 12.17).

All of this is very much a work in progress. Some actors are large and relatively sophisticated, e.g. The Nature Conservancy and timber companies keen to burn to dampen fuels. Most, however, are small or

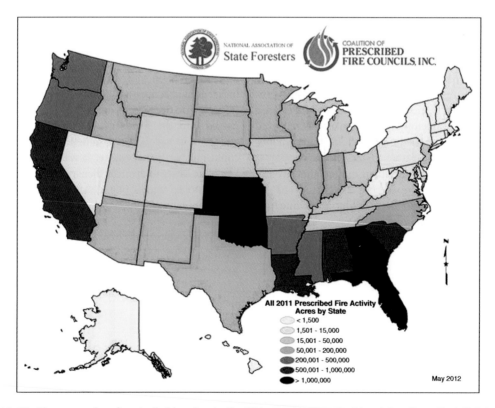

Figure 12.17 The geography of controlled burning in the USA, roughly organized by states (From Mark Melvin, 2012). Reproduced by permission of the Coalition of Prescribed Fire Councils, Inc.

isolated and must cooperate to gain the knowledge, equipment and legal sanction to burn. As the competing processes sort themselves out, the national landscape will become dappled and jumbled. The simplicity of a single-theme driver of fire history will splinter. No one, however, should lament the loss of conceptual clarity when it comes with better practices on the ground.

12.5.2.7　The Big Burn

All of these expressions of human-influenced fire, even the most mammoth megafires, pale before the magnitude of industrial fire, itself the most potent of the pathologies of the pyric transition. There is little prospect that it will shrink soon. Nations seek the transformation as rapidly as possible.

New sources of fossil fuel are being developed briskly. While not limitless, they appear unbounded for centuries. Controlled combustion remains the primary means to power, and fossil fuels, the principle source of combustion. The effluent of the outcome is the big dump of pollutants that has so overloaded the Earth system and is unhinging even our climate.

As the magnitude of the Anthropocene becomes more apparent year by year, accelerating through a cataract of history, it is worth pausing to recall that its core is fire. Global warming – climate change and much of the rest of the global change agenda – are the result of a radical mutation in how humans manipulate combustion. Many observers have commented on how global warming will affect fire on the land, and a few have brooded over how landscape fire might affect global warming, but fewer still have sought to link these through their shared cause. The driver of the Anthropocene, its informing principle, has been anthropogenic fire. It is the torch in the hands of humanity that continues to be the lever that governs the machinery.

It is easy to understand why, because of its malfeasance with industrial combustion, humanity's fire practices overall might become suspect and tainted with illegitimacy. However, the claim that anthropogenic fire is somehow unnatural is hard to accept. Fire is what humanity does. It is what we do that no other species can. Managing fire remains the signature of our ecological agency and, in a genuine sense, our ecological duty to Earth. The issue is not whether we must manage fire, but how well or how poorly we will do it.

Further reading

Cohen, J. (2008). The Wildland-Urban Interface Problem. A Consequence of the Fire Exclusion Paradigm. *Forest History Today (Fall)*: 20–26.

Ellis, S., Kanowski, P., Whelan, R. (2004). *National Inquiry on Bushfire Mitigation and Management*. Commonwealth of Australia

Goldammer, J.G. (2002). Fire Situation in Indonesia, *International Forest Fire News* **26**, 37–45.

Pyne, S.J. (2007). *Awful Splendour: A Fire History of Canada*. University of British Columbia Press.

Pyne, S.J. (1997). *Vestal Fire: An Environmental History, Told through Fire, of Europe and of Europe's Encounter with the World*. University of Washington Press.

Radeloff, V.C., Hammer, R.B., Stewart, S.I, Fried, J.S., Holcomb, S.S., McKeefry, J.F. (2005). The Wildland-Urban Interface in the United States. *Ecological Applications* **15**(3), 799–805.

Steffen, W., Grinevald, J., Crutzen, P., McNeill, J. (2011). The Anthropocene: conceptual and historical perspectives. *Philosophical Transactions of the Royal Society A* **369**, 842–867.

Wildland Fire Leadership Council. (2011). *U.S. National Cohesive Strategy for Wildland Fire Management*.

Chapter 13

Fire management

13.1 Introducing integrated fire management

Fire management has become a distinct – often a stand-alone – enterprise of modern societies. It became necessary when the pyric transition began dissolving traditional means of living on the land and, with them, inherited means of applying and withholding fire. Formal fire management substitutes for traditional fire practices; what had been encoded in social mores and ways of life must now be identified, rationalized and reassembled. Progressive thinking today insists that fire is not simply something to exclude but to manage, and that management should be integrated.

In its usual definition, integrated (or integral) fire management (IFM) refers to a full-spectrum programme that includes prevention, preparedness (training, fuels projects), prescribed burning, active suppression and post-fire rehabilitation. It recognizes that all the parts of a programme must fit together. Preventing fires is cheaper than fighting them, but prevention cannot succeed unless the capacity exists to suppress those that do occur. Also, suppression will work best when it is pre-planned and the landscape fashioned in ways to enhance it. What IFM emphasizes is the widely held recognition among fire officers that co-existing with fire means more than hurling crews and aircraft at flames, however attractive this may be as political theatre.

Moreover, integrated fire management should also mean that programmes, policy and practice must synchronize with their sustaining society. It must fit with how a people actually live on the land, or wish to live. It must mesh with politics, mores and knowledge, particularly for societies that have broken the old links that long bound their ways of living with their ways of using fire. An integrated programme must find or forge new bonds; it must invent research mechanisms to create the knowledge where traditional lore has been lost or does not apply; it must devise policies to substitute for inherited and long-adapted customs; and it must construct a governance structure by which such invented norms may be applied with a sense of social legitimacy. Such an undertaking goes far beyond the usual scope of 'fire', whether as protection, management, or crisis response.

Most often, 'fire management' means protection from wildfire – a necessary task. A wildfire is an emergency that threatens lives, property, or ecological assets. Society must counter it and compensate for the losses it imposes. A more robust notion, however, is that fire management means protecting against fires you do not want and promoting those you do. It means developing the techniques necessary to apply and withhold fire but, more profoundly, it means engaging with the sustaining society and its culture to determine ends and means – which fires are deemed good and which bad, what methods of management are judged appropriate, and what costs and rules of engagement are acceptable. It means determining the

Fire on Earth: An Introduction, First Edition. Andrew C. Scott, David M.J.S. Bowman, William J. Bond, Stephen J. Pyne and Martin E. Alexander.
© 2014 John Wiley & Sons, Ltd. Published 2014 by John Wiley & Sons, Ltd.

why of fire management as well as its how. It means deciding just what the 'problem' truly is.

13.2 Two realms: managing the pyric transition

The pyric transition divides fire management into realms, as it does fire on Earth overall. There is one regimen characteristic of those societies that have not made the transition, another for those who have, and yet another for those still undergoing it. Since the untouched are few (and fading), the larger question of fire management means how to make the transition and how to establish a working system once the transition has more or less passed.

13.2.1 *Amid the transition*

Societies in the transition have the tougher challenge. They must, typically, fight off an eruption of bad fires, which requires a capacity they do not have, while not extinguishing all good fires, which they still require. The deep driver here is the need to envision – and to the extent possible, to enact – the kind of landscape they wish to have when they finally pass through the transition. Such tasks lie far beyond the grasp of fire organizations. No fire organization can control broad patterns of land use, internal migrations, economic shifts to manufacturing or services, inflation or currency exchange crunches. They can only deal with their epiphenomena as these appear in woods, fields, pastures and wilds.

Understandably, consultants, fire officers and officials will want to install 'modern' systems of protection or management as soon as possible. The wisest counsel, though, may be for fire agencies to build up local resources to cope with local problems, while understanding that the present circumstances, the rolling crises, are ephemeral. They will last perhaps 50–60 years – a span of several working generations – but are not a permanent condition. As lands divide and gel into the geographies they will have on a more permanent footing, separate policies and practices should evolve with them.

A sense of urgency, however, will likely evolve to create national programmes in which sparse and expensive resources can be shared. Mostly, this means fire suppression technologies such as aircraft. It also means creating a civil society for fire management, since no apparatus of the state can cope alone with all the varied fire needs of a place. This is a matter of legitimacy as much as effectiveness.

Since fire management even in advanced countries is largely reactive, it is quixotic to assume that an emerging fire economy will have more than marginal control over its own development. That, however, is the need.

13.2.2 *After the transition*

What fire management system is appropriate after the transition depends on what kinds of lands emerge and what a society wishes those lands to be. The primary distinction will probably hinge on public and private lands, each of which will have different requirements for fire management.

Generally, private lands will continue to undergo a transition from open to closed combustion, and fire agencies will assist by enforcing codes and fighting fires that result. Two exceptions are likely. In subtropical lands devoted to extensive agriculture, particularly grazing, controlled burning will persist as a means of refreshing grasses and protecting against wildfire. Examples include forest plantations and, especially, grazing such as occurs in Australia's Northern Territory. The other exceptions are privately owned landholdings that serve a public purpose, such as the preservation of species and habitats. An example would be lands overseen by The Nature Conservancy. Where fire is a part of a preserved historic landscape, it needs to continue. Elsewhere, flame will recede from fields, forests and cities. The pyric transition will, by default and determination, work out its illimitable logic.

Public lands such as forests, parks or wildlife refuges present a different problem. They reside, at least in principle, outside the pyric transition – that was the point in their creation. However, if they occupy places that are intrinsically fire-prone, their fires will not seep away through attrition by substitution and suppression, as with most private lands. Moreover, they are subject to different styles of management; they are, by their nature, political entities, subject to causes and answerable to constituencies perhaps well removed from their particular setting. The most extensive of these holdings tend to be relics of a colonial era, which is a further complication. Wildland fires will preferentially collect in such places, if only because, by definition, these are where the wildlands are.

That is the post-pyric geography in abstraction. In practical terms, the countryside tends to fragment into three landscape types. One is the rural or working landscape, as reshaped by a fossil-fuel-stoked industrialization. That reconstruction may still be in progress or may have achieved a certain quasi-stable accommodation. A second is protected landscapes. These are places reserved to shield nature from most human usage, except the most benign (such as hiking, birdwatching, or picnicking). Some may be outright wildland, or even legally constituted wilderness or biosphere reserves. A third involves the urban recolonization of rural lands, which places suburbs and strip development adjacent to or within fire-prone settings.

Each of these three geographies has its own suite of preferred practices. On working landscapes, the pyric transition rubs fire away, except in subtropical locales where burning is necessary to maintain grazing or beat down brush encroachment. On wild landscapes, the process historically begins with fire protection, since most such lands were reserved in order to shield them from the extravagance of wildfire during the transition. They then struggle to reincorporate an appropriate regimen of burning. With the peri-urban landscape, the tendency is to adopt urban fire standards and all-risk services. There is, in brief, no single policy, practice or programme to embrace them all.

13.3 Strategies

What 'fire management' mostly means today is wildland fire management on public lands, or on those patches of private lands held for public purposes, such as ecological servicing. Some places lack any disposition to fire. If these places burnt in the past, it was because people created the circumstances and kindled the fires. Unless protecting the cultural landscapes that resulted are the reason behind the preserve, fire management will mean keeping fire out.

However, most such reserves are fire-prone, and their fire administration means installing and maintaining the right regime. In broad terms, four options or strategies are possible: lighting, fighting, leaving

alone and reshaping. Each has its advantages and its liabilities. While the issues seem clear in principle, reality is murkier. Wildland fire management mixes, blurs and fudges. The ancient military adage that no plan survives contact with the enemy is certainly true of wildland fire. In almost all instances, proper management will mean finding a suitable mix among the options, adjusted to particular sites. In the US, policy now allows for various responses to even a single fire – a recognition that fire can have multiple characteristics along different parts of its perimeter.

Because of its extensive record, American experience can provide examples to illustrate the principles under discussion. This will keep the four options within a single national framework, which helps to make them choices available to a common society rather than choices among societies. This chapter will conclude with examples from elsewhere in the world to illuminate the ways that certain strategies come to define national programmes. All illustrate that the core of fire management is land management – how a people live on their land. The firefight is an emergency response, like a post-trauma surgery, not a means by which to govern or live sustainably in fire-prone settings.

13.3.1 Strategy 1: leave to nature

13.3.1.1 Overview

One option is to do nothing, or to do as little as possible. The idea has practice, policy and often philosophy behind it. Practice suggests that removing human agency is easier, cheaper and safer than fighting fire; policy, that the human presence may be at odds with goals of naturalness; and philosophy, that human intervention is generally unwise and rarely successful. For fire management these understandings translate into programmes to grant wildland fire as much room to roam as possible (Rollins et al., 2007).

The practice notes that it is expensive, often dangerous and frequently unnecessary to fight fires in remote settings, and that suppression brings environmental costs on its own in the form of scarring fire lines, intense burnouts and waters polluted with chemical retardants. Not attacking fires that cannot spread outside of their political reserves is sensible; it saves money and advances ecological goals. So does a variant, which backs off from frontline fighting and allows a fire more space to free-burn and a suppression organization a chance to halt spread on its preferred terms, where natural barriers can leverage its strength.

Policy matters in places dedicated to nature protection or legal wilderness, where active intervention violates the spirit of the reserves if not their charter. Whether the intervention takes the form of suppression or prescribed burning matters little. The breach is the human hand doing what should be nature's work. In reality, nature reserves are varied – not even a spectrum, so much as a constellation. Some aspire to be as pristine as possible. Others, devoted to particular species or habitats, may require active tinkering, if only because their scale will not allow wide-ranging processes such as seasonal migrations and fires to free-roam. Intervention is conceived as a surrogate that is less damaging than doing nothing.

The philosophy is largely one attracted to non-anthropogentric values such as biocentricity, ecocentricity or notions of deep ecology. It begins with the argument that the record of humanity's environmental tweaking in wildland settings contains few successes. Overwhelmingly, the chronicle shows many failures to enhance landscapes according to environmental norms, often resulting in degradation and mercilessly subject to unintended consequences. Such observations hold, whether or not the intervention is based on the best available science, since science is always incomplete and tends toward a reductionism that frequently aggravates, rather than alleviates, damages. Each failed intervention invites another (this formulation, of course, assumes an older style of administration in which science proposes and management applies; concepts like 'adaptive management' aspire to temper the process and leave it less vulnerable to the inevitable unexpected consequences).

All in all, with or without a deeper philosophical grounding, tolerating more fire where possible can advance an agency's mission and make better use of its limited resources. It is a case of practice shaping policy, and both trumping philosophy. Where done well, the strategy is part of a larger plan for administering land and fire that explains why, where and how more fire will be accepted. It may specify prescriptions. It will likely include active monitoring, either by air or on the ground. Ideally (though rarely in practice), it will include an evaluation of both the

process of accommodating and of fires' effects. Partisans may dismiss the need for post-fire assessments on the grounds that whatever happens is, by definition, for the good, but breakdowns in operations will likely put the public in a different mind.

Even where not implemented, the tolerance for more fire may affect other aspects of an integrated programme. It may, for example, encourage prescribed fire as a necessary alternative. It might also promote less aggressive suppression by encouraging more indirect attacks and lightening the imprint of suppression. From an administrative perspective, relying on lightning fire deflects culpability should the fire go wrong or outcomes differ from what is expected. There is no agent responsible (or accountable) for either cause or consequence.

13.3.1.2   Evaluation

The strategy, like all the others, has its downsides, the most spectacular and politically dangerous of which involve escapes. A fire may bolt out of its reserve and threaten timber concerns, protected but fire-sensitive habitats or even towns. Or, if flame does not escape, smoke might, swirling far from the originating landscape and causing accidents or having deleterious effects upon public health. Such breakouts lead to calls for direct action to halt the unwanted burning. At this point, with the fire large, containment is onerous and expensive. The costs to public support for policies may be even more exorbitant than operational expenditures.

Even without such ruptures, the programme, if done properly, is not free. It requires monitoring, which may go on for weeks. It needs an infrastructure capable of fighting escape fires, although these expenses are often hidden within the suppression organization. There may be a demand by the public for evaluation, which can alter not only future practices but an entire fire programme. The environmental consequences can only be known after the fact; this is a faith-based ecology nicely suited to support a philosophy of wilderness or naturalness, but it is not necessarily adept at shielding threatened species or habitats. If the strategy is loosed onto a legacy of disrupted landscapes, free-burning fire will most likely not behave as it did before or recreate former landscapes.

Mostly, this strategy requires considerable space. Geographic space in that fire, to work its range of influences, needs room to respond. Intellectual space in that it requires a willingness to stand aside and, without knowing the full consequences, let nature burn as it will. Political space as a tolerance for a pluralism of fire practices. (Revealingly, the official term has consistently proved unstable, morphing from 'let-burn' to 'prescribed natural fire' to 'wildland fire use' or 'resource benefit fire', which partly reflects evolving notions of what the practice really is and how it ought to be done as well as concern that the public know the agency is still 'in control'.) All in all, the strategy works best in remote, protected, mountainous fire-prone but self-contained places that have not experienced a serious rupture in their fire histories.

13.3.1.3 Selected examples: from proof of concept to national policy

As a legitimate option, the strategy originated in the USA with the two great national parks in the Sierra Nevada of California, Yosemite and Sequoia-Kings Canyon. Both had high-elevation landscapes intended for management as wilderness, broken into forest and rock, and were ideal for self-regulating lightning-kindled fires. Excepting a sharp break in 1989, when national directives compelled all programmes to be rebooted, the scene has moved from experiment to established practice.

The most intensively studied area is the Illilouette Creek basin, which was incorporated into Yosemite's wilderness fire zone in 1973. Over the next 40 years, the Illilouette experienced 157 fires, 28 of which exceeded 40 hectares in area. A dramatic series of maps reveals how the burns fashioned a dynamic jigsaw puzzle of scars. Some areas have re-burnt more than once, while some have yet to burn. The actual consequences of any fire depend on weather and fuel, which correlates with time since last burn. The burn record is striking compared to the previous 40 years, in which suppression kept all large fires out, but 40 years is still a short slice of fire history. How future fires will interact with those that have so far occurred will surely both fascinate fire officers and frustrate simplistic models of fire and fuel (van Wagtendonk, 2007; Figure 13.1).

A variant has evolved in Alaska, where towns are scarce and the backcountry large. While it was still a territory, a fire protection system was established under the Bureau of Land Management. As a result of the Alaska National Interest Lands and Conservation Act of 1980, the state underwent a division of ownership – and hence of fire responsibilities – between the State of Alaska, native corporations and the federal government, which consolidated the fire programmes of its various agencies into the Alaska Fire Service. The full suppression era had proved expensive and never especially effective, and so it became ephemeral.

The outcome was a division of the vast landscape into four zones, each of which had a different priority. Sites of high value (including cities and villages) earned 'critical protection', which meant the most effective response possible. Other sites were ranked for 'full protection', 'modified protection' or 'limited protection'. Each downward grade allowed more room for free-burning fire. Wilderness and other nature preserves, lands with little economic or ecological value, or places remote and costly to access, all adopted features of natural fire programmes in which observation replaced suppression.

Such programmes have rebounded back to the Lower 48. They have demonstrated conditions under which fire might be allowed something closer to its natural role, and they have granted fire officers experience in handling such events. Current federal fire policy now allows for fire managers to respond to a single fire in a variety of ways ('take appropriate strategic responses'), one of which is to back off and allow fire to wend through complex landscapes, while concentrating efforts on more easily defended perimeters and on hardening critical assets such as towns and perimeters.

13.3.2 Strategy 2: replace wild fire with tame fire

13.3.2.1 Overview

A second strategy accepts that fire will occur, or it is in the interests of people to make it occur, and that the appropriate method is for people to carry out the burning themselves. They do so by seizing control of an existing fire regime and burning the land on their own terms, usually pre-emptively. Alternatively, they introduce fire to places where it is possible but is not naturally present, in order to enhance those features of a landscape they seek. They substitute tame fire for wild fire. The term is shorthand for that most hoary of anthropogenic fire practices: setting fires where and when people wish them.

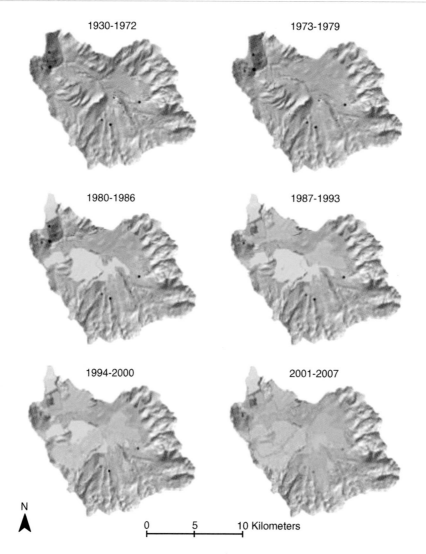

Figure 13.1 Fire in the Illilouette Creek basin, Yosemite National Park (From van Wagtendonk, 2007). Reproduced by permission of Jan W van Wagtendonk.

'Prescribed fire' is thus a capacious and ambiguous term. It can mean setting broadcast burns under commercial plantations, firing slash piles, lighting fallow and field straw or even letting wildfires burn (under a set of guidelines). While the term itself dates from the 1930s, the practice is ancient; it is, in fact, the most elemental of humanity's fire arts. What has changed is its setting: the physical landscapes to which it is applied, the social (often bureaucratic) contexts within which it nestled and by which it is constrained, and the intellectual setting in which it is re-stated in overtly scientific language.

People and torch reincarnate into more modern avatars. Instead of a firestick, practitioners reach for a drip torch or a helitorch. Instead of a hunting ground or a berry patch, they burn logging slash, over-thick understoreys and the habitats of threatened species. Instead of shielding their territories from nature's wildfire, they may seek to extend nature's diminished reach by leveraging its burns with their own.

But fire's setting also includes social and political contexts. Prescribed fire must operate within a legal regime that assigns culpability, determines rights and obligations and may restrict or promote its use. Traditional norms that once governed who could burn, and when, have been replaced by modern ones. Instead of elders or a village headman, institutions have established formal policies, publishing guidelines for conducting burns and creating pro-grammes for training and certification. On private lands, too, regulations govern burning.

Even in places where legal concepts of strict liability apply, there may be fire bans on critical days, and escape fires will result in lawsuits or arrests (escape fires can be defined as vandalism, trespass or arson). On public estates, some countries may grant pre-scribed burning its own identity, separate from sup-pression, complete with its own crews and funding.

13.3.2.2 Evaluation

The benefits are obvious: they are what controlled burning has always wrought. They bring fire where it is needed, and in a regime that suits human interests. While the techniques are many, they flourish better in some situations than in others.

They work most smoothly where fire routinely exists and where its absence results in rapid and obvious changes. Mostly, these are subtropical regions in which, without fire, woody species overwhelm open spaces and may promote high-intensity burns over surface ones, and in which no surrogate methods exist to sustain the landscape in the manner that people wish. Bulldozers and wood-chippers may successfully remove fuel, but they cannot catalyze the range of ecological effects that fire achieves and upon which the ecosystem depends. Not least, a prescribed fire may serve as a surrogate for free-burning fire where matters of scale or a legacy of land management prohibits letting fires free-range.

Prescribed fire works best when it is embedded in a persistent culture of burning, where people regard fire as a natural feature of the landscape and anthropo-genic burning as a customary right, and where they have kept fire in some form on the land. In such circumstances, as landscapes and societies modernize, the actual conditions under which prescribed fire must occur will morph. However, people continue to regard anthropogenic burning as normal and proper and they may take special political measures to ensure its continuance. They may, for example, grant a landowner a legal 'right to burn'. They may place the burden of proof on those who object to prescribed fire, rather than on those who use it.

The converse is equally true. Examples include places where fire is not intrinsic to the native ecology; where fire's absence does not result in obvious and immediate changes; where technological surrogates to

fire's desired effects exist (e.g. reducing fuels); where a culture of burning has expired or has never existed; and where legal protection is shaky. In such condi-tions, where society views fire with suspicion and agencies lack special funding for burning, prescribed fire will struggle. Its liabilities, such as escaped flames, wanton smoke and operational costs, will seem more critical than its nominal advantages.

Modern landscapes, too, have less capacity to absorb escapes. People cannot migrate from burnt to unburnt sites, extensive fire-dependent ranching and farming cannot compete economically with industrial versions, and people who no longer rely on open combustion are less tolerant of neighbours who do.

Other difficulties assert themselves. Setting a fire, as opposed to accepting one from nature, identifies a clear chain of agency, which translates into a descent of legal liability. In Western legal systems, someone who starts a fire bears a responsibility distinct from that of someone who 'manages' fire started by light-ning, accident or arson. Prescribed fire, moreover, requires special funding and cannot rely on emergency or supplementary accounts such as those available for firefighting. Furthermore, prescribed fire may not yield the full range of outcomes that a less 'controlled' burn might, and too many outcomes are, in fact, not very predictable. Their capacity to control and to predict consequences is usually less than agencies care to admit. Also, before prescribed burning can do the work expected, it may require complex and expensive preparations.

In brief, the difficulties of prescribed burning are many. The practice will likely survive where it does ecological work required by the land under consider-ation, where it enjoys social support as part of a community's cultural identity, or where economics favours extensive over intensive management. Mostly, though, modern versions struggle because the tech-niques are not those that, in the past, made prescribed fire the treatment of choice. Under a traditional regime, practitioners typically set small burns across a large area over a long period of time. By contrast, modern requirements favour set-pieces in the form of designated sites and times, with written guidelines. These lack the flexibility and nimbleness that made burning into a kind of foraging; by administrative design and out of legal concerns, they resemble instead a kind of mirror image of suppression. That is not how anthropogenic fire became pervasive across Earth or how it might flourish in the coming era.

13.3.2.3 Selected examples: Myakka River and Babcock-Webb WMA, Florida

Probably no place in the industrial world has developed prescribed burning as thoroughly as Florida. The reasons are several, among them the ability to burn nearly all year round, a long cultural tradition of burning, the creation of a collaborative system for burning and the threat of horrific wildfire if prescribed fire does not replace it. Well into the 20th century, the saying circulated that Florida burns twice a year. When federal agencies began granting exceptions to its suppression-only policy, Florida was the point of departure. Today, some 80% of the prescribed fire conducted by the National Park Service occurs on two Florida parks.

What is striking, however, is the way in which the state of Florida has also participated and has, in fact, become a lynchpin. While Florida assumed responsibility for fire protection in most counties through the Department of Forestry, the state also acquired lands for parks and nature reserves, and the state soon found it had to manage fire, which meant prescribed burning. Two examples are particularly relevant. Myakka River State Park underwent a familiar cycle, in which fire protection replaced the historic burning that had characterized the Big Prairie until it became necessary to restore fire through prescription, while the Babcock-Webb Wildlife Management Area never completely shed its fire heritage and segued seamlessly into the modern era. Myakka River thus shows prescribed fire used for restoration, Babcock-Webb for maintenance (Pyne, 2011a).

Florida developed late, mostly after World War II, and then along its coasts. The interior remained as open range, burnt as often as the land could take fire, sometimes twice a year. What ranchers missed, lightning kindled. In 1934, interest in preserving patches of native vegetation combined with a movement to create the Myakka River state park and forest. The park became the flagship for a later system of protected sites.

The authorities followed the common scenario for such places. They developed camp grounds, laid out roads, drained swamps and instigated fire control. Inadvertently, they denied the park the two processes that had shaped it – floods and fires. By 1969, the unintended consequences had become significant problems, and the Florida Park Service found itself the only agency in the state that had not adopted prescribed burning. It converted and undertook to reinstate fire in order to bring the raucously overgrown landscape back to something resembling the Big Prairie of pre-settlement times.

It proved more onerous to put fire back in than it had been to take it out, since conditions had changed. Ecologically, woods and rough had replaced grasses; palmetto prairies had rooted tenaciously; roads and canals had upset the park's hydrology; creeping urbanization created neighbours less tolerant to spill-over burns. The old regime had been perpetuated by ranchers tossing wooden matches from horseback and pickup truck, but the new order demanded costly preparations. Pre-burn treatments included mastication and herbicides. Fire's restoration took almost as many years as its removal had. By 2010, Myakka River State Park had more or less achieved what reinstatement it could (Figure 13.2).

Figure 13.2   A. Fire removed, an impenetrable rough in Myakka State Park.
B. Fire restored, emulating the 'big prairie' savannas and pine groves in the mid-19th century. (Photos courtesy of S.J. Pyne.)

Babcock-Webb Wildlife Management Area evolved out of ranches and, during its early decades, ranchers leased the site, ran cattle and burned. The site never experienced a compulsory era of fire exclusion. The place remained a working landscape, with hunters and recreationists replacing herders. As public use intensified, however, the authorities sought to systematize the inherited laissez-faire burning by establishing burning blocks from roads and plough lines. Initially, they burned every 2–3 years, then they ratcheted down the frequency to 1–2 years. The torch had passed, but fire had remained and there was no wrenching, costly reformation required to have fire perform its ancient duties.

Although a smoke column from one would be visible to the other, Myakka River and Babcock-Webb experienced dramatically different fire histories. Paradoxically, Myakka River, intended to preserve 'original natural Florida', received far more attention (and suffered from well-intentioned but misguided efforts to exclude fire), while Babcock-Webb, left to rudely folk treatments, came through the transition with its indigenous landscape more intact. However, the deep lesson might be that it is easier to keep fire on the land and tweak its use than to remove it and then attempt to restore it.

13.3.3 Strategy 3: change the combustibility of the land

13.3.3.1 Overview

Reshaping the land to promote the fires you want and prevent those you do not want is also an ancient strategy, as well as an inevitable one, since burning by itself remakes landscapes (and their fuels). What matters here is that the reverse is true as well. Altering the landscape changes the kind of fires that occur. Those fires may be deliberate or accidental, natural or anthropogenic – it matters only that people shape the settings within which they must burn and so modify those fires and their effects. Prescribed fires will do as they are intended, but wildfires will be easier to control and damages will be less.

The possible arrangements make a constellation of scatter points. At one corner lies the aboriginal fire economy of firestick farming, in which a synergy of hunting, foraging and burning remake the land to support more of the same, and shield inhabitants from threatening wildfires. At another corner resides farming outright, in which burning is almost completely constrained within a cultivated landscape. This, of course, is the honoured basis behind most fire management in Europe.

The advantages to the strategy are obvious. People are more likely to get fires they want (or can live with) by deliberation than by default. No one would question, for example, the determination to remove brush or encourage fire-flushed pasture within a herding economy, or to clear around houses to prevent fire spreading from woods to cabin. In overtly humanized landscapes, where people have clearly arranged the parts, there is little issue with further arranging. The sticking points come with wildlands.

The intent of protected wildlands by such means as nature preserves, parks, wildlife refuges, wilderness and biosphere reserves is that they are more or less natural. Applying an agronomic model to wildlife may violate the spirit, if not the legal constitution, of a refuge. Clear-cutting swathes through a park may lessen its propensity for crown fires but could destroy the landscape that was the point of preservation. Sending in busy hands, with or without chainsaws and ORVs, lessens the solitude and sanctity of a wilderness. Such places exist not to manage fire, but to have fire management advance their fundamental goals for sustaining ecological integrity, outdoor recreational opportunities or just open space.

So, while the possible practices are many, they are typically partial rather than comprehensive. That is, they are most likely to appear as fuel breaks, selective treatments around houses or valued but fire-sensitive sites, logging slash disposal or protecting reserve borders. Historically, officers often burnt around the boundaries early in the dry season ('early burning') in order to prevent fires from sweeping into the forest from outside. Today, the problem in industrial societies is more often to prevent fires within the reserve from propagating outside. Moreover, changing a landscape for fire can happen through fire, as much of the logic behind prescribed fire is to rearrange or reduce fuels. In some instances, this means landscape-scale treatments. Most often, however, it refers to targeted patches of problem 'fuels', either to abate them or to redirect the way that fire burns on the landscape at large.

Redesigning the landscape makes sense to harassed fire officers, but it may also destroy the values of the reserve. It can also often rile the public into protest, which can compromise the fire programme overall.

Since fire works most effectively when combined with other practices, what techniques are appropriate – what sort of synergies might be permitted – is often at the core of public controversies over what kind of fire management is acceptable.

In 2000, the USA adopted a National Fire Plan that put fuels management at its nuclear core, with the idea of a quelling what appeared to be a plague of large, high-intensity fires that were burning outside their historic range. In effect, fire management (and fire ecology) were simplified into fuels management. This had the additional benefit (so it seemed) of aligning policy with the most fully developed of the fire and forestry sciences, that is, silviculture and fire behaviour.

Critics voiced concern that reserved landscapes were not treasured for their hydrocarbon stocks but for their scenic allure, their biodiversity and their ecological services. A fire programme had to reconcile with such values – not simply manipulate fuels or fire behaviour. The dominant language of fire science proved inadequate to this task, and field officers have turned to mechanical treatments for what is, at heart, a biological issue. Fuels were not a summary index of landscapes, but a surrogate for one. Policy followed the best available science, not the real task, which required some capacity to undertake biological controls or ecological engineering.

13.3.3.2 Selected example: the Flagstaff model

By the 1990s, the montane forests of the American West were widely regarded as an ecological mess. They were overgrown with small trees, stagnant, infested with mistletoe, beetles and budworm, and were obese with fuels. They were ripe for wildfire. A century of settlement had removed the indigenous burners, stripped grassy understoreys through overgrazing, broken forest structure by big-tree logging and had actively excluded fire. Routine controlled fires were no longer set and wildfires were attacked. The Blue Mountains of north-eastern Oregon became the poster child for a putative crisis of 'forest health'. However, a prescription for restored vigour emerged from the ponderosa pine of Arizona's Mogollon Rim.

Here was the largest contiguous stand of ponderosa pine in the world. Around Flagstaff, which housed the Ecological Restoration Institute (affiliated with Northern Arizona University) and hosted the Fort Valley Research Forest (the oldest in the US Forest Service), the dynamics of disruption were well

documented and understood. In 1870, pine savannas had carried routine surface fires through and around clumps of big trees. By 1990, however, grazing, selective logging and fire exclusion had allowed a thickening of young trees to create a continuous escalator of combustibles from fallen needles draped over windfall to upper-limbs of old-growth giants. The surface was paved with woody debris. Biodiversity plummeted. When fires returned, as they inevitably did, they burnt more intensely and blasted out thousand-acre cavities in the canopy. Such fires were well outside the evolutionary experience of the ponderosa. If uncontrolled, they would erase the pine as ruthlessly as clear-cutting.

The proposed solution was an adaptive silviculture that sought to restore the structure and processes of the pre-settlement forest. It relied on the aggressive thinning of small-diameter trees (a kind of woody weeding) that would allow for the reintroduction of surface fire without risking blowups. Slashing and burning would, together, unwind the biotic tangle that had knotted into an unmanageable and impoverished landscape and would allow nature to reassemble itself into something like its former regime. The prescriptive package became known as the 'Flagstaff model' (Friederici, 2003; Figure 13.3).

It soon became apparent, however, that technical advice was only half the equation. The other half was social, which meant reinstating a structure for discussion and restoring processes for political decision. Not everyone wanted to see saws back in the woods, fearing that they meant the reinstatement of a logging industry that would strip out the remaining big trees. Few observers trusted the Forest Service, or forestry in general, to do now what they had failed to do in the past. Many critics, including scientists, worried that the Flagstaff model, grounded in a particular landscape, would expand into places where it was ill-suited – sites like lodgepole pine forests naturally prone to crown fires, or designated wilderness for which wholesale manipulation was inappropriate.

The mechanics of environmental review could make decisions a tedious undertaking, and when the Flagstaff model was used to underwrite national initiatives such as the Healthy Forests Restoration Act, which sought to limit appeals by environmental activists, it roused further suspicion that the ecological silviculture was an intellectual smokescreen for more devious conspiracies. There was also the question, perhaps fatal, of cost. Without a market for small-

Figure 13.3   Ponderosa pine forests at Fort Valley Experimental Forest, Flagstaff, Arizona, USA.

A. Before treatment.
B. After thinning and burning.

(Photos courtesy of S.J. Pyne.)

diameter wood, thinning was too simply too expensive to scale up from demonstration plots to forest-wide projects.

What kept advancing the process, however, was the spectre of big fires. With a long drought cycle, record fires began scouring out swathes of forest. It was difficult for any group to see benefits in such conflagrations. Slowly, markets for small-wood products emerged. With passage of the Collaborative Forest and Landscape Restoration Act in 2009, funding became available to undertake million-acre projects over a 20-year calendar. Their essence was to restore fire resilience to the Mogollon Rim, and their working prescription was the Flagstaff model.

13.3.4 Strategy 4: suppress

13.3.4.1 Overview

Fire prevention, fire suppression, fire exclusion – the concept comes with many overlapping terms, but they all embrace a strategy intended to prevent unwanted fires from igniting or, if started, to stop them from spreading. In extreme cases, the intention is to abolish fire altogether.

As with all matters pertaining to fire, context determines practical significance. The defining feature is whether or not a landscape is naturally prone to burning. If not – if there is no natural basis for fire or if there is no created cultural landscape that depends on fire and which people wish to perpetuate – then more or less complete suppression is possible. The mechanics of the pyric transition work their magic through substitution and suppression. Where fire operates as a tool, other tools that are not based on open burning can replace it. Where, in such settings, fires break out, they can be beaten back and extinguished with minimum costs and little environmental damage. While 'fire suppression' remains as a shorthand for the sum of these processes, 'fire exclusion' is a more appropriate expression.

The difficulty comes when such experiences are extrapolated into fire-prone environments. Here, suppression costs rise because the firescape encourages burning. Ecological costs escalate because the land is deprived of a process to which it has become adapted, and political costs worsen because expectations, or what might constitute a social contract, are at odds. The firefight is a great set-piece of modern times, but it is not a strategy for long-term administration. It more resembles a declaration of martial law, as though the fire were a riot that needs quelling. It is not a basis for governance. In settings dedicated to quasi-natural conditions, in which landscapes are disposed to fire or have known fire for a long time, continued callouts for suppression simply create an ecological insurgency that can lead to a nasty cycle of bigger fires and more costly firefights. Eventually, the fire regime itself changes.

Still, the ability to contain unwanted fires is essential for most fire management programmes. A grassland scheduled for three-year rotational burning cannot afford to have a wildfire sweep over it out of sequence. A restoration project committed to reshaping the firescape to keep fire on the surface does not want a high-intensity wildfire while it is under way. A public estate cannot afford wildfires that rush out and threaten surrounding communities.

In this sense, a suppression force is the equivalent of a constabulary. What matters is getting one proportional to the task, and not to turn the land management organization into the equivalent of a police state. A paramilitary force may make useful political theatre, but it will convey a false sense of security and success.

As wildland fire administration has evolved, 'suppression' (or 'fire control') has come to mean many things; it absorbs a gamut of actions. It is well established that the surest strategy to contain a fire is to strike it early and hard. For many decades, as it was building up a national firefighting force, the US adopted a '10 am policy' that stipulated that every fire should be controlled by 10 am on the morning following its report. The strategy intended to hold both the area burnt and the costs to a minimum. As a device for installing a first-order firefighting system, the approach has merit, not unlike a forced stimulation of industry in other sectors.

However, as blowback increased, the country adapted 'suppression' to include options that allowed for looser standards. Suppression might be 'modified', 'limited' or even restricted to 'observation'. It might mean attacking one flank of a fire and letting other flanks burn freely. It might involve backing off to a natural barrier like a lake or a stony crestline and burning out. It might mean a regime of close observation or monitoring that segues into variants of 'let burning'. It can be applied with minimal impact instead of brute force. In brief, suppression (or fire control) does not demand a single response but can be adapted to circumstances. The ultimate goal is that the fire behaves in ways acceptable to managers. If the fire exceeds those bounds, it must be contained or pushed back. That is the job of suppression.

Suppression is thus less a strategy than a support operation for other strategies. Often, because it has public visibility and can protect social assets like rural communities or a timber berth, it can amass more equipment, crews and resources generally than other tasks of fire management, and so can help to establish an infrastructure on which the fire programme as a whole relies. Where fire is abundant, however, or conditions favour it, the folk saying holds that crews are not putting fires out, they are only putting them off.

13.3.4.2 Conundrum of the big fire

One aspect in the political economy of wildfire deserves special attention: the 'big fire', or the 'big fire season' – the breakout fire that overwhelms fire-fighting capabilities, sends costs ballistic and blasts over the countryside (and sometimes towns). Such episodes tend to deform an entire fire management system and can wipe out, in one rush, all the successes achieved by patient application over decades. As the crisis boils over, rules get suspended, budgetary constraints blow away and a hard-won integration of practices into a coherent system disaggregates as money, political attention and materiel flow into the firefight. Big fires can act on a fire community as a declaration of war does on a country.

The concern continues beyond the actual fire or fire complex. Because no local agency can cope, but must request help from others, the reach of the fire broadens and may become national, or even international. During the crisis, power rises up and concentrates. These emergency arrangements may subsequently be institutionalized and, by such means, the needed ability to address exceptional events can deform the entire workings of a fire programme. The big fire can, in short, challenge and distort the governance structure for fire management overall.

There are ample historic illustrations. In each, it is not enough for fires to be big; they must interact with the political and social context and become significant, and those that do can catalyze major reforms. The Great Fires of 1910 in the United States shaped policy for over 50 years and continue to influence how America responds to wildfire. The 1939 Black Friday fires in Australia set into motion an Australian style of fire management that shaped research, policy and operations. The 1972 fires outside Moscow led to reforms throughout the Soviet Union. The 1979–82 fires in Canada catalyzed its national structure of wildland fire management. The 1998 fires that savaged Mesoamerica led to the modern national programme in Mexico.

Programmes begun in response – in reaction – to a crisis can be hard to retrofit to other purposes later, which is why suppression, once formalized, persists so stubbornly. Also, nationalizing fire agencies can dampen, or even drive out, a civil society of fire practice, which is what must ultimately support a sustainable programme.

The big fire has haunted wildfire fire programmes since they commenced their modern form in the late 19th century, and no-one has solved the problem. In all likelihood, there is no technical solution. On a short-term basis – a season, a period of extreme fire

weather – a build-up of suppression forces operating under a 10 am-style mission may succeed in keeping fire starts from metastasizing into monster burns. Eventually, however, one or more will escape, and these may become so large that they consume all of the area previously protected. The wisest approach is to apply all the techniques of fire management, adjusted to the particular landscape, so that the 100-year fire, when it occurs, will not destroy everything of value – not only tangible assets, but the character of the fire programme. What is unlikely to succeed is to increase fire suppression forces incrementally in the delusion that all potential big fires can be prevented.

Yet, in this regard – the need for multiple responses, for varied tasks, for a pluralism of practice – the big fire is not so much an exception as an exaggerated version of what fire management requires overall. Almost never does one kind of fire dominate, and almost nowhere does one strategy alone succeed. Wildland fire management does not mean applying the latest scientific discovery with the newest technology to advance a single mission. Rather, it involves reconciling the messy, oft-conflicting values of society with the endless shape-shifting properties of fire on lands that, in turn, morph as fire and land use and institutions change. It means mixing the strategic options in a proportion that yields the fire regime determined best for a particular site. Universal mandates, like the 10 am policy, can serve to create a fire programme or to colonize a landscape with fire institutions. However, they cannot govern one.

13.3.4.3 Select example: Los Angeles County, Southern California

When the landscape in question is a cityscape, there is little option other than suppression; prescribed fire means the demolition of derelict buildings. No one is going to let urban fire free-burn on the theory that this is how cities used to renew themselves. The interesting question is how municipalities that include open space (if not outright wildland) organize. The answer seems to be that the wildland fire agency will assume the shape of urban fire services.

Nowhere has this hybrid found a more sophisticated expression than in Southern California, where a dense and sprawling cityscape crowds against public wildland or fast-converting rural countryside. The story of how that mosaic of jurisdictions, purposes and practices has evolved is a cameo of the US national narrative, only more compressed in space and time, so much so that it seems to be a fire entity all its own. In 2011, Los Angeles County had a fire budget of $900 million. Its Forest Service neighbour, the Angeles National Forest, had the largest fire budget in the national forest system and the three heaviest districts for fire calls. Its state partner, CalFire, has the largest state fire programme in the country. For all, the pressures point to fire suppression within the context of an all-hazard emergency service.

The keystone is Los Angeles County Fire Department. The county had a major municipal fire service with the City of Los Angeles, but its bailiwick reached across 4000 square miles and over the mountains to the north, and in 1911 it created a Forestry and Fire Warden Department to extend fire protection to its backcountry and rural hamlets. Both city and country developed rapidly and, beginning in the 1940s, explosively.

The forestry division of LACFD diverged from its counterpart in the Angeles National Forest. The US Forest Service focused on its mountain wildlands and then began fretting over its urbanizing periphery. The LACFD focused on its urban core, while trying to absorb a fire-prone wildland fringe. Of the two models, the urban was the stronger. Each big fire year then came with almost metronymic predictability and boosted both the demand for better fire protection and the funds to enact it. Beginning in WWII, a California fire plan created a master mutual-aid agreement that promoted cooperation beyond the old alliance between Los Angeles County and its national forest (Pyne, 2011b).

Still, major fires – what came to be called fire sieges – could blow away even the proudest of regional fire departments. When that actually happened, in 1970, a research programme commenced, with the intention to allow the many fire services to coordinate during emergencies. By 1976–77 the outcome, what later evolved into the incident command system, went operational, headquartered in Los Angeles County. Meanwhile, the US Forest Service beefed up its capabilities and the State of California upgraded its Office of Emergency Services and its capabilities. What had been a minor trend – the migration of urban models to counties under contract – became a defining one.

When the smoke lifted, LACFD was an urban fire service. Fewer than 20% of its alarms involve fires of any kind. The forestry division has shrunk to a relict appendage relative to its urban mission (although, with 42–43 'badged foresters', it remains large by any

other measure). LACFD acts as the fire department for 58 of the county's 88 municipalities. It has responsibility for the county's unincorporated areas, which it protects under contract from CalFire. It is the regional coordinator for the California Emergency Management Agency. It has mutual aid agreements with Santa Monica Mountains National Recreation Area and the Angeles National Forest. It is one of two departments qualified for international deployment for urban search and rescue. It one of the five largest fire departments in the US.

When proponents and critics argue for or against the move to make wildland fire into an all-hazard emergency service, it is the LACFD that serves as an exemplar. Ultimately, people control fire by controlling the landscape. If that landscape is urban, or urbanizing, or destined for future urbanization, then fire management becomes urban fire protection and may increasingly evolve away from fire altogether.

13.4 Institutions: ordering fire

13.4.1 Fire management operations

Ideas are one thing, but implementing them is another. In the past, institutions tended to be customary – part of an everyday fabric of life and work. Applying and withholding fire on the land followed traditions encoded from long experience. However, as the pyric transition broke those patterns, as imperialism forced a new order onto societies, as novel landscapes emerged, as modern societies became more complex, or all in combination, then new arrangements had to arise. Formal institutions replaced customary ones. Explicit policies supplanted unwritten mores. Bureaucratic guidelines overwrote embedded cultural expectations. Often, these had to be invented, and fire agencies had to establish research programmes to supply what inherited experience no longer could.

Even so, institutions for the management of wildland fire are full of anomalies and eccentricities. Very few were created for fire management from their inception, but were tweaked and adapted from organs intended for other tasks (most evolved out of bureaus established by foresters, for example, and still bear the birthmarks of that origin). So, too, procedures and programmes dedicated to fire suppression continue to influence the more full-spectrum fire management of their successors. Changes in purposes and practices that had once emerged out of the pushes and pulls of long history had to be conducted by formal processes of political debate.

Fire management, in brief, requires institutions of governance, and these will vary by scale and context. They will assume forms appropriate not only to the lands under administration but to the societies that create and staff them. The history of fire management is largely the narrative of how, at considerable cost and effort, such reforms have occurred.

13.4.1.1 Single agency models

The possible models are many, ranging from the organization of suppression forces on individual fires to international organizations committed to disaster assistance and collective scientific research. A few principles seem to persist throughout, however, and one is the need to reconcile a fire agency with its larger institutional context. What, exactly, is the mission of the agency, how should it be organized to achieve this mission and how does a fire programme interact with other institutions? Another is the impact of the big event – that is, a large fire or fire complex or a major research programme – which the local or individual agency cannot by itself support. Who should direct and finance undertakings beyond the capacity of a local unit? Such questions are not merely matters of policy but of politics – of how a society chooses to organize itself.

When most wildland or rural fire agencies were established, their mission was to fight wildfire. Although they could field dedicated crews, few did, instead relying on part-time employees, or an agency 'militia', or on outside gangs of workers drafted only during the period of emergency. This meant that, although firefighting was a distinct operation, it was embedded within its host agency. Fire programmes in forestry bureaus behaved rather as city fire departments did, or as a military operation, as a somewhat stand-alone service that fought fires. Few such organizations remain today; as land use has morphed, so have the agencies responsible for dealing with fire on them. The trend shows a strong bifurcation, on the one hand toward further integration with land management, and on the other toward integration with urban-style services. Each reflects a different perspective on what the mission of a fire agency should be.

The first of these perspectives plants fire operations squarely within a larger context of land management, in which the fire programme seeks to limit unwanted fires and to promote desired ones. Its autonomy is limited. Firefighting (or fire lighting, for that matter) is not an isolated undertaking that trumps every other activity. At a deeper level, this vision reflects an understanding that fire control resides in controlling fire's setting.

Ultimately, a solution requires fully intertwining fire management with land management. Each needs the other. Borneo does not have a peat fire problem; it has a problem with converting tropical peat lands to palm oil plantations, and this manifests itself as fire. California does not have an exurban fire problem; it has a problem pushing housing developments into marginal lands, which places them along a burning border.

The second perspective identifies the principal fire problem with wildfire's impact on human communities. The primary purpose of a fire agency is to protect lives and property, but it will tend to take on other emergency tasks in addition to fire. It will, in other words, evolve much as urban fire departments have done, into an all-hazard emergency service. It will become for the peri-urban landscapes of industrial societies what city fire departments have become for metropolitan areas. The wildland periphery becomes absorbed into an urban fire regime.

Each approach has vices and virtues. The fire-as-land-management model struggles to cope with specialty skills and the politics of dealing with its peri-urban fringe. It worries that such fires drain money and attention away from its core mission, which is wildland management. The fire-as-emergency-service model can excel at protecting society from fires that it does not want in its exurban enclaves, but stumbles to apply the fires it does want in its wildlands or rural countryside. It seeks to further its goals by a forward strategy of pushing protection further into the backcountry, so that fewer fires can escape to threaten the perimeter or engulf a countryside partially reclaimed by houses.

Each strategy reflects a distinct understanding of what fire protection means. For the first, it ultimately resides in fire's environment. Fire management (and control over wildfire) derives from good land management. For the second, it comes from the ability to stop fires or to repair rapidly the damages that fire or other hazards inflict on society. It is increasingly obvious that neither model can do both tasks well. The bifurcation in fire management thus emulates the bifurcation of land use away from a broadly rural countryside.

The trend globally is to move toward the all-hazard fire service. A few places where the frontier between wildlands and cities has become more or less distinct and stable – notably Southern California in the US – have managed to invent fire services that can occupy that 'I-zone'. However, most cannot. The ideal solution is to create fire agencies appropriate to their task, which means separate operations for each geographical setting. The likely outcome, though, is to force existing agencies to accommodate a variety of duties at which, during extreme events, they are certain to fail.

13.4.1.2   Multiple agency models

The above observation may serve as a general expectation. Whatever their design, few agencies can stand by themselves, because there will always be some big event that overwhelms the system. The solution is to pool resources during such episodes.

In countries where multiple organizations have overlapping or parallel responsibilities, the trend is to sign mutual assistance pacts, to arrange for common dispatching centres and to organize transfers of crews, aircraft and materiel during major outbreaks. A regional or national coordination centre that can oversee fire suppression resources over a large area has become the norm. The arrangement is best developed in North America, where the US, Canada and Mexico have national centres, along with treaties among themselves, to allow transfer across country borders.

The movement of hardware, however, requires software if it is to be used smoothly. After the 1970 fires demonstrated the difficulties of coordinating among the many fire agencies in Southern California, researchers devised what became the ICS, an operational platform for allowing different entities to cooperate effectively on fires and to make decisions about how to allocate resources during multiple-fire incidents. Further developments led to the National Incident Management System, which subsequently migrated from the Forest Service to the Federal Emergency Management Agency and expanded from firefighting to all-hazard emergency responses. The model has now gone international.

During the past 50 years, the apparatus for an international approach to fire has matured, with mutual aid treaties, study tours, joint fire research projects, etc. Most arrangements have been bilateral between countries. The USA included fire within its international aid (and disaster relief) programmes. The UN Food and Agricultural Organization (FAO) has encouraged, through its forestry component, attention to fire statistics, sponsored fire consulting and promoted guidelines. The UN International Strategy for Disaster Reduction has a Wildfire Advisory Group. A Global Fire Management Center in Freiburg, Germany has overseen a composite of regional networks into a planetary one. Several international fire conferences occur annually.

To sum up, wildland fire has become the purview of a global guild. It has a distinct character as an occupation and as a collective community across many countries. Its sub-communities have begun assembling manuals for 'best practices'.

13.4.1.3   Private-land agencies

Fire services have become so identified with public safety, or with public lands, that it hard to appreciate the extent to which their transfer into the public sphere is a modern phenomenon. Previously, fire management was the responsibility of landowners and citizens generally. By removing fire from vernacular life, however, industrialization segregated fire protection from everyday society.

However, integrated fire management is impossible without working relationships with citizens other than the fire services proper. In developing countries, where fire problems typically emanate from agricultural burning, it makes more sense to organize volunteer bushfire brigades and regulate burning, than to create dedicated public bureaus. In developed countries, having ceded most responsibilities to public agencies, a counter-reformation is under way to return some rights and responsibilities to civil society. Besides, the volunteer fire department remains a significant feature of rural life and is unlikely to be replaced by more costly full-time departments.

The USA again exhibits an interesting spectrum of examples. The private sector manufactures tools and protective gear that once belonged only to public agencies; it conducts scientific research; it boasts consultants and contractors (often retired public fire officers); and it fields private services.

Prescribed fire associations have sprung up to assist landowners to do their burning. In places, legislation explicitly recognizes a right to burn and provides some legal protection, subject to a process of permitting or authorization. Volunteer fire departments are being integrated into a national system of standards. Some NGOs, such as The Nature Conservancy, do considerable amounts of burning on their lands and integrate into fire management networks at local, regional and national levels. In fact, such organizations can serve as brokers between the general public and governmental agencies. However, only the largest and wealthiest institutions and estates can afford a stand-alone fire programme. Almost all will seek out associates or subsidies from the state or both (Figure 13.4).

But that is true generally. Even the largest state-sponsored fire agencies can rarely operate without help from others. All must apply for mutual aid, whether for firefighting, fire lighting, or fire preventing. The trend is so obvious that it is almost invisible. Gone are the founding days when one agency might rule hegemonically; whatever happens now will occur within a framework of collaboration. No one can fight big fires by themselves. No one can do complex, multi-disciplinary research without cooperation among disciplines and agencies. No one can conduct prescribed

Figure 13.4 (A) Burning accomplished by The Nature Conservancy, almost 1.5 million acres by 2010 (Data from TNC). (B) A TNC prescribed burn in Florida (Photo: Parker Titus, Reproduced with permission of The Nature Conservancy).

burning in isolation, without at least agreement with, and preferably assistance from, neighbours. The dynamics of governance – the ways that people, cultures and political institutions construct the options available for management – matter at least as much as the particulars of physical geography. Together, collaboratively, they shape fire's behaviour.

13.4.2   Fire research

Institutions run on information and, where accumulated folk lore is gone, knowledge must be manufactured. This is the task of fire research. From its onset, modern fire management has placed research among its cornerstones. Whatever it did would be based on science. It assumed that fire protection, like forestry, was a species of engineering.

This simple vision proved quixotic. Just as fire management meant more than fighting fires, so fire management involved more than applying the latest scientific discovery through newly devised tools. It was about inventing institutions, about reconciling society's wishes with what was possible on the land, about debating what constituted a proper fire regime. So, too, fire research has (slowly) evolved from silviculture and applied science into a broader gamut of scholarship.

The classic themes of fire research derive from the historic needs of fire suppression. This was government science sponsored to solve the problem of wildfire on the public estate. At issue were matters like fire behaviour, fire danger rating and fire effects, both environmental and economic. Eventually, more attention went into fire ecology, fire economics and what might be termed fire sociology, and even operations research into decision-support guidelines. But the strictures that kept research in the sciences have mostly held (Figure 13.5).

Societies with wildlands had to invent fire-specific research as they did fire-specific agencies. They might farm some tasks out to universities, if the universities had disciplines capable of absorbing them. Mostly this meant forestry (and rangeland) schools, since fire has claimed no particular academic discipline of its own. However, forestry schools, even when expanded into resource management, proved far too limited. Serious countries have accordingly established facilities to conduct the research they need or to place special research centres within universities. The major sources of fire science are now the USA, Canada, Australia, the European Union and, though diminished from previous peaks, South Africa and Russia.

The models are many, each adapted to the political structure and land ownership of the host country. Most rely on a mixed economy of research: some funding through the agency, some by a national facility, some by universities, some through a general funding authority. Recent decades have witnessed a shift toward more diverse arrangements and fewer dedicated laboratories, but also have seen a trend toward fire themes that relate to national interest topics such as global change. More tellingly, perhaps, the old relationship between research and management has shifted from master and servant and to

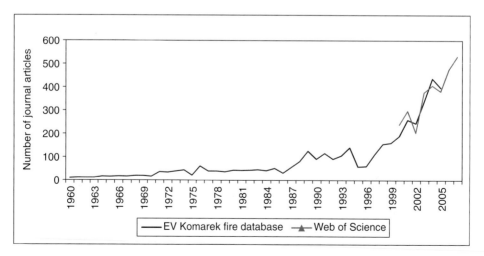

Figure 13.5 Scientific publications over the past 50 years, as recorded in the Komarek Fire Bibliography and Web of Science (Data courtesy of Peter Frost).

that of joint collaborators. Fire research can no longer assume a command-and-control model in which research identifies goals and management applies them. Even in principle research cannot direct operations, make decisions and determine purposes. It can, rather, advise and audit.

Fire agencies need knowledge as much as money. Research can oil the gears of operations, streamline the process of decision-making and measure achievements in the field against aspirations. Still, fire management is a guild, not a profession, and the knowledge that agencies lost when they wiped away indigenous practices, they can reacquire through new, if constructed, experiences. Humanity coped with landscape fire for thousands of years without computers, pumps, decision charts and nomex.

In a deeper sense, what a commitment to science does is to help bond fire management to its sustaining society. It connects it to the technology, the institutions, the high culture and the wider worldview of a larger culture. It forces a group that tends to think with its hands to think with its head. It keeps fire management – a very practical undertaking – from being isolated in an occupational ghetto.

13.5 Ideas: conceptions of fire

Ideas have consequences; not least, as science commands more and more of intellectual life, it influences

how a culture imagines fire. The relationship is, again, mutual. How a society defines fire is not just something that derives from its science. Its fire science also derives from how that society conceives of fire. If it imagines fire as a matter of scientific definition, it will turn to science for answers. If it imagines fire as a political concern, it will defer to politics. Such notions matter, because how a society defines fire and its problems will influence what kinds of solutions it proposes.

Overwhelmingly, in modern societies, science is seen as the means to identify fire's problems and to propose solutions. However, even within that scientific worldview, different conceptions of fire are possible. Three models – what might aptly be termed paradigms – are especially pertinent. Each grounds its understanding in a distinct discipline and, accordingly, proposes different remedial actions. Intriguingly, each is capable of absorbing the others, and each offers a comprehensive conception of how fire works in the world. The choice between them depends not on properties inherent to fire, but on the ways humans can understand those features and what questions they wish to ask. To simplify the choices, label the models as physical, biological and cultural (Figure 13.6).

13.5.1 The models as world views

The physical model defines fire as a chemical reaction shaped by its physical surroundings. The zone of

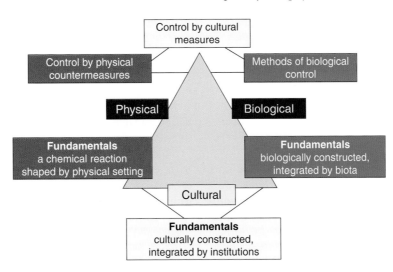

Figure 13.6 Conceptual models of fire: an intellectual fire triangle.

combustion – in which hydrocarbons are oxidized – has properties set by such parameters as fuel loads, fuel arrangements and fuel moisture content, as well as the flow of oxygen. The zone of burning moves around the landscape according to winds, terrain and caches of combustibles. Know that fire behaviour, and you know fire's effects on landscapes and societies. Fire ecology is the study of how that wave of burning alters ecosystems. Fire sociology is the study of how humans have defended themselves against unwanted fires and how they have turned fires into tools.

The biological model defines fire as a biological construction, a product of the living world. Life creates the fuels and the oxygen that combustion requires; life, in the character of humanity, supplies most of the ignitions that spark fire; and life, through ecological and evolutionary processes, shapes the biomass in which burning occurs. The chemistry of combustion is, after all, a *bio*-chemistry. When it happens in cells, it is called respiration, and when on landscapes, fire. Biology, in other words, underwrites fire behaviour and defines the ways in which people can manipulate fire to yield the landscape effects they want.

The cultural model relocates fire within a social context. It says that fire is a cultural construction, an interaction between nature and society. It notes that the choice of a fire model is a cultural decision; that science is itself a cultural invention; and that all of the concerns that motivate social interest are culturally defined. Whether fire is a problem (or is an asset) is a cultural call. No solution will succeed unless it satisfies social expectations, and this will depend on how society defines the issue. The physical and biological models only have meaning insofar as they make sense to a society. The other models exist not because they are embedded in nature, but by virtue of their usefulness to people.

13.5.2   *The models as management*

Each models proposes a different suite of solutions. The physical model offers physical countermeasures. Slashing and burning, rearranging blocks of hydrocarbons, controlling ignition, fighting flames with cooling agents, hardening assets against radiant heat and ember attacks – these are the treatments of choice. Fire ecology translates into the 'ecology' of fuels. Fire protection for settlements means early warning systems, an infrastructure for suppression and community evacuations. Treat fire as though it were an earthquake or a tsunami.

The biological model looks to measures of biological control, a fire equivalent to integrated pest management. It notes that fire serves ecological purposes, that ecological engineering is an appropriate way to rearrange living biomass, which is more than lumps of hydrocarbons. It might conceive of fires as biological processes – the essential task is not to abolish them but to ensure that they do the proper work expected, and that the locus of action is the matrix of biomass. Bad fires represent broken biotas, or diseases that occasionally become plagues because of a collusion of geographic and social factors. Treatments might look to public health measures such as quarantining, vaccinating and emergency remediation. Physical measures may compromise biological values.

The cultural model sees the problem as matching ideas and institutions with socially perceived problems. Bad fires – or bad fire regimes – result from cultural practices such inserting fire where it should not be, removing it where it should be, and generally misaligning social goals with natural processes. Breakdowns are akin to riots or insurgencies, or to political quarrels that result from mis-defining fire and what people can do with it. Fire institutions are out of step with how fire actually works or how people wish it to work. Solutions might take such forms as education, sponsored research, new legislation, revamped policies and agencies to apply them. It may mean changing society as much as changing firescapes.

Which is best? It depends. For dealing with a fire that is blowing and going, the physical model is ideal. For overseeing a firescape, a landscape rife with ecological goods and services, the biological model is better able to reconcile means with ends. For managing a fire programme, the cultural model is clearly superior. A wise fire agency will support all and select what, for any particular issue, best answers its needs.

Too often, however, the physical model dominates to the exclusion of the others. The reasons are historical and intellectual. Fire science was funded to assist with fire control, and physics seemed an (apparently) surer epistemology. Management follows the best science and the most pressing crises, even if that conception speaks only to a fraction of the problems before it, and if ongoing wildfires are but a fraction of what a full-spectrum fire management programme needs.

13.6   Fire management: selected examples

Integrated fire management is difficult. Every strategy, every institution, every idea, every operation has its disappointments, breakdowns, unexpected outcomes and outright failures. However, exemplars also exist that seem to be succeeding, and that – for their time – match place and practice with their larger sustaining society.

As with all matters pertaining to fire, exemplars are experiments. What follows are some of the most interesting experiments today. Since the USA supplied illustrations for our analysis of fire strategies, selections will come from elsewhere on Earth. Each must reconcile natural conditions with human land use and the politics that must negotiate between wind, fuel and evolutionary history and cultural values, social norms and economics. Common to all is the recognition that fire needs to be managed and that some outside stimulus – a law or treaty, a funding mechanism, an institution – seems necessary to catalyze reform.

13.6.1   Banff National Park, Canada

Banff National Park sprawls over 6641 km^2 along the eastern slope of the Canadian Rockies. Established in 1885 Banff is the flagship of Parks Canada, and it continues to serve as an exemplar for problems and strategies and as a source of inspiration and apostles. Nowhere has this proved more true than with fire (White and Fisher, 2007; Figure 13.7A, B).

Historically, Banff experienced a lot of fires, big and small, across seasons. The evidence is written in its vegetation and, in more recent times, in the chronicle of explorers and early park wardens. The record reached a peak around the 1880s as the park was proclaimed and as a transcontinental railroad rushed through the Bow Valley. The park did what prevailing doctrines urged: it built up tourism facilities, while protecting the surrounding natural estate by fire protection, predator control and the culling of elk. Burning declined sharply, to the point that fire essentially disappeared as an ecological presence. Forests metastasized. Banff suffered what critics believed was a fire famine.

By the 1960s, a new ethos began to inform park management that emphasized a doctrine of natural regulation. Attention shifted from hotels, golf courses, divided highways, ski resorts and other recreational developments, and toward redeeming the natural environment. Predator control ceased, elk culling ended and efforts were introduced to restore fire. The belief was that, by removing a disrupting human presence, nature would find its appropriate balance.

With regard to fire, Banff's physical parameters clearly dominated the scene. Amid such monumental scenery and obvious climatic influences, humanity's role was evidently trivial, so managers should stand to the side as fully as possible. The less intervention, the better. Since the only fire science was physical science, the prevailing models of fire seemed to support this perspective. New guidelines were promulgated in 1979, which stipulated as a goal that Canadian national parks should maintain 50% of their long-term fire cycle.

However, picking up the other end of the stick introduced its own imbalances. It appeared that nature set few fires. Lightning kindled plenty west of the Rockies, but not across the divide. The remarkable fire load apparent up through the 1880s was the product of anthropogenic burning. Some – like that associated with the railroad – were clearly excessive and damaging, but the fires that had sculpted the landscape had come from people. Moreover, restoring fire did not, as expected, rejuvenate extensive aspen stands. Instead, they sparked browsing by large elk herds that threatened to eliminate many stands altogether.

Thus, simply allowing natural fire to free-roam did not work, because there were too few fires. On the other hand, prescribe-burning, while it might get the mandated hectares burnt, did not yield the biological results that the programme sought. Meanwhile, continued development in the Bow Valley compromised Banff's standing as a World Heritage Site and pushed people against a lodgepole pine forest that swept continuously across the valley and hillsides, where it had previously existed in patches and stringers.

Parks Canada commissioned a Banff Bow Valley Study to address these issues. The Banff conclusions were codified into a new management plan in 1997 that essentially required the park to embrace complexity and to manage it adaptively. 'Complexity' meant incorporating many disciplines, not just natural science. It meant involving many publics, not just park staff and research scientists, and embedding fire management within a fuller ecological matrix, not just aligning flame to fuel and climate. It meant reinstating something like the historic population

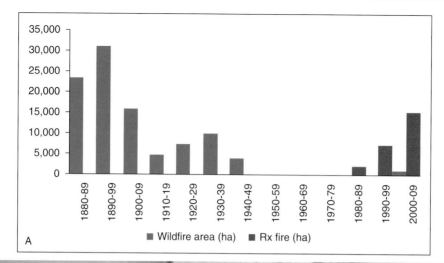

Figure 13.7   The fall and rise of fire at Banff National Park.

A. The earliest figures are most likely above historic averages because of railroad construction. Then they fall with fire control and prevention programs. Finally, fire returns, almost all of it through prescribed burning (Data from Banff National Parks, Parks Canada).

B. A stand-replacing prescribed fire, using terrain to help guide fire behavior (Photo from Randy Komar).

dynamics of Banff's fabled big fauna and, more controversially, it meant addressing the role of humans in more realistic ways. Thanks for the Canada Parks Act of 2000, which mandated ecological integrity as the goal of management, selective interventions were possible.

The diagnostic aspen was a bellwether. Renewal could not happen without controlling the elk population, which meant reintroducing predators, notably wolves. This knocked the elk population down, but required other interventions, since wolves were not protected outside the park and elk alone could not sustain the ideal landscape. Other fauna entered the mix; most recently, the park is reintroducing bison.

However, wolves and grizzly bears were not the only top-end predators – people were. The big animals, part of Banff's charisma, had to reconcile with the human populations – resident, seasonal and First Nation. The use of fences and special overpasses to span the highways emulated old migration patterns. The town site (and bordering properties) had to be protected against wildfire by logged fuel breaks. Also, fires had to be lit, often following preparatory thinning; by 2004, some 170 km² had been kindled. Philosophically, the strategy committed Banff to an enduring human presence as an active agent in the landscape. A diaspora of Banff fire managers has subsequently propagated attitudes and techniques – and a feeling of success – throughout Parks Canada.

13.6.2   Kruger National Park, South Africa

Located on nearly a million hectares of bushvelt and savanna in northeast South Africa, Kruger has long been the crown jewel of regional conservation. It was gazetted as a wildlife preserve in 1898 and proclaimed a national park in 1926. Since then, fire has remained an indelible presence and an enduring management concern. In 1954, Kruger initiated a multi-year fire experiment site, which continues to this day and is the longest-established anywhere. However, since its founding as a park, it has undergone some seven management regimes, a history that offers an unparalleled experiment in administrative policy and practice (Van Wilgen *et al.*, 2004; Figure 13.8).

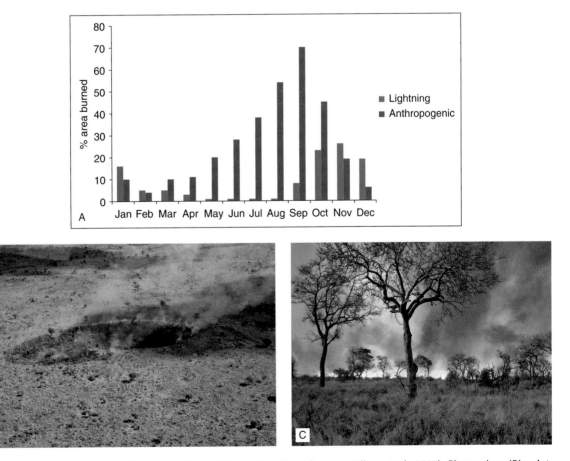

Figure 13.8   Annual rhythm of burning at Kruger National Park (Data from van Wilgen *et al.*, 2007). Photos show (B) point ignition method now favoured by managers, and (C) spreading fire in bushveldt (Photos courtesy of B.W. van Wilgen).

The early wardens recognized that fire was inevitable in savannas, since they could do little to stop it and it was useful for big game, however much academic science might disparage it. With park status, fire became an object of formal administration, bounded by policies. From 1926 to 1947, the park formalized the old system, accepting fires that it could not have prevented and selectively igniting burns in old grass to promote forage. This was regarded as insufficiently grounded in science, however. With a change in national politics, administrators from 1948–1956 attempted to reduce fire's frequency to less than five years, or even remove it altogether. The policy did not abolish burning but the abrupt shift in regime wrecked havoc with wildlife. From 1957–1980, controlled burning was reinstated but this time on a more rigorous basis. The park was divided into over than 400 blocks, bordered by fuel breaks, and each block was burnt every three years roughly with the onset of spring rains. Eventually, the practice was deemed too mechanical – an agronomic model, as it were.

Ecological science became fascinated with diversity amid communities and processes, and conservation thinking followed suit. From 1981–1991, managers introduced variability into the block-burning, according to rainfall (which determined fuel) and seasons. More land was burnt, along with some summer burns, following very wet seasons (50%) than after dry ones (20%). Then, from 1991–2001, the park experimented with a 'natural' (or natural regulation) regime, in which it would allow lightning fires to burn freely, while suppressing anthropogenic fires. Staff dismantled about half the fuel break system. In 1996, about a quarter of the park burnt after a lightning fire bust. The mix of fires was again deemed inappropriate and the costs of suppressing burns exorbitant.

In 2001, policy again reformed. This latest iteration tolerates lightning-kindled fires, which constitute 10% of starts, but suppresses some fires where they are unwanted for one reason or another and sets fires where, according to several metrics, the park believes the landscape needs more fire. To encourage further variability in fire intensity, it prescribes burns by point ignition and, to match fuel and flame, it emulates lightning starts, which it permits to burn until it reaches a monthly area-burnt goal. Fire officers are also seeking to move more of the burning into early-season phases. They operate under a philosophy of adaptive management.

This is a fascinating narrative, a stele of conservation philosophy over a century. Cynics might observe that the park had, after 60 years of research, come back to the stance its first wardens had taken. Critics might whinge that by setting burnt-area goals – a carrying capacity for fire, as it were – the park was culling wildfire instead of wildlife. Conservation strategists might ponder the awkward relationship between scientific research and management practice. However, despite a few wobbles and a short misguided attempt to exclude fire, amid a political landscape that had effectively undergone two revolutions and lay adjacent to decades of civil war in Mozambique, the park had accomplished its most critical task. It had kept usable fire on the land, a mix of combustion, along with Kruger's resident flora and fauna.

The lessons are several and continue to evolve. What the Kruger story suggests is that the extremes are dangerous, that neither fire exclusion nor lightning alone can work, and that the attempt to push fire regimes into one polarity or the other is itself upsetting. The veldt can take a lot of fire, and the more mixed the burning, the better. One concern is that the statistics only record the physical measures of fire, such as ignitions and burnt area; they do not speak to the fire ecology of the park. The assumption is that if enough is burnt well, the veldt will thrive. The reality is that fire's effects reflect an interaction with everything else on the land, especially with grazers, which help determine the amount of grass to burn, and the predators, which tweak the populations of grazers. Elephants particularly interact with fire to sculpt the woods. Even porcupines have been shown to influence fire, by making gnawed trees, rich with pitch, more vulnerable to surface flame.

The other issue is people. Southern Africa is a hearth for hominins, and humanity has interacted with fire since our origins. It seems as difficult to exclude anthropogenic fire as it does lightning. In fact, fire continued throughout the protection era because people kept setting them, and not always according to programmes (refugees from Mozambique started many during the natural-regulation era, for example) or according to the prescriptions of scientific investigators (the science kept changing). Remarkably, the evidence suggests that the amount of land burnt has remained more or less constant, varying with wet and dry cycles and not with changing climates of opinion. What theories of ecology and concepts of conservation had changed were the attributes of the park's fire regimes – seasonality, frequency, intensity.

What scientific theories still struggle to incorporate is the role of people as keystone creatures in fire ecology. Paradoxically, that has meant ignoring the role of managers and of the scientists advising them. By clearly demonstrating their influence Kruger might stand as a model generally.

13.6.3   La Sepultura Biosphere Reserve, Mexico

La Sepultura embraces 167 310 ha in the mountains of Chiapas in Mexico's far south. In 1995, a year after the Zapatista uprising, a presidential decree established a preserve under the terms of UNESCO's Biosphere Reserve programme. This includes a core zone dedicated to preservation (13 759 ha) and a wider buffer zone, within which reside 23 000 people organized into 47 communities, of which 35 are ejidos and 80 ranches. Some 3.5 million people reside within the reserve's zone of influence. Fire is among the land's primary ecological processes and its oversight is a defining feature of the reserve's brand of collaborative management (Huffman, 2010; Figure 13.9).

To simplify, the fire scene is one in which traditional use must reconcile with the interests of the modern state, and local values must reach an accommodation with global environmental concerns. Famous for its biodiversity and endemism – the reason for its designation as a biosphere reserve – La Sepultura has become no less an expression of institutional diversity. The State of Mexico has declared it a Protected Natural Area, administered through the Comisión Nacional de Áreas Naturales Protegidas and, in 2009, Mexico adopted a national policy on fire use, with standards overseen by the Secretaría de Medio Ambiente y Recursos Naturales. The site conforms to UNESCO Biosphere Reserve standards. The Nature Conservancy, an international NGO, has facilitated research and planning with collaboration from Mexican universities. Also, of course, there is a host of local communities, each with their own structures for governance. All view fire in the reserve through somewhat different prisms.

Anthropogenic fire has been on the landscape since the advent of people. For at least 5000 years, people have burnt in the service of maize cultivation, a swidden system known as *milpa*. The Spanish conquest in the 16th century introduced livestock and ranches (*agropecuaria*), which prior practices adapted to and absorbed. In the mid-19th century, the

Mexican state created a forestry bureau, which (like such agencies everywhere) sought to control fire, but its script never reached Chiapas.

By the late 20th century, the lands that became La Sepultura were routinely – necessarily – burnt to sustain small-plot milpa and agropecuaria and other minor purposes. All in all, those fires affected an estimated 83% of forested land in Chiapas. Almost certainly, the chronic churning wrought by human land use, with fire as a catalyst, helped to account for the extraordinary ecological variety that made La Sepultura worth the state's effort to protect.

A significant fraction of such burns escaped control, however. Usually, these were swallowed up in the larger landscape as little more than nuisances. In 1998, during a record-shattering drought, they became a scourge. It was a fire year of record for Mexico overall. In Chiapas, wildfires burnt some 89 000 ha, not sparing even tropical forests normally exempt from fire. If La Sepultura were to accomplish its purpose, it required some mechanism that would allow the local people to continue their fire-catalyzed agriculture while safeguarding the reserve.

In 2005, La Sepultura announced a planning programme for fire management that would synchronize these various institutions and interests in collaborative and 'integral' ways. In essence, the project sought to formalize traditional knowledge and usage in ways transparent to outsiders, including the state. It did not prohibit indigenous fire practices, but attempted to shape them to promote values of interest to the state and the global environmentalist community, while embedding those practices within a system of oversight. The outcome was, in the good sense of the term, a compromise.

Each side both got and gave up. The indigenous practitioners saw their long experience with fire recognized and inserted into written plans, and they received official legitimacy for such usage. Critically, they remained the agents of fire's application on the ground. The institutions responsible for the reserve, in turn, gained some control over the process by requiring burn authorizations. They could direct fire away from sensitive areas and, in theory, reduce the number of escape fires from burning done at inappropriate times and places. They could, within limits, regulate traditional use, and their authority had the force of law.

Much as La Sepultura seeks to reconcile nature preservation with agricultural practice, so it must

Figure 13.9 Burning in La Sepultura Biosphere Reserve, Chiapas, Mexico.

A. Daytime burn that shows typical landscape and fuels

B. The much preferred night-time burning, in which evening humidity recovery dampens the fire.

(Photos courtesy of Mary Huffman).

harmonize ways of knowing and doing and of governing the ceaseless negotiations between them. To those versed in fire management on public wildlands, the outcomes may seem peculiar. The experiment unfolded amid an ongoing insurrection. Researchers have sought to identify and re-code traditional understanding into the conceptual language of modern fire science. However, the La Sepultura story is more typical of fire management on Earth than sites like Yellowstone National Park. Even protected places display an enormous diversity of institutional arrangements, and they require fire management suited to their circumstances, not derived from abstract first principles.

13.6.4 West Arnhem Land Fire Abatement Project, Australia

Arnhem Land is a bulbous peninsula in the centre of northern Australia. It has long been a land informed by two geographic contrasts. One, of landforms, pairs a central sandstone plateau with coastal lowlands. The other, of climate, pits wet and dry seasons. For over 40 000 years Aboriginal peoples have moved seasonally between plateau and lowland with those rhythms, burning the landscape as they travelled and creating an immense, relatively stable and biodiverse biota as a result.

Since the 19th century, a cultural contrast has remade this scene. As Europeans moved into the tropical savannas, Aborigines declined, some from outright violence, many from disease and the dissolution of traditional society. Ranchers and large cattle stations replaced local indigenous groups and hunter-gathering. The fire regimes shifted accordingly. Aboriginal practice had begun burning small select areas as soon as fuels became available. Later fires burnt against them as the landscape dried, and they burnt along routes of travel (songlines) as peoples moved seasonally to access other habitats. It was a complex choreography that inscribed a matrix of fire, one that expanded or contracted as circumstances permitted. Lightning, effective at the first tremors of the rainy seasons, burnt what remained.

This arrangement broke down with European colonization. The agents of burning left, fuels amassed, and what fires remained went feral. Meanwhile, ranchers imposed another regime by burning towards the end of the dry season, which promoted hotter and vaster fires. A biotic tapestry that had been woven by particular patterns of fire began to unravel. However, recognizing the ecological value of the countryside, Australians proclaimed a substantial patch as Kakadu National Park between 1979 and 1991, and this subsequently became a World Heritage Site.

Park officials sought to control what they regarded as a damaging regime of burning, but experience soon revealed fire exclusion as not only hopeless but damaging. When a 1992 Supreme Court decision recognized Aboriginal land rights on Crown land and some national parks, experiments evolved to incorporate indigenous fire lore into a hybrid programme of management. Fire management found ways to broker modern science with traditional lore, and old practices with modern technology – what became known as the 'two toolkits' strategy. By 2004, that example was enlarging to embrace Aboriginal lands outside the park proper.

The 1997 Kyoto Protocol redefined this historical dialectic between the burning regimes of the two cultures by incorporating industrial combustion, as measured by greenhouse gas emissions, into the narrative. No longer could fire management be measured simply in terms of ignitions and area burnt, but as carbon stored and released. Of course, attention focused, as intended by Kyoto, on the burning of fossil fuels. However, in national carbon budgets, agricultural and wildland burning now had to be entered into ledgers. Open burning could augment or diminish a country's carbon accounts.

Tentatively, Australia has moved toward adoption of a Carbon Pollution Reduction Scheme and continues to debate how open burning might be a part. With or without formal agreement, the country has been trending toward some kind of carbon trading scheme.

So, when Conoco Phillips built a natural gas processing facility in Darwin Harbour, an inspired concept emerged to link the disruption to Earth's atmosphere caused by burning fossil biomass with that caused by changing fire regimes amid surface biomass, and to search out a common solution. Was it possible, by restoring older regimes on the countryside, to offset the carbon that would be released by the products of the Darwin LNG facility? The West Arnhem Land Fire Abatement (WALFA) Project was the outcome (Russell-Smith *et al.*, 2009; Figure 13.10).

WALFA drew on the experiences at Kakadu and on a number of governmental organizations for research

Figure 13.10   Burning in the West Arnhem Land Fire Abatement Project.

A. Aerial ignition along traditional routes.
B. Detail of patch burning characteristic of early season fires.

(Photos courtesy of Jeremy Russell-Smith).

support, logistics and legitimization. Critical, however, was an agreement with the DLNG to contribute AU\$ 1 million a year for 17 years in order to reduce annual carbon emissions from a baseline value; the goal was 100 000 tons of CO_2 equivalent. The contract provided both funds and national purpose. Along with contributions from government and NGOs, the project could proceed. The strategy was to reintroduce early season burning according to designs that would eliminate late season burns, particularly wildfires. By tweaking the fire regime, research suggested that carbon could be (and, in fact, was being) stored on the land. Along the way, the project helped to preserve elements of Aboriginal culture and yielded jobs for Aborigines as burners.

Its architects recognize the uniqueness and fragility of the experiment, but plans are under way to expand the model across other portions of northern Australia. In one way or another, the need to reconcile the two grand realms of combustion (open and closed) and Earth's two great economies (nature's and humanity's) will, in the future, drive other such undertakings.

13.6.5 Västernorrland County, Sweden

In pre-industrial times, northern Sweden had a typical fire profile for a boreal forest, with roughly 1% burnt a year. Scots pine, on drier sites, burnt perhaps every 40–60 years, and the wetter spruce patches much longer. Around towns like Umeå, swidden agriculture was common and pastoralists burnt to promote forage for goats and cows (see Figure 11.12). Further inland, an extensive Saami pastoralism, based on reindeer, discouraged burning in order to spare the lichens needed for winter range. The border between the two economies, however, was usually burnt to make a biotic cordon sanitaire. In both arrangements, the forest displayed a patchiness: the pines were mixed in age; pitchy snags and stumps abounded; dead wood dappled the forest floor.

The migration of industrial forestry north through the Baltic broke this pattern. In river basin after river basin, the large pine were clear-felled and floated to mills, and the slash left behind to fuel conflagrations. Beginning in the 1870s, large fires swept the landscape almost decadally. In 1888, Umeå burnt. When it was rebuilt, it was planted with silver birch to serve as a fuel break, in recognition that a new fire regime had emerged.

The big burns continued through 1914. By then, Sweden had begun enacting a series of forest laws that steadily transformed its swidden farms into tree farms. The legislation tamed logging, intensive silviculture replaced wholesale cutting, cultivation cleaned up the woods and removed the dead wood that had propelled the worst fires. Forestry became a species of farming. Following European examples, the landscape reflected social models. In industrial Sweden, wildfire had no place but neither, increasingly, did fire of any kind. Wildfire had spread with untrammelled logging; fire exclusion spread with systematic reconstruction of Swedish society and land. Like Umeå, the country rebuilt in ways that would prevent the threat of conflagration and the perceived wastage of natural resources.

Yet protection had its costs. The forests of Västernorrland became a commercial duopoly of Scots pine and Norway spruce, with no more variety than a wheat field. Some species, such as fungi that depended on dead wood and certain beetles that required burnt wood, shrivelled into insignificance. Very little of the landscape belonged in nature reserves – mostly patches that for one reason or another had escaped logging and conversion. Where forestry was less intrusive, spruce and pine reproduction began to flood the understorey. The wild and the natural, it seemed, had little value either to Swedish consciousness or politics, which valued social security and economic productivity.

Modern environmentalism, however, has impressed itself on the country, not with the same ferocity as first-wave industrial logging, but sufficiently to promote notions of biodiversity. One response is the standard one: to create nature reserves and to encourage fire within those small settings to stimulate the dead and the burnt wood that critical species require and, more grandly, to recreate something of the pre-industrial forest. It has helped that the University of Umeå has a forestry school, which has evolved into a forest ecology department, and that its faculty took an interest in fire ecology. For many years, it was virtually the only such group in northern Europe. With patience and forbearance, ideas about suitable fire regimes have moved from the laboratory into the field (Rydkvist and Eriksson, 2009; Figure 13.11).

In 1993, Västernorrland burned 35 ha in Jämtgaveln nature reserve, followed in 1995 and 1996 with prescribed burns of 23 and 19 ha. All were in areas that had been 'heavily logged', and so were not representative of

Figure 13.11   Prescribed fire around Västernorrland, Sweden.

A. Burning in Holmen forests (Photo courtesy of David Rönnblom).
B. Burning in Långskidberget nature reserve, just west of Västernorrland (Photo courtesy of Anders Granström).

early conditions. Neither were they likely to yield the biodiversity sought and, indeed, they did not. Still, this was a breakthrough event because it was the first time a Swedish jurisdiction had sponsored such a fire.

In 1999, the programme expanded to include a prescribed fire in unlogged forest, specifically 120 ha in Helvetesbrännans reserve. The burn served as a proof-of-concept test that subsequently inspired other counties to consider a similar programme. In 2004, the programme enlarged its ambitions in an attempt to restore a forest to pre-industrial conditions and found a collaborator in Mid-Sweden University at Sundsvall. This time, the field operations required thinning of invasive spruce before re-introducing fire (the spruce had increased in density 4–5 times beyond historic conditions).

From 2004–2006, six prescribed fires burned 150 ha. The experimental project became, again, an exemplar for other reserves, not only intellectually but operationally, since the small units demanded relatively intensive preparations, relied on control methods other than mineral-soil fuel breaks and had created new dead wood without burning up all the old. The research programme set as a new goal to burn into spruce.

By North American standards, the experiments looked like boutique burning. However, Sweden did not have the scale of landscapes to absorb slop-overs, could not rely on a network of like-minded collaborators and neighbouring landowners, and did not enjoy the kind of system that underwrote premier prescribed-fire programmes. Instead, it offered another variant that grew out of its national experience. It carried prescribed fire for ecological purposes into private lands – not lands acquired for preservationist purposes as with The Nature Conservancy, nor commercial forest lands burnt to reduce fuel, but industrial forests burnt to promote biodiversity.

The leverage comes from the Swedish Forest Stewardship Council's standards for forest certification, which incorporates specific guidelines for burning. Specifically, FSC forest certification requires an annual burn equivalent to 5% of the area cut. If the logged stand has 15% of its volume intact, then the company can multiply its actual burnt area by 1.5; if more than 30%, then by 2; and if it burns in uncut stands, by 3. The programme, in brief, balances cutting and burning to create quasi-natural patches.

Interestingly, parts of Sweden made the transition from swidden to industrial logging by incorporating slash burning into its operations, a practice that disappeared as the country underwent its pyric transition. Burning logging slash marks a return to ancient habits, some of which had created the landscapes that modern Sweden desires (Swedish FSC, 2011).

Today, all the major Swedish forestry companies comply with FSC standards. Effectiveness varies, as one might expect, with companies, personalities and counties. Critics complain that certification has not fashioned the environmental conditions intended and that, in instances, it provides cover for the bad old ways. However, it is a bold effort to move environmental burning out of the ghettos of nature preserves and, as experience and beneficial outcomes accrue, it will surely spread.

The deeper difficulties lie where they always do – in reconciling nature and culture. The prime month for burning is July, which coincides with the national holiday season, and this puts labour out of sync with the job to be done. Also, the vision of an ideal society does not easily align with the natural materials available to make it happen. For some peoples, 'nature protection' means shielding landscapes from unwonted human meddling, which is to say, protecting nature from people. For others, it means buffering society against the vagaries of nature – in other words, protecting people from nature. How Sweden defines those competing visions will determine what kind of fire regime it has in the future.

Further reading

Friederici, P. (ed.) (2003). *Ecological Restoration of Southwestern Ponderosa Pine Forests*. Island Press.

Huffman, M.R. (2010). *Community-Based Fire Management at La Sepultura Biosphere Reserve, Chiapas, Mexico*. Dissertation, Colorado State University.

Pyne, S. (2011a). 'Slow Match, Fast Flames'. http://firehistory.asu.edu/slow-match-fast-flames (posted February 2011).

Pyne, S. (2011b). 'Imperium in imperio: Los Angeles County Fire Department'. http://firehistory.asu.edu/imperium-in-imperio-los-angeles-county-fire-department (posted July 2011).

Rollins, M., Morgan, P., Stephens, S., Holden, Z. (2007). *Final Report. Historical Wildland Fire Use: Lessons to be Learned from Twenty-five Years of Wilderness Fire Management*. JFSP Project Number 01-1-1-06.

Russell-Smith, J., Whitehead, P., Cooke, P. (eds.) (2009). *Culture, Ecology, and Economy of Fire Management in*

References and further reading for part three

Achard, F., Eva, H.D., Mollicone, D., Beuchle, R. (2008). The effect of climate anomalies and human ignition factor on wildfires in Russian boreal forests. *Philosophical Transactions of the Royal Society B: Biological Sciences* **363**(1501), 2331–2339.

Amiro, B.D., Todd, J.B., Wotton, B.M., Logan, K.A., Flannigan, M.D., Stocks, B.J., Mason, J.A., Martell, D.L., Hirsch, K.G. (2001) Direct Carbon Emissions from Canadian Forest Fires, 1959–1999 *Canadian Journal of Forest Research* **31**, 512–525.

Anderson, K. (2006). *Tending the Wild: Native American Knowledge and the Management of California's Natural Resources*. University of California Press.

Archibald, S., Roy, D.P., Van Wilgen, B., Scholes, R.J. (2009) What limits fire? An examination of drivers of burnt area in Southern Africa. *Global Change Biology* **15**, 613–630.

Balée, W. (1994). *Footprints of the Forest: Ka'apor Ethnobotany – the Historical Ecology of Plant Utilization by an Amazonian People*. Columbia University Press, New York, NY.

Bartlett, H.E. (1955). *Fire in Relation to Primitive Agriculture and Grazing in the Tropics: Annotated Bibliography*. University of Michigan Botanical Gardens.

Batchelder, R.B., Hirt, H.F. (1966). *Fire in Tropical Forests and Grasslands*, Technical Report ES, 23. U.S. Army Natick Laboratories.

Blackburn, T.C., Anderson, K. (eds.) (1993). *Before The Wilderness: Environmental Management by Native Californians*. Malki Press.

Boucher, D. (1991). *Ride the Devil Wind: A History of the Los Angeles County Forester & Fire Warden Department and Fire Protection Districts*. Fire Publications, Inc.

Boura, J. (1998). Community Fireguard: creating partnerships with the community to minimize the impact of bushfire. *Australian Journal of Emergency Management* **13**, 59–64.

Boyd, R. (1999). Strategies of Indian Burning in the Willamette Valley. In: Boyd, R. (ed.) *Indians, Fire, and Land in the Pacific Northwest*, pp. 94–139. Oregon State University Press.

Bradshaw, R.H.W., Tolonen, K., Tolonen, M. (1997). Holocene Records of Fire from the Boreal and Temperate Zones of Europe. In: Clark, J.S. *et al. Sediment Records of Biomass burning and Global Change*, pp. 347–365. NATO ASI Series I, Vol. 51. Springer-Verlag.

Braudel, F. (1972). *The Mediterranean and the Mediterranean World in the Age of Philip II*. 2 vols. Translated by Sian Reynolds (Harper and Row; see Vol. 1, pp. 85–102 for transhumance).

Cary, G., Lindenmayer, D., Dovers, S. (eds.) (2003). *Australia Burning: Fire Ecology, Policy and Management Issues*. CSIRO Publishing, Collingwood.

Cochrane, M.A. (2009). *Tropical Fire Ecology. Climate Change, Land Use, and Ecosystem Dynamics*. Springer-Praxis, Heidelberg, Germany/Chichester, UK.

Cohen, J. (2008). The Wildland-Urban Interface Fire Problem: a Consequence of the Fire Exclusion Paradigm. *Forest History Today* Fall issue, 20–26.

Conklin, H.C. (1961). The Study of Shifting Cultivation. *Current Anthropology* **1**, 27–61.

Davidson, E.A., de Araújo, A.C., Artaxo, P., Balch, J.K., Foster Brown, I., Bustamante, M.M.C., Coe, M.T., DeFries, R.S., Keller, M., Longo, M., Munger, J. W., Schroeder, W., Soares-Filho, B.S., Souza, C.M. Jr., Wofsy,

Fire on Earth: An Introduction, First Edition. Andrew C. Scott, David M.J.S. Bowman, William J. Bond, Stephen J. Pyne and Martin E. Alexander.
© 2014 John Wiley & Sons, Ltd. Published 2014 by John Wiley & Sons, Ltd.

S.C. (2012). The Amazon basin in transition. *Nature* **481**, 321–327.

Delcourt, H.R., Delcourt, P.A. (1997). Pre-Columbian Native American Use of Fire on Southern Appalachian Landscapes. *Conservation Biology* **11**(4), 1010–1014.

Ekstam, U., Aronsson. M., Forshed, N. (1988). *Ängar. Om naturliga slåttermarker I idlingslandskapet.* LTs Förlag, Stockholm, Sweden.

Eldredge, I. (1911). Fire Problems of the Florida National Forest. *Proceedings of the Society of American Foresters* 164–171.

Ellis, S, Kanowski, P, Whelan, R (2004). *National Inquiry on Bushfire Mitigation and Management.* Commonwealth of Australia.

Estany, G., Badia, A., Otero, I., Boada, M. (2008). Socio-ecological Transformation from Rural into a Residential Landscape in the Matadepera Village (Barcelona Metropolitan Region), 1985–2008. *Global Environment* **5**, 8–38.

Evans, E.E. (1940) Transhumance in Europe. *Geography* **25**, 172–180.

Fernow, Bernhard, *et al.* (1912). *The Forest Conditions of Nova Scotia.* Commission of Conservation, Ottawa, Canada.

Forestry Department, Food and Agriculture Organization of the United Nations. (2006a). *Fire Management: Review of international cooperation.* Working Paper FM/18E.

Forestry Department, Food and Agriculture Organization of the United Nations. (2006b). Fire Management. *Voluntary Guidelines. Principles and strategic actions.* Working Paper FP/17/E.

Forestry Department, Food and Agriculture Organization of the United Nations. (2007). *Fire Management – global assessment 2006.* FAO Forestry Paper 151.

Friederici, P. (ed.) (2003) *Ecological Restoration of Southwestern Ponderosa Pine Forests.* Island Press, Washington, DC.

Fuhlendorf, S.D., Engle, D.M., Kerby, J., Hamilton, R.G. (2009). Pyric Herbivory: Rewilding Landscapes through the Recoupling of Fire and Grazing. *Conservation Biology* **23**(3), 588–598.

Gabriel, H.W., Tande, G.F. (1983). *A Regional Approach to Fire History in Alaska.* BLM-Alaska Technical Report 9.

Gammage, B. (2008). Plain Facts: Tasmania Under Aboriginal Management. *Landscape Research* **33**(2), 241–254.

Gammage, B. (2011). *The Biggest Estate in the World: How Aborigines Made Australia.* Allen and Unwin.

Goldammer, J.G. (2002). Fire Situation in Indonesia. *International Forest Fire News* **26**, 37–45.

Goodsblom, J. (1992). *Fire and Civilization.* Penguin Press, London.

Gorte, R.W. (2005). *Wilderness: Overview and Statistics. Congressional Research Service, Report to Congress.* Order code RL31447, CRS.

Granstrom, A., Niklasson, M. (2008). Potentials and limitations for human control over historic fire regimes in the boreal forest. *Philosophical Transactions of the Royal Society B: Biological Sciences* **363**, 2353–2358.

Grübler, A., Nakićenović, N. (1996). Decarbonizing the Global Energy System, *Technological Forecasting and Social Change* **53**, 97–110.

Guyette, R.P., Mjuzik, R.M., Dey, D.C. (2002). Dynamics of an Anthropogenic Fire Regime. *Ecosystems* **5**, 472–486.

Guyette, R.P., Spetich, M.A., Stambaugh, M.C. (2006). Historic fire regimes dynamics and forcing factors in the Boston Mountains, Arkansas, USA. *Forest Ecology and Management* **234**, 293–304.

Hallam, S.J. (1975) *Fire & Hearth: a study of Aboriginal usage & European usurpation in south-western Australia.* Australian Institute of Aboriginal Studies, Canberra, Australia.

Hecht, S., Cockburn, A. (2011). *The Fate of the Forest: Developers, Destroyers, and Defenders of the Amazon,* updated edition. University of Chicago Press, Chicago, IL.

Hirsch, K.G., Pinedo, M.M., Greenlee, J.M. (1996). *An International collection of wildland-urban interface resource materials.* Information Report NOR-X-344, Northern Forestry Centre.

Huffman, M.R. (2010). *Community-Based Fire Management at La Sepultura Biosphere Reserve, Chiapas, Mexico.* Ph.D. dissertation, Colorado State University, CO.

Ingram, R.P., Robinson, D.H. (1998). Evolution of a Burning Program on Carolina Sandhills National Wildlife Refuge. In: Pruden, T.L., Brennan, L.A. (eds.) *Fire in Ecosystem Management: Shifting the Paradigm from Suppression to Prescription,* pp. 161–166. Proceedings, 20th Tall Timbers Fire Ecology Conference. Tall Timbers Research Station.

International Forest Fire News, Vol. 1–41 (United Nations Publication, 1990–present).

Johannessen, Carl L. Davenport, W.A., Millet, A., McWilliams, S. (1971). The Vegetation of the Willamette Valley. *Annals of the Association of American Geographers* **61**, 286–302.

Jones, R. (1969). Fire-stick Farming. *Australian Natural History* **16**, 224–228.

Kilgore, B.M. (2007). Origin and History of Wildland Fire Use in the U.S. National Park Service, *George Wright Forum* **24**(3), 92–122.

Komarek, E.V. (1964). The Natural History of Lightning, pp. 139–183. In: *Proceedings, Tall Timbers Fire Research Conference* **3**. Tall Timbers Research Station.

Kull, C.A. (2004). *Isle of Fire: The Political Ecology of Landscape Burning in Madagascar.* University of Chicago Press, Chicago, IL.

Kunstadter, P. (1978). Subsistence Agricultural Economies of Lua' and Karen Hill Farmers, Mae Sariang District, Northwestern Thailand. In: Kunstadter, P., Chapman, E.C., Sabhasri, S. (eds.) *Farmers in the Forest. Economic Development and Marginal Agriculture in Northern*

Thailand, pp. 74–133. University Press of Hawaii for the East-West Center: Honolulu, HI.

Lewis, H.T. (1985). Burning the 'Top End': Kangaroos and Cattle. In: Ford, J. (ed.), *Fire Ecology and Management in Western Australian Ecosystems*. WAIT Environmental Studies Group Report No. 14. Western Australia Institute of Technology.

Lewis, H.T. and Theresa A. Ferguson. (1988). Yards, corridors, mosaics: how to burn a boreal forest, *Human Ecology* **16**(1): 57–77.

Melvin, Mark. (2012) *National Prescribed Fire Use Survey Report*. Technical Report 01-12. Prescribed Fire Councils, Inc.

Mistry, J., Berardi, A., Andrade, V., Krahô, T., Krahô, P., Leonardos, O. (2005). Indigenous fire management in the Cerrado of Brazil: The case of the Krahô of Tocantins. *Human Ecology* 33, 365–386.

Myers, R.L. (2006). *Living with Fire – Sustaining Ecosystems and Livelihoods through Integrated Fire Management*. Global Fire Initiative. The Nature Conservancy.

Myers, R.L., Ewel, J.J. (eds.) (1990). *Ecosystems of Florida*. University of Central Florida Press, Orlando, FL.

Nakićenović, N. (1996). Freeing Energy from Carbon. *Daedalus* **125**(3), 95–112.

NASA. MODIS website: http://modis.gsfc.nasa.gov/.

Nature Conservancy (2007). *Fire, Ecosystems, and People. Threats and Strategies for Global Biodiversity Conservation*. Global Fire Initiative.

Newton, D.E. (2002). *Encyclopedia of Fire*. Oryx Press, Westport, CT.

Normark, E., Rothpffer, C. (2011). *Guidelines for Sustainable Forest Management*. Holmen Skog, Sweden.

Pereira, J.S., Pereira, J.M.C., Rego, F.C., Silva, J.N., Silva, T.P. (eds.) (2006). *Incêndios Florestais em Portugal*. ISA Press, Lisbon, Portugal.

Pivello, V.R. (2011). The use of fire in the cerrado and Amazonian rainforests of Brazil: past and present. *Fire Ecology* **7**(1), 24–39.

Pyne, S.J. (2007). *Awful Splendour: A Fire History of Canada*. University of British Columbia Press.

Pyne, S. (2001). *Fire: A Brief History*. University of Washington Press.

Pyne, S. (2012). *Fire: Nature and Culture*. Reaktion Books.

Pyne, S. (2011b). *Imperium in imperio*: Los Angeles County Fire Department. http://firehistory.asu.edu/imperium-inimperio-los-angeles-county-fire-department (posted July 2011).

Pyne, S. (2008). Passing the Torch: Getting a Grip on an Age of Megafires. *American Scholar* **77**(2), 22–33.

Pyne, S. (2011a). Slow Match, Fast Flames. http://firehistory.asu.edu/slow-match-fast-flames (posted February 2011).

Pyne, S.J. (1997). *Vestal Fire: An Environmental History, Told through Fire, of Europe and of Europe's Encounter with the World*. University of Washington Press.

Radeloff, V.C., Hammer, R.B., Stewart, S.I, Fried, J.S., Holcomb, S.S., McKeefry, J.F. (2005). The Wildland-Urban Interface in the United States. *Ecological Applications* **15**(3), 799–805.

Rodríguez, I. (2006). Pemon Perspectives of Fire Management in Canaima National Park, Southeastern Venezuela. *Human Ecology* **35**(3), 331–343.

Rodríguez-Trejo, D.A., Martínez-Hernández, P.A., Ortiz-Contla, H., Chavarría-Sánchez, M.R., Hernández-Santiago, F. (2011). The Present Status of Fire Ecology, Traditional Use of Fire, and Fire Management in Mexico and Central America. *Fire Ecology* **7**(1), 40–56.

Rollins, M., Morgan, P., Stephens, S., Holden, Z. (2007). *Final Report. Historical Wildland Fire Use: Lessons to be Learned from Twenty-five Years of Wilderness Fire Management*. JFSP Project Number 01-1-1-06.

Román-Cuesta, R.M., Gracia, M., Retana, J. (2004). Fire Trends in Tropical Mexico. A Case Study of Chiapas. *Journal of Forestry* **102**, 26–32.

Russell-Smith, J., Whitehead, P., Cooke, P. (eds.) (2009). *Culture, Ecology, and Economy of Fire Management in North Australian Savannas. Rekindling the Wurrk Tradition* (esp. pp. 287–312). The West Arnhem Land Fire Abatement (WALFA) project: the institutional environment and its implications. Tropical Savannas Cooperative Research Centre.

Rydkvist, T., Eriksson, A-M. (2009). Prescribed Fire as a Restoration Tool and its Past, Present and Future Use in Västernorrland County, Sweden. *International Forest Fire News* **38**, 4–11.

Sahlin, M. Report. (2011). *Under the Cover of the Swedish Forestry Model*. Swedish Society for Nature Conservation.

Smil, V. (1994). *Energy in World History*. Westview Press, Boulder, CO.

Steensberg, A. (1993). *Fire-Clearance Husbandry. Traditional Techniques Throughout the World*. Poul Kristensen, Herning.

Steffen, W., Grinevald, J., Crutzen, P., McNeill, J. (2011). The Anthropocene: conceptual and historical perspectives. *Philosophical Transactions of the Royal Society A* **369**, 842–867.

Stein, S.M., McRoberts, R.E., Mahal, L.G., Carr, M.A., Alig, R.J., C, S.J., Theobald, D.M., Cundiff, A. (2009). Private Forests, Public Benefits: increased housing density and other pressures on private forest contributions. Gen. Tech. Report PNW-GTR-795.

Stewart, O.C. (Lewis, H.T., Anderson, M.K., eds.) (2002). *Forgotten Fires. Native Americans and the Transient Wilderness*. University of Oklahoma Press, Norman, OK.

Suomen Antropologi **4**. Special issue on swidden cultivation (1987).

Swedish Board of FSC (2009). *Swedish FSC Standard for Forest Certification*.

Tall Timbers Research Station. (1962–2012). *Tall Timbers Fire Ecology Conferences, Proceedings.*

van Wagtendonk, J.W. (2007). The History and Evolution of Wildland Fire Use. *Fire Ecology Special Issue* **3**(2), 3–17.

van Wilgen, B.W., Biggs, H.C. (2011). A critical assessment of adaptive ecosystem management in a large savanna protected area in South Africa, *Biological Conservation* **144**, 1179–1187.

van Wilgen, B.W., Govender N., Biggs, H.C. (2007). The contribution of fire research to fire management: a critical review of a long-term experiment in the Kruger National Park, South Africa. *International Journal of Wildland Fire* **16**, 519–530.

van Wilgen B.W., Govender N., Biggs, H.C., Ntsala D., Funda X.N. (2004). Response of Savanna Fire Regimes to Changing Fire-Management Policies in a Large African National Park. *Conservation Biology* **18**(6), 1533–1540.

van Wilgen, B.W., Biggs, H.C., O'Reagan, S.P., Mare, N. (2000). A fire history of the savanna ecosystems in the Kruger National Park, South Africa, between 1941 and 1996. *South African Journal of Science* **96**, 167–178.

Ward, D., Sneeuwjagt, R. (1999) Believing the Balga. *LANDSCOPE* **14**(3), 10–16.

Weise, D.R., Martin, R.E. (tech. coords.) (1995). *The Biswell Symposium: Fire Issues and Solutions in Urban Interface and Wildland Ecosystems.* Gen. Tech. Report PSW-GTR-158. U.S. Forest Service.

White, C.A., Fisher, W. (2007). Ecological Restoration in the Canadian Rocky Mountains: Developing and Implementing the 1997 Banff National Park Management Plan. In Price, M. (ed.) *Mountain Area Research and Management: Integrated Approaches.* Earthscan, London.

White, C.A., Pengelly, I.R., Zell, D., Rogeau, M-P. (2004). Restoring Heterogeneous Fire Patterns in Banff National Park, Alberta. In: Taylor, L., Zelnik, J., Cadwallander, S., Hughes, B. (eds.), *Mixed Severity Fire Regimes: Ecology and Management, Symposium Proceedings*, pp. 255–266. Washington State University Cooperative Extension Service and the Association for Fire Ecology, Pullman, WA.

White, C.A., Perrakis, D.D.B., Kafka, V.G., Ennis, T. (2001). Burning at the Edge: Integrating Biophysical and Eco-Cultural Fire Processes in Canada's Parks and Protected Areas. *Fire Ecology* **7**(1), 74–106.

Wildland Fire Leadership Council. (2011). U.S. National Cohesive Strategy for Wildland Fire Management.

Williams, J.T., Hyde, A.C. (2009). The mega-fire phenomenon: Observations from a coarse-scale assessment with implications for foresters, land managers, and policy-makers. Proceedings from the Society of American Foresters 89th National Convention (SAF).

Williams, J., Albright, D., Hoffmann, A.A., Eritsov, A., Moore, P.F., De Morais, J.C.M., Leonard, M., San Miguel-Ayanz, J., Xanthopoulos, G., van Lierop, P. (2011). *Findings and Implications from Coarse-Scale Global Assessment of Recent Selected Mega-fires, 5th International Wildland Fire Conference (9–13 May 2011).* Sun City, South Africa.

Wrangham, R. (2009). *Catching Fire: How Cooking Made Us Human.* Basic Books.

Zackrisson, O. (1976). Vegetation dynamics and land use in the lower reaches of the river Umeälven. *Early Norrland* **9**, 7–74.

Zalasiewicz, J., Williams, M., Haywood, A. & Ellis, M. (2011). The Anthropocene: a new epoch of geological time? *Philosophical Transactions of the Royal Society A: Mathematical, Physical and Engineering Sciences* **369**, 835–841.

PART FOUR

The Science and Art of Wildland Fire Behaviour Prediction

Photo

Outdoor experimental fires have provided an important source of data on free-burning wildland fire behaviour. This image shows a CSIRO fire behaviour documentation team in action on an experimental fire in a heath fuel complex in South Australia in March 2008 as part of Project FuSE, a Bushfire Cooperative Research Centre venture. (photo from David Bruce. © Bushfire CRC).

Fire on Earth: An Introduction, First Edition. Andrew C. Scott, David M.J.S. Bowman, William J. Bond, Stephen J. Pyne and Martin E. Alexander.
© 2014 John Wiley & Sons, Ltd. Published 2014 by John Wiley & Sons, Ltd.

Preface to part four

What does wildland fire behaviour involve, and what makes it an important topic? Can we predict it? If we can, how well do we achieve this challenge? How does one use fire behaviour knowledge in wildland fire management? Answers to these questions will be found in the following pages.

In this fourth and final part of the book, we focus on the fundamentals of vegetation fire behaviour from the standpoint of understanding heat transfer mechanisms, methods of measurement, description and characterization and ways of forecasting probable fire behaviour in relation to the three components of the fire environment (namely fuels, weather and topography). This information, in turn, serves as a means to support the development of practical ways of living with and managing ignitions of planned and accidental origin.

Part Four consists of three chapters which effectively cover the 'theory and practice' of wildland fire behaviour prediction. In Chapter 14, the physical processes, controlling influences and characteristics associated with the behaviour of free-burning fires are examined in light of what has been learned from conducting experimental fires in the laboratory and field, as well as from observing prescribed fires and wildfires, including extreme fire phenomena. In Chapter 15, the historical evolution of decision aids, different types of models and systems for predicting wildland fire behaviour and fire danger are recounted, along with the methodologies involved in model evaluation and preparation of various types of fire behaviour forecasts. Finally, in Chapter 16, applications of fire behaviour knowledge in five distinct areas of fire management are discussed (i.e. fire suppression, firefighter safety, community fire protection, fuel management and fire effects prediction).

The references contained in the bibliography and in the 'Further reading' sections at the end of the individual chapters should enable the reader to explore any particular topic presented here in greater depth. The references included within these sources may also be consulted for further information.

Chapter 14

Fundamentals of wildland fire as a physical process

14.1 Introduction

Many regions of the globe are prone to the occurrence of wildfires (Krawchuk *et al.*, 2009). Consequently, we need to learn how to better coexist with wildland fire, always remembering that no radically new concept in fire suppression can be anticipated. Suppression costs, at least in the United States and Canada, have skyrocketed in the past three decades, while the adverse impacts of wildfire on the land and communities worldwide has reached epidemic proportions. Furthermore, many of the world's ecosystems and their inhabitants are dependent on fire for their continued existence.

Drysdale (2011) has stated that: 'Further major advances in combating wildfire are unlikely to be achieved simply by continued application of traditional methods. What is required is a more fundamental approach which can be applied at the design stage ... such an approach requires a detailed understanding of fire behaviour'.

The term 'fire behaviour' is commonly regarded as the manner in which fuel ignites, flame develops, and fire spreads and exhibits other related phenomena. In his 1951 seminal publication, *Fire Behaviour in Northern Rocky Mountain Forests*, Jack S. Barrows (1951) outlined the basic concepts involved in the practice of prediction or forecasting wildland fire behaviour that are as valid today as they were more than 60 years ago.

As Barrows' schematic diagram illustrates (Figure 14.1), the process of judging fire behaviour requires the systematic analysis of many factors which can be divided into a five-step process:

Step 1: *Basic knowledge*. The foundation for judging probable fire behaviour must rest on basic knowledge of the principles of combustion:
- What is necessary for combustion to occur?
- What causes the rate of combustion to increase or decrease?
- How may combustion be reduced or stopped?

Step 2: *Forest knowledge*. Three basic factors in a forest area – weather, topography, and fuels – are important indicators of fire behaviour.

Step 3: *Aids and guides*. Several aids and guides are available to assist in evaluating weather, topography and fuels.

Step 4: *Estimate of situation*. The probabilities for various patterns of fire behaviour are systematically explored through an estimate of the situation, based upon the combined effects of weather, fuels and topography.

Step 5: *Decision*. The end product of the fire behaviour analysis is a decision outlining when, where and how to control the fire, and spelling out any special safety measures required.

Fire on Earth: An Introduction, First Edition. Andrew C. Scott, David M.J.S. Bowman, William J. Bond, Stephen J. Pyne and Martin E. Alexander.
© 2014 John Wiley & Sons, Ltd. Published 2014 by John Wiley & Sons, Ltd.

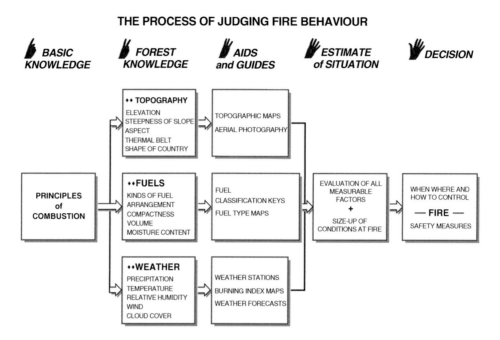

THE PROCESS OF JUDGING FIRE BEHAVIOUR

Figure 14.1   Judging wildland fire behaviour requires the systematic analysis of many factors (from Barrows, 1951).

These guidelines are applicable to both the control and use of planned or unplanned ignitions. They would also be valid in designing simulation studies involving the prediction of wildland fire behaviour.

A working knowledge of the fundamental principles, and some of the advanced concepts, concerning the dynamics and phenomenology of free-burning fires is considered essential to a greater understanding and appreciation of the more applied aspects of predicting wildland fire behaviour and of applying such information to practical problems and issues. As Adams and Attiwill (2011) have succinctly stated: 'The fundamentals of bushfire are not difficult to understand. A fire starts when there is a source of ignition and a supply of fuel. What happens thereafter is variable and complex, depending on the moisture conditions, nature, quantity and distribution of the fuel, on topography, and on the weather.'

In this first instalment of Part Four, Chapter 14, we summarize the background information pertaining to Barrows' steps 1 and 2 above, setting the stage for the chapters that follow. Chapter 15 is devoted to covering steps 3 and 4. Finally, in Chapter 16, we effectively address step 5 and more.

The International System (SI) of units is used exclusively throughout Part 4. Table 14.1 has been provided to help make conversions.

14.2   The basics of combustion and heat transfer

Wildland firefighters, for example, may be far more interested in practice than theory. Nevertheless, some time must be devoted to the principles of combustion and heat transfer as a prelude to the study and analysis of wildland fire behaviour.

14.2.1   Combustion chemistry

Wildland fires are the result of the combustion or burning of vegetative fuels, and these fires involve both chemical and physical processes (see also chapter 1, Part One). The fuel or plant material that burns, or is consumed, in a wildland fire is one of the two products of the photosynthetic process (Davis, 1959).

Table 14.1 List of International System (SI)-to-English unit conversion multiplications factors and formulae for the more common wildland fire-related quantities. To convert an SI unit to an English unit, multiply by the conversion factor or use the formula given in second column (e.g. 10.0 cm × 0.394 = 3.94 in). To convert an English unit to an SI unit, multiply by the inverse conversion factor or use the formula given in the right-hand column (e.g. 10 in × 2.54 = 25.4 cm).

SI unit	Conversion factor or formula	English unit	Inverse conversion factor or formula
Centimetre (cm)	0.394	Inch (in)	2.54
Degree Celsius (°C)	5/9 (°F − 32)	Degrees Fahrenheit (°F)	(9/5 °C) + 32
Grams per cubic centimetre (g/cm^3)	62.4	Pounds per cubic foot (lb/ft^3)	0.016
Hectare (ha)	2.47	Acre (ac)	0.405
Kilograms per cubic metre (kg/m^3)	0.0624	Pounds per cubic foot (lb/ft^3)	16.0
Kilograms per square metre (kg/m^2)	0.205	Pounds per square foot (lb/ft^2)	4.88
	4.46	Tons per acre (T/ac)	0.224
Kilojoules per kilogram (kJ/kg)	0.430	Btu per pound (Btu/lb)	2.32
Kilometre (km)	0.621	Mile (mi)	1.61
Kilometres per hour (km/h)	0.621	Miles per hour (mi/h)	1.61
Kilowatts per metre (kW/m)	0.289	Btus per second per foot (Btu/s-ft)	3.46
Kilowatts per square metre (kW/m^2)	0.088	Btu/(s-ft^2)	11.3
Metre (m)	3.28	Feet (ft)	0.305
Metres per minute (m/min)	3.28	Feet per minute (ft/min)	0.305
Metres per minute (m/min)	2.98	Chain per hour (ch/h)	0.335
Percentage slope (%)	tan^{-1} (% Slope ÷ 100)	Degree slope (°)	(tan °Slope) × 100
Reciprocal centimetre (cm^{-1})	2.54 in^{-1}	Reciprocal inch (in^{-1})	0.394
Tonnes per hectare (t/ha)	0.446	Tons per acre (T/ac)	2.24

Photosynthesis is the chemical process or reaction by which carbon dioxide (CO_2) and water (H_2O), coupled with the sun's energy, are combined gradually to produce organic matter and life-giving oxygen (O_2), the former being comprised principally of cellulose and other carbohydrates. This can be expressed in a general way by the following chemical equation:

$$CO_2 + H_2O + \text{Solar Energy} \rightarrow \text{Cellulose} + O_2$$

Fire, however, can suddenly reverse the process, releasing the thermal energy produced by combustion of the fuel due to an external heat source, as illustrated in this expression:

$$\text{Cellulose} + O_2 + \text{Heat} \rightarrow CO_2 + H_2O \\ + \text{Thermal Energy}$$

In this sense, combustion and photosynthesis are chemically similar. However, in contrast to photosynthesis, combustion is a rapid, chaotic chemical oxidation-type process. A kindling or fuel ignition temperature, which acts as a triggering mechanism or catalyst, must be reached to initiate the combustion process.

The decay or decomposition of plant material can accomplish the same end result as fire, only much more slowly. Generally, the heat released is not noticeable. When the production of plant matter exceeds the rate of decomposition, the net result is an accumulation of fuels, which increases the potential for sustained fire propagation and intensity.

Three basic elements in proper combination are necessary for combustion and flame production to occur – fuel to burn, heat to ignite it and oxygen (i.e. air) to support the process. The dependence of all three of these factors interacting simultaneously leads to the popularization of the 'fire triangle' concept (Figure 14.2A) (see also Figure 1.1, Chapter 1, Part One). Combustion is considered as a chemical chain reaction, because heat

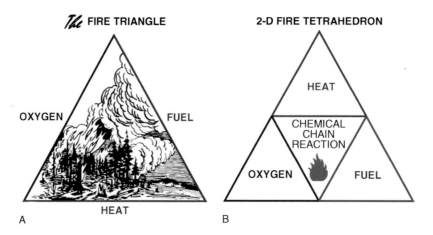

Figure 14.2   (A) The fire triangle or combustion triangle (from Barrows, 1951). (B) A two-dimensional version of the fire tetrahedron (adapted in part from Curl, 1966).

produced by the reaction then acts as a catalyst, which further increases the rate of reaction.

While the fire triangle still remains the basic link in the chain reaction of combustion, the 'fire tetrahedron' (Figure 14.2B) incorporates the uninhibited chemical chain reaction component of combustion and reflects the flame development process more accurately (i.e. the heat produced is greater than the heat required for maintaining combustion and results in a continuing process following ignition, provided all components of the fire triangle are present and the fuel remains in a combustible state).

A tetrahedron is a solid figure with four triangular surfaces. It was chosen over the square because, like the triangle, each side or 'leg' is in contact with each of the other sides. These symbolic models serve as a simple reminder of the ingredients necessary for ignition and combustion (or fire) to exist – namely oxygen, fuel and heat. Once initiated, the combustion is sustained by means of the heat-producing chemical reaction (Figure 14.2B). If any one of these four components is absent, eliminated or not in proper balance, then combustion and, hence, a fire, cannot occur or ceases to continue.

14.2.2   Combustion phases

Most observers of wildland fires would commonly think of combustion as a simple two-stage process.

Flaming combustion is characterized by the movement of a visible flame through a fuel bed. Smouldering or glowing combustion is generally associated with the residual burning of fuels after the actively spreading fire front has passed.

Combustion in wildland fires actually consists of at least four more or less distinct but overlapping phases (Figure 14.3). While the entire process is exceedingly complex and dynamic, the four phases can be described in general terms:

Preheating or pre-ignition combustion phase: unburnt fuel ahead of the advancing flame front is heated and raised to its ignition temperature. Water vapour is driven to the surface of the fuels and expelled into the surrounding air. As the fuel's internal temperature rises, cellulose and other compounds begin to decompose, releasing combustible organic gases and vapours.

Flaming combustion phase: the flammable gases escaping from the fuel surface are ignited in the presence of oxygen, and energy in the form of heat and light are produced. The flaming combustion phase can involve both active (solid flame zone) and secondary (discontinuous flame zone) flaming, depending on the composition of the fuel bed.

Smouldering combustion phase: the concentration of combustible vapours above the fuel is too small to support a persistent flame. Gases and vapours condense, appearing as visible smoke as they escape into the atmosphere.

Figure 14.3 A wind-driven surface fire in a lodgepole pine stand near Prince George, British Columbia, illustrating the various phases of combustion (from Lawson, 1972).

Glowing combustion phase: most volatile gases have been driven off. There is little visible smoke. The carbon remaining in the fuel is oxidized and continues to produce significant heat. Only embers are visible – no flame.

One would have little difficulty recognizing each of these phases or stages when the rate of combustion is slow. However, video taken within the plots burnt during the International Crown Fire Modelling Experiment in Canada's Northwest Territories has provided an entirely new means of visualizing the four phases of combustion experienced in a rapid spreading, wildland fire (Taylor *et al.*, 2004).

14.2.3 Parts of a spreading fire

For the purposes of communication, the anatomical parts of a wildland fire pattern are depicted in Figure 14.4A. It is common to talk about the head and front, flanks (e.g. right/left, east/west), and rear or back of a fire. There may also be wind-blown spot fires out ahead of the main advancing fire front. The fire perimeter represents the entire edge or boundary around the fire. In turn, the area enclosed by the fire perimeter represents the fire area or burnt area.

Unburnt islands may be found within the main body of the fire. Pockets or bays may also be found along the perimeter in addition to fingers – all of which result in a highly irregular perimeter.

These features of a fire can be caused by any or all of the components of the fire environment (e.g. surface and ground fuels in a lowland area being too moist to burn as is most often the case with peat bogs). Knowing what fuels did not burn is just as important, if not more so, than knowing those that did.

14.2.4 Principles of heat transfer and fire spread

Van Wagner (1983) emphasized that certain universal principles are applicable to all wildland fires, regardless of whether they are spreading through live and/or dead fuels:

1 There must be sufficient fuel of appropriate size and arrangement for the fire to burn in and through.
2 This fuel must be sufficiently dry to support a spreading combustion reaction.
3 There must be an agent of ignition.

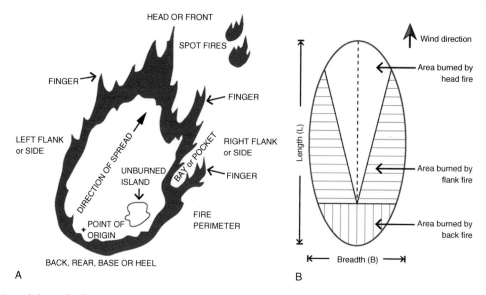

Figure 14.4 Schematic diagrams illustrating: (A) the parts of a free-burning, wind-driven wildland fire (after Moberly *et al.*, 1979); (B) a simple elliptical fire growth model (after Van Wagner, 1969), with the point of ignition or origins located at the junction of the four area growth zones. Reproduced with permission of the Canadian Institute of Forestry–Institut forestier du Canada.

In general terms, the ideal prerequisite for a spreading fire to occur is a continuous layer of fine surface fuels (e.g. conifer needles, hardwood leaves, grass, lichen, moss, finely divided shrubs, or other lesser vegetation). If such a condition does not exist, the likelihood of fire spreading from the point of ignition will be small.

While the fuels must also be dry enough, it is difficult to specify a moisture content threshold. This is commonly termed the 'moisture of extinction' (Rothermel, 1972) and this has been found, from laboratory and field studies, to depend on the quantity and spatial arrangement of the fuels and the wind strength (Wilson, 1985; Burrows *et al.*, 2009). However, it has been observed generally that fire spreads poorly or not at all when surface litter and cured grass moisture contents are above 30% (oven-dry weight basis), whereas shrubby fuels or conifer foliage may support fast-spreading crown fires at live fuel moisture contents of 100% or greater.

The process of ignition can be caused by natural or anthropogenic sources. A cloud-to-ground lightning strike is the most common natural source of ignition. Wildfires initiated by humans can be the result of:

1   accidents or human carelessness (e.g. wind-blown firebrand from a campfire or debris burning,

discarded cigarette, fireworks celebration, arcing power lines); or
2   arson (i.e. the malicious setting of a fire with matches, incendiaries such as flares, etc.).

There are many ignition sources associated with unwanted human-caused wildfires. A person using a hand-held drip torch, for example, to prescribe-burn an area also constitutes a deliberate (but, by definition, lawful) ignition source (see Chapter 1, Part One), unless by chance an escape occurs, in which case it is then regarded as a wildfire.

Regardless of how complex the behaviour of a fire is to describe, there are two limiting criteria common to all spreading fires in live and/or dead wildland fuels (after Van Wagner, 1983):

1   The fire must transfer enough heat to the unburnt fuel to dry it out and raise it to ignition temperature by the time the flame front arrives.
2   The moving flame front must consume enough fuel produce a continuous solid flame.

The term 'heat transfer' refers to the physical processes by which the heat produced by combustion is transmitted from burning to unburnt fuels. There are three principal mechanisms of heat transfer involved in the spread of wildland fires (Figure 14.5):

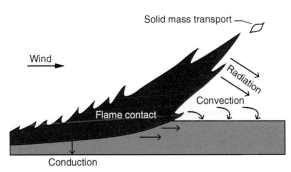

Figure 14.5 Schematic diagram illustrating the mechanisms of heat transfer involved in a wind-driven surface on level terrain (after Rothermel, 1972).

Convection: the transfer of heat by movement of a fluid (i.e. liquid or gas), including by actual direct flame contact. This is the natural movement of hot air and combustion products upwards in the absence of any appreciable wind and/or slope.

Radiation: transfer of heat in straight lines at the speed of light from hot particles of matter (solid, liquid or gas) to cooler regions in the fuel layer or its surroundings.

Conduction: the transfer of heat through matter from a region of high temperature to a region of lower temperature.

These three mechanisms of heat transfer can be seen in the kitchen. For example, steam rising from a boiling pot of water on the stove (convection), heat or warmth from an electrical stove's burner (radiation), and the gradual warming of the metal handle on a frying pan used in cooking (conduction).

Rising convective currents play a major role in the pre-heating of elevated fuels (e.g. lower portion of tree crowns), along with radiation. Convection and thermal radiation are the major contributors to the pre-heating of fuels of heading fires spreading with wind, upslope or both. Thermal radiation is the dominant mechanism for fires backing into the wind and/or downslope. There is much debate about the relative roles of convection versus radiation in heat transfer and fire spread. Of these two, the dominant mechanism is determined by the fire environment and fire dynamics – namely, the fire's energy output rate. Difficulties associated with measuring heat or thermal fluxes in wildland fires (i.e. the rate of heat energy transfer) have made settling the issue difficult.

Radiant heat transfer is manifested in the form of the heat felted from the visible flames. Convective heat transfer, on the other hand, is not visible to the naked eye. This has important safety implications for both firefighters and members of the general public alike, who may unknowingly be subjected to an invisible, but potentially lethal, dose of superheated air.

Because of their porous nature, wildland fuels are poor conductors of heat, and as a result, conduction plays a minor role in the flaming propagation of wildland fires. However, conduction is the main heat transfer mechanism in the burning of large logs and other heavy fuels such as deep organic layers (Figure 14.5).

A fourth means of heat transfer, which is primarily dependent on convection, termed 'solid mass transport' – or 'spotting' in the wildland firefighter's vocabulary – involves live or active firebrands being carried upwards in the convective currents above a fire and subsequently deposited varying distances downwind of the main advancing flame front (Figures 14.4, 14.5 and Figure 14.6). A similar situation can occur in mountainous topography, when burning material such as a log or tree cone rolls downhill (i.e. via gravity).

14.3 The wildland fire environment concept

The prediction of wildland fire behaviour is considered a prerequisite for the safe and effective control and use of fire. However, this requires a solid understanding of the fire environment concept. The fire environment and its relationship to fire behaviour is a vast and complex subject (Barrows, 1961).

The fire environment represents the surrounding conditions, influences and modifying forces that

Figure 14.6 Sequence of photos taken during the afternoon of August 22, 2005 near Coimbra, Portugal, showing some of the complexities involved in free-burning wildland fire behaviour. (A) An advancing wildfire in a maritime pine forest spotting into an opening (local time: 16:39:57), followed by (B and C) the coalescence of spot fires (respectively 16:41:14 and 16:41:28) and (D) merging with the main flame front (16:41:38), resulting in a greatly increased flame height. The elapsed time between photos (A) and (D) was 105 seconds (from Alexander and Cruz, 2013a). Reproduced with permission of Elsevier.

determine the inception, growth and subsequent behaviour of a free-burning wildland fire (Country-man, 1972). The fire environment can be represented by an equilateral triangle (Figure 14.7A) with the lower two sides or 'legs' representing the fuel and topography components, and the top side represent-ing the air mass (i.e. a widespread body of air that is approximately homogeneous in its horizontal extent, particularly with respect to temperature and mois-ture) or simply the weather. It is the current state of the fire environment and their interrelationship with one another, and with fire itself, that determines the behaviour of a fire at any given moment.

From a fire behaviour standpoint (Figure 14.7B), the fire environment can be separated into two broad classes – 'open' or 'closed' (Countryman, 1966). In urban fires involving buildings (i.e. a closed fire environment), the weather or outside influences and topographic effects play a less dominant role compared to fuel influences. However, with wildland fires, the distinction between the two types of fire environments is not so obvious. While the environ-ment outside of the buildings in a wildland area is not considered confined (i.e. open), a surface fire burning inside a dense conifer forest stand may, for example, be regarded as a closed fire environment.

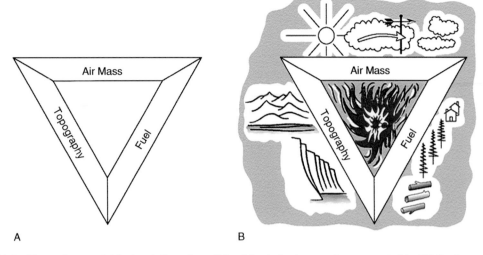

Figure 14.7   The environmental factors influencing wildland fire behaviour can be represented by (A) the fire environment triangle (from Countryman, 1966), with each side representing the three major components – fuel, topography and air mass or weather. The extent of the environment affecting a fire also changes with the size and characteristics (e.g. intensity) of the fire itself. By adding fire to the fire environment triangle, we can represent this relationship symbolically with (B) the fire behaviour triangle (from Countryman, 1972).

14.3.1   *Fire weather and climate*

A working knowledge of fire weather meteorology is essential to the understanding of wildland fire behaviour. A good introductory reference book on meteorology is Ahrens (2008).

Weather is the state of the atmosphere surrounding the Earth. We commonly think of weather in very simplistic terms, such as:

- How warm or cold is it?
- Is it raining and, if not, how sunny will it be?
- Will winds be light or strong?

While our immediate interests are in what is happening at ground level, the conditions for several kilometres above Earth's surface are equally as important, although not nearly as obvious.

Because the atmosphere is constantly changing, so too is the weather. The changes are caused by differential heating at the Earth's surface, resulting from the planet's rotation around the sun. The amount of heating varies with the steepness and aspect of the ground surface, the time of day (e.g. day or night), the season of the year and the closeness to the Earth's equator.

The Earth's surface responds differently to the heating by the sun during daylight hours. For example, water bodies are heated much less than land masses, and land covered with vegetation is heated less than bare ground surfaces. The reverse occurs at night. The differential heating resulting from differences in latitude and water versus land masses sets up circulation by convection, creating large-scale weather patterns.

Even after the passage of more than 40 years, Schroeder and Buck (1970) is still considered the classic, authoritative information source on the subject of fire weather meteorology. The term 'fire weather' is generally regarded as, collectively, those weather elements that affect the likelihood of a fire starting, the fire's behaviour and impact and its suppression difficulty. Fire weather also plays a significant role in prescribed fire planning and execution, in terms of achieving the desired fire effects while minimizing the chances of an escape.

The influence of a weather element on ignition and fire behaviour potential can be direct, as with the effect of wind speed on fire spread, and/or indirect, as is the case with fuel moisture and fuel temperature (Table 14.2). The term 'fire climate' represents the composite pattern, or integration over a period of time, of the fire weather elements that affect fire occurrence, fire behaviour, fire suppression and prescribed fire use. The fire climate of an area defines the

Table 14.2 A list of common surface weather elements or variables and the primary manner in which they indirectly and directly influence the behaviour of free-burning wildland fires.

Weather element	Influence(s)	Supplementary comments
Air temperature	Drying effect on the moisture content of dead fuels, both in the short-term and cumulative; alters fuel temperature.	Common measure: degrees Celsius (°C). Synonymous with dry-bulb temperature.
Relative humidity	Drying and wetting effects on dead fuels, both in the short and long term.	Commonly referred to as 'RH'. Unit of measure: percent (%). Often used as a direct surrogate for the moisture content for fine deal fuels.
Wind speed	Alters the angle of the flame front so that flames are driven into the unburnt fuel; determines spotting distances.	As measured at the international standard height of 10 m in the open or 10 m above closed vegetative canopy. Common measure: kilometres per hour (km/h).
Wind direction	Determines the direction of fire spread; in some situations, a relatively sudden and major change in wind direction can very quickly turn the flank of fire into a wide head fire (e.g. cold frontal passage).[1]	The direction *from* (not *to*) which the wind is blowing. Common measures: degrees (°) or cardinal direction (e.g. north, north-east, east, south-east, south, south-west, west and north-west).
Precipitation	Wetting effect on dead fuels, both in the short and long term.	Both the amount in millimetres (mm) and duration (e.g. hours) of precipitation events are important. A prolonged dry spell or drought can acerbate a fire situation but is not a prerequisite for a major incident.
Cloudiness	Short-term drying effect on dead fuels and on altering fuel temperature.[2]	Cloud amount expressed as a percent (%) or in eights (e.g. 3/8).
Atmospheric stability	An unstable atmosphere causes winds to be turbulent and gusty, leading to erratic fire behaviour; thunderstorm activity increases leading to strong downdraft winds and the possibility of lightning fire ignitions.	Whether an atmosphere is 'neutral', 'stable' or 'unstable' depends on whether the temperature decrease with altitude is equal to, less than or greater than the dry adiabatic lapse rate (i.e. about 1°C per 100 metres).

[1]Light winds (i.e. <5 km/h) are commonly associated with variable wind directions.
[2]Clouds serve to block direct solar radiation from reaching the ground surface. Some cloud types can locally forewarn of impending critical fire weather conditions. See *Fire Weather and Fire Behavior Impacts of Clouds* (http://www.wildfirelessons.net/Additional.aspx?Page=299).

length and severity of the fire season (Pearce *et al.*, 2011). Fire climate statistics can give a general picture of what one might expect in regard to a range of fire weather patterns in the future. Fire weather climatology is a valuable tool in both wildfire and prescribed fire planning.

14.3.2   Fuels

As Brown and Davis (1973) have so eloquently noted, 'The ignition, build-up, and behaviour of fire depends on fuels more than any other single factor. It is the fuel that burns, that generates the energy with which the fire fighter must cope, and that largely determines the rate and level of intensity of that energy.'

A wildland fire is not capable of starting and spreading if there is not sufficient fuel of the right kind available. No matter what the current weather conditions may be, if there is no flammable fuel available, then there can be no fire!

The composition and extent of vegetation types in a given geographic locale is, in the broadest sense, a reflection of the area's climate, geomorphology and land use history. Fire climate thus influences the quantity and condition of wildland fuels. For example, wildland fires are rare in arid climates, as the fuel is normally too sparse to sustain a propagating fire. Conversely, in some tropical rainforests of the world, the vegetation and climate is too moist to support combustion. However, there are exceptions to these two extremes, when above-average rainfall occurs in the former case or by man-made disturbances, coupled with unprecedented drought, in the latter (see chapter 1 Part One and chapters 6 and 10, Part Two).

Natural vegetation or live wildland fuels can be readily altered by certain meteorological phenomena (e.g. ice, snow, hail and wind storms, frost), including multiple drought years (Buck, 1951), insect attacks (e.g. mountain pine beetle, spruce budworm), pathogens (e.g. sudden oak death, dwarf mistletoe) and even fire itself (e.g. crown scorching after a fire has burnt through

Figure 14.8 Profile of a forest, illustrating the location and composition of four commonly recognized fuel layers or stratum, based principally on the characteristic type of fire activity associated with each of them (after Brown and Davis, 1973). Reproduced with permission of McGraw-Hill.

the surface fuels), in ways that can increase or decrease fire behaviour potential. Introduced or invasive plant species can also greatly increase the fire susceptibility of a landscape (Brooks *et al.*, 2004). Notable examples include the proliferation of cheatgrass in the western United States, gorse in New Zealand, and the stands of melaleuca in Florida (see chapter 6, Part Two).

Wildland fuels vary widely in their distribution, physical characteristics and effects on fuel moisture, wind and fire behaviour. A means of stratification on the basis of their vertical distribution and general properties is necessary for effective communication. Based on their location and effects on fire behaviour, four more or less distinct strata are generally recognized: ground, surface, ladder and crown fuels (Figure 14.8), although other classification schemes have been developed (e.g. Gould *et al.*, 2011). Collectively, these strata would constitute a forest fuel complex or fuel type which may either be simple (i.e. a single strata) or complex (i.e. involving multiple strata) in nature.

Ground fuels normally support smouldering or glowing combustion in ground or subsurface fires. Surface fuels are responsible for propagating surface fires and, in turn, initiating crown fires. Ladder or bridge fuels (e.g. understorey conifer reproduction, tall shrubs, needle drape, tree lichens and bark flakes on tree boles) provide for vertical continuity between surface and crown fuels, thus contributing to the ease of single or group tree torching and the onset of crowning. Crown

fuels represent the standing and supported combustibles in the overstorey of a forest stand (or tall shrub field) that are not in direct contact with the ground, and which are generally only consumed in crown fires.

The delineation of surface and crown fuel stratums is generally regarded as occurring approximately two metres above the ground fuels, with ladder or bridge fuels representing the transition zone between surface and crown fuels (Brown and Davis 1973). In certain vegetation types, such as shrubland and logging slash, the ladder and crown fuels will not exist, while in some grasslands and shrublands, there may also be no true ground fuel layer.

14.3.2.1 Physical fuel properties

Fire behaviour depends not only on fire potential at one location, but also on a range of associated factors that include the distribution and characteristics of the individual and collective elements comprising the fuel complex (Table 14.3).

The extrinsic properties of vegetation fuel complexes generally have a far greater influence on fire behaviour dynamics than the intrinsic physical and chemical properties that are characteristic of the fuels themselves (e.g. mineral, heat and chemical content, particle density, thermal conductivity). The main extrinsic fuel properties include the quantity, size and shape, arrangement, continuity or pattern.

Table 14.3 Physical fuel properties affecting various elements of wildland fire behaviour potential and their relative influence as outlined in 1975 by Hal E. Anderson, USDA Forest Service, Northern Forest Fire Laboratory, Missoula, Montana (from Alexander, 2007).

Fuel properties (dimensionless)	Ignitability		Fire spread		Energy release							Physical obstruction	
	Ignition probability	Spot fire ignition	Linear	Size	Fire intensity	Flame height	Scorch height	Crowning potential	Firebrand generation	Fire duration	Fire persistence	Ground	Aerial vegetation
PARTICLES:													
Size (diameter length)	X	X	X	X	X	X	X	X	X	X	X	X	X
Shape (geometric factor, surface area volume)	X	X	X	X	X	X	X	X	X			X	
Density (weight/volume)	X	X	X	X	X	X	X	X	X	X	X	X	
BEDS:													
Load (weight/unit area)	X		X	X	X	X	X	X		X	X	X	X
Depth (thickness)			X	X	X	X	X	X		X	X	X	X
Continuity: Vertical (length)		X					X	X					X
Horizontal (length)	X	X	X	X				X				X	
Live/Dead ratio	X	X	X	X	X	X			X				X
Extent (% of land area)		X		X								X	X

Quantity: fuel load represents the oven-dry weight of material per unit area. The recommended SI units for fuel load are kg/m^2 for general usage and t/ha for large-scale mental image (Van Wagner, 1978); note that $1.0 \ kg/m^2$ equals 10 t/ha. The amount of fuel present directly affects a fire's energy output rate. The total fuel load is not nearly as important as the 'available fuel' (i.e. the quantity of material that would be consumed in the flaming zone).

There is undoubtedly a lower limit, below which spreading combustion is not possible (i.e. a minimum fuel load for surface fire spread), although this is difficult to indentify precisely for a given fuel complex unless deduced from experimental burning or other means (e.g. Burrows, 1994; Lavoie, 2004). However, based on past experience, it is generally possible to express such a threshold in terms of the number of years since the last fire (e.g. Rogers, 1942), especially in short-interval fire regime fuel types such as grass, eucalypt forest, Mediterranean climate shrublands and south-eastern US pine forests. This is not so readily discernible in longer-interval fire regime fuel type complexes such as the high-elevation spruce-fir forests in the Rocky Mountains of western North America.

Size and shape: because the heat required to raise the temperature of a fuel particle to ignition must go through the fuel's surface, those fuels with a high surface area to volume ratio (cm^{-1}) burn more readily than those with a lower value. The size (i.e. diameter and length) and geometric shape (e.g. irregular versus angular) of a fuel particle will determine the ratio of its surface area to its volume. Small, flat fuels have a greater surface area to volume ratio than larger round particles. A fuel's surface area to volume ratio also affects the rate of drying and wetting in dead fuels. Thus, the finer the fuel particle, the better the overall burning characteristics. The shape of fuel particles is also an important factor in determining the aerodynamic features of potential firebrand materials associated with spotting.

When describing fuel size, it is common to use terms such as 'fine' or 'medium' and 'heavy' or 'coarse' fuels, even though precise criteria have not been established, at least not globally anyway. Dead fuels less than about pencil thickness (≈ 0.6 cm) in diameter are almost completely consumed over a wide range of burning conditions.

Arrangement: the manner in which fuels are arranged can have a profound influence on wildland fire behaviour. Fuel arrangement involves the vertical (and, to a certain extent, the horizontal) distribution of combustible fuel particles. The rate of combustion or burning efficiency increases as the amount of air space in a fuel bed increases, up to an optimum spacing. A highly compacted fuel bed would have a lower potential fire spread rate. At some point, the fuels become too sparse or too compacted to support further fire spread.

One simple measure of fuel arrangement is bulk density. Numerically, this is equal to the fuel load divided by depth of the particular fuels (e.g. forest floor, roundwood surface fuels, canopy foliage) and is expressed in kg/m^3 for general usage and in g/cm^3 for small-scale mental image (Van Wagner, 1978). The canopy base height, describing the distance between surface and crown fuels, is a useful measure of fuel arrangement in conifer forest stands.

Continuity or pattern: while fuel arrangement should be regarded as pertaining to a fire's behaviour at any given moment (i.e. small scale), fuel continuity refers to consistency or lack of fuels at a much larger scale with respect to a fire's probable path. Horizontal fuel continuity determines whether a fire will spread for any significant distance. Fuel moisture content level can also determine whether there are fuels available to support fire spread. When fuels ahead of the fire cease to be uniform and 'continuous', and become patchy or 'discontinuous' due to an unburnable ground surface (e.g. bare soil, rock, gravel), then the threshold for fire spread is largely controlled by the wind speed (Figure 14.9). Discontinuous fuel types, like the pinyon-juniper woodlands of the western USA, are oftentimes viewed as being especially volatile when this wind speed threshold is exceeded. There is no easy metric for fuel continuity other than percent vegetative or fuel cover.

14.3.2.2 Fuel moisture content

Fuel moisture content represents the amount of water present in the fuel expressed as a percentage of the fuel's oven-dry weight, and it is conceivably the most important parameter affecting fire behaviour as it determines fuel ignitability (Figure 14.10), rate of combustion and the amount of fuel consumed. The moisture content of fuels will influence whether they act as a heat source or as a heat sink, either intensifying or diminishing the level of fire behaviour. The influences controlling the moisture content of 'live' and 'dead' fuels is decidedly different (Nelson, 2001).

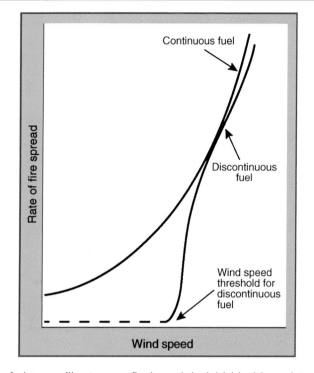

Figure 14.9 Discontinuous fuel types will not carry a fire beyond the initial ignition point until a particular wind speed threshold is reached or exceeded. This is in contrast to a continuous fuel type, where fire spread is possible even under calm or light wind conditions and steadily increases with increasing wind speed (after Gill *et al.*, 1995).

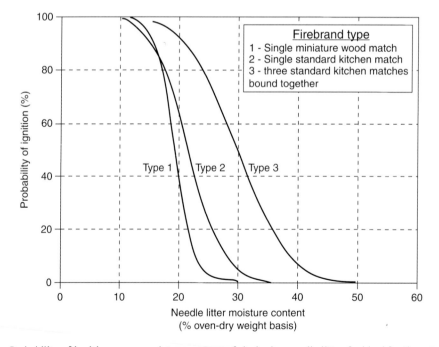

Figure 14.10 Probability of ignition versus moisture content of slash pine needle litter fuel bed for three types of flaming firebrands of different sizes (adapted from Blackmarr, 1972).

The moisture content of dead fuels, such as the forest floor layer and dead-down woody surface fuels, is determined by the wetting and drying processes associated with past and present weather conditions on an hourly, daily or seasonal basis. Just how rapid changes in moisture content occur is determined by the characteristics of the fuel, such as size and age (Nelson and Hiers, 2008). Numerous models have been developed for predicting dead fuel moisture content directly from observed weather elements, such as dry-bulb temperature, relative humidity, rainfall, solar and terrestrial radiation, dew and wind speed.

Fine fuels, such tree lichens, react very quickly to the minute-by-minute changes in relative humidity. This is in contrast to deep, compact organic matter or large logs, which may require substantial rain or prolonged dry spells for their moisture status to change appreciably.

Seasonal changes in living vegetation can have a major effect on the condition of both live and dead fuels with respect to moisture content and, in turn, wildland fire behaviour. The most pronounced and readily observable example is the degree of curing that occurs in annual and perennial grasslands during the fire season (Cheney and Sullivan, 2008), i.e. the proportion of cured and/or dead plant material in a grassland fuel complex, expressed as a percentage of the total fuel mass. In northern climes, distinct changes in the flammability in the surface fuel strata are also readily apparent in hardwood and mixed hardwood/conifer forests during the cured stage in the spring, following snow-melt, compared to the 'green-up' of the understorey and overstorey vegetation in early summer (Alexander, 2010). A further change occurs in autumn, as the hardwood trees and understorey shrubs drop their leaves and the senescence or dying of the herbaceous vegetation advances.

The moisture content of most living vegetation, especially perennials (i.e. trees, woody shrubs and certain species of grasses), is controlled by the species physiology and time of year and, generally, has very little to do directly with current weather conditions. Drought and heat wave conditions, however, can contribute to plant moisture stress (Fahnestock, 1962). The moisture content of living vegetation is highest at the time of the emergence of new plant tissue, and gradually declines as the growing season progresses. Lichens and mosses, on the other hand, react to changes in weather conditions in the same manner as fine, dead fuels, as do herbaceous plants (i.e. annual grasses, herbs and forbs), once they have reached a cured state.

As drought conditions intensify, the differences in dead fuel moisture between different topographic aspects becomes less discernible. The same can be said for extended rainy periods.

14.3.3 Topography

Topography, or the 'lay of the land', is an important factor in determining fire behaviour in mountainous country or complex terrain (Campbell, 2005). With the exception of the mechanical effect of slope steepness (or 'power of the slope') on rate of fire spread (Figure 14.11) (i.e. in bringing the fuels closer to the advancing flame front), the influence of topography on wildland fire behaviour in mountainous country depends largely on how it alters both meso-scale and micro-scale meteorological variables (e.g. air temperature, relative humidity, wind speed and direction, insolation) and how these variables, in turn, affect dead fuel moisture content and wind speed near the ground (Table 14.4).

Elevation, aspect and slope steepness influence the local climate, so have a long-term effect on the vegetation, including species composition and, thus, the nature of the wildland fuels in the area. Latitude is also a governing factor. Elevation general controls the timing of snow-free cover in the spring, the length of fire season, plant phenology (e.g. curing of grasses), and the severity of daily and seasonal fire danger.

14.3.4 Variations and interactions

As a fire front grows and spreads across the landscape, it can encounter considerable spatial and temporal variation in the components of the fire environment, and especially so in mountainous country. Technically, with respect to time, topography can be considered as static. However, in rugged terrain, it can change greatly in horizontal space. Except for the moisture content of dead fuels, fuel properties change so slowly that they can be considered as static for a given fire. However, like topography, there can also be wide variations in fuels spatially.

Figure 14.11 A high-intensity fire advancing upslope (60% or 31°) as part of the experimental burning in native shrub fuels carried out on March 15, 2005 near Lake Taylor on the South Island of New Zealand during Project FuSE, a Bushfire Cooperative Research Centre venture (photo Fraser Townsend, Scion, New Zealand. Reproduced with permission of Bushfire CRC).

On the other hand, fire weather conditions in some regions of the globe are constantly changing in both time (from day to day, hour to hour and even minute to minute) and space. A distinct diurnal cycle in air temperature and relative humidity is a well-known fire weather pattern in many regions of the world (Figure 14.12).

As the air temperature increases during the daylight hours, the relative humidity will decrease to its daily minimum near mid-to-late afternoon, coinciding with the maximum air temperature for the day. The relative humidity will then gradually increase to its daily maximum just before sunrise, when the air temperature is at a minimum. Exceptions to this general pattern do occur from time to time (e.g. Mills, 2008).

The wildland fire environment is an integration of the effects of all its components. Because the components are closely interrelated, changes in one group of factors can cause changes in others. Clear-cut harvesting of a forest stand, for example, causes significant changes in both the fuel structure and availability, as well as the fire weather conditions, given that the forest canopy directly influences fuel moisture and wind conditions inside a forest stand (Countryman, 1956, 1960).

There is a dangerous tendency to use stereotypical descriptions to quickly size-up wildland fire environments, fire climates or fuel types as to their apparent benign fire behaviour potential. Certain forest cover types are regarded as perpetual 'asbestos' or 'rain' forests, incapable of supporting high-intensity fires. For example, young and semi-mature trembling aspen stands in the summer, following leaf-out of the understorey and overstorey vegetation, are not usually fire-prone, even under extremely dry conditions.

However, mature and over-mature trembling aspen stands exhibiting large accumulations of dead-down woody fuels and a dense conifer understorey can be quite flammable, even during mid-summer. Also, particular environments are not often viewed as capable of sustaining a high-intensity fire. For example, mention Prince

Table 14.4 List of topographic characteristics and the manner in which they directly or indirectly influence the behaviour of free-burning wildland fires.

Topographic characteristic	Description of influence(s)	Supplementary comments
Slope steepness	As angle of the slope increases, the advancing flames are tilted closer to the unburnt fuels lying upslope of the advancing fire front, even in the absence of wind, which serves to enhance the convective and radiant heating, leading to a higher rate of fire spread.[2]	Numerically, slope steepness represents the 'rise' in elevation over the 'run' in horizontal distance. This is commonly expressed as percentage (%) slope in North America but in degrees (°) slope in Australia, for example.[1]
Elevation	Weather elements change with a rise in elevation. Air temperature generally decreases and relative humidity increases as does precipitation, the net result being an increase in dead fuel moisture.	Expressed as above mean sea level (m).
Aspect	Aspect refers to the direction a slope is facing. Along with the degree of slope steepness, aspect determines the duration and level of solar radiation, thereby determining surface fuel temperatures and the drying regime of dead forest fuels via air temperature and relative humidity.[3]	Referred to variously as slope aspect, slope exposure, direction facing or simply 'slope'. Most commonly expressed in terms of four (i.e., north, east, south and west) or eight (i.e. add north-east, south-east, south-west and north-west) cardinal directions.
Configuration or shape of the country	In narrow canyons or ravines, the radiation from the burning of one slope can lead to increased drying and preheating of the opposite slope, thereby making it highly susceptible to spot fire development. A fire in a steep, narrow chute or chimney virtually 'erupts' due to the simultaneous effects of slope steepness (including slope attachment[4]), enhanced uphill air flow and cross-canyon radiant heating.	Canyons with long, shallow slopes do not have near the same impact on spotting and rate of fire spread potential.
Barriers to fire spread	These can be either natural or man-made. Examples include rock slides, bare mineral soil (e.g. ploughed field, established firebreak, prepared fireguard), roads and water bodies.[5] Each of these barriers are devoid of fuels and thereby directly limit fire spread.	The effectiveness of a topographical barrier being breached depends on the width of the barrier and the level of fire behaviour, in terms of flame dimensions and spot fire distances at the time the advancing fire meets the barrier.

[1]A 1% slope means a rise or fall of 1.0 m of elevation at a horizontal distance of 100 m. A 100% slope would equal 45°. When slope steepness exceeds 60–70%, flames tend to bathe or attach to the slope directly, even in calm winds, and fire behaviour can become very intense and unstable (Van Wagner, 1977b).
[2]The chances of burning material rolling downhill and initiating upslope fire runs also increases as slope steepness increases.
[3]In the northern hemisphere, south and west aspects receive more direct solar radiation (given that they are nearly perpendicular to the sun's rays) than north and east aspects (which are oriented more parallel to the sun's rays and are thus shaded during more of the day because of the sun's angle during the heat of the afternoon). Countryman (1966) provides a good example of the diurnal variation in surface air temperatures with aspect. In the southern hemisphere, it is just the opposite: north aspects are drier and warmer than south aspects.
[4]As slope steepness increases, the flames tend to lean more and more toward the slope surface, gradually becoming attached, the result being a sheet of flame moving roughly parallel to the slope (Van Wagner, 1977b). It appears that 50% slope represents a key threshold for 'slope attachment', but the actual value will differ, depending on the prevailing wind strength as well as on the fuel type characteristics. There is no universal agreement on slope steepness classes or adjectives. However, Barrows (1951), for example, considers 0–20% as 'gentle', 21–40% as 'moderate', 41–60% as 'steep' and >60% as 'very steep'.
[5]In some areas, the height of the water table (Williams, 1954; Fahnestock, 1959) can determine whether the area constitutes an effective barrier to fire spread.

Edward Island, Canada's smallest province, in the context of wildland fire, and most Canadian fire managers would immediately think of a landscape dominated by potato fields. Yet, as demonstrated during 'seven days in May' of 1986, crown fires can readily occur in the island's white spruce plantation forests. Therefore, such broad generalizations or preconceived notions are dangerous and should be avoided in order to prevent being surprised by an unexpected fire behaviour-related event.

14.3.5 Methods of monitoring and assessment

Weather conditions at and above the Earth's surface are monitored by a network of surface weather and upper air monitoring stations, supplemented by satellite or remote sensing imagery (see chapter 1, Part One). Forecasts of weather conditions for varying time periods are derived from these monitoring activities coupled with computer-based, numerical weather prediction models that are based on scientific understanding of

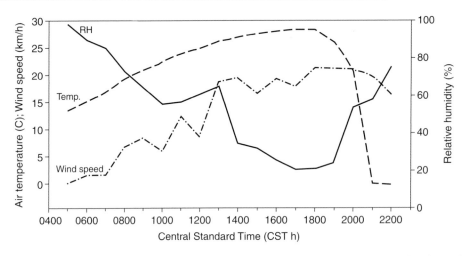

Figure 14.12 Diurnal trend in several fire weather elements as recorded on July 9, 1980 at the airport in Flin Flon, Mantioba (after De Groot and Alexander, 1986). At 1400 hours, the Chachukew Fire in nearby east-central Saskatchewan breached its containment line and advanced nearly 6 km in the next five hours as a crown fire. Reproduced with permission of Canadian Forest Service Publications (Natural Resources Canada).

atmospheric processes. A synoptic weather chart or map provides a spatial representation of the actual or forecasted state of the atmosphere at an instant in time.

Two types of fire weather forecasts are common. A zone or area fire weather forecast is regularly issued during the fire season for particular geographical regions. These are typically delineated on the basis of fire climatic and/or administrative considerations. Spot fire weather forecasts are issued on request to fit the time, topography, and weather for a particular fire location, and these are more detailed, timely and specific than a zone or area forecast (Cuoco and Barnett, 1996).

Fire management agencies have also established their own fire weather station networks and system of fire weather forecasting. Manually operated weather stations have largely been replaced by remote automated weather stations (RAWS) (Finklin and Fischer, 1990). Belt weather kits were introduced in the late 1950s to monitor on-site weather conditions (i.e. air temperature, relative humidity, wind speed and direction) before and during wildfires and prescribed fires, but these have now started to be replaced by small, hand-held electronic instruments that complete the same measurements. Still, simple tools like the Beaufort scale for estimating wind strength based on visual indicators (List, 1951), and conventional observations of cloud cover and type, still have great value.

Tremendous strides have been made over the years to increase fuel sampling efficiency. Labour-intensive, destructive sampling methods, while still

required in certain situations, have been largely eliminated. Fuel properties can be determined directly in the field using various non-destructive inventory techniques (Brown et al., 1982), or indirectly through prediction, which requires known quantitative relationships between fuel properties and vegetation characteristics (e.g. canopy bulk density can be predicted from stand characteristics such as height and basal area or density). Estimation generally requires a combination of both inventory and prediction. Field measurements made of forest floor depth or grass fuel height can be converted to fuel loads, provided that previous sampling has established depth- or height-weight relationships.

In spite of the increased efficiencies, fuel sampling can still be a very tedious, time-consuming and expensive task. The needs of fire researchers and fire managers can also be quite different. The fuel photo series approach, for example, offers a simple means of quickly obtaining an estimate of certain fuel characteristics in situations where a high level of detail and accuracy is not considered necessary (Lavoie et al., 2010). The fuel hazard rating schemes used in Australia offer a similar capability (Gould et al., 2011).

Over the years, field personnel have devised simple, yet expedient, means of judging fine dead fuel moisture by simply twisting a conifer needle or snapping a twig (Burrows, 1991). Forced-air drying ovens, while still a mainstay in many fire research circles, have given way to rapid, electronic fuel moisture analyzers that are capable of determining moisture content directly in the field in a matter of seconds (rather than having to wait 24 hours for oven-drying to be completed). In addition, satellite remote sensing of dead and live and fuel moistures appears to hold great promise for the future.

While topographic features can be readily identified in the field, the observer is limited by his or her position on the landscape. Slope steepness can be visually estimated in the field or measured more accurately with a clinometer. At the same time, aspect can also be judged visually and more precisely assessed using a compass. Topographic maps and aerial photos were considered major innovations for fire managers, as were digital terrain models in more recent times. Nowadays, both managers and researchers alike have *Google Earth* capability, allowing for the ready assessment of topography from a seemingly unlimited number of perspectives.

Many significant advances in the monitoring and assessment of the fire environment have been made directly or indirectly over the past 65 years or so. These include (but are not limited to) satellites, weather radar, geographical information systems (GIS), portable upper air sounding stations, electronic weather stations, hand-held global positioning system (GPS) devices and Light Detection And Ranging (LiDAR) optical remote sensing, in addition to the enormous advances made in both the software and hardware aspects of computer technology. Today's ability to detect and track wildfire activity from satellites has no doubt far exceeded the original vision (Singer, 1962) (see chapter 1 Part One).

If a fire manager or fire researcher from the 1960s awoke today from a long Rip Van Winkle-like slumber, they would no doubt be amazed at the current state of the fire environment intelligence-gathering system. However, in spite of the technology and modelling advances that have been made in meteorology, for example, the general difficulty of monitoring and forecasting winds and rain on an hourly or daily spatial basis still persists, as does trying to forecast general weather trends out to longer than approximately six days. Limitations in what is possible continue to exist, even given certain innovations like ensemble predictions derived from operational weather forecasting models, which seek to identify the expected 'spread' of weather conditions and assess the probability of particular weather events.

Even the most advanced currently available fuel mapping system (e.g. LANDFIRE) cannot accurately estimate the very simplest of crown fuel characteristics (i.e. canopy base height). Consider, too, Van Wagner's (1990) claim that 'the problem of how to describe a fuel complex in terms that would permit a physical deduction of how fire would spread through it has so far proved intractable'. Nothing has appreciably changed in the intervening 20-plus years to suggest that this statement is no longer valid.

14.4 Characterization of wildland fire behaviour

Because there are many aspects to wildland fire behaviour, there are, in turn, many quantitative measures to describe the behaviour of free-burning wildland fires (Burrows, 1984a). Wildland fires are also commonly described in qualitative terms and, thus they have come to be referred to as 'cool' versus 'hot' and 'light' versus 'severe'. Descriptive terms, such as 'smouldering', 'creeping' and 'running' (i.e. a rapidly spreading surface or crown fire with a well-defined head) are often used to characterize fire spread, whereas other aspects of fire behaviour are often referred to as spotting, torching or candling (Figure 14.13).

The most basic features of a wildland fire are that:

1 it spreads;
2 it consumes fuel; and
3 it produces bright light, heat, buoyancy and smoke.

It is therefore useful to think about the behaviour of wildland fire in terms of how fast it travels, where it is headed, what it looks like – including its size and shape (e.g. flame dimensions) – and what it feels and sounds like from some distance. The colour of smoke produced by a fire can even provide a crude indication of its general behaviour (Table 14.5).

14.4.1 Types of wildland fires

Wildland fires are commonly recognized on the basis of the fuel layer(s) controlling their propagation (see Figure 1.8, chapter 1, Part One):

- Ground fire or subsurface fire.
- Surface fire.
- Crown fire.

Figure 14.13 Sequence of photos depicting the 'candling' of a spruce bark beetle-killed white spruce tree in the Tracy Avenue Fire near Homer, Alaska on May 1, 2005. The elapsed time was approximately 1 minute, 23 seconds (photos Wade W. Wahrenbrock, Kenai Peninsula Borough, Soldotna, Alaska).

Table 14.5 Smoke colour as a visual approximation of forest fire behaviour (after Burrows, 1984a).

Smoke colour	Fuel moisture status	Relative fire intensity/vigour
Dense white	Very moist	Mild
Grey	Moist	Mild to moderate
Black	Dry	High
Copper-bronze	Very dry	High to severe

Figure 14.14 A free-burning active crown fire spreading through a lodgepole pine stand ($\approx$20 m tall) in the boreal forest region of central Alberta, near the community of Swan Hills on August 23, 1981. The head fire rate of spread and fire line intensity were estimated to be about 15 m/min and 11 250 kW/m respectively. Note that the 'wall of flame' extends well above the top of the tree canopy (photo Martin E. Alexander).

Large fires in conifer forest fuel types will contain evidence of all three types of fire, even when a large fire could be regarded on the whole as a crown fire.

Ground or subsurface fires spread very slowly, with no visible flame and sometimes with only the occasional wisp of smoke. Heading surface fires can spread with the wind (Figure 14.3) and/or upslope (Figure 14.11), and backing surface fires burn into the wind or down-slope. A crown fire is dependent on a surface fire, both for its initial emergence and its continued existence. Thus, a crown fire advances through both the surface and fuel canopy fuel layers, with the surface and crown fire phases more or less linked together as a unit or a 'wall of flame' (Figure 14.14).

Ground or subsurface fires do not occur in grasslands unless there is an underlying layer of peat. Crown fires, effectively, do not occur in grasslands, low brush fields, clear-cut logging slash, blow-down fuel beds and the vast majority of hardwood stands. However, this is not to suggest that very high-intensity surface fires are not possible in these vegetation and fuel types. Crown fires can occur in many of the Australian native eucalypt forest types, and also in tall shrubland

fuel complexes (Figure 14.15), such as chaparral in southern California and fynbos shrublands in South Africa (see Figure 1.7, chapter 1, Part One).

14.4.2 Point and line source fire growth following ignition

Two basic types of fire ignition sources are recognized in wildland fire management. A point source fire is commonly thought of in connection with a single, accidental or otherwise unplanned ignition such as a lightning-ignited fire start (Figure 14.16), escaped campfire or as a spot fire ignition (associated with either a wildfire or as used in prescribed burning).

A line source fire is associated with deliberate ignition (using a hand-held drip torch or an aerial drip torch slung below a helicopter), as employed in a prescribed burning operation or as part of a suppression firing operation on a wildfire. There are several variations of both the point and line source fires (Figure 14.17). A key attribute of the elliptical shaped fire pattern is the length-to-breadth (LB) ratio (i.e. the

Figure 14.15   High-intensity flame front associated with experimental burning in native shrub fuels, carried out on March 16, 2005 near Lake Taylor on the South Island of New Zealand during Project FuSE, a Bushfire Cooperative Research Centre venture (photo Stuart A.J. Anderson, Scion, New Zealand, Reproduced with permission of Bushfire CRC).

Figure 14.16   Photo sequence of the (A) discovery, (B) onset of crowning and (C and D) the initial run of the Washington Creek Fire (299) near Fairbanks, Alaska, on June 21, 2004 (photos Frank V. Cole, Alaska Division of Forestry, Fairbanks, Alaska).

Figure 14.17   Flank-fire ignition pattern used in burning wheat stubble near Perkins, Oklahoma, on June 24, 2005 (from Weir, 2009; photos by Jay Kerby).

total length of fire based on forward and back spread distances relative to its maximum width or breadth). For example, the LB ratio for the idealized elliptical shaped fire portrayed in Fig. 14.4B is approximately 2.0.

In the absence of wind and slope, an incipient fire will develop a roughly circular shape (i.e. LB = 1.0), provided the fuels are uniform (Figure 14.18A). If the wind direction is relatively constant, the fire will be quite elongated (Figure 14.18B) but, if it does vary, the LB will be much less than if winds were more unidirectional (Figure 14.18C).

Fires ignited as a single line source will reach their maximum potential very quickly after ignition, in contrast to a point source fire, which can take considerably longer to develop (Figure 14.19).

A fire originating from a single ignition point will steadily increase its forward rate of advance with elapsed time, eventually reaching an approximate average or 'quasi-steady state' for the prevailing environmental conditions (Box 14.1). The period of time required for a fire to attain this idealized equilibrium-like level of fire behaviour is highly variable.

From experimental fires carried out in various grassland types in the Northern Territory of Australia (Cheney and Sullivan, 2008), it was found that, for a given moisture content, this rate depended on the wind speed and the head fire width, which may be as great as 200 metres. In general, grass fires took 30 minutes to reach their maximum rate of advance. However, they could take from as little as 12 minutes to more than an hour, depending on the fluctuations in wind direction, which determined the overall width of the head fire. In forest fuels, fires can take even longer to attain a 'quasi-steady state' propagation.

Figure 14.18 Fire perimeters recorded at two-minute intervals for point-source ignition experimental fires conducted on level ground: (A) under near calm in-stand wind conditions (0.3 km/h) at a fuel moisture content of 7.0%; (B) with in-stand winds of 3.2 km/h with a constant direction at a fuel moisture content of 4.3%; and (C) with in-stand winds of 1.8 km/h but of variable direction at a fuel moisture content of 9.0% (from Curry and Fons, 1938), in (D) ponderosa pine stands at the Shasta Experimental Forest in northern California (photo John R. Curry, USDA Forest Service).

14.4.3 Rate of fire spread

Rate of fire spread refers to the speed or velocity at which a fire extends its horizontal dimensions, expressed in distance per unit of time. For general use, m/min is recommended or, for a long-term and/or large-scale mental image, km/h (Table 14.6). The same units are used for expressing the rate of fire perimeter increase.

Ground fires spread by smouldering combustion, which is a very slow process – perhaps in the order of

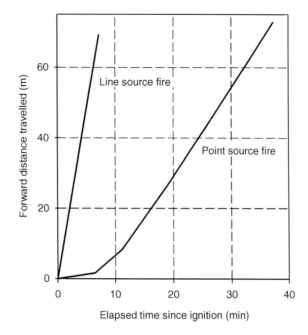

Figure 14.19 Comparison of forward spread distances versus elapsed time for the 'head' of a point-source and a line-source fire, based on simultaneous ignitions carried out in plots measuring ≈80 × 80 m in size within a plantation of slash pine – palmetto-gallbery located within the Dixon Memorial State Forest in Ware County, Georgia (after Johansen, 1987). This was accomplished by initiating the point source fire with a match once the strip or 'line of fire' had been created with a single hand-held drip torch.

just a few centimetres per hour. In comparison, surface and crown fires spread by flaming combustion, which is faster by at least an order of magnitude higher. The threshold spread rate for a creeping surface fire, with barely visible flames, is around 0.1 m/min. Spread rates of head fires burning in surface fuels under dense forest canopies, seldom exceed 5–10 m/min without some form of canopy fuel involvement, but can reach up to 25 m/min in open grown forest types. Crown fires in closed-canopy conifer forest fuel types generally spread at rates of 15–30 m/min. Fires in fully-cured grasslands are easily capable of exceeding the spread of surface and crown fires in forests.

The maximum possible spread rate for fires backing into the wind or downslope generally does not exceed 1.0 m/min. Interestingly, there is no agreement within the fire behaviour research community on whether fires backing downslope spread at a rate less than or equal to the zero slope case. Similarly, there is disagreement in the wildland fire science literature on whether fires backing into the wind increase or decrease their rate of advance with increasing wind speed, or maintain a rate similar to the case of zero wind.

14.4.4   Fire shape and size

The shape or overall pattern of a free-burning wildland fire in homogeneous fuels, while perhaps highly irregular in detail, generally resembles an ellipse (Figure 14.4B), although the reason for this has not be deduced theoretically. Elliptically shaped fires have been reproduced in wind tunnel laboratory experiments (Fons, 1946). Wind-driven fires, in particular, can quite often be represented by an elongated ellipse that corresponds roughly to the outline of the burnt area, provided the wind direction does not vary widely. Other fire shapes are seen in discontinuous fuels.

Assuming an elliptical fire shape, a rough estimate of the size of a wildland fire can be made in terms of area and perimeter on the basis of its combined forward and backward spread distances and LB ratio, based on the mathematical formulae for an ellipse

Table 14.6 A suggested classification scheme for the rate of spread (ROS) of wildland fires (adapted from Alexander and Lanoville, 1989).

Descriptive term	ROS (m/min)	ROS (km/h)	General type of fire or fire activity
Very slow	<0.1	<0.01	Ground or subsurface fires.
Slow	0.1–1	0.01–0.1	Backing fires into the wind or downslope.
Moderately slow	1–3	0.1–0.2	Heading surface fires.
Moderately fast	3–10	0.2–0.6	Vigorous surface fires. Vertical fire development in conifer forests.
Fast	10–18	0.6–1.1	Onset of crowning in dense conifer forests. Highly vigorous surface fires in open fuel types.
Very fast	18–25	1.1–1.5	Fully developed crown fires.
Extremely fast	>25	>1.5	Conflagrations.

Source: Canadian Forest Service Publications (Natural Resources Canada). Reproduced with permission.

Box 14.1 Description of fire behaviour and growth from an experimental point source ignition fire in a black spruce stand in north-eastern Alberta, Canada (after Kiil, 1975). Courtesy Canadian Forest Service Publications, Natural Resources Canada.

Fire started at 1300 hours and was allowed to spread freely until head fire reach prepared fire guard. Head fire rate of spread increased steadily from about 0.6 m/min during the first three minutes to about 2.4 m/min at ten minutes following ignition, with the fire having burned an area of 0.05 ha at this time.

Torching of individual trees and crown-to-crown ignitions became increasing frequent 10–15 minutes following ignition, with firebrands being carried 60 m ahead of the fire front. By this time, head fire rate of spread was about 6.6 m/min and appeared to be approaching an equilibrium intensity. Flame length and depth of the fire front fluctuated considerably with wind speed, but averaged about 5 m and 3 m, respectively, in the ten minutes following ignition. Fire area at 15 minutes after ignition was estimated to be 0.17 ha.

By contrast, the backfire spread at a steady rate of 0.5 m/min into the prevailing wind, flame length and depth of the fire front averaging 1.0 and 0.8 m, respectively.

Head fire intensity increased steadily from 1050 kW/m after five minutes following ignition, to 4200 kW/m ten minutes later. Backfire intensity remained constant at 250 kW/m.

(Equations 14.1 and 14.2 – Box 14.2). The LB ratio is a function of wind speed and broad fuel type (Table 14.7). Such relationships can be used to estimate on-site winds from observations of LB ratios of active fires and post-burn patterns, assuming the flanks of a wildfire are not influenced by any suppression action or barriers to fire spread.

Box 14.2 List of equations commonly used for making calculations of wildland fire behaviour characteristics, the associated symbols and abbreviations, and units.

Equation number	Equation
14.1[a]	$A = (\pi \div (4 \times \mathrm{LB})) \times \mathrm{FSD}^2) \div 10\,000$
14.2[a]	$P = \pi \times (\mathrm{FSD} \div \mathrm{LB}) \times (1 + 1 \div \mathrm{LB}) \times (1 + [(\mathrm{LB} - 1) \div (2(\mathrm{LB} + 1)]^2)$
14.3[b]	$t_r = D \div R$
14.4[b]	$D = R \times t_r$
14.5[c]	$I_B = (H \times w_a \times R) \div 60$
14.6[c,d]	$L = 0.0775\, I_B^{0.46}$
14.7[e,f]	$I_o = (0.010 \times \mathrm{CBH} \times (460 + 25.9 \times \mathrm{FMC}))^{1.5}$
14.8[e,g]	$R_o = 3.0 \div \mathrm{CBD}$
14.9[h]	$\mathrm{CBD} = \mathrm{CFL} \div \mathrm{CD}$
14.10[d,i,j]	$\Delta T = (3.85 \times I_B^{2/3}) \div z$
14.11[d,k]	$h_s = 0.1483 \times I_B^{2/3}$

Key

A	Fire area (ha)
CBH	Canopy base height (m)
CBD	Canopy bulk density (kg/m³)
CD	Crown depth (m)
D	Flame depth (m)
FMC	Foliar moisture content (% oven-dry weight basis)
FSD	Fire spread distance – heading and backing components (m)
h_s	Crown scorch height (m)
H	Net low heat of combustion (kJ/kg)
I_B	Fireline intensity (kW/m)
I_o	Critical surface fire intensity for initial crown combustion (kW/m)
L	Flame length (m)
LB	Length-to-breadth ratio
P	Fire perimeter (m)
R	Rate of fire spread (m/min)
R_o	Critical minimum spread rate for active crowning (m/min)
t_r	Flame front residence time (min)
w_a	Amount of fuel consumed (kg/m²)
z	Height above ground (m)
ΔT	Temperature rise above ambient air conditions (°C)

Reference

[a] Forestry Canada Fire Danger Group (1992)
[b] Fons et al. (1963)
[c] Byram (1959a)
[d] Alexander (1982)
[e] Van Wagner (1977a)
[f] Van Wagner (1993)
[g] Alexander (1988)
[h] Cruz et al. (2003)
[i] Thomas (1963)
[j] Van Wagner (1975)
[k] Van Wagner (1973)

If the environmental conditions remain unchanged, the fire perimeter will increase linearly with time (Van Wagner, 1965). On the other hand, the rate of area growth will increase quadratically (i.e. total area burnt increases as the square of time). Any fire which remains out of control for days or weeks is likely to be subject to considerable changes in wind direction and in other fire weather elements, fuel complex differences and topographic features (e.g. lakes and other barriers to fire spread). Such variations in the fire environment will determine the shape of a fire and its final size, which sometimes can exceed more than one million hectares (Rothermel et al., 1994).

14.4.5   Fuel consumption

Research studies and operational experience have shown that the ignition, behaviour, immediate impacts and long-term effects of fire are determined in large part by the level of fuel consumption. The quantity of fuel consumed in a wildland fire varies widely, depending on the fuel or vegetation type (Stocks and Kauffman, 1997). Fuel consumption can vary over two orders of magnitude from 0.1 kg/m² (representing, for example, a gentle surface fire consuming the leaf litter in a hardwood stand in the spring or fall), to more than 10 kg/m² (associated with the high-intensity burning of a heavy blow-down fuel complex).

In addition to the weight per unit area of fuel consumed, other measures of fuel consumption include the percent fuel reduction relative to the pre-fire quantity. In the case of dead-down woody surface fuels, this could be by the roundwood diameter size class. For thick organic layers, 'depth of burn' is a common metric used to describe the actual vertical reduction (i.e. in cm). In can also be expressed as a percentage reduction of either the pre-fire depth or load.

Table 14.7 Length-to-breadth (LB) ratio for elliptical fire shapes on level terrain as a function of broad fuel type and 10-metre open wind speed (adapted from Taylor et al., 1997).

Fuel type	10-m open wind speed (km/h)										
	0	5	10	15	20	25	30	35	40	45	50
					LB						
Forest	1.0	1.1	1.5	2.0	2.6	3.3	3.8	4.4	5.0	5.6	6.1
Grassland	1.0	2.3	3.2	3.5	4.4	4.9	5.3	5.7	6.1	6.4	6.8

Source: Canadian Forest Service Publications (Natural Resources Canada). Reproduced with permission.

The degree of fuel consumption in any given fuel complex depends first of all on the pre-fire fuel load. In fuel types composed exclusively of fine fuels, such as fully cured grass, total fuel consumed literally equals the pre-fire fuel load. In fuel types exhibiting an organic layer, the fuel consumption of this component is related in large part to the fuel moisture content and its distribution, but other fuel properties (e.g. bulk density and mineral or inorganic content) also play a role. Dead-down woody surface fuel consumption is largely dictated by piece diameter. In any analysis of fuel consumption, the type of fire must also be considered. In a conifer forest stand, if the surface fire is sufficiently intense to induce crowning, there will be a certain degree of consumption of the canopy fuels, as well as ground and surface stratum.

It is difficult, under field conditions, to determine precisely the amount of fuel consumed in the flaming combustion phase versus the smouldering and glowing combustion phases of a wildland fire. As a result, certain assumptions are adopted and estimates are made. This is easily done with fuel types like fully cured grasslands, but it can be problematic in other situations that involve heavy fuel loads of varied composition. Persistent smouldering and glowing combustion are commonly associated with heavy woody surface fuel concentrations and/ or deep organic layers and low fuel moisture conditions, following passage of the active flaming fire front.

14.4.6 Flame front dimensions and characteristics

Despite the obvious fluctuations seen in the flames of steadily moving wildland fires, it is nevertheless possible to recognize that there is a height, depth, length and angle to the flaming fire front (Figure 14.20). The flame height represents the average vertical extension of the flames, as measured from the ground surface; the occasional flame flashes that rise above the general level of flames are typically not considered. In turn, the flame depth represents the width of the zone within which the continuous flaming occurs behind the leading edge of the fire front. The flame length, on the other hand is the distance from the flame height and the midpoint of

the flame depth at the ground surface. The flame angle is the angle formed between the flame at the fire front and the ground surface, whereas the flame tilt angle is the angle formed between the fire front and the vertical.

Flame length and flame height are considered equal, or near so in the case of level terrain and calm conditions. In this situation, the flame tilt angle would be 0° and the flame angle 90°.

The amount of radiation received at some distance ahead of the flame front is dictated by the size and orientation of the flames (Sullivan *et al.*, 2003), as illustrated in Figure 14.21. For comparative purposes, the radiant heat flux at a distance of two metres from flames one metre tall is around 2.0 kW/m^2. A person standing at such a distance from the flame front would find the pain unbearable on bare skin after about 30 seconds.

The thermal environment of a wildland fire is perhaps best characterized by the signature it leaves in the form of a time-temperature trace as the moving flame front passes by a given point (Wotton *et al.*, 2012). Such a graph represents the 'thermal pulse' of a wildland fire associated with the various stages of spreading combustion (Figure 14.22).

The duration of flaming combustion is commonly represented by the flame front residence, usually defined as the time required for the flame depth zone to pass a given point in the fuel bed surface (Equation 3 – Box 14.2). Flame front residence times vary according to the fuel type characteristics, as determined by fuel bed structure, such as available fuel load, compactness and average fuel particle size and density. For example, surface heading fires in pine stands typically have flame front residence times of 30–60 seconds, in contrast to 5–15 seconds in grasslands and 1–2 minutes in logging slash fuel beds. In turn, estimates of flame depth can be made using rate of fire spread using one of these nominal residence times (Equation 4 – Box 14.2). There is, quite surprisingly, a general paucity of data on flame front residence times in some important fuel types (e.g. southern California chaparral, Gambel oak in the central south-western region of the United States).

The total length of time that fuel will continue to smoulder and glow after passage of the flaming front is correspondingly much larger. In grasslands, this time might only be a minute or less. However, in

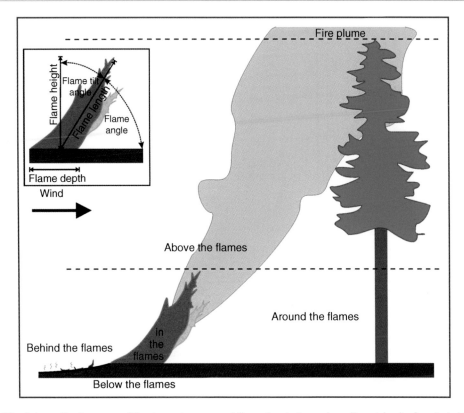

Figure 14.20 Schematic diagrams of fire impact zones and flame front dimensions (insert box) of a wind-driven surface fire on level terrain (adapted from Alexander (1982) and Burrows (1995)).

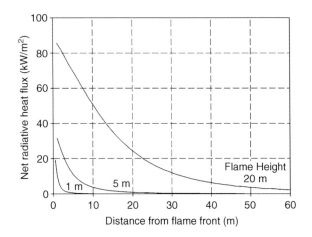

Figure 14.21 Radiant heat flux on a receiving surface, as a function of flame height and distance from the flame front (after Ryan and Koerner, 2011). For comparison's sake, peak sunlight on a clear, bright summer's day is around 1.0 kW/m^2. The threshold of pain for radiant heat on bare skin for most individuals after eight seconds would 6.4 kW/m^2 and after three seconds it would be 10.4 kWm^2(Drysdale, 2011).

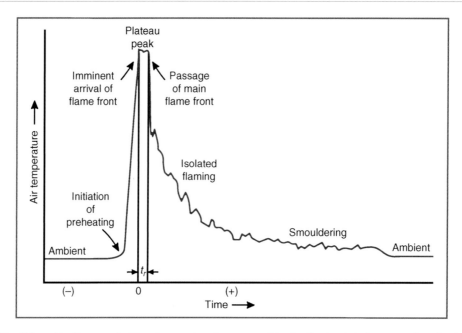

Figure 14.22 Schematic diagram of a time-temperature trace or profile at a fixed point above a surface fire in relation to the passage of the flaming front and denoting the flame front residence time (from Alexander, 1998).

heavy fuel situations, it can be quite variable, depending on the fuel quantity remaining and level of dryness, and considerably longer when the burnt area is large (e.g. 10–30 minutes or even more). Many fire managers, after several years of experience with ground or sub-surface fire activity in a given area, have been able to relate relative fire danger indexes to smouldering and glowing potential and, in turn, develop guidelines for themselves (e.g. Melton, 1989).

14.4.7   Fireline intensity

The maximum temperature experienced in the flames of wildland fires are around 1000°C and are judged to occur as readily in a single burning pine needle as crowning forest fire (Van Wagner and Methven, 1978). The difference between these two extremes is the energy output or release rate.

In 1959, a physicist with the US Forest Service, George M. Byram, introduced his concept of fireline intensity, which has proven highly valuable in characterizing the behaviour of wildland fire for more than 50 years. Fireline intensity constitutes the rate at which the combustion of fuel produces heat per unit of time per unit length of fire front, regardless of its depth. Numerically, it is equal to the product of the heat yield of the burnt fuel, the quantity of fuel consumed and the linear rate of fire spread (Equation 5 – Box 14.2). For practical purposes, the fuel heat yield can be considered a constant. Thus, a given fireline intensity value can be attained by different combinations of fire spread rate and fuel consumption (Figure 14.23A).

Expressed in terms of kW/m, fireline intensity represents the energy output (i.e. kJ/s) being generated from a 1-m wide strip extending from the fire edge back through the flame depth. Fireline intensity can vary over an exceedingly wide range, from as low as 10 kW/m (where surface fires are just barely able to sustain themselves) to in excess of 100 000 kW/m (for major conflagrations). Byram (1959a) considered that 'by moving briskly one could step over the fire without getting burned' at an intensity of 200 kW/m, and that at 3500 kW/m, 'the roar of the flames would be accompanied by occasional explosive and whistling sounds'. Fully developed crown fires commonly exhibit fireline intensities greater than 10 000 kW/m.

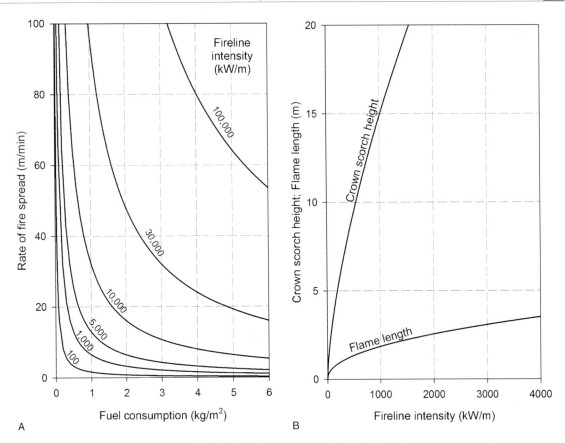

Figure 14.23 Graphical representation of: (A) fireline intensity as a function of rate of fire spread and fuel consumption, assuming a net low heat of combustion of 18 000 kJ/kg (after Alexander and Cruz, 2012c). Reproduced with permission of CSIRO Publishing; and (B) Byram's flame length-fireline intensity and Van Wagner's crown scorch height-fireline intensity relationships (adapted from Alexander and Cruz, 2012b). Reproduced with permission of the Canadian Institute of Forestry/l'Institut forestier du Canada.

Byram showed that fireline intensity was related to flame length (Figure 14.23B). Many members of the wildland fire community have generally assumed that Byram's flame length-fireline intensity relationship (Equation 6 – Box 14.2) is universal in nature, but several other fire researchers have since come to derive similar equations for structurally different fuel types (Alexander and Cruz, 2012a).

Fireline intensity can be calculated for any point on the fire perimeter. The fastest spreading part of a fire is regarded as the head and the back the slowest, the flanks being intermediate between the two. It is for this reason that the flames at the front of a moving wildland fire are the greatest in size and the smallest flames are at the rear (Figure 14.16B).

The proportions of the area and perimeter of an elliptically shaped fire can be determined for any given fireline intensity class (Table 14.8).

14.4.8 Crown fire initiation and propagation

Van Wagner (1977a) considered that the onset of crowning in a conifer forest stand occurs when the intensity of a surface fire – as defined per Byram's fireline intensity – attains or exceeds a certain critical value. He believed that the initiation of crowning is primarily dependent on the canopy foliar moisture content and canopy base height (Equation 7 – Box 14.2). The greater the values of these two variables, the greater the values of surface fire intensity required to initiate a crown fire. The flames of a surface fire do not necessarily have to reach or extend into the lower canopy fuel layer to induce crowning.

Van Wagner judged that, for crown fire propagation to proceed horizontally, the fire would need to

Table 14.8   Example of the proportional area and perimeter by fireline intensity class for an elliptical shaped fire with a length-to-breadth ratio of 3.0: 1 and a 'peak' or maximum head fireline intensity of 5000 kW/m (from Catchpole *et al.*, 1992). Such distributions can be calculated for any fire characteristic that is dependent on fireline intensity (e.g. flame length, crown scorch height).

Fireline intensity (kW/m)	Percentage of total:	
	perimeter length	area burnt
<500	25	7
500–1000	32	19
1000–1500	18	17
1500–2000	9	13
2000–2500	5	10
2500–3000	3	7
3000–3500	2	6
3500–4000	2	5
4000–4500	1	6
4500–5000	2	10

Source: Canadian Forest Service Publications (Natural Resources Canada). Reproduced with permission.
Note: Perimeter values do not total 100 due to rounding.

meet or exceed a critical minimum spread rate for a solid flame front to develop and maintain itself within the canopy fuel layer (Equation 8 – Box 14.2). He considered this condition to be largely dependent on the canopy bulk density – a function of an estimate of the available fuel load and depth of the canopy layer (Equation 9 – Box 14.2). Thus, if following initial crown ignition, a fire does not grow and attain a sufficiently high enough spread rate (as determined by the canopy bulk density), the development of a true crown fire is viewed as unlikely. The robustness of Van Wagner's (1977a) simple criterion for fully developed crown fire spread has come to be validated in several different conifer forest types.

If burning conditions were severe enough to initiate crown combustion, but then not sufficient to maintain continuous flame in the crown space, Van Wagner considered this type of fire a passive crown fire. Otherwise, when both the initiation and propagation requirements were met, the fire was considered an active crown fire (Figure 14.14).

14.4.9   Fire impact zones

There are clearly many descriptors or characteristics of free-burning wildland fire behaviour. These include, but are not limited to, the type of fire (ground, surface, crown), linear rate of advance, flame front dimensions (length, height and depth), fireline intensity, flame front residence time and smoulder time, just to name a few that have been discussed in this chapter. Many of these have direct, practical application in both the control and use of fire from operations and planning perspectives, including interpreting ecological response and damage potential.

These fire behaviour descriptors also determine the immediate physical or acute impacts of wildland fires, which give rise to the resulting ecological effects. For example, under calm wind conditions, the temperature profile in the thermal or convective plume above a surface fire is a function largely of fireline intensity (Equation 10 – Box 14.2). The flame with flame front residence time, this will dictates the height of crown scorch, which, in turn, determines the probability of tree mortality. Van Wagner (1973) was the first fire researcher to publish formally on the fact that crown scorch height could be linked to fireline intensity (Figure 14.23B). Many in the wildland fire community have come to consider his crown scorch height-fireline intensity relationship (Equation 11 – Box 14.2), as a generic one, while others have developed similar equations for specific tree species and fuel situations.

Thus, a fire manager or researcher who wishes to link fire effects to fire behaviour (Burrows, 1995), may find it useful to think in terms of the following fire impact zones be it with respect to plant response or soil heating (Figure 14.20):

1 around the flames;
2 below the flames;
3 in the flames;
4 above the flames;
5 behind the flames.

Such a framework will allow the focus to be on selecting the most appropriate fire behaviour descriptor(s) for use as independent variables, to correlate a specific ecological effect of interest or in calibrating a biophysical process-oriented model. For example, the degree of soil heating experienced in a forest stand during a fire will depend to a large extent on the thickness of the remaining duff layer (Van Wagner, 1970).

14.5 Extreme wildland fire behaviour phenomena

A common axiom in wildland fire management is that approximately 95 percent of area burnt by wildfire is generally caused by less than five percent of the incidents (USDA Forest Service, 1966). It is this relatively small number of wildfires that end up exhibiting extreme or severe fire behaviour. While the term 'extreme fire behaviour' is used by both urban or structural and wildland fire services, the characteristics are usually different, although at times they share certain commonalities (e.g. rapid or abrupt changes in fire spread and flame size).

In all cases, the phenomena involved are exceedingly complex. As a result, a number of myths or fallacies regarding extreme wildland fire behaviour have emerged over the years (e.g. Cheney and Sullivan, 2008). We also have a tendency in the wildland fire community to use metaphors to help simplify the complexity of the physical processes involved, which can sometimes be misleading (e.g. 'the fire erupted like a volcano').

Although he offered no specific definition, Byram (1954) is believed to have been the first to mention 'extreme fire behaviour' regarding wildland fires. Some have complained that the term has no meaning or value, yet it is a deeply entrenched phrase within the lexicon of the wildland fire management community. While precise threshold values have yet to be suggested, extreme fire behaviour is usually accepted as involving one or more of the following

characteristics – with full recognition that there are many interconnections:

• Continuous crowning.
• Extremely fast rates of fire spread.
• High fireline intensities and large flames.
• Prolific spotting.
• Presence of large fire whirls and other vortices.
• Well-established convection column.

It is common for fires exhibiting extreme fire behaviour phenomena to behave in an apparently erratic and, sometimes, dangerous manner. Predictability is very difficult at times, as such fires are not only affected by the fire environment, but often exercise some degree of influence over their own environment. *Because of the high degree of unpredictability associated with extreme fire behaviour, one should come to expect the unexpected* (Box 14.3). Noteworthy is the fact that these characteristics and features of extreme fire behaviour can occur on both small and large fires (Byram, 1959b).

14.5.1 Continuous crowning

The term 'crowning' refers to both the ascent of flames into the crowns of trees and spread through the forest canopy (Figure 14.14). When a conifer forest stand crowns, fuel in addition to that at the surface is consumed; available crown fuels consist primarily of needle foliage, mosses and lichens, bark flakes and small woody twigs. This additional canopy fuel consumed by a crown fire, combined with a doubling of the rate of spread, leads to a sudden escalation in fireline intensities and flame height; in turn, this change in fire behaviour results in greater spot fire distances and at least a four-fold jump in area burnt with time. Is it any wonder that some wildland fires just seem to literally 'blow up'?

One of the earliest uses of the term 'blow-up', in a scientific sense, was by Brown (1937). This occurred in his discussion of the fire behaviour associated with the Blackwater Fire in the Shoshone National Forest in north-eastern Wyoming, in which 15 firefighters perished on August 21, 1937.

The remarkable or staggering power of crown fires is often times seen in the aftermath of burning. Large areas of timber sometimes will have been blown down, suggesting fire-induced winds of 130–190 km/h.

Box 14.3   Byram's (1954) facts regarding extreme wildland fire behaviour and exceptions to the rule.

1   Most severe fires, and a considerable number of blow-ups, occur during the middle of the afternoon on sunny days. On such days, the atmosphere is often turbulent and unstable to a height of several thousand feet. However, some of the worst forest conflagrations in the United States have either occurred at night or have reached the peak of their intensity at night (usually between sundown and midnight). At this time, the lower layers of the atmosphere (up to $\approx$150 m or more) are usually stable.

2   Some of the worst western fires in the past 15 years (i.e. 1939 to 1953) have been in rough country, which might indicate that topography is a dominating factor. On the other hand, there have been conflagrations, such as those that occurred in the Lake States many years ago, which burnt in nearly flat or rolling country. Some of the conflagrations have been compared to 'tornadoes of fires'.

3   An intense fire may occasionally spread rapidly across slope or downslope at night in the general direction of the cool downslope winds. Yet this same rapid downslope spread may happen in the middle of the afternoon when the surface winds, if any, would be upslope. Fires have travelled across drainages (upslope and downslope) as though these did not exist.

4   Turbulence in the atmosphere seems to be closely related to extreme fire behaviour, yet on a large proportion of warm, sunny days, the atmosphere is unstable. Often, fires do not build up to extreme intensity on such days.

5   Many intense fires have been accompanied by high winds, but some of the most dangerous and erratic fires have burnt when the wind speed was not especially high.

6   High temperatures and low relative humidity accompany a large proportion of severe fires, but some of the most intense and rapid-spreading fires have burnt when the temperature was low and falling. The fires in the East and South-east in the fall of 1952 are examples.

7   Prolonged periods of drought and dry weather show a strong correlation with intense hot fires, but the Brasstown fire in South Carolina in March 1953 burnt only a week after nearly two inches of rain had fallen on ground well charged with winter rainfall. However, both burning index and build-up index were high on this day.

8   The amount of fuel available to a fire is an important factor in its behaviour. At times, the effect of an increase in quantity of fuel on fire intensity appears to be considerably greater than would be expected from the actual fuel increase itself. For example, doubling the amount of fuel might increase the apparent intensity four or five times.

9   Arrangement, as well as quantity, of fuel is important. Extreme fire behaviour seems most likely to occur in dense conifer stands. Intense fires also build up in stands of evergreen brush, and in the South can readily cross swamps if the brush is dense enough.

10   On those fires to which one would be most likely to apply the term 'blow-up' (owing to the sudden and often unexpected build-up of turbulent energy), there is an obvious and well developed convection column that may extend high into the atmosphere.

11   Large fires exhibiting extreme behaviour have been known to put up convection columns to a height of $\approx$760 m or more. Since about 70 percent of the total air mass is below the tops of such convection columns, these fires have literally pierced the atmosphere. They are volume phenomena and have storm characteristics like certain other disturbances in the atmosphere. This, in part, seems to explain why they do not conform to the 'rules' of fire behaviour. These 'rules' are based on the far more frequent ordinary fire, which is pretty much a surface phenomenon.

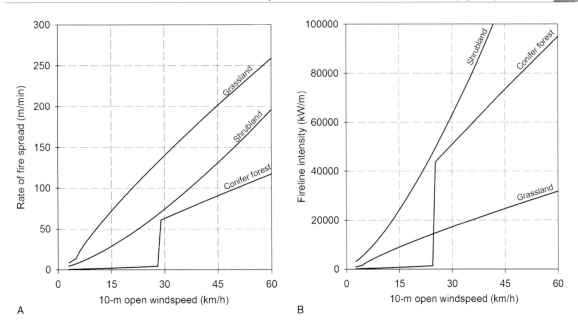

Figure 14.24 Graphical representations of (A) rate of fire spread and (B) fireline intensity as a function of wind speed for three broad fuel complexes. The following environmental conditions are assumed: slope steepness, 0%; ambient air temperature, 30°C; relative humidity, 20%; foliar moisture content for conifer forest, 110%; and shrub live fuel moisture content, 75%. The vertical 'kink' in the conifer forest curves represents the point of surface-to-crown fire transition (adapted from Alexander and Cruz, 2013c).

In other cases, relative largely tree stems (40–50 cm diameter at breast height) have been 'snapped' off at heights of 3.0 to 4.5 above ground.

One aspect of crown fire phenomena that has yet to be adequately resolved is the existence of an independent crown fire. Van Wagner (1993) regarded his concept of an independent crown fire (in which a fire's propagation is assumed to advance entirely within the crown fuel layer, unsupported by an accompanying surface fire) was dubious – especially on level terrain. However, it may have value in certain types of closed-canopy conifer forest stands on steep slopes as a short-lived phenomenon, although not as a steady-state means of fire propagation.

14.5.2 Extremely fast rates of fire spread

Crown fires in conifer forests can exceed spread rates of 100 m/min under the influence of strong wind speed conditions (Figure 14.24A). The fastest spreading crown fire run in a conifer forest documented to date occurred in South Australia during the Ash Wednesday fires of February 16, 1983. The Mount Muirhead Fire was reported to have spread through radiata pine plantations at approximately 208 m/min, or 12.5 km/h for a period of about one hour (Keeves and Douglas, 1983).

Fires spreading in grasslands and shrublands can easily advance twice as fast as crown fires, especially on steep slopes, even with moderately strong winds. For example, it was found, during the major run of the South Canyon Fire on July 6, 1994 in western Colorado (Butler *et al.*, 1998), that the flame front advanced up slopes greater than 55% at a rate in excess of 360 m/min (21.6 km/h). Several fires occurring in fully-cured grasslands on flat topography, such as the Wangary Fire of 10 January 2005 in South Australia (Cheney and Sullivan, 2008), have been documented as moving at upwards of 30 km/h for short periods of time!

Wildfires in forests have been known to make major sustained runs of 30–65 km in one direction over flat and rolling to gently undulating terrain in less than a single ten-hour burning period. Such was the case of the Lesser Slave Lake Fire in central Alberta, which advanced 64 km in ten hours on May 23, 1968, through several coniferous forest types (Kiil and Grigel, 1969). Similar types of incidents have occurred elsewhere (e.g. Cruz *et al.*, 2012).

Given that grass fires have the potential to spread at nearly twice the rate of a crowning conifer forest fire, they are capable of covering the same distance in half the time. Fires in grasslands and forests driven by winds of 50 km/h – exhibiting LB ratio ∼6.5 (Table 14.7) – are thus capable of advancing 60 to 70 km in a single burning period, and burning over an area of at least 50 000 ha with a perimeter length of 140 km in the space of just five and ten hours respectively.

14.5.3 High fireline intensities and large flames

Given the extremely fast rates of fire spread, high fireline intensities are to be expected, as well, in turn, as large flames. Wind-driven fires or fires burning up steep mountain slopes through dry and plentiful fine fuels are quite capable of attaining fireline intensities exceeding 30 000 kW/m (Figure 14.24B). In such situations, it is quite common for people to use the cliché that the fire 'sounded like a freight train or a jet plane' passing by or approaching.

Efforts to estimate flame heights of crown fires objectively are complicated by the fact that sudden ignition of unburnt gases in the convection column can result in flame flashes that momentarily extend some 100 m or more into the convection column aloft. One such flame flash that extended almost 200 m above the ground was photographically documented during the Ash Wednesday fires that plagued south-eastern Australia on February 16, 1983 (Sutton, 1984). Such flashes can easily result in overestimates of average flame heights, which usually range from about 15–45 m on high-intensity crown fires (Byram, 1959b). Average flame heights of crown fires are thus generally regarded as being about two to three times the stand height (Figure 14.14). This is in contrast to wind-driven grass fires, where average flame heights seldom exceed 4.0 m.

14.5.4 Prolific spotting

High-density, short-range spotting up to ≈100 m is a common feature in certain wildland fire situations

(Figure 14.5) and represents close to a maximum distance with grass fires, regardless of the wind conditions, unless woody firebrand material are present (e.g. from shrubs or scattered trees). Lower density, intermediate- to medium-range spotting upwards to 2 km is a common occurrence with high-intensity surface and crown fires. In continuous fuels, short and intermediate distance spot fires are generally overrun by the main advancing flame front before they are able to develop sufficiently to increase's a fire's overall rate of advance and, in turn, initiate a 'leapfrog'-like effect (Figure 14.25).

Spot fire density will gradually increase as the moisture content of fine dead fuels steadily decreases from 10% down to critically dry levels as low as 2–3%. This raises the possibility of numerous spot fire ignitions rapidly coalescing and creating a firestorm or an 'area ignition' effect (Arnold and Buck, 1954).

Prolific spotting emanating from high-intensity, wind-driven fires will allow for the main advancing head effectively to bypass areas that would otherwise not burn. High-density spotting or 'ember attack' associated with the wildfire that descended upon Slave Lake in central Alberta, Canada, from the outlying forested areas on May 15, 2011, contributed to the destruction of a third of the town's residential properties and businesses (Sweeny, 2012). This particular incident represents an excellent example of a fire effectively 'following the path of least resistance'. Barrow (1945) was one of the first to enunciate the role that spotting plays in the ignition of houses assaulted by wildfires.

Longer range spot fire distances of 6–10 km and even slightly greater have been reported in shrublands and conifer forests. For example, during the Great Peshtigo Fire of October 8, 1871 in north-eastern Wisconsin, eyewitnesses describe firebrands falling upon the decks of ships located 11 km from the Lake Michigan shoreline (Haines and Kuehnast, 1970).

The spotting potential of Australia's native eucalypt forests is unparalleled in terms of both density and distance as a result of the abundance and aerodynamic properties of the tree bark (McArthur, 1967). Spot fire distances of 30–41 kilometres were, for example, authenticated during the Black Saturday fires that occurred in Victoria, Australia, on February 7, 2009 (Cruz *et al.*, 2012), yet again reconfirming this notorious reputation.

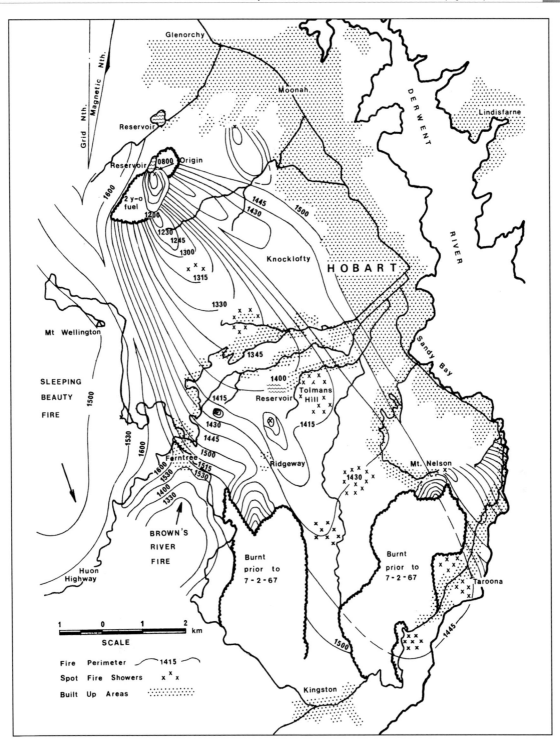

Figure 14.25 Progress map showing the perimeter expansion and spotting activity associated with the major run of the Hobart Fire on February 6, 1967, Tasmania, Australia (from Cheney, 1976). Reproduced with permission of Taylor & Francis.

14.5.5 Presence of large fire whirls and other vortices

Fire whirls or vortices are spinning columns of hot and gases rising vertically up from a fire and carrying aloft smoke, debris, flame and firebrands (Figure 14.26). They can range from one metre to several hundred metres in diameter and can involve the entire fire area, or only hotspots within or outside of the main perimeter of a fire (Forthofer and Goodrick, 2011). Horizontal vortices may also form during high-intensity burning.

Many fire whirls are small and short-lived. However, when they become large, they are capable of attaining tornado-like wind velocities and of potentially inflicting serious damage and injury. Just such an incident took place on the Indians Fire in the Los Padres National Forest in southern California on June 11, 2008 (Moore, 2008). Several firefighters became entrapped by a large fire whirl which was estimated to be approximately 300 metres in diameter. One firefighter sustained serious burn injuries but recovered.

Large fire whirls may remain relatively stationary, although they sometimes travel considerable distances –

Figure 14.26 Fire-induced tornadic whirlwind associated with the Polo Fire that occurred on March 7, 1964, near Santa Barbara, California, as recorded by a Santa Barbara News-Press photographer (from Pirsko and Sergius, 1965). This tornado-like fire whirlwind cut a 1.6 km long path, injuring four people, destroying two houses, a barn, and four automobiles, and wrecking a 100-tree avocado orchard. The fringe of the fire whirlwind passed over a fire truck and sucked out the rear window of the vehicle. A firefighter standing on the rear platform of the vehicle was pulled up vertically, so that his feet were pointing to the sky while his hands clasped the safety bar. A small piece of plywood was rammed 7.6 centimetres into an oak tree. Pirsko and Sergius (1965) regarded the Polo Fire whirlwind as 'a classic case of extreme fire behaviour'.

upwards of 14 kilometres from the main fire in one instance. Their occurrence is commonly associated with an unstable atmosphere, light to moderate winds and large heat sources, created by dry and plentiful fuel concentrations or terrain configurations that concentrate heat from the fire, such as the lee side of a ridge.

14.5.6 Well-established convection column

All wildland fires exhibit some form of a convection column comprised of hot gases, smoke, firebrands and other combustion by-products that rise up above the combustion zone of a fire. The most obvious indication of a fire's convection column, as seen from a distance, is the rising, billowing smoke plume or wedge (Kerr *et al.*, 1971).

With strong winds at the surface that steadily increase with height, a fire's convection column is tilted much 'like a hockey stick', as some would suggest (Figure 14.27A). In this case, the kinetic energy of the wind field exceeds the convective energy of the fire. Smoke, embers and convection heat are transferred directly to the fire front, creating immediate and enormous safety issues for the public and emergency responders who are directly downwind of the fire's forward assault.

Provided that weather, fuel and/or topographic conditions do not change appreciably, wind-driven or wind-dominated fires will continue to spread generally in one direction unabated. Therefore, within this context, their progression is predictable. The 1956 Dudley Lake Fire in northern Arizona (Schaefer, 1957), the 1967 Sundance Fire in northern Idaho (Anderson, 1968), the 1968 Lesser Slave Lake Fire in central Alberta, and the 1990 Canyon Creek in western Montana (Goens, 1990), represent four examples of the better known and documented cases of this conflagration type of fire.

When the winds at the surface are light and remain so, or even gradually decrease with height, the atmosphere is unstable and high fuel loads are present, wildland fires are capable of producing towering, active and well-developed convection columns with a strong vertical component and a characteristic inside-out rolling motion resulting from the tremendous buoyancy caused by such an intense and generally large heat source (Figure 14.27B). When the winds aloft are exceptionally strong, then the potential for long-distance spotting greatly increases.

When the convective energy of the fire exceeds the kinetic energy of the wind field, it is common for these convection-dominated fires to form pyro-cumulus (pyroCu) clouds above them (Figure 9.11, chapter 9, Part Two). With top heights reaching 5000 to 10 000 m above ground, they are like their thunderstorm cousins, capable of triggering cloud-to-ground lightning strikes (Gill, 1974) and downdraft winds with unknown trajectories. In contrast to the wind-dominated case, strong fire-induced updrafts draw the smoke into the convection column, and the flank and front of the fire are often visible.

The difficulty of predicting the behaviour of high-intensity, convection-dominated type of fire is the uncertainty associated with their general direction or path of spread. The 1974 Dryden-18 Fire in northwestern Ontario that was fed by high blow-down fuel loads (Stocks, 1975), the Shoshone Fire that burned in Yellowstone National Park in northwestern Wyoming during the 1988 fire season (Rothermel, 1991a), and the 2001 Chisholm Fire in

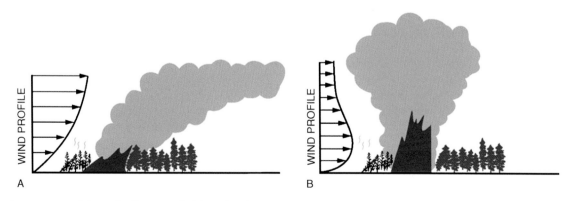

Figure 14.27 Schematic diagrams of: (A) a wind-driven crown fire with winds increasing with height; (B) a convection-dominated crown fire with light winds aloft (adapted from Rothermel, 1991b). Reproduced with permission of the Society of American Foresters.

central Alberta (Quintilio *et al.*, 2001), all of which occurred in exceptionally dry, conifer forests, represent three very divergent cases of extreme fire behaviour resulting from convective dominance.

In addition to their direct effects at the Earth's surface, convection-dominated fires can also lead to pyrocumulonimbus (pyroCb) as opposed to PyroCu cloud development (Fromm *et al.*, 2010). PyroCbs are capable of transporting smoke directly into the upper troposphere and lower stratosphere regions of the atmosphere (i.e. ~7–14 km above the ground in the northern hemisphere) (see Figure 1.38, chapter 1, Part One). This region of the Earth's atmosphere plays an important role in global climate.

14.6 Field methods of measuring and quantifying wildland fire behaviour

The methods used in the measurement and quantification of fire behaviour in an outdoor environment will depend on the size of the fire and also the circumstances (i.e. whether it is an experimental fire, an operational prescribed fire or a wildfire). Safety and logistical issues will limit what is possible with respect to wildfires and, to a large extent, most operational prescribed fires. Aside from these considerations, funding, time and number of personnel available will influence what would be 'nice to do' and what is realistic.

No comprehensive review and summary of the various methodologies for the measurement and quantification of fire behaviour associated with experimental fires and operational prescribed fires yet exists. However, useful information and ideas can be found throughout the literature; the fire monitoring handbook developed by the USDI National Park Service (2003) constitutes a good starting point. The methodologies employed in major experimental burning projects warrant serious study. These include the Annaburroo project in grass in Australia's Northern Territory (Cheney *et al.*, 1998) in grass (1986), the International Crown Fire Modelling Experiment (ICFME) in Canada's Northwest Territories (Stocks *et al.*, 2004) in jack pine-black spruce forest (1995–2001), Project Vesta in Western Australia (McCaw *et al.*, 2012) in dry eucalypt forest (1996–2002) and Project Ngarkat in South Australia (Cruz *et al.*, 2013) in mallee-heath shrubland (2006–2008).

While the principles used in the monitoring and documentation of wildfire behaviour are similar to those used on experimental and operational prescribed fires, the methods are considerably different. A recent paper by Cruz *et al.* (2012) constitutes a good example of the present approaches used in completing a case study of wildfire behaviour. Sampling rates on experimental fires and operational prescribed fires can be predetermined, based on the purpose of the study (e.g. how frequently will wind speed and fire spread be recorded?). With wildfires, opportunistic observation is more likely the situation.

A distinct advantage on experimental and operational prescribed fires is the opportunity to conduct pre-fire and post-fire fuel sampling in order to determine the available fuel consumption – a variable required for calculation of fireline intensity. This is a much more difficult task to accomplish in a wildfire setting, although not impossible. The advance lead time is also hugely advantageous for instrumentation setup to gather data on other characteristics of fire behaviour. For example, it would not be possible, in most cases, to lay out plastic sheeting downwind of a rapidly advancing wildfire in order to study spot fire densities. However, this could be done with relative ease on experimental fires and operational prescribed fires. In the case of wildfires, one might have to rely on other sources that were in place prior to the fire's occurrence (e.g. trampoline or plastic covering for a boat).

The specific method selected for determining rate of spread on an experimental fire or operational prescribed fire will often depend on the type of ignition, the size of the area to be burnt, the anticipated spread rate and fireline intensity and the purpose of the study. It may be quite possible to use the same technique on both types of fires. For example, the forward spread distances of the line- and point-source ignition fires depicted in Figure 14.19 were marked with one-gallon paint can lids at one-minute intervals by field observers. However, another technique or a modification would be required to obtain a record of the point-source fire's perimeter development as a function of time (Figure 14.18A-C).

Wildland fire behaviour research has benefited greatly from technological advances. Seemingly antiquated methods of documenting, observing, measuring and recording data (Figure 14.18D) on rate of spread, flame dimensions and other aspects of fire behaviour (e.g., spotting) have been replaced or augmented by developments in electronics and photography (Figure 14.28). For example, thermocouples have been used in wildland fire research for at least 75 years. The recording devices of the past were exceedingly cumbersome to install in a field setting. Present-day data logging and storage capability has made the task considerably easier.

Figure 14.28   Thermal image of a prescribed fire in a ponderosa pine-Jeffery pine stand on the Eagle Lake District of the Lassen National Forest, California. The image is from a MikroScan 7200 Thermal Imager. It illustrates the spatial variability in soil and stem heating associated with a duff mound burning experiment. The duff mound around the base of the tree in the foreground had been left unraked. The tree directly to the right of this had had the duff mound raked away from its base (dark zone around tree). The lighter or brighter the colour, the higher the surface temperature (photo James R. Reardon, USDA Forest Service, Missoula, Montana).

One of the latest breakthroughs in instrumentation came in 1997, when Jim Kautz, a photographer with the US Forest Service's Missoula Technology and Development Center, developed an 'in-fire' video camera box for the ICFME experimental crown fires, based on a suggestion by the author of this chapter (MEA), who served as one of the project research coordinators for the Canadian Forest Service at the time. This innovation has brought a whole new dimension to the study of wildland fire behaviour. Another more recent development has been successful use of small, unmanned aircraft to monitor experimental field fires.

14.7   Towards increasing our understanding of wildland fire behaviour

Wildland fires are highly variable, multi-dimensional phenomena and are not always easily observed,

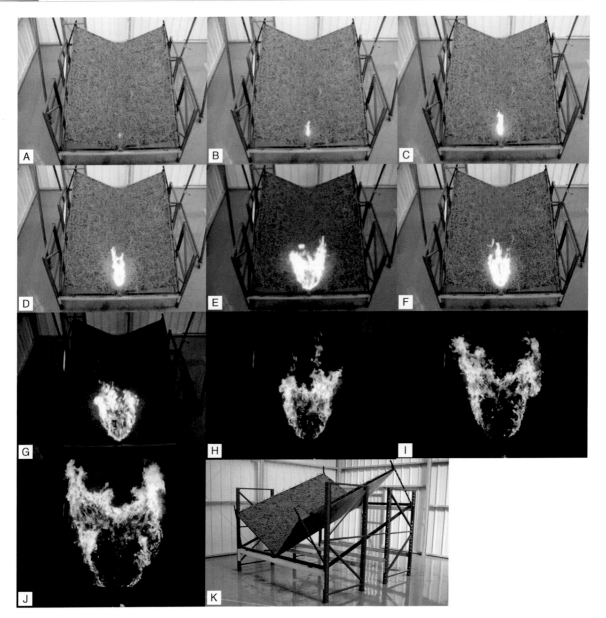

Figure 14.29 Experimental fire conducted on a specially constructed burning table described by Viegas and Pita (2004), designed to simulate fire behaviour in V-shaped canyons as carried out at the Forest Fire Research Laboratory located in Lousã, Portugal, and operated by the Association for the Development of Industrial Aerodynamics (ADAI)). The fuel is comprised of maritime pine needle litter, with a moisture content of 11% (oven-dry weight basis) and load of 0.6 kg/m^2 Reproduced with permission of Domingos X. Vavier, Departamento de Engenharia Mecânica, Universidade de Coimbra, Portugal. http://www.uc.pt/fctuc/dem/.

monitored, explained or documented. Nevertheless, research carried out in the field and the laboratory (Figure 14.29) coupled with operational experiences over the years, has offered us insights into the physical processes involved in the behaviour of free-burning wildland fires, both qualitatively and quantitatively in many cases. Research has also

provided concepts and principles that have proved useful in the understanding wildland fire behaviour.

Some, but not all, of the remaining knowledge gaps and basic research needs pertaining to the fundamentals of fire behaviour have been alluded to throughout this chapter. Unfortunately, a comprehensive problem analysis on the subject does not presently exist, and, in general, is beyond the scope of this chapter and book, although Alexander and Andrews (1989) provide a good outline. Nevertheless, two specific areas deemed worthy of concentrated study will be touched on briefly.

In comparison to the research that has been undertaken into the behaviour of single-point or line-source fire ignitions, there has been very little scientific field experimentation associated with interacting flame fronts involving far more complex ignition patterns (Finney and McAllister, 2011). Yet, the merging of flame fronts, whether from lines or points, is common on prescribed fires and in burn-out and backfiring operations associated with the containment of wildfires. The increased flame heights associated with the junction zones of such merging are a well-known outcome (Figure 14.15). What is probably less appreciated are the differences in fuel consumption associated with multi-point or line-source ignitions, compared to single-point or fully-developed line fires and the type of ignition device used. Codifying the behaviour and characteristics of merging flame fronts from multi-point or line-source ignition fires is quite likely to require a completely different set of input considerations and output metrics that has yet to be fully explored.

There is no question that a great deal of knowledge has been acquired about various aspects of extreme fire behaviour from wildfire observation and case study documentation. While very useful data has been gathered (e.g. fire spread rates under critical fire weather conditions), the 'snapshot' approach that has unfolded has also caused many a misunderstanding (Alexander, 2009a). Coupled with this are the many lost opportunities to gain additional information because of seemingly higher priorities. No doubt, the occasional 'windfall' observation will prove useful. However, a clear understanding of the dynamics involved in high-intensity wildfires in different fire environments likely will come only from the systematic study of a well-funded and dedicated group involving both fire research and operations staff over a long time frame in order to gain the experience necessary to make significant contributions. To date, the closest we have come to realization of this goal was the efforts of the fire behaviour documentation team that operated out of the Southern Forest Fire Laboratory at Macon, Georgia, from the late 1950s to the early 1970s.

These are but two examples of major subject areas that will require a significant change in how fire behaviour research is conducted in the future. It is fully expected that our understanding of wildland fire behaviour will continue to evolve in the light of knowledge generated from field studies, laboratory experiments, and simulation modelling. At the same time, there is a real value in promoting efforts that critically analyze and comprehensively synthesize our existing knowledge as a basis for new studies (Trevitt, 1989).

Further reading

Albini, F.A. (1993). Dynamics and modeling of vegetation fires: observations. In: Crutzen, P.J., Goldammer, J.G. (eds.), *Fire in the environment: The ecological, atmospheric, and climatic importance of vegetation fires*, pp. 39–52. John Wiley & Sons, Chichester, UK. Environmental Sciences Research Report ES 13.

Alexander, M.E., Stefner, C.N., Mason, J.A., Stocks, B.J., Hartley, G.R., Maffey, M.E., Wotton, B.M., Taylor, S.W., Lavoie, N., Dalrymple, G.N. (2004). *Characterizing the jack pine-black spruce fuel complex of the International Crown Fire Modelling Experiment (ICFME)*. Natural Resources Canada, Canadian Forest Service, Northern Forestry Centre, Edmonton, Alberta. Information Report NOR-X-393. 49 p.

Andrews, P.L. (1996). Part 1. Fire environment. In: Pyne, S.J., Andrews, P.L., Laven, R.D. (eds.), *Introduction to wildland fire*. 2nd edition, pp. 3–168. J. Wiley & Sons, Inc., New York, NY.

Arnaldos, J., Navalón, X., Pastor, E., Planas, E., Zárate, L. (2004). *Manual de ingeniería básica para la prevención y extinción de incendios forestales*. [Basic engineering handbook for wildfires prevention and suppression]. Mundi-Prensa, Madrid, Spain. (In Spanish.)

Babrauskas, V. (2003). *Ignition handbook: principles and applications to fire safety engineering, fire investigation, risk management and forensic sciences*. Fire Science Publishers, Issaquah, Washington.

Baker, W.L. (2009). *Fire ecology in Rocky Mountain landscapes*. Island Press, Washington, D.C. 605 p.

Byram, G.M. (1957). Some principles of combustion and their significance in forest fire behavior. *Fire Control Notes* **18**(2), 47–57.

Catchpole, W. (2002). Fire properties and burn patterns in heterogeneous landscapes. In: Bradstock, R.A., Williams, J.E., Gill, A.M. (eds.), *Flammable Australia: The fire regimes and biodiversity of a continent*, pp. 49–75. Cambridge University Press, Cambridge, UK.

Cheney, N.P. (1981). Fire behaviour. In: Gill, A.M., Groves, R.H., Noble, I.R. (eds.), *Fire and the Australian biota*, pp. 151–175. Australian Academy of Science, Canberra, Australian Capital Territory.

Cheney, N.P. (1990). Quantifying bushfires. *Mathematical and Computer Modelling* **13**(12), 9–15.

Cheney, N.P., Gill, A.M. (eds.) (1991). *Conference on bushfire modelling and fire danger rating systems proceedings*. CSIRO Division of Forestry, Yarralumla, Australian Capital Territory.

Chuvieco, E. (ed.) (2003). *Wildland fire danger estimation and mapping: The role of remote sensing data*. World Scientific Publishing Company Pte. Ltd., Singapore.

Cochrane, M.A., Ryan, K.C. (2009). Fire and fire ecology: Concepts and principles. In: Cochrane, M.A. (ed.), *Tropical fire ecology: Climate change, land use and ecosystem dynamics*, pp. 25–62. Praxis Publishing, Ltd., Chichester.

Cottrell, W.H., Jr. (2004). *The book of fire*. 2nd edition. Mountain Press Publishing Co., Missoula, MT.

Cruz, M.G., Alexander, M.E. (2012). Evaluating regression model estimates of canopy fuel stratum characteristics in four crown fire prone fuel types in western North America. *International Journal of Wildland Fire* **21**, 168–179.

DeBano, L.F., Neary, D.G., Ffolliott, P.F. (1998). *Fire effects on ecosystems*. John Wiley & Sons, New York.

De Groot, W. J., Alexander, M.E. (1986). Wildfire behavior on the Canadian Shield: A case study of the 1980 Chachukew Fire, east-central Saskatchewan. In: Alexander, M. E. (ed.), *Proceedings of the third Central Region Fire Weather Committee scientific and technical seminar* pp. 23–45. Environment Canada, Canadian Forestry Service, Northern Forestry Centre, Edmonton, Alberta.

Glickman, T.S. (ed.). (2000). *Glossary of meteorology*. 2nd edition. University of Chicago, Chicago, Illinois.

Goldammer, J.G., de Ronde, C. (eds.). (2004). *Wildland fire management handbook for sub-Sahara Africa*. Global Fire Monitoring Center, Freiburg, Germany and Oneworldbooks, Cape Town, South Africa.

Gorski, C.J., Farnsworth, A. (2000). Fire weather and smoke management. In: Whiteman, C.D. (ed.), *Mountain meteorology: Fundamentals and applications*, pp. 237–272. Oxford University Press, New York.

Hollis, J.J., Gould, J.S., Cruz, M.G., Doherty, M.D. (2011). *Scope and framework for an Australian fuel classification*. A Report for the Australasian Fire and Emergency Service Authorities Council. CSIRO Ecosystem Sciences and Climate Adaptation Flagship, Bushfire Dynamics and Applications, Canberra, Australian Capital Territory. CSIRO ePublish ID: EP113652. 90 p.

Hollis, J.J., Matthews, S., Anderson, W.R., Cruz, M.G., Burrows, N.D. (2011). Behind the flaming zone: Predicting woody fuel consumption in eucalypt forest fires in southern Australia. *Forest Ecology and Management* **261**, 2049–2067.

Johnson, E.A. (1992). *Fire and vegetation dynamics: Studies from the North American boreal forest*. Cambridge University Press, Cambridge.

Johnson, E.A., Miyanishi, K. (eds.). (2001). *Forest fires: Behavior and ecological effects*. Academic Press, San Diego, California.

Lavoie, N., Alexander, M.E., Macdonald, S.E. (2007). *Fire weather and fire danger climatology of the Fort Providence area, Northwest Territories*. Natural Resources Canada, Canadian Forest Service, Northern Forestry Centre, Edmonton, Alberta. Information Report NOR-X-412. 51 p.

Lillquist, K. (2006). Teaching with catastrophe: Topographic map interpretation and the physical geography of the 1949 Mann Gulch, Montana wildfire. *Journal of Geoscience Education* **54**, 561–571.

McCaw, L., Cheney, P., Sneeuwjagt, R. (2003). Development of a scientific understanding of fire behaviour and use in south-west Western Australia. In: Abbott, I., Burrows, N. (eds.), *Fire in ecosystems of south-west Western Australia: Impacts and management*, pp. 171–187. Backhuys Publishers, Leiden, the Netherlands.

Omi, P.N. (2005). *Forest fires: A reference handbook*. ABC-CLIO Inc., Santa Barbara, CA.

Ottmar, R.D. (2013). Wildland fire emissions, carbon, and climate: Modeling fuel consumption. *Forest Ecology and Management*, doi.org/10.1016/j.foreco.2013.06.010

Rosenfeld, D., Fromm, M., Trentmann, J., Luderer, G., Andreae, M.O., Servranckx, R. (2007). The Chisholm firestorm: Observed microstructure, precipitation and lightning activity of a pyro-cumulonimbus. *Atmospheric Chemistry and Physics* **7**, 645–659.

Rothermel, R.C., Rinehart, G.C. (1983). *Field procedures for verification and adjustment of fire behavior prediction*. U.S. Department of Agriculture, Forest Service, Intermountain Forest and Range Experiment Station, Ogden, Utah. General Technical Report INT-142. 25 p.

Simard, A.J., Haines, D.A.. Blank, R.W., Frost, J.S. (1983). *The Mack Lake Fire*. U.S. Department of Agriculture, Forest Service, North Central Forest Experiment Station, St. Paul, Minnesota. General Technical Report NC-83. 36 p.

Simeoni, A. (2013). Experimental understanding of wildland fires. In: Belcher, C.M. (ed.). *Fire phenomena and the earth system: An interdisciplinary guide to fire science*, pp. 35–52, John Wiley & Sons, Ltd., Chichester.

Van Wagner, C.E. (1972). Duff consumption by fire in eastern pine stands. *Canadian Journal of Forest Research* **2**, 34–39.

Vines, R.G. (1981). Physics and chemistry of rural fires. In: Gill, A.M., Groves, R.H., Noble, I.R. (eds.), *Fire and the Australian biota*, pp. 129–149. Australian Academy of Science, Canberra, Australian Capital Territory.

Weir, J.R. (2009). Conducting prescribed fires: A comprehensive manual. Texas A&M Univ. Press, College Station. 194 p.

Weise, D.R., Wright, C.S. (2013). Wildland fire emissions, carbon and climate: Characterizing wildland fuels. *Forest Ecology and Management*, doi.org/10.1016/j.foreco.2013.02.037

Werth, P.A., Potter, B.E., Clements, C.B., Finney, M.A., Goodrick, S.L., Alexander, M.E., Cruz, M.G., Forthofer, J.M., McAllister, S.S. (2011). *Synthesis of knowledge of extreme fire behavior: Volume 1 for fire managers*. U.S. Department of Agriculture, Forest Service, Pacific Northwest Research Station, Portland, Oregon. General Technical Report PNW-GTR-854. 144 p.

Zimmerman, T. (2013). Information technology and the work of managing fires. *Wildfire* **22**(2), 14–18.

Chapter 15

Estimating free-burning wildland fire behaviour

15.1 Introduction

Anyone who has personally observed the behaviour of a free-burning wildland fire for the first time will attest to the wide variations in the speed, direction and size of the flame front. There would appear to be very little rhyme or reason to the way a fire behaves. It all seems to add up to erratic behaviour without any logical explanation.

Some fire behaviour researchers feel that most wildland firefighters base their expectations of how a fire will behave largely on experience and, to a lesser extent, on fire behaviour guides (Burrows, 1984b). We see, for example, in the 2002 movie *Superfire*, where there is an explanation by Dr. Simon Orr (played by actor Latham Gaines) of Caltech about how a satellite-based, computerized decision support system can predict a fire's behaviour. One of the firefighters then points out that they 'just ask Joe' what a fire is going to do. 'Joe' as it turns out, is their veteran smokejumper foreman Joe Nighttrail (played by actor Wes Studi), who doesn't need a computer to know where a fire is going.

Over the years, many wildland firefighters have developed a remarkable skill at predicting fire behaviour, particularly if they have worked long enough in an area to become familiar with the fire environment features and associated fire behaviour (Box 15.1).

Experienced judgment can, however, have limitations, depending on the nature of the experiences and what one remembers about them and documents. This brings to mind comedian Steve Wright's comment that 'everyone has a photographic memory, some just don't have film'.

Still, others, like Williams and Rothermel (1992), maintain that the best chance for success in predicting wildland fire behaviour requires a mix of fire experience with analytical modelling methods. Explicitly trusting the output of models can be just as dangerous as relying exclusively on limited, experienced judgment.

The goal of wildland fire behaviour research is to provide simple, timely answers to the following types of questions, given an actual ignition or a simulated fire occurrence, for any specified set of fuel, weather, and topographic conditions (after Luke and McArthur, 1978):

- What will be the head fire rate of spread?
- What will be the area, perimeter length, and forward spread distance after one hour, two hours, three hours, and so on?
- Will it be a high-intensity or low-intensity fire?
- Will it be a crown fire or a surface fire?
- How difficult will it be to control and extinguish?
- Will mechanical equipment and/or air tankers be required to contain the fire, or can the fire be handled safely and effectively by ground crews?

Fire on Earth: An Introduction, First Edition. Andrew C. Scott, David M.J.S. Bowman, William J. Bond, Stephen J. Pyne and Martin E. Alexander.
© 2014 John Wiley & Sons, Ltd. Published 2014 by John Wiley & Sons, Ltd.

The old-timer among firefighters is often inclined to forget that knowledge of fire behaviour is not the knowledge of instinct. He gathered this knowledge for himself bit by bit from personal experiences as the years progressed. Sometimes he has difficulty in passing that knowledge along to others, for the very good reason that he fails to realise how much can remain unknown to the man who has never had the opportunity to observe similar events personally.

To demonstrate this lack of knowledge due to lack of experience, a middle-aged man tells a little story about himself of an event that happened more than 20 years ago. That day, the young fellow easily reached an observation point near the fire before an older ranger arrived. There before him was a rolling inferno of flames such as he had never before seen. Fascinated and frightened, he told himself that all the power of Man could never stop this fire. The old ranger wheezed up, rolled himself a cigarette and mumbled to himself, "The head will run into that old burn in half an hour and by sundown the wind will die and we'll cold trail her." Then he turned slowly to a messenger and said, "Joe, go phone headquarters and tell them the fire is under control."

The young man's misjudgement was built upon fear and ignorance. Ignorance is simply a lack of knowledge. The older man certainly had jumped too far to a conclusion, but he had made an estimate of the situation and formulated a plan of action with quiet confidence based upon the knowledge of long experience. Such experience requires the passage of years and the observation of many errors. Errors of judgment are generally very costly.

- Will the mop-up efforts require more time than normal?
- Is there a possibility of it 'blowing up'? If so, will the blow-up produce a towering convection column or have a wind-driven smoke plume?
- What will be the spotting potential – short- or long-range?

- Are fire whirls and/or other types of wildland fire vortexes likely to develop? If so, when and where?

The supporting work undertaken by fire behaviour researchers to codify the relationships between fire behaviour variables and environmental conditions, and then produce the tools that enable one to answer these kinds of questions at least adequately enough for most fire control and fire use, constitutes the 'science' of predicting fire behaviour. In turn, the 'art' of fire behaviour prediction represents the artful application of the science (Figure 15.1), including meaningful communication of the information to different audiences.

15.2   A historical sketch of wildland fire behaviour research

In what year did wildland fire behaviour research start? Prior to the early 1920s, wildland fire behaviour-related research was largely limited to analyses of fire weather conditions or fire report data carried out by meteorologists and operational foresters on a part-time basis (Potter, 2012). The most notable individuals were the team of Stuart Bevier Show and Edward I. Kotok, representing operations and research respectively, who were stationed in California.

The world's first full-time forest fire researcher, Harry T. Gisborne, a forester by training, was appointed to the US Forest Service's Northern Rocky Mountain Forest Experiment Station on April 1, 1922. He and other station researchers would undertake many fundamental studies of fire weather and fuel moisture. It is undoubtedly a testament of Harry Gisborne's quintessential pioneering spirit that he died of a heart attack on November 9, 1949, while engaged in a field inspection of the steep and difficult topography of the fatal Mann Gulch Fire near Helena, Montana, in which 13 firefighters had lost their lives. Gisborne hiked up those high-elevation canyon walls that November day, despite knowing full well that he had heart problems. Hardy (1983) provides an excellent accounting of the Gisborne era of forest fire research.

Up until the early 1950s, there were probably less than 20 people worldwide who were dedicated to the study of wildland fire behaviour. Additional fire research staff would come to be hired at the Priest

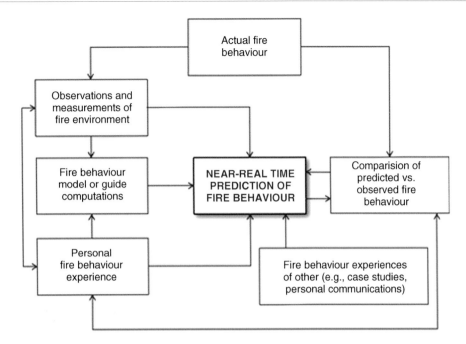

Figure 15.1 Flow diagram illustrating that the 'science and art' of wildland fire behaviour prediction. This encompasses the coupling of practical knowledge, professional judgment and fire behaviour experiences (including local knowledge) with the computational tools produced by fire research (from Alexander and Cruz, 2013b). Reproduced with permission of the Canadian Institute of Forestry/l'Institut forestier du Canada.

River Forest Experiment Station in northern Idaho and at other forest experiment stations across the United States in years that followed, including most notably George M. Byram at the Appalachian Forest Experiment Station in Asheville, North Carolina, in the mid-1930s. A Division of Forest Fire Research was established in 1950 within the US Forest Service's national office in Washington, DC, with Arthur A. Brown as its first director.

A forest fire research program was initiated in Canada in 1925 by James G. Wright, a civil engineer with the Dominion Forest Service. His student assistant from the summer of 1928, Herbert W. Beall, received a permanent appointment in 1932 after graduating from the University of Toronto forestry school (Van Wagner, 1987b; Beall, 1990).

Research into aspects of fire behaviour in the taiga or boreal forest region of the Union of Soviet Socialist Republics began in the 1930s (Artsybashev, 1984). Notable Soviet fire researchers of the time include Valentin G. Nesterov and A.A. Molchanov.

Alan G. McArthur's appointment in 1953 as a bushfire researcher in the Commonwealth of Australia's Forestry and Timber Bureau represented a first for

that country. The Commonwealth Scientific and Industrial Research Organization (CSIRO) eventually added fire research staff, as did some Australian states, most notably Western Australia, Victoria, New South Wales, Queensland and Tasmania (Luke and McArthur, 1978).

15.2.1 Experimental fire behaviour field studies

The first known field research into rate of fire spread was undertaken by Show (1919), a silviculturist with the US Forest Service, who documented the growth of small, experimental point-ignition source fires in the ponderosa pine forest type of northern California at the Feather River Experiment Station during the summers of 1915–1917. It is of interest to note that in the last article by Harry Gisborne, published posthumously in 1950, he describes how he worked with Bevier Show to undertake various ignition tests with cigar and cigarette stubs back in 1922.

The work initiated by Show in ponderosa pine was continued in the 1930s at the Shasta Experimental

Forest (see Chapter 14, Figure 14.18D) by John R. Curry and Wallace L. Fons, a forester and mechanical engineer, respectively. Similar experimental fires were undertaken in Mississippi in the longleaf pine-slash pine type, beginning in the mid-to-late 1930s, specifically to investigate rate of fire spread in relation to fire detection by lookout towers which had also been examined in the northern California study.

A number of experimental burning studies were carried out in the south-eastern states by US Forest Service fire research and others at several locations, beginning in the late 1940s and continuing well into the 1990s, as part of a broader look at prescribed fire use on a rotational basis in different southern pine types. Experimental fires were also undertaken in oak-chaparral in Arizona and aspen-northern hardwood stands in the Lake States in the late 1960s and early 1970s.

Other notable fire behaviour-related research studies undertaken in the US following World War II include Operation Firestop, carried out at Camp Pendleton in southern California in 1954–1955 to examine advanced firefighting methods (Cermak, 2005). A study of logging slash fire behaviour that also examined the effects of slash aging was carried out by forester George R. Fahnestock and others at the Priest River Experimental Forest in northern Idaho during the 1950s. A study of mass fire behaviour called Project Flambeau (1964–1967), involving the simultaneous burning of numerous large debris piles, was carried out at desert sites in Nevada and California under the technical leadership of Clive M. Countryman. This also involved participation from Australia, Canada, and England.

In Canada, Wright (1932) had devised methods of estimating fuel moisture content and a two-minute test fire procedure, initially developed and tested at the Petawawa Forest Experiment Station in eastern Ontario, for assessing the forest fuel flammability that would come to be used at 11 field stations across country for the next three decades. The resulting database, comprised of some 20 000 fire test observations in many different vegetation and fuel types, is still proving useful.

The experimental burning eventually extended to the deliberate initiation of crown fires using a line-source ignition in small (0.2 ha) plantation plots of red pine at the station by Charles E. Van Wagner, beginning in the early 1960s (Place, 2002). This, in turn, led to the burning of experimental plots up to 5 ha in size

in other conifer forest types, logging slash and trembling aspen stands, beginning in the mid to late 1960s and continuing for the next two to three decades.

Following his appointment, Alan McArthur very quickly initiated a very impressive program of experimental burning in Australia's grassland and eucalypt forest fuel types, involving point-source ignition fires allowed to burn from 30 minutes to upwards of an hour (Figure 15.2). The work was initially focused in the Australian Capital Territory and New South Wales but was then extended to Western Australia. More than 800 experimental fires alone were documented in the dry eucalypt forest.

Subsequently, fire researchers in Western Australia undertook to develop their own experimental burning program, using McArthur's techniques to support their prescribed fire use practices in native forests and exotic pine plantations. Several of the state and territorial forest and fire services, as well as CSIRO, have since undertaken numerous experimental burning projects.

As wildland fire behaviour research expanded globally, other countries and international entities, such as the Food and Agriculture Organization of the United Nations (e.g. Peet, 1980), recognizing the value of undertaking experimental fires in the field as a means of developing local fire behaviour guides, began to invest the time and effort to do so. This included, most notably but not limited to, Brazil, South Africa, Kenya, Portugal, Spain, New Zealand, France, Scotland and Turkey (Figure 15.3), involving point- and line-source ignition fires in grasslands, shrublands, and pine forests.

15.2.2 Laboratory test fires

In time, experimental fires were also being carried out in a laboratory environments (Figure 15.4) as a supplement to the field studies of fire behaviour. In spite of issues of scale, some laboratory fire experimentation has proven useful as part of the development of operational fire behaviour models (e.g. in studying the relative effect of slope steepness on rate of fire spread).

The first experimental fires carried out in a wind tunnel were made in 1938 by Wally Fons. Interestingly, the specially constructed wind tunnel for the study of fire behaviour was located outdoors at the Shasta Experimental Forest in northern California, as

Figure 15.2 A point-source ignition experimental fire in dry scherophyll eucalypt forest at Black Mountain, Australian Capital Territory, being monitored and documented by a bushfire research crew from the Commonwealth of Australia's Forestry and Timber Bureau during the 1960s (from 1965 Annual Report of the Forestry and Timber Bureau).

opposed to inside a building. Fons (1963) also initially led Project Fire Model (1959–1966), an experimental study of model fires conducted by the US Forest Service. Indoor experimental burning significantly escalated with the creation by the US Forest Service of the three national forest fire laboratories dedicated to the study of forest fires at Macon, Georgia, in 1959 (USDA Forest Service, 1993), at Missoula, Montana,

Figure 15.3 Experimental crown fire spreading through a 22-year-old calabrian pine plantation (planted at 3×2 m spacing) in north-western Turkey, August 4, 2007. Average tree height was 8 m. The plot measured 40×80 m. This fire was conducted as part of the experimental burning program being carried out under the leadership of Dr. Ertugrul Bilgili, Faculty of Forestry Karadeniz Technical University, Trabzon, Turkey (photos Bülent Sağlam, Faculty of Forestry, Artvin Çoruh University, Artvin, Turkey).

Figure 15.4 Laboratory test burning being carried out as part of the study of fire behaviour at the Harrison Experimental Forest, Desoto National Forest, Mississippi, October 17, 1937. Under controlled conditions, fuel of various moisture contents is burnt with different amounts of wind movement to determine the effect of fuel moisture content and wind on rate of spread and fuel consumption (photo T.T. Koharn, USDA Forest Service, Southern Forest Experiment Station).

in 1960 (Smith, 2012) and at Riverside, California, in 1963 (Wilson and Davis, 1988).

The novelty of possessing a wind tunnel has passed, as several organizations now have their own such facilities (Figure 15.5) – although, admittedly, the level of sophistication varies widely. In recent years, some groups have constructed vertical wind tunnels for the purpose of studying airborne firebrands.

Laboratory investigation of fire behaviour has not been limited to just wind tunnels. To the list we can also add: open combustion chambers, where a variety of fuels have been burnt, including Christmas trees up to five metres in height; burning tables that can also be tilted to various levels of slope steepness, including the simulation of canyon walls (see Chapter 14, Figure 14.29); firebrand generators; and a variety of devices (e.g. cone calorimeter, epiradiator) used for testing the flammability of fuel particles.

The study of certain aspects of wildland fire behaviour in a laboratory environment has not been limited just to forestry research organizations. For example, some of the work undertaken by Dr. Phillip H. Thomas, a physicist with the former Fire Research Station at Borehamwood, England (Read, 1994), although primarily focused on urban or structural fires, led to advances in the understanding of wildland fire spread, flame size and convection column temperatures. In recent years, the Fire Research Division of the US National Institute of Standards and Technology (NIST) have been undertaking wildland-urban interface fire related testing in their Large Fire Laboratory located on the NIST campus in Gaithersburg, Maryland.

15.2.3 Wildfire monitoring and case studies

The notion of using wildfires as a source of fire behaviour research data continues to be extensively utilized (Figure 15.6). Several different approaches have been taken over the years but, basically, it breaks down into active observation or post-mortem reporting (e.g. fatality fire or board of review investigation). The later would consist of follow-up detective work by fire research or fire control staff, while the former would involve a single dedicated person or, more typically, a team effort (Alexander and Thomas, 2003a, 2003b).

Figure 15.5 A laboratory test fire in progress within the CSIRO pyrotron located in Canberra, Australian Capital Territory. The pyrotron is a 25 m long, fireproof wind tunnel with a glass-panelled observation area design to monitor the burning fuel beds under controlled conditions (Sullivan *et al.*, 2013) (photo Andrew L. Sullivan, CSIRO Ecosystem Sciences and Climate Adaptation, Canberra, Australian Capital Territory).

Figure 15.6 Scene of the Butte City Fire in south-eastern Idaho during its initial major run on July 1, 1994, during which time it advanced 19 km in 6.5 hours, burning an area of 8300 ha (from Butler and Reynolds, 1997).

Table 15.1 Reliability rating for weather, fuel, and rate of spread information for wildfires in dry eucalypt forests (from Cheney *et al.*, 2012). Earlier on, Cheney *et al.* (1998) developed a similar rating scheme for wildfires in grasslands.

Rating	Weather	Fuel	Rate of spread
1	Nearby meteorological station or direct measurements in the field.	Fuel characteristics inferred from a fuel age function developed for the particular fuel type.	Direct timing of fire spread measurements by observers.
2	Meteorological station within 50 km of the fire.	Fuel characteristics inferred from a visual assessment of nearby unburnt forest.	Reliable timing of fire spread by a third party.
3	Meteorological station > 50 km of the fire, reconstruction of wind speed for fire site.	Fuel characteristics inferred from a fuel age curve for a forest type of similar structure.	Reconstruction of fire spread with numerous cross references.
4	Spot meteorological observations near the fire.	Fuel characteristics typical of equilibrium level in a dry sclerophyll forest.	Doubtful reconstruction of fire spread.
5	Distant meteorological observations at locations very different from fire site.		Anecdotal or conflicting reports of fire spread.

Source: Reproduced with permission of Elsevier.

Harry Gisborne is believed to have published the first wildfire case study in his description of the Quartz Creek Fire, which occurred on the Kaniksu National Forest adjacent to the Priest River Experimental Forest during the summer of 1926. He continued the active monitoring of wildfires for several more years (e.g. the 1929 Half Moon Fire in Glacier National Park), arguing vehemently that such an approach could, in fact, yield useful fire behaviour data (Gisborne, 1929).

Wildfire monitoring has not been without its controversy however. In the late 1930s, the Southern Forest Experiment Station of the US Forest Service operated a crew specifically dedicated to studying the behaviour of free-burning wildfires. This included their documentation of the 1938 Honey Fire that occurred on the Kisatchie National Forest in north-central Louisiana. The research crew's case study reporting on this fire still stands as one of the classic case studies, yet they were roundly criticized for not attempting to suppress the fire when they had the chance to, in spite of the fact that they had been given prior approval to focus strictly on fire behaviour documentation (Alexander and Taylor, 2010).

The opportunity to document the characteristics of fast-spreading, high-intensity wildfires generally offer the only means of acquiring data toward the extreme end of the fire behaviour spectrum. Generally, this is not possible unless, by chance, an 'escape' from an experimental fire or operational prescribed fire occurs.

Rate of spread data gleaned from wildfire case studies is used in both the building of models and in their testing. The reliability and completeness of the information gathered on rate of spread, wind speed and fuels from wildfires can, however, vary widely

(Table 15.1). In some situations, for example, one could have acquired quite reliable and detailed information on a fire's rate of spread by continuous monitoring from the air and ground, but the weather station network may, in turn, be extremely sparse, with perhaps observations only being recorded once an hour.

15.2.4 Operational prescribed fires

Like the occasional wildfire, fire researchers have, at one time or another, made good use of the opportunities to gather certain types of data from prescribed fires (i.e. fires deliberately set to accomplish specific resource management objectives such as ecological restoration, wildlife habitat improvement or hazardous fuels reduction) (Alexander and Thomas, 2006). However, it would appear that this was not a common fire research practice until the mid-1950s.

The Prescribed Burn Fireclimate Surveys, carried out in 1956 and 1957 on controlled burns of 160 to 400 ha in size and conducted for wildland brush control in central California (Schroeder and Countryman, 1960), constituted one of the first such efforts. These landscape-scale fires provided an opportunity to carefully observe each fire's behaviour in relation to closely monitored fire weather conditions at several locations around each burn unit. Since then, fire researchers have gathered information on various aspects of free-burning fire behaviour such as rate of fire spread, flame dimensions, fuel consumption, spot fire distances, fire whirl activity and smoke column heights.

15.2.5 Individual fire report data

A good many wildland fire management agencies require that, for annual statistical purposes, a report form be completed on every wildfire occurrence. While the content of these forms varies widely, many ask for information on the fire's general behaviour (e.g. rate of spread at the time of initial attack). Fire behaviour researchers have, from time to time, made good use of such reports as a source of fire behaviour data for model building or testing (e.g. Haines *et al.*, 1986). Alan McArthur, for example, was one individual who made use of such data in developing the relations between fire behaviour and the fire environment embedded in his fire danger meters. Similar information may also be filed on prescribed fires.

Even a simple metric like final fire size can have value as a measure of fire behaviour (Figure 15.7), even if there is no information on fire perimeter at times between ignition and containment. At one time,

the US Forest Service looked to correlate information obtained from their individual forest fire report data with weather conditions and fuel types as a means of developing guides to estimating certain fire behaviour characteristics (e.g. rate of fire spread). As Brown (1959) notes, 'We now know that was expecting too much'.

In spite of their general low reliability, individual fire report data have been used as means of evaluating model performance and deriving general trends. For example, it was found from an analysis of such information, for wildfires occurring in the National Forests of the California Region of the US Forest Service, that fires typically double their spread rate for about every 30% increase in slope steepness. This result has, in turn, been compared to the findings from field studies and laboratory investigations, as a form of a check on the latter two methods in determining the mechanical effects of slope steepness on rate of fire spread.

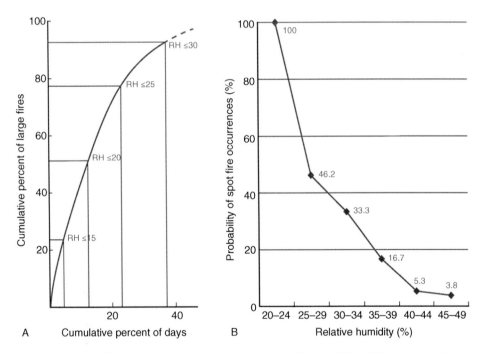

Figure 15.7   Two examples illustrating the association between relative humidity (RH) in relation to measures of fire incidence. (A) Cumulative percent of 'large fires' (>121 ha) versus cumulative percent of days within RH classes based on 266 class E fires that occurred over a ten-year period (1950–1959) in Georgia, USA (from Krueger, 1961). (B) The percent probability of spot fires occurring as a function of RH classes observed on 99 prescribed fires conducted from 1996 to 2002 in Oklahoma, USA (from Weir, 2004). Both Krueger (1961) and Weir (2004) found that an RH value of 25% constituted a critical threshold. However, such benchmark values do vary regionally.

15.3   Models, systems and guides for predicting wildland fire behaviour

Formal recognition that a prediction of probable fire behaviour is essential to developing strategies and tactics for dealing with unwanted wildland fires extends back nearly 100 years. DuBois (1914) recognized that some form of a systematic method of assessing the inception, spread and difficulty of controlling a wildfire was needed. Roy Headley, a former Chief of Fire Control in the US Forest Service (1919–1942), remarked in his unpublished book manuscript titled *Re-thinking Forest Fire Control* (completed in 1943) that such a system would 'allow a man in charge of a going fire to be less of a gambler and more of a manager'.

Coert DuBois's vision was, of course, eventually realized. His foresight came to be refined in the years that followed by the likes of Harry T. Gisborne, Lloyd G. Hornby, Charles C. Buck and Jack S. Barrows, just to name a few of the other pioneering visionaries. The degree of sophistication that has come to be achieved is quite remarkable, considering that the level of support for wildland fire behaviour research pales in comparison to other disciplines in the physical sciences.

15.3.1   The formative years

The earliest forms of assessing fire behaviour potential consisted simply of the direct touching of the fuels in the field to gauge the level of dryness and, hence, the current fire danger level. This approach was obviously somewhat subjective. Relative humidity was soon seen as a far more objective measure of fire potential, and this still has value today (Hoffman and Osborne, 1923; Countryman, 1971).

However, it was not until the early 1930s that the first rigorous methods of fire danger assessment were developed based on fire weather observations and measurements of fuel moisture. In 1930, the first of many forest fire danger meters developed by Harry Gisborne for the US Northern Rocky Mountain Region became available for use. In addition, Jim Wright's *Forest-Fire Hazard Tables for Mixed Red and White Pine Forests, Eastern Ontario and Western*

Quebec Regions was published as Forest-Fire Hazard Paper No. 3 by Canada's Dominion Forest Service in 1933. A forest fire weather and fire danger rating research programme was initiated by the Forests Department of Western Australia the following year (Wallace and Gloe, 1938).

Fire danger rating systems were the primary form of gauging wildland fire behaviour potential from the mid-1930s up until the early 1960s. The outputs were typically qualitative or relative numerical values. An excellent accounting of the early developments in the field of fire danger rating can be found in Davis (1959), later updated by Brown and Davis (1973).

15.3.2   Forms of decision aids and field guides

Numerous mathematical models, computerized decision support systems and field guides have since come to be developed for predicting wildland fire behaviour. Some of the models are made available as easy-to-use decision support tools. For more complex models, the complexity is typically – but not always – buried out of sight in the form of prepared tables, or in graphical computational aids such as nomographs or nomograms, slide-rule devices (Figure 15.8) and computer programs.

Computer calculation of wildland fire behaviour for operational and research purposes began to take hold in the early-to-mid 1970s. The use of programmable pocket or hand-held calculators proved popular in the late 1970s and the first half of the 1980s. Computer-based calculation of wildland fire behaviour has steadily grown since that time, as a result of technological advances in software and hardware that have evolved, including the application of geographic information system (GIS)-based fire growth modelling. Among field-going firefighters, however, there still seems to be a thirst for simple, manual methods of computing fire behaviour characteristics.

Regardless of the particular type of fire behaviour tool (e.g. fireline notebook or smartphone), it is imperative that users understand and appreciate the underlying theory that supports the field application. Otherwise, the tool becomes the proverbial 'black box' (Figure 15.9).

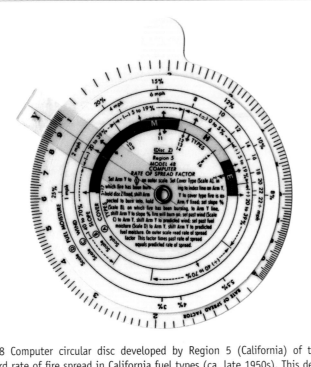

Figure 15.8 The Model 48 Computer circular disc developed by Region 5 (California) of the USDA Forest Service for computing heading or forward rate of fire spread in California fuel types (ca. late 1950s). This device, which continued to be used up until the late 1970s, is undoubtedly based on the analysis of individual fire report data carried out by Abell (1940).

Figure 15.9 Schematic diagram illustrating the notion of a 'black box', in which the inputs and outputs are known but knowledge of the internal workings are not (adapted from http://en.wikipedia.org/wiki/File:Blackbox.svg).

15.3.3 Types of fire behaviour models

The core component or engine of fire behaviour models are typically distinguished into two main categories: empirical; and physical or theoretical. Hybrids involving each type also effectively exist (e.g. semi- or quasi-physical or empirical models). Sullivan (2009a,b) has published the most recent review of models and modelling systems for predicting surface fire behaviour.

Both types of models have their place. A fire behaviour model derived experimentally from relatively local situations will almost invariability outperform a theoretical or physics-based model for the same given locale, but might have limited application in other areas with similar but distinctly different fuel complexes. Even when the empirical method of scientific investigation is employed in the development of a fire behaviour guide or model, knowledge garnered from physical theory and laboratory experiments is often combined with field evidence obtained from experimental fires, operational prescribed fires and/or wildfires.

Physics-based models are formulated on the basis of the chemistry and physics of combustion and heat transfer processes involved in a wildland fire (Morvan, 2011). They range in complexity from models for calculating rate of fire spread based solely on the radiation from the flaming front to three-dimensional models coupling fire and atmospheric processes. Examples of the later include FIRETEC, FIRESTAR, and the Wildland-urban interface Fire Dynamics Simulator (WFDS).

Physics-based models hold great promise in being able to advance our theoretical understanding of wildland fire dynamics. Considering their computational requirements (typically in excess of ten hours for a small scale simulation), it is not envisioned that physics-based fire behaviour models would replace their operational counterparts in the foreseeable future. For the time being, they should be viewed as research models and not for operational prediction of wildland fire behaviour or for other fire and fuel management applications (Alexander and Cruz, 2013a).

15.3.4 The two solitudes in wildland fire behaviour research

The term 'two solitudes' was originally used in reference to a perceived lack of communication and, moreover, to a lack of will for communication between Anglophone and Francophone people in Canada. This was popularized by Hugh MacLennan's 1945 novel entitled *Two Solitudes*. Van Wagner (1971) suggested that, perhaps, a similar situation exists with respect to the conduct of forest fire behaviour research – namely, a field or empirical approach versus a laboratory and/or theoretical approach.

Van Wagner compared the difficulties and advantages of indoor versus outdoor research on forest fire behaviour and concluded that theoretical and small-scale modelling of forest fires was so difficult as to render it unlikely that achieving the desired practical endpoint could be possible; the observation and study of both wildfires and outdoor experimental fires had the best chance for success. It was possible to maintain control over certain influencing variables, such as wind and dead fine fuel moisture. However, when it came to conducting laboratory test fires, there were scaling problems and limitations to the fireline intensities that would not be achievable. Outdoor experimental fires offered the distinct advantage of being able to observe real-world conditions but, even so, when it came to time for burning, researchers were still at the mercy of nature.

Fifteen years later, he acknowledged that there was a tendency for both approaches to converge to similar practical states. He also believed that, with a subject as complex as fire behaviour, pure scientific logic was simply not enough. His thinking was that fire behaviour researchers 'had better be something of artist as well as a scientist' (Van Wagner, 1985).

Some fire behaviour researchers have suggested that the data from empirical studies could eventually lead to improvements in both laboratory-based and theoretically-based models. This has already happened to a certain extent. We have seen for example, theoretical or physics-based modellers make good use of findings obtained from experimental fire field studies, such as the importance of head fire width on rate of spread. In turn, traditional empiricists are increasingly looking upon physics-based fire behaviour models as a possible means of solving particular fire dynamics questions (e.g. what is the relative effect of conifer foliar moisture content on crown fire rate of spread?).

In her investiture speech as the 27th Governor-General of Canada in September 2005, Michaëlle Jean specifically stated that, 'the time of "two solitudes" had finished' in Canada. While there are signs that this could be the case when comes to wildland fire behaviour research in the not too distant future, there is still a considerable gulf between the two factions. In fact, there is not even any agreement on the effect of slope steepness on the rate of fire spread among the empirical and quasi-physical models (Figure 15.10).

15.3.5 Current operational fire danger rating and fire behaviour modelling systems

As mentioned previously, the primary leaders in the field of fire danger rating and fire behaviour modelling have traditionally been Australia, Canada and the USA. The models and systems for rating fire danger and predicting fire behaviour developed in these three countries have come to be applied and/or adapted by many other countries. However, this is not to suggest that other countries have not made significant contributions in their own right. Russia, for example, has a number of learned scientists who have written much on the theoretical aspects of wildland fire behaviour. Some of this work has been translated into English (e.g. Grishin, 1997) or has appeared in English language journals. However, the work has yet to find its way into practical application.

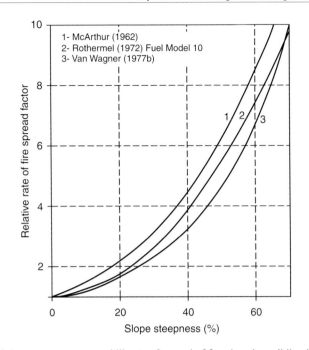

Figure 15.10   The effect of slope steepness on uphill rate of spread of free-burning wildland fires in the absence of wind according to Australian (McArthur, 1962), Canadian (Van Wagner, 1977b) and American (Rothermel, 1972) sources (after Alexander, 1998).

15.3.5.1   Australia

The systems of rating fire danger and models used for predicting fire behaviour in Australia are empirically based. The fire danger rating system originally developed for Australia by Alan McArthur in the late 1950s consists of separate indices for dry eucalypt forests with fine fuel quantities averaging 12.5 t/ha, and open grasslands consisting of fairly dense stands of improved pastures carrying a fuel load of 4.5 t/ha (Luke and McArthur, 1978). The indices integrate the combined effects of fine fuel moisture content (determined directly from air temperature and relative humidity) and wind speed. Either the Keetch-Byram Drought Index or the Mount Soil Dryness Index, coupled with the number of days since rain and the amount, are used to determine the proportion of fine fuel available for combustion in the forest fire danger index. An assessment of the degree of curing is required as input in the grassland fire danger index.

Both the forest and grassland fire danger index numbers are directly related to ignition probability and forward rate of fire spread, while suppression difficulty is based on a scale of 1 to 100. A value of 100 represented the 'near worst possible' fire weather

conditions that occurred during the Black Friday fires in the state of Victoria on January 13, 1939. While a value of 100 had been exceeded on several occasions in the past (e.g. Ash Wednesday fires of February 16, 1983), as a result of the Black Saturday fires of February 7, 2009 in Victoria, Australia (Gellie *et al.*, 2013), other fire danger rating categories (e.g. catastrophic) have been suggested for cases when a fire danger index of 100 or greater is reached.

Alan McArthur did devise the basic fire danger index so that it could be adjusted for slope steepness (Figure 15.10) and variable fuel quantities to give the head fire rate of spread (km/h) and flame height (m), as well as the average (as opposed to the maximum) spotting distance (km) in forests and the elliptical fire area (ha) versus elapsed time since ignition in grasslands. While McArthur initially presented his two fire danger rating systems in the form of tables, ever since 1962, these have appeared in circular or rectangular slide-rule form. It is noteworthy that McArthur's work represented the very first quantitative guides to predicting wildland fire behaviour anywhere in the world, making it a rather unique fire danger rating system. His models have now been encapsulated into the

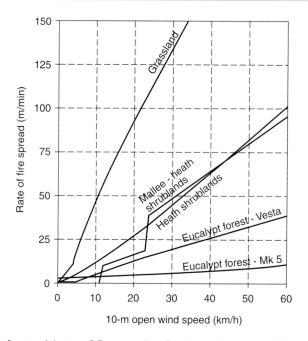

Figure 15.11 Comparison of potential rate of fire spread as function of wind speed for various Australian fuel types, assuming a slope steepness of 0% and fine dead fuel moisture of 7% (Patterned after Cruz and Gould, 2010).

PHOENIX RapidFire fire growth simulation model (Tolhurst *et al.*, 2008).

Australian bushfire researchers have elected to keep forecasts of fire danger rating separate from fuel type-specific predictions of fire behaviour. Fire danger ratings are intended to reflect the severity of fire weather conditions for a broad, standard fuel type. Fire behaviour predictions involve more exacting assessments of fuel structure, fuel moisture, topography and wind speed inputs (Figure 15.11).

The *Forest Fire Behaviour Tables for Western Australia*, or 'Red Book', as it is known locally, has gone through a number iterations since the 1960s. The last major update was authored by Sneeuwjagt and Peet (1998). This guide is also empirically based and is intended primarily for predicting head fire rate of spread for use in prescribed burning operations. Five native hardwood forest and two exotic pine plantation fuel types are currently recognized.

In the past 25 years or so, improvements have been made on Alan McArthur's original work on predicting fire spread in continuous grasslands and in dry eucalypt forests, including examination of the fuel age-load effect on fire spread rate (Gould *et al.*, 2007). Empirically based fire behaviour models have now been developed for woodlands and eucalypt forests with a grass understorey, spinifex/hummock grasslands, buttongrass moorlands, mallee-heath shrublands and heathlands. The Australia bushfire community has recently started the process of formulating a national fire danger rating system.

15.3.5.2 United States of America

Richard C. Rothermel (1972), with the support of a team of technicians and other scientists, developed a mathematical model for predicting surface fire rate of spread and line-fire intensity, based on a judicious mixture of physical theory and parameterized using laboratory test fires and a few documented wildfires in grasslands. Rothermel's model has had a profound effect on wildland fire behaviour training, planning and operations. For a historical perspective on its development and subsequent use, (see Steen, 2007).

The basis of operation involves a 'fuel model' which constitutes a description of key surface fuel bed variables required for the solution of his mathematical model, such as load, depth, particle size, etc.. The initial field application involved 13 stylized or static fuel models (Anderson, 1982) and has since been expanded to 53 fuel models (Scott and Burgan, 2005). In other words, the fuel descriptors are pre-specified. Both classifications considered a broad range of vegetation or fuel types (e.g. grasslands, shrublands, forest stands, and slash). There is also the option to develop custom fuel models.

The relative effect of slope steepness on fire spread rate (Figure 15.10) in Rothermel's model is considered to be dependent on the fuel bed's packing ratio (i.e. the proportion of fuel bed space occupied by fuel). The model assumes a quasi-steady state rate of spread from the moment of ignition (i.e. it does not consider the initial build-up from the time ignition to this state). The model prediction of rate of fire spread enables other fire behaviour characteristics to be computed (e.g. fireline intensity, flame length and the elliptical fire area and perimeter length).

Methods exist for adjusting the 6.1 m open wind speeds to a mid-flame height as required by the Rothermel (1972) model (Andrews, 2012). Auxiliary procedures also exist for estimating dead and live fuel moistures for use with this fire spread model.

Rothermel (1972) readily acknowledged that his surface fire model was not applicable to predicting the behaviour of crown fires, because the nature and mechanisms of heat transfer between the two spread regimes were quite different. Nineteen years later, however, he did deduce a simple, 'first approximation' statistical model for predicting crown fire rate of spread applicable to the US Northern Rocky Mountains, based on a correlation between the observed spread rates of crown fires and corresponding predictions of his surface fire rate of spread model (Rothermel, 1991a).

The Rothermel (1972) surface fire model forms the basis for the vast majority of guides and computerized decision support systems for predicting fire behaviour in use today in the United States and in other countries (e.g. South Africa), including the 1978 and 1988 versions of the US National Fire-Danger Rating System (NFDRS) (Deeming *et al.*, 1977; Burgan, 1988). The NFDRS is intended for fires that are spreading without spotting or crowning, although it was assumed that experience with the system would enable users to identify the critical levels of fire danger where such fire behaviour is likely to occur.

There are three main outputs from the NFDRS. The first two are the Spread Component (SC) and Energy Release Component (ERC), representing the fire spread rate (expressed in ft/min) and energy release, respectively. The SC and ERC are combined to give the Burning Index (BI), the resultant value of which is equal to ten times the flame length (ft) based on Byram's (1959a) relationship to fireline intensity (e.g. a BI of 30 represents a flame length of 3 ft).

A total of 20 fuel models are recognized, as well five percent slope classes. The moisture content of the live fuel and four size classes of dead fuels are calculated from weather observations and the previous day's value.

The present editions of the NFDRS evolved from the original 1972 version, which was also based in part on the Rothermel (1972) surface fire model. None were considered as fire behaviour prediction systems *per se*. As Rothermel (1983) remarked, it was not until Frank Albini 'let the genie out of the bottle with publication of his book of nomographs in 1976' that a field method of obtaining outputs from his surface fire model for the 13 original fuel models existed, without the use of a computer (Figure 15.12). Since then, a number of other manual methods have been devised.

Since the late 1990s, a number of existing and newly developed computerized decision-support systems have either separately implemented or linked to Rothermel's surface and crown rate of fire spread models with Van Wagner's crown fire transition and propagation criteria. These include both stand-scale and landscape-scale fire behaviour modelling systems:

- BehavePlus
- FARSITE
- NEXUS
- Fire and Fuels Extension to the Forest Vegetation Simulator (FFE-FVS)
- FlamMap
- Fuel Management Analyst (FMA) Plus

These fire modelling systems are used extensively for fire operations, planning, and research purposes. Andrews (2007, 2013) provides excellent overviews regarding the utility of all the systems listed above except the last one. The most recent additions to the family of US fire behaviour modelling systems that rely upon Rothermel's surface and crown rate of fire spread models include ArcFuels (Ager *et al.*, 2011) and the Wildland Fire Decision Support System (WFDSS) (Noonan-Wright *et al.*, 2011), a web-based system used for operational decision making on wildland fires.

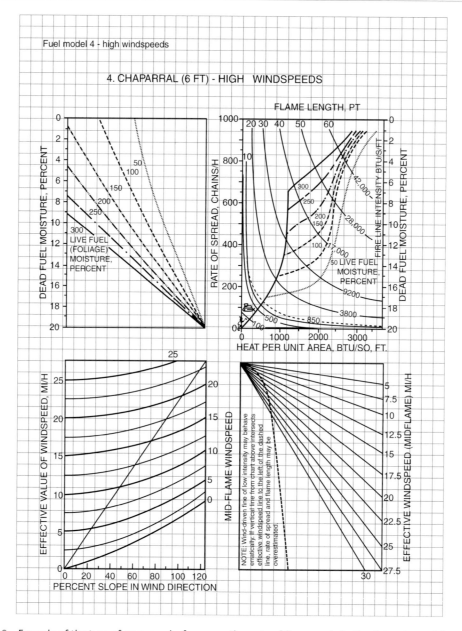

Figure 15.12 Example of the type of nomographs for computing rate of fire spread, fireline intensity and flame length, as developed by Frank A. Albini (from National Wildfire Coordinating Group, 1992). This particular one is for fuel model 4 – chaparral (Anderson, 1982) – at high wind speeds.

15.3.5.3   Canada

The Canadian Forest Fire Danger Rating System (CFFDRS), as developed by the Canadian Forest Service (CFS) Fire Danger Group, consists of two major subsystems or modules – the Canadian Forest Fire Weather Index (FWI) System and the Canadian

Forest Fire Behaviour Prediction (FBP) System (Taylor and Alexander, 2006). The distinction in the CFFDRS between weather-based fire danger rating and fire behaviour prediction is, thus, seamless.

The FWI System has been in use across Canada since 1970 (Van Wagner, 1987a). It provides numerical ratings of relative fire potential for standard fuel type (i.e. mature jack or lodgepole pine stand) on level terrain, based solely on past and present fire weather observations. The FWI System consists of six standard components, including three fuel moisture codes that follow daily changes in the moisture content of three classes of forest fuel with different drying rates (i.e. litter, duff, and deep organic matter), known as the Fine Fuel Moisture Code (FFMC), Duff Moisture Code (DMC), and Drought Code (DC), respectively. The other three components are relative numerical indexes of fire behaviour, known as the Initial Spread Index (ISI), Buildup Index (BUI) and Fire Weather Index (FWI), representing rate of fire spread, amount of fuel available for combustion and fireline intensity, respectively. In all cases, the higher the code or index value, the drier the fuel – or the more severe, the fire behaviour.

In contrast to the FWI System, the FBP System accounts for the variability in actual fire behaviour amongst fuel types, taking into account slope steepness, in absolute terms. A partial, interim edition was released in 1984, followed by the first complete edition

in 1992 (Forestry Canada Fire Danger Group, 1992), followed by the publication of theFBP System 'red book' field guide (Taylor *et al.*, 1997).

The forerunner to the FBP System was originally conceived as a series of regionally developed burning or fire behaviour indexes. The first one, entitled *A Burning Index for Spruce-Fir Logging Slash with Guidelines for Their Application*, authored by John Muraro of the CFS's Pacific Forest Research Centre in Victoria, British Columbia, was published in 1971.

The FBP System is largely empirical in nature, based in part on correlations between indexes of the FWI System and observed fire behaviour data generated from outdoor experimental burning, monitoring of operational prescribed fires and wildfire case studies. The system provides quantitative estimates of head fire spread rate (m/min), fuel consumption (kg/m^2), fireline intensity (kW/m) (Figure 15.13) and type of fire description. With the aid of an elliptical fire growth model, it gives estimates of fire area (ha), perimeter (m) and

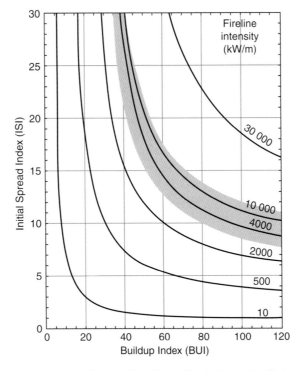

Figure 15.13 Head fire intensity class graph for Canadian Forest Fire Behavior Prediction System fuel type C-6 (conifer plantation) on level terrain as a function of the Initial Spread Index and Buildup Index components of the Canadian Forest Fire Weather Index System (slope steepness, 0%; foliar moisture content 120%; and canopy base height 7 m). The shaded area represents the region of intermittent crown fire activity.

perimeter growth rate (m/min), as well as flank fire and backfire behaviour characteristics.

The FBP System includes very simplistic functions for the acceleration in rate of fire spread for a point source ignition to a quasi-steady-state equilibrium. Emphasis is placed on the influences of fire weather (i.e. fuel moisture and wind) on potential fire behaviour for a given fuel type and the mechanical effects of slope steepness (Figure 15.10).

Fuel types in the FBP System are viewed as 'identifiable associations of fuel elements of distinctive species, form, size, arrangement and continuity that will exhibit characteristic fire behaviour under defined burning conditions'. Seventeen discrete fuel types are currently recognized. These include coniferous, deciduous and mixed wood forest stands, as well as coniferous logging slash and open grasslands. The number of fuel types is a reflection of currently available fire behaviour knowledge, rather than being a definitive fuel type classification scheme.

The FBP System forms the basis for a major component of PROMETHEUS – the Canadian wildland fire growth simulation model (Tymstra et al., 2010). The system constitutes the 'fire behaviour engine' for several computerized decision support and analysis systems used in wildland fire management decision making in Canada, including the Canadian Wildland Fire Information System (CWFIS), Spatial Fire Management System (sFMS), Probabilistic Fire Analysis System (PFAS), Burn-P3 (Probability, Prediction, and Planning) Model and the Canadian Fire Effects Model (CanFIRE).

15.3.5.4 New Zealand

Following a formal review of potential options, New Zealand adopted in 1980 the FWI System, with some minor adaptations, as the basis for a national system of rating fire danger based on fire weather severity, after having used a fire danger meter from the southeastern United States for a little more than 30 years. In the early 1990s, they elected also to adopt the overall framework and philosophy of the CFFDRS (Pearce and Clifford, 2008). This led to the initial development in 2008 of the New Zealand Fire Behaviour Prediction System, based in part on the Canadian FBP System combined with the experimental burning and wildfire documentation carried out locally over the past 15 years or so.

Both a field guide (Pearce et al., 2012) (dubbed the 'orange book' and patterned after the Canadian FBP System 'red book') and computer software have been produced, and appropriate modifications have been made to the New Zealand option in the PROMETHEUS fire growth model. The latest innovation by the Scion rural fire research group is the application of SmartPhone technology to fire behaviour prediction.

15.3.5.5 Other fire behaviour predictive models systems and guides

Several other models and systems are worthy of mention. For a complete listing of other quantitative methods to the operational prediction of wildland fire behaviour see Alexander and Cruz (2013b).

Crown fire initiation and rate of spread. The Crown Fire Initiation and Spread (CFIS) software system is a suite of empirically based models for predicting crown fire behaviour, founded largely on a reanalysis of the experimental fires carried out as part of developing the Canadian FBP System (Alexander et al., 2006). The main outputs of CFIS are as follows:

- Likelihood of crown fire occurrence or initiation.
- Type of crown fire (passive crown fire or active crown fire) and its associated rate of spread.
- Minimum spotting distance required to increase a crown fire's overall forward rate of spread.

The system depends on a minimal number of inputs: air temperature and relative humidity (to compute fine dead fuel moisture), 10 m open wind speed, an estimate of the available surface fuel load, the canopy base height and, finally, canopy bulk density. The crown fire initiation model can have as its input the above environmental variables, or can use several FWI System components.

CFIS is considered most applicable to free-burning fires that have reached a pseudo-steady state burning in live, boreal or boreal-like conifer forests found in western and northern North America (i.e. it is not directly applicable to insect-killed or otherwise 'dead' stands). Furthermore, the models underlying CFIS are not applicable to prescribed fire or wildfire situations that involve strong convection activity as a result of the ignition pattern. Level terrain is assumed, as the CFIS does not presently consider the mechanical effects of slope steepness on fire spread rate.

The crown fuel ignition model (CFIM) developed by Cruz et al. (2006a, 2006b) is a good example of the tendency to see the merging of empirically and physically based approaches. CFIM is a quasi-physical model developed to predict the onset of crowning based on fundamental heat transfer principles. A series of sub-models, that take into account surface fire characteristics along with canopy fuel properties, is used to predict the ignition temperature of canopy fuels above a spreading surface fire.

Pine Plantation Pyrometrics (PPPY) is a new modelling system developed to predict fire behaviour in Australian industrial pine plantations over the full range of burning conditions in relation to proposed changes in fuel complex structure from fuel treatments (Cruz *et al.*, 2008). The system comprises a series of sub-models, including CFIM and elements of CFIS, which describe surface fire characteristics and crown fire potential in relation to the surface and crown fuel structures, fuel moisture contents and wind speed.

Maximum spotting distance. Albini (1979) developed a physically based model for predicting the maximum spotting distance from single or group tree torching that covers the case of intermediate-range spotting of up to perhaps 1.5–3.0 km. He also developed similar models for burning piles of slash or 'jackpots' of heavy fuels and wind-aided surface fires in non-tree canopied fuel complexes, such as grass, shrubs and logging slash.

More recently, an alternative predictive system has been put forth for estimating the maximum spotting distance from active crown fires as a function of the firebrand particle diameter at alighting, based on three inputs, namely: canopy top height; free flame height (i.e. flame distance above the canopy top height); and the wind speed at the height of the canopy (Albini *et al.*, 2012). Determining whether a given ember or firebrand will actually cause a spot fire must still be assessed separately, based on the probability of ignition.

15.4 Limitations on the accuracy of model predictions of wildland fire behaviour

Fire behaviour models and related decision support systems should obviously be sensitive to those parameters known to affect fire behaviour, such as variations in live and dead fuel moistures, wind speed and slope steepness, among others. As Kessell *et al.* (1980) pointed out over 30 years ago, models 'are only as good as the data, understanding, assumptions, and mathematics that go into their construction' (Figure 15.14). It is easy to forget that fire behaviour models and fire behaviour modelling systems are mechanical schemes that, in all likelihood, cannot produce an exact representation of a natural phenomenon like wildland fire.

All fire behaviour prediction tools will produce results that do not agree exactly with observed fire behaviour. In some instances, the disagreement can be quite significant, with large and consistent under-prediction trends (Figure 15.15). *Overestimates of fire behaviour can be easily readjusted without serious consequences, but underestimates of fire behaviour can be disastrous* (Table 15.2). The most important source of error in any particular prediction of fire behaviour may be difficult to determine, regardless of the whether the model systematically over- or under-predicts.

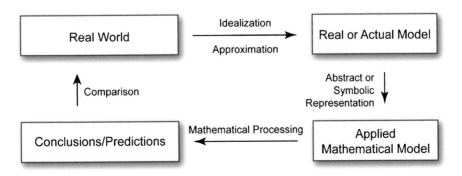

Figure 15.14 Flow chart illustrating the basic steps involved in computer simulation modelling (after Kessell *et al.*, 1980).

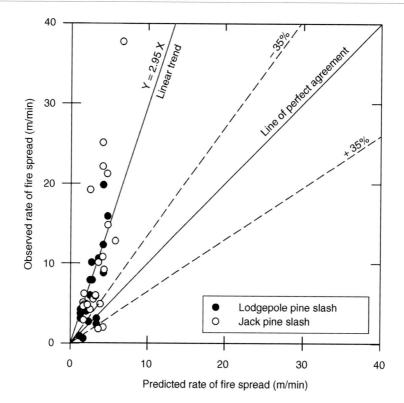

Figure 15.15 Observed rates of spread for experimental fires in cured, needle-bearing lodgepole pine logging slash in Alberta and cured, needle-bearing jack pine logging slash in north-eastern Ontario, versus predictions from Rothermel's (1972) surface fire rate of spread model for fuel model 12 – medium logging slash as described by Anderson (1982). The dashed lines around the line of perfect agreement indicate the ±35% error interval (after Alexander and Cruz, 2013b). Reproduced with permission of the Canadian Institute of Forestry/l'Institut forestier du Canada.

Table 15.2 The scope of quantitative wildland fire behaviour prediction (adapted from Rothermel, 1980).

Fire situation	Intended use	Resolution		Relative usefulness/ value	Ease of prediction accuracy	Impact of inaccurate prediction
		Timeframe	Area			
Possible	Training	Long-term	Not applicable	Moderate	Extremely to very easy	Minor or minimal
	Long-range planning (e.g. preparedness system development)	Yearly/ seasonal	State/province/ territory	Good	Easy to moderately easy	Significant
Potential	Short-term planning (e.g. daily fire assessment	Daily/weekly	Forest/district	Very good	Moderately difficult to difficult	Serious
Actual	Near real-time (e.g. automated dispatch, project fires, escaped fire situation analysis)	Minutes to hours	Stand or site-specific	Excellent	Very to extremely difficult	Critical

Source: Reproduced with permission of the Society of American Foresters.

In real-world applications of fire behaviour models or modelling systems, prediction accuracy is also dependent upon the skill and knowledge of the user, including the correct assessment of the fire environment and current fire status. The Campbell Prediction System (CPS) is an example of a practical way to use on-site assessments of a fire's history and its environment for determining a fire's probable behaviour in order to develop safe fire suppression strategies and tactics (Campbell, 2005). As the CPS slogan states, *Learn from the Past . . . Predict the Future*. The CPS is based on a combination of scientific research and practical knowledge, logic and its own terminology. A software application of CPS is currently under development. In the CPS, observations of current observed fire behaviour serve as the baseline for future predictions of fire behaviour.

15.4.1   Three sources of error

Albini (1976) considered that the main sources of error in fire behaviour predictions were due to the lack of model applicability, a model's internal inaccuracy and errors associated with input values. While much progress has been made in wildland fire behaviour modelling during the intervening years, these same basic principles still remain valid.

15.4.1.1   Model applicability

If one applies a model or a modelling system to a situation for which it was not intended to be used, the error associated with the prediction can quite possibly, in turn, be very large. Most rate of fire spread models have the following kinds of limitations and should not be expected to predict what they do not pretend to represent:

1   The fuel complex is assumed to be continuous, uniform and homogeneous. The more the actual fuel situation departs from this idealized assumption, the more likely the prediction will not match the observed fire behaviour.

2   Some models assume that the fuel bed is a single layer and is contiguous to the ground. In other words, there is no distinct gap between fuel layers (e.g. a forest stand with ground/surface fuels and crown or aerial fuels).

3   Fire spread by short-range spotting (flying embers or firebrands) is not accounted for by laboratory or theoretically based rate of fire spread models, or by fire behaviour modelling systems that rely upon such models. Some empirically based models or modelling systems such as the Canadian FBP System and Rothermel's crown fire rate of spread model indirectly include the influence of short- and intermediate-range spotting on rate of fire spread, although it is not expected that these models will be able to account for the effect of short-range spotting over the full spectrum of possible burning conditions.

4   The influence of fire whirls and similar extreme, fire-induced vortices on the rate of spread or growth of a free-burning wildland fire are not modelled. Site-specific predictions of vertical and horizontal vortex activity in wildland fires are not yet possible, although it is generally known where they are most likely to occur.

15.4.1.2   Model input data

Predictive models must be sensitive to those parameters known readily to affect fire behaviour, such as wind speed, dead fuel moisture and slope steepness, among others. If these input data are not accurate then, in turn, model output can be in error significantly. Given the non-linear nature of the various relationships found in free-burning wildland fire behaviour, model output may be highly sensitive to a particular parameter over one range of values and quite insensitive to that same parameter over a different value range. Rate of fire spread models, for example, are comprised of power or curvilinear functions of wind strength, slope angle and fuel moistures (Figure 15.16).

As a result of this non-linear nature in fire behaviour, it is often difficult to make a valid quantitative statement about the relationship between input data accuracy and output accuracy. As such, the model in question must be used to establish its requirements for data accuracy, considering the range of values of the variables used for input.

The greatest challenge faced by a fire model user in a predictive situation is the accurate estimation of representative input values. The old saying 'garbage in, garbage out' (or the newer one 'garbage in, gospel out') (GIGO) applies equally well to computer-based fire behaviour predictions.

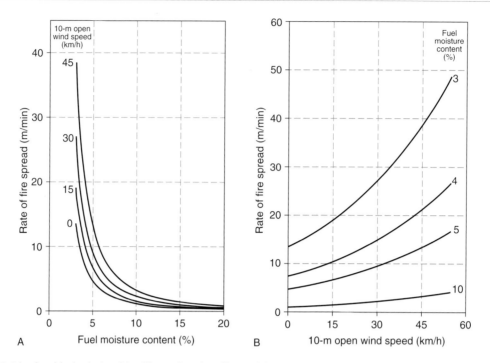

Figure 15.16 Graphical relationships illustrating the effects of fuel moisture content and wind speed on the heading or forward rate of fire spread in dry scherophyll eucalypt forest on level terrain (adapted from McArthur, 1967).

A deterministic approach for fire behaviour prediction assumes best estimates of input conditions to represent the fire environment. Nonetheless, the fuel complex is not uniform, continuous or homogeneous. Nor is the wind speed and direction constant, the slope steepness uniform or the moisture content of dead and live fuels the same from place to place, especially in complex, mountainous terrain. Yet, methods and guides to predicting fire behaviour generally assume such idealized burning conditions.

If standard techniques and procedures are strictly adhered to, the error component arising from uncertainty in input data is reduced to acceptable levels. If no direct measurements or observations are made, inaccurate forecasts are used or predictions are based solely on 'guesstimates', then the error associated with the input data will be the dominant error source.

15.4.1.3 Model relationships

Wildfires, being unpredictable as to their timing and location, and often occurring in remote locations, are seldom ideal subjects for conventional instrumentation and measurement. Furthermore, some aspects of wildfire behaviour are difficult to monitor and, thus,

to document precisely (e.g. maximum spot fire distances).

In the absence of a long-term, concerted effort to monitor and document wildfire behaviour systematically, data to test theoretical or empirical model formulae against actual wildfire behaviour must accumulate slowly from opportunistic high-quality observations. As a result, model testing or evaluation continues to be limited largely to laboratory experimental fires (or ones that are laboratory-like but conducted outdoors), operational prescribed fires or outdoor experimental fires. However, even general qualitative observations of wildland fire behaviour observations have value (Hester, 1952).

Albini (1976) considered that the causal relationships between the driving variables and fire behaviour models in most models must be viewed as weakly tested, semi-empirical in nature and subject to exception. Experimental fires carried out in plots of very uniform fuel complexes involving grasslands, shrublands and conifer forest stands has shown that there will be some degree of unexplained variation. Even with laboratory fires involving replicated or reproducible fuel beds, there can be as much as ±20% unexplained variation in observed spread rates. Given the inherent natural variation in wildland fire behaviour, Albini suggested that model builders considered models successful if the relationships predict fire behaviour within a factor of two or three over a range of two or three orders of magnitude.

15.4.1.4 Analyzing model performance

Model evaluation based on independent datasets is an important part of the model development process. Cruz and Alexander (2013) recently undertook a comprehensive analysis of 49 fire spread model evaluation datasets involving 1278 observations in seven broad fuel type groups. They found that the mean percentage error varied between 20% and 310%, and that only three out of 49 studies had average errors below 25%. They also found that that the concept of an exact prediction of rate of fire spread is an elusive one. Assuming that an exact prediction is one in which the error is less than $\pm 2.5\%$ of the observed rate of spread, only 3% of the predictions (i.e. 38 out of 1278) were found to be within this interval.

On the basis of their analysis, Cruz and Alexander (2013) suggested that from strictly a research perspective, an error threshold of $\pm 35\%$ would constitute an acceptable level of performance for model predictions of rate of fire spread (Figure 15.15). However, until some of the limitations of model applicability are resolved by further research, improvements in the accuracy of model relationships beyond the current level are unlikely to increase the overall accuracy of model predictions of fire behaviour.

It is worth noting that physics-based models have yet to be compared against a set of outdoor experimental fires involving any more than four observations – a statistically negligible number. Without the calibration and evaluation of these models against a dataset covering a broad range of fuel, weather and fire behaviour conditions, it is uncertain as to what the bounds or limits of their application are.

15.5   The wildland fire behaviour prediction process

Judging the quality of fire behaviour forecasts or predictions solely on the basis of the outcome can

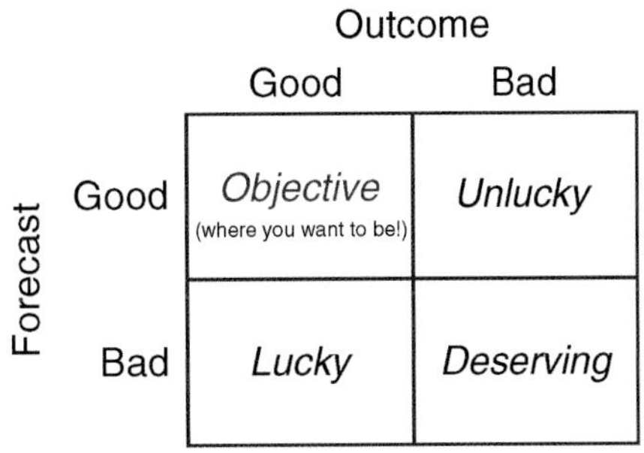

Figure 15.17   The two-by-two fire behaviour forecast or prediction matrix illustrates that even good forecasts can sometimes have unlucky outcomes (from Alexander and Thomas, 2004). The objective of wildland fire behaviour forecasting or prediction is to produce a good forecast and, in turn, a good outcome (i.e. the forecast or prediction closely matches what actually happened).

be hazardous. Just by chance, a good prediction or forecast can sometimes have a bad outcome, and a bad or poor prediction or forecast can sometimes result in a good outcome (Figure 15.17).

Predicting or forecasting wildland fire behaviour involves a good many uncertainties. Most people under stress use intuition and other heuristics to deal with uncertainty. Like good decision-making, good forecasting or prediction of wildland fire behaviour can be learned. However, as with all learning, one must be prepared to pay the price through concentration and hard work. For most of us, good decision-making skill does not come naturally.

In their book *Decision Traps: The Ten Barriers to Brilliant Decision-Making and How to Overcome Them*, Russo and Schoemaker (1989) examine the common pitfalls for decision-makers. They are also considered valid for fire behaviour specialists and others making predictions and forecasts of potential wildland fire behaviour. Excellent or effective decision-making involves making our way through ten common decision-making traps (see Table 15.3). These traps have been grouped into five distinct elements:

1   *The 'meta-decision' decision* (Decision Trap 1). Before we make a specific decision, we must make a decision about the decision itself (i.e. think about the general nature of the decision that is going to be made).
2   *Framing or structuring the question* (Decision Traps 2 and 3). In framing, good decision-makers think about the viewpoint from which they and

Table 15.3 The ten most dangerous decision traps to avoid in wildland fire behaviour forecasting or prediction (adapted from Russo and Schoemaker, Simon & Schuster, 1989).

Decision trap number, name and explanation	Some simple possible remedies and helpful hints
1. Plunging in: Beginning to gather information and reach conclusions without first taking a few minutes to think about the crux of the issue you are facing or to think through how you believe decisions like this one should be made.	Pause for a moment and ask yourself: • What is the crux or primary difficulty of this situation? • Can you daw on feedback from related predictions you have faced in the past?
2. Frame blindness: Setting out to solve the wrong problem because, with little thought, you have created a mental framework for your decision that causes you to overlook the best options or lose sight of important objectives. **3. Lack of frame control:** Failing to consciously define the problem in more ways than one, or being unduly influenced by the frames of others.	'Framing' means defining what must be decided and determined in a preliminary way, and what criteria would cause one to prefer one option over another, thus: • Try and remain open-minded. • Challenge yourself; know when to reframe. • Consider what is important and what is not.
4. Overconfidence in your judgment: Failing to collect key factual information because you are too sure of your assumptions and opinions. **5. Short-sighted shortcuts:** Relying inappropriately on 'rules of thumb', such as implicitly trusting the most readily available information or anchoring too much on convenient facts (e.g. it always begins to rain rather heavily beginning on about August 15).	• Appreciate the human propensity to be overconfident; know what you do not know. • Be careful with ideas and data that confirms, rather than challenges, your beliefs. • Be careful with your use of rules of thumb. • Never consider a rule of thumb as sacred. • Know enough about why a rule of thumb works to be able to predict when it will fail.
6. Shooting from the hip: Believing you can keep straight in your head all the information you have discovered, and therefore 'winging it' rather than following a systematic procedure when making the final choice. **7. Group failure:** Assuming that, with many smart people involved, good choices will follow automatically, and therefore failing to manage the group decision-making process.	• Fight against using intuition alone; people who do so are usually suffering from 'information overload'. • Fight against 'group think'; this occurs when group members striving for unanimity override the motivation to appraise alternative courses of action realistically.
8. Fooling yourself about feedback: Failing to interpret the evidence from past outcomes for what it really says, either because you are protecting your ego or because you are tricked by hindsight. **9. Not keeping track:** Assuming that experience will make its lessons available automatically, and therefore failing to keep systematic records to track the results of your decisions and failing to analyze these results in ways that reveal their key lessons.	• Acknowledge that mistakes are inevitable; the person who never makes a mistake is unlikely to accomplish much. • Periodically list your failures – if the list is short, be suspicious. • Appreciate the fact that experience provides only data, not knowledge. • Consider the value of wildfire case studies.
10. Failure to audit your decision process: Failing to create an organized approach to understanding your own decision-making, so that you remain constantly exposed to the preceding nine decision traps.	• Evaluate your use of time. • Practice 'reflection in action' (i.e. the ability to think and ponder something while you are still doing it).

others will look at the issue, and they decide which aspect they consider important and which they do not. Thus, good decision-makers inevitably simplify the world, while constantly being wary of their simplifications.

3 *Information gathering intelligence* (Decision Traps 4 and 5). This involves seeking out the knowable factors and reasonable estimates of 'unknowable facts' that will be needed to make the decision.

4 *Coming to conclusions* (Decision Traps 6 and 7). Sound framing and good intelligence do not necessarily guarantee a wise decision. A systematic approach forces one to examine many different aspects of the problem, and it often leads one to make better decisions than spending hours of unorganized thinking would.

5 *Learning or failing to learn from experience* (Decision Traps 8, 9 and 10). Everyone needs to establish a

system for learning from the results of past decisions. This usually means keeping track of what you expect would happen, systematically guarding against self-serving explanations and then making sure you review the lessons your feedback has produced and apply the results the next time a similar situation comes along. In an operational setting, taking a few minutes each day to sit down and look back is an important part of the process.

It is worth noting that the use of rules of thumb is a very common practice in wildland fire management circles (e.g. Mitchell, 1937). They can help you make quick decisions when problems are too complex and there is insufficient time for a truly analytical solution. Too often, however, people explicitly trust rules of thumb, especially those based on experience, as if they were certainties, and they fail to recognize when these rules should be used with wariness or not at all.

The 'crossover' concept is a common rule of thumb used in the boreal forest regions of Canada as a simple means of indicating the potential for extreme fire behaviour (Lawson and Armitage, 2008). It occurs during the course of the diurnal cycle on those days when the relative humidity in percent is less than or equal to the air temperature in degrees Celsius; this implies a dead fine fuel moisture content of 8–9% or less. For example, the conditions for crossover are met when the air temperature reaches 28°C and the relative humidity is 26%, but not when it is 20°C and 30%. Note that this rule of thumb does not work when the air temperature is expressed in degrees Fahrenheit, nor is it very differentiating in semi-arid or arid climates, when the conditions for crossover would be met virtually every day of the fire season. While the crossover concept definitely has value, the tendency has been to consider it the only time that extreme fire behaviour is possible (i.e. there is no consideration for wind strength or the level of dryness in medium and heavy fuels).

15.5.1 Assumptions

Predictions of wildland fire behaviour invariably involve assumptions. This includes the underlying assumptions associated with the model or guide used to compute fire behaviour characteristics (see, for example, Box 15.2). The assumptions upon which a model or guide is based, and the range in environmental conditions and fire behaviour upon which they are considered valid, need to be carefully defined and frequently rechecked. One should be wary of models and guides for which the limitations on applicability are not spelled out in practical terms.

Box 15.2 List of basic assumptions (adapted from Davis and Dieterich 1976) associated with a model for predicting rate of fire spread (ROS) in Arizona oak chaparral developed by Lindenmuth and Davis (1973).

- Litter or dead twigs 0.635 cm or less in diameter but reasonably dry (15% moisture content or less) for reliable predictions. Usually this will be reached with five days of seasonably warm weather, April through October, and with ten days during the rest of the year.
- Winter rainfall records representative of the area for which the ROS values are desired (past wildfires, going wildfires, or prescribed burns) must be available. These records make it possible to select one of the three climatological conditions that best describe past winter rainfall in order to estimate leaf moisture.
- The burning experiments on which the model was based were conducted in pure shrub live oak (*Quercus turbinella* Greene), and oak leaves were used for determining leaf moisture content.
- Slope was not included as a factor in the model, because the experimental burns were conducted on essentially flat terrain (obviously, the fire manager would not ignore slope as a factor in predicting ROS but, because the interactions of slope and wind with fuel are complex, they will have to use their best judgment in 'accounting' for slope as they attempt to interpret the meaning of the predicted ROS value).
- ROS is expressed in terms of feet per minute – forward spread with the wind, or radial spread with no wind.

Most fire behaviour models or guides also include the following general types of assumptions (after Taylor *et al.*, 1997):

1   The model or guide is applicable to certain fuel conditions.
2   The fuels are uniform and continuous.
3   The fuel moisture values used are representative of the fire site.
4   The topography is simple and homogeneous.
5   Wind speed is constant and unidirectional.
6   The fire is free-burning and unaffected by fire suppression activities.

In rolling to gently undulating terrain, it is often assumed that the surging and stalling of a fire as it climbs and descends slopes averages out (i.e. slope steepness is assumed to be equal to zero). While a whole host of assumptions specific to a particular prediction or forecast of fire behaviour may be invoked, they are seldom publicly stated except for the date and time interval that they are deemed valid for, although they are nonetheless documented.

Fire behaviour forecasts, much like forecasts of severe summer weather, assume that fire environment conditions are favourable for the development of, or that potential exists for, a fire or fires to exhibit a certain level of fire behaviour. Whether the situation materializes or not is dependent on a whole host of factors (e.g. correct assessment of current fire status, a proper appraisal of fuels, weather and topography). If the forecast wind speed used in the prediction does not occur, then one should not be surprised that there could be a considerable departure from the actual fire behaviour that did happen.

15.5.2   *General procedures*

The general procedures involved in predicting wildland fire behaviour, whether it be for a given day (i.e. a single burning period) or over the course of weeks or even months, involves three primary components:

1   a means of evaluating the fire environment inputs (e.g. fuel type, fuel moisture, wind speed, slope steepness);
2   a means of calculating the two basic descriptors of fire behaviour, namely rate of fire spread and fireline intensity;

3   methods of interpreting other characteristics of fire behaviour, including fire growth with time.

A schematic diagram of how information flows through one particular system of predicting wildland fire behaviour is shown in Figure 15.18.

15.5.2.1   Short-term projections

Rothermel (1983) described, in his seminal publication *How to Predict the Spread and Intensity of Forest and Range Fires*, the typical steps taken in the process of predicting wildland fire behaviour. These can be summarized as follows:

1   *Assess the past and present fire situations.* What has the fire done before you were able to observe it and what is it doing now? What were the attendant environmental conditions?
2   *Determine critical areas.* Identify those values at risk or hazardous fuels where predictions are needed most.
3   *What information is needed and when?* The prediction needs to be timely and must include the pertinent information relative to the situation (e.g. the fire is not controllable and will reach the community in less than five hours on a 2.5 km wide front).
4   *Estimate inputs.* Appraising the fuels, weather and topography requires skill. If you are not experienced with the situation and/or area you find yourself in, try to find an experienced and knowledgeable local person who can advise you.
5   *Calculate fire behaviour.* Use the model and/or guide that is most appropriate for the situation at hand.
6   *Interpret the outputs.* Apply an elliptical pattern for newly reported starts or pick several points to project in cases of larger, irregularly shaped fire perimeters. Give an indication of whether extreme fire behaviour is likely to occur.
7   *Further fire assessment.* In the latter stages of a fire (i.e. following initial containment), focus on the possibility of excursions or breaching of control lines due as dictated by weather changes.

As Rothermel (1983) points out, estimates of potential fire behaviour 'can be made in an amazingly short time when the procedures are understood well enough to recognize the simplifying assumptions that

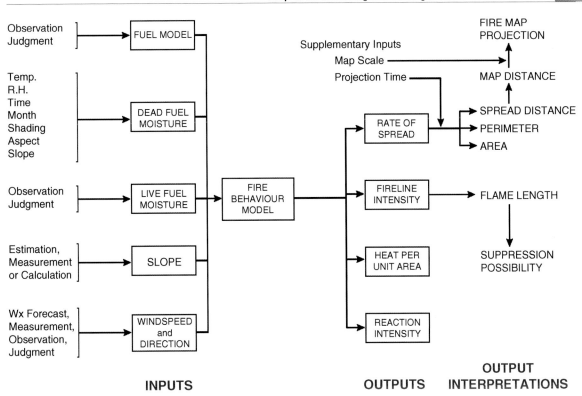

INPUTS **OUTPUTS** **OUTPUT INTERPRETATIONS**

Figure 15.18 Information flow associated with the Rothermel (1983) system of predicting wildland fire behaviour (from Rothermel, 1983).

can be made while still retaining the significant factors'. A very simple form of fire behaviour prediction, borrowed from meteorology and termed 'persistence forecasting', assumes that what the weather is doing now is what it will continue to do. This is used in predicting wildland fire behaviour, especially in cases when time is at a premium.

It is not uncommon to hear reports that the models or software for predicting wildland fire behaviour 'don't work'. Such was the case during the 2012 fire season in Colorado, for example (Behar, 2012). As a species, humans have short memories. The same sentiments were expressed during the 1988 fire season in the Greater Yellowstone Area. There will nevertheless be times on actual fire incidents where the fire behaviour model or modelling system does not match the fuel situation and even conventional adjustment procedures will fail to resolve the problem. In such cases, the best approach, as Beighley and Bishop (1990) found, is to: 'Begin with careful observation, continue with analysis of what is seen, and apply what is learned'.

In contrast to the case of the wind pushing the fire is one general direction, any fire which remains burning beyond its initial burning period is likely to be subjected to considerable changes in wind direction and other fire weather elements, fuel type differences and topographical variations, all of which collectively determine the direction, speed, growth, etc. on any particular day. Under these circumstances, a large portion of the fire perimeter can remain active for an extended time period, thus requiring multiple projections of fire growth.

Rothermel's (1983) guide remains a valuable reference on the principles of projecting the spread and growth from the uncontrolled edge of a large, irregularly shaped free-burning wildland fire, even though manual methods have largely been replaced by computerized decision support systems (Pearce, 2009). However, those individuals who learned the original 'hands on' techniques and, in turn, applied them operationally, probably have a far better appreciation of the limitations of fire modelling systems like FARSITE, PROMETHEUS and PHOENIX RapidFire, because they understand the theory underlying the 'black box'.

15.5.2.2   Long-range projections

Formal demands for long-range projections of fire growth began to take place in the mid-1980s as fire management in the US steadily moved into an era of managing fire 'with' time as opposed to 'against' time (Mohr *et al.*, 1987). Considerable experience was gained in developing techniques and methodologies for doing so as a result of the long-duration fires that occurred in the Greater Yellowstone Area during the 1988 fire season and in the years that followed. Long-range projections tend to be done by teams comprised of fire behaviourists, fire weather meteorologists and other technical specialists. Rothermel (1998) provides a good summary of his experiences and ideas on the processes involved in long-range projections of fire growth.

While there are some similarities between day-1 forecasting or prediction of fire potential, compared to multi-day or long-range projections, there are obviously major differences (i.e. bigger scale and longer time frame to be considered). The crux of the problem is that, given continuous fuels, long-range fire growth depends strongly on the ensuing weather. Also, because weather cannot be forecasted reliably beyond 5–7 days, the question then becomes: what can be done?

In long-range fire growth projections, a good deal of emphasis is placed on climatological information, as opposed to focusing on tomorrow's weather forecast. The fuel type mosaic and the characteristics of the individual types, along with the topographic features of the broader landscape, must also be considered, as well as the values at risk. Daily fire progression maps from recent wildfires and reconstructions of past wildfires (e.g. Tymstra, 2014) can also prove valuable as a means of vindicating long-range fire growth simulations.

Large fires are inevitably the result of major runs associated with strong wind events. The task then becomes just how many such events will occur from some point in the fire season until the end of the fire season? The task of ascertaining this kind of information was initially based on personal estimates after consulting climatological records or expertise. In recent years, a number of computerized decision support systems like PFAS, RERAP (Rare Event Risk Assessment Process) and the Fire Spread Probability (FSPro) component of WFDSS have aided or, in some cases, it appears as if they have replaced the general thought process. These types of decision support systems generally rely upon climatological data coupled with fire spread and growth models.

15.5.3   Means of dealing with uncertainty

In the continuing desire for 'better' predictions, it is easy to lose sight of the fact that model predictions are only a guide and that perfect, real-time prediction of wildland fire behaviour will probably never be achievable. There are limits to what can be accurately predicted.

Wildland fire behaviour predictions are inevitably fraught with uncertainty. To date, the operational use of fire behaviour models has largely followed a deterministic approach to a simulation or prediction, based on the notion of providing the best possible estimates of the input variables. One of the limitations of this approach is that it fails to provide any indication of the uncertainty surrounding the model predictions and, hence, fails to quantify the forecast predictability. Some individuals have endeavoured to compensate for this fact by considering 'most probable' and 'worst case' scenarios in both short-term and long-range projections.

The application of multiple or ensemble predictions, and data assimilation methods have the potential to reduce the uncertainty in fire behaviour forecasts. Cruz (2010) demonstrated the application of a simple Monte Carlo-based ensemble method to incorporate weather input uncertainty into the prediction of grassland rate of fire spread. In this case, the modelled outputs did not improve the general fit statistics but provided complimentary information, such as error bounds and probabilistic outcomes, which extended the range of questions that can be answered by fire behaviour models (Figure 15.19). Ensemble forecasting has also been extended to geographical information system (GIS)-based fire growth modelling used in long-range projections (e.g. PFAS, FSPro).

15.5.4   Gauging the potential for extreme fire behaviour

Extreme fire behaviour is considered to represent a level of fire activity that precludes any direct

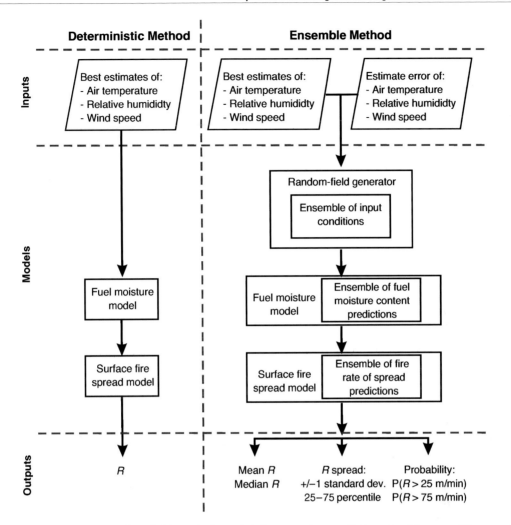

Figure 15.19 Information flow for a (left) purely deterministic approach to predicting rate of fire spread R, versus (right) an ensemble procedure to predict rate of fire spread R, including confidence intervals and probability density functions (after Cruz, 2010). Reproduced with permission of CSIRO Publishing.

suppression action by conventional means, because of the danger posed by the speed and intense heat of the moving flame front. In the broadest sense, the conditions favourable for the development of extreme fire behaviour have been known for a very long time and should come as no real surprise. They include:

- Continuous fine fuels in sufficient quantity and arrangement, both vertically and horizontally.
- A dry spell of sufficient length to reduce the moisture content of dead fuels to a uniformly low critical

level, coupled with high ambient air temperatures and low relative humidity.
- Strong prevailing winds or steep slopes.
- An unstable atmosphere.

Study of past conflagrations (i.e. large, fast-moving wildfires exhibiting many or all of the features associated with extreme fire behaviour) inevitably reveals that their development constitutes a simple case of all the 'planets aligning'. In other words, all of the conditions mentioned above occur simultaneously. Insect- and disease-killed stands can exacerbate matters (Kuljian and Varner, 2010; Page *et al.*, 2013b).

The transition between a slow-moving, low-intensity surface fire to a fast-moving fire with geometrically large flames can occur quite quickly. The borderline between a surface fire and a crown fire in a conifer forest can be such a delicate balance that a slight change in one of the key variables during peak burning during the day may be enough to trigger a quiet, gentle surface fire into an erratic, high-intensity crown fire (McArthur *et al.*, 1966).

While many of the guides and models to predicting fire behaviour are capable of identifying when to expect extreme fire behaviour, confirmation should be sought out, particularly where it relates to dead and live fuel moisture levels at a broader landscape scale. The threshold conditions for extreme fire behaviour have come to be established from local study of wildfire activity in relation to fire weather and fire danger ratings (e.g. Thorpe, 1999; Rothermel, 2000; Curcio, 2009).

The potential for extreme fire behaviour cannot be deduced from one single number or variable. Consider, for example, the Haines Index, which is an indicator for large fire growth potential based on the degree of stability and moisture (e.g. relative humidity) of the lower atmosphere. While correlations have been found between the Haines Index and rate of fire spread (Werth and Ochoa, 1993), it cannot be used directly to predict fire spread rate, given that wind speed is not a factor in its computation. A maximum index value of 6 is just as possible in the winter with continuous snow cover as it is in the middle of the summer fire season.

Extreme fire behaviour is often regarded as unusual and/or unexpected. As for it being unexpected, it is easy to forget that no two fires are alike. Furthermore, wildland fire behaviour readily exemplifies the principle of 'equifinality' (i.e. in open systems, a given end state can thus be reached by many different means). Nevertheless, this general feeling or attitude towards extreme fire behaviour may very well be symptomatic of an improper assessment of the fire environment, a lack of monitoring of the changing fire behaviour and fire weather conditions, and not applying existing guides to the assessment of fire behaviour. It could also be a reflection of experience level, including how the person in charge uses their intuition in predicting a fire's behaviour (Box 15.3).

> **Box 15.3   On the place of experienced judgment in the prediction of wildland fire behaviour (from Gisborne, 1948).**
>
> 'For what is experienced judgment except opinion based on knowledge required by experience? If you have fought forest fires in every different fuel type, under all possible different kinds of weather, and if you have remembered exactly what happened in each of these combinations of conditions, your experienced judgment is probably very good. But if you have not fought all sizes of fires in all kinds of fuel types under all kinds of weather, then your experience does not include knowledge of all the conditions.'

15.6   Specialized support in assessing wildland fire behaviour

The first specialists in wildland fire behaviour began to appear on the scene in the late 1950s. A designated fire behaviour specialist was part of the US Forest Service overhead teams associated with the large fire organization structure that would be put in place whenever a campaign or project fire occurred. The creation of this position came about as a result of a recommendation contained in a report prepared by a task force established by the Chief of the US Forest Service in 1957 *to recommend action to reduce the chances of men being killed by burning while fighting fire* following the deaths of 79 firefighters on 16 tragedy fires that had occurred on national forest lands in the previous 20 years.

The fire behaviour specialist was deemed responsible for:

1 identifying unusual fire conditions that may develop or exist because of weather, fuel, topography or a combination of all three of these factors;
2 forecasting probable fire behaviour and interpreting its effects on the control job and the safety of fireline personnel.

US Forest Service fire research staff stationed at the California Forest and Range Experiment Station were instrumental in pioneering the use of the position on going wildfires. This included the likes of Craig C. Chandler, Clive M. Countryman, Mark J. Schroeder and Carl C. Wilson.

The first use of the fire behaviour specialist position occurred in southern California in 1958 (Chandler and Countryman, 1959). Interestingly enough, the fire behaviour forecast form has not changed since that time. By at least the 1961 fire season, the term 'fire behaviour specialist' had been replaced by 'fire behaviour officer' or FBO, and this was, in turn, replaced by 'fire behaviour analyst' or FBAN in the early-to-mid 1980s, with widespread adoption of the Incident Command System for wildfire management. A fire weather meteorologist has also generally been assigned to any wildfire involving a fire behaviour specialist, FBO or FBAN in order to provide on-site fire weather forecasting support. Canada began to deploy FBOs on large fire overhead teams in the early 1980s. The formal use of FBANs has extended to parts of Europe (Molina *et al.*, 2010) and also into Australasia in recent years.

Beginning in the mid-1980s, a new form of fire behaviour intelligence gathering and analysis began to occur in the US Northern Rocky Mountain region and gradually expanded throughout the country as well as internationally – first in Canada and later in Australia. This took the form of fire behaviour service centres that were temporarily 'activated and staffed during critical multiple fire situations to improve fire safety and fire suppression decisions' (Bushey and Mutch, 1990).

A fire behaviour service centre would typically include an FBAN and a fire weather meteorologist, located within a permanent fire coordinating centre as opposed to a fire camp or incident command post, in order to deliver technical fire behaviour support for a large geographical area or region. This has proved highly beneficial, considering that most firefighter fatalities as a result of burn-overs and entrapments have commonly occurred in the initial attack and extended attack stages of fire suppression, in which cases there would not be a dedicated FBAN in place as there would be on a large fire incident.

One of the latest developments in the field of fire behaviour specialization has been the creation of 'predictive services' (PS) units at the US National Wildland Fire Coordination Center in Boise, Idaho, and the various regional centres (Winter and Wordell, 2009). PS units are organized very much like fire behaviour service centres and are responsible for producing both short- and long-term products that integrate fire weather and fire danger information for decision support purposes, as well as providing fuels and fire behaviour advisories and 'red flag' watches and warnings. All of the units are staffed with one or more fire weather meteorologists. Some units have a full-time wildland fire analyst, as well as a specialist in fire behaviour as needed.

Sadly, permanently appointed specialists in wildland fire behaviour have yet to reach the same level of recognition and status as other aspects of wildland fire management in regards to pre-suppression, suppression and fire use (e.g. fire prevention coordinator, air attack supervisor, prescribed fire manager). However, a few Canadian fire management agencies have seen fit to create a provincial or territorial fire behaviour specialist position within their organization since 1981.

In spite of these examples of specialized support, assessing fire behaviour potential in some form or another is, in fact, the personal responsibility of everyone involved with wildfires and prescribed fires, regardless of their position, training, education and experience – not just those who specialize in this activity. This includes the entry-level firefighter, up through the ranks to the incident commander involved in the containment of a large fire. In addition to field personnel, it also includes other fire management personnel who may not be directly involved in a fireline position (e.g. dispatchers, district or regional fire management officers).

15.7 Looking ahead

Predicting free-burning wildland fire behaviour is a complicated task. Long-distance spotting, development of fire whirls and simultaneous ignition over large areas as a result of high-density spotting and the ensuing mass ignition effect, are just some of the fire behaviour characteristics that are difficult to predict with any degree of accuracy as of yet.

The wildland fire management community suffers from a 'silver bullet' syndrome when it comes to the prediction of fire behaviour. Fire research is expected to come up with new technologies and/or a new model to solve theirs problem, resulting in the belief that they do not have to think about their problem any further. Unfortunately, there are no 'quick fixes' or 'silver bullets'. The ability to artfully predict or forecast fire behaviour can only be mastered through continual study, observation and practice.

Rothermel (1987) perceptively pointed out the paradox faced by the developers of wildland fire behaviour models in addressing the needs of various fire behaviour practitioners:

- The models and systems are not accurate enough.
- The models and systems are too complicated.

The attempted resolution of either one of these problems worsens the other. Presumably, what is required at the field level are rough but reliable decision aids for predicting wildland fire behaviour. However, regardless of the type of model or system, fire behaviour researchers have a professional responsibility to ensure that their products have been adequately evaluated and field tested before they are applied to specific fire and fuel management tasks (Wade, 2011).

Wildland fire research has largely concentrated on the physical aspects, or the 'science', of wildland fire behaviour prediction for nearly 100 years now. Very little research or even thought has been given to the 'art' of fire behaviour prediction (Weick, 2002), which is currently only taught at the 'School of Hard Knocks' (Newbould, 2005). Fire operations and fire behaviour training would undoubtedly benefit from research into the art of predicting fire behaviour from the standpoint of the humanities and social sciences, in the same manner as it has benefited from the physical sciences. This represents an especially critical need, in light of the manner in which computer technology has come to dominate the field of wildland fire behaviour prediction so heavily.

Further reading

Alexander, M.E. (2008). *Proposed revision of fire danger class criteria for forest and rural areas of New Zealand.* 2nd edition. National Rural Fire Authority, Wellington, in association with the Scion Rural Fire Research Group-Christchurch, New Zealand.

Beck, J.A. (1995). Equations for the forest fire behaviour tables for Western Australia. *CALMScience* **1**, 325–348.

Beverly, J.L., Wotton, B.M. (2007). Modelling the probability of sustained flaming: Predictive value of Fire Weather Index components compared with observations of site weather and fuel moisture conditions. *International Journal of Wildland Fire* **16**, 161–173.

Chandler, C., Cheney, P., Thomas, P., Trabaud, L., Williams, D. (1983). *Fire in forestry. Volume I: Forest fire behavior and effects.* John Wiley & Sons, New York.

Fujioka, F.M., Gill, A.M., Viegas, D.X., Wotton, B.M. (2009). Fire danger and fire behavior modeling systems in Australia, Europe, and North America. In: Bytnerowicz, A., Arbaugh, M.J., Riebau, A.R., Andersen, C. (eds.), *Wildland fires and air pollution*, pp. 471–497. Elsevier B. V., Oxford. Developments in Environmental Science Volume 8.

Hardy, C.C., Hardy, C.E. (2007). Fire danger rating in the United States of America: An evolution since 1916. *International Journal of Wildland Fire* **16**, 217–231.

Lee, B.S., Alexander, M.E., Hawkes, B.C., Lynham, T.J., Stocks, B.J., Englefield, P. (2002). Information systems in support of wildland fire management decision making in Canada. *Computers and Electronics in Agriculture* **37**, 185–198.

Pastor, E., Zarate, L., Planas, E., Arnaldos, J. (2003). Mathematical models and calculation systems for the study of wildland fire behavior. *Progress in Energy and Combustion Science* **29**, 139–153.

Planas, E., Pastor, E. (2013). Wildfire behaviour and danger ratings. In: Blecher, C.M. (ed.), *Fire phenomena and the earth system: An interdisciplinary guide to fire science*, pp. 53–76, John Wiley & Sons, Ltd., Chichester.

Reifsnyder, W.E., Albers, B. (1994) *Systems for evaluating and predicting the effects of weather and climate on wildland fires.* 2nd edition. World Meteorological Organization, Geneva, Switzerland. Special Environmental Report 11. 34 p.

Schlobohm, P., Brain, J. (comps.) (2002). *Gaining an understanding of the National Fire Danger Rating System.* National Interagency Fire Center, National Fire Equipment System, Boise, Idaho. Publication NFES 2665. 71 p.

Sharples, J.J., McRae, R.H.D., Weber, R.O., Gill, A.M. (2009). A simple index for assessing fire danger rating. *Environmental Modelling & Software* **24**, 764–774.

Sullivan, A.L., McCaw, W.L., Cruz, M.G., Matthews, S., Ellis, P.F. (2012). Fuel, fire weather and fire behaviour. In: Bradstock, R.A., Gill, A.M., Williams, R.J. (eds.), *Flammable Australia: Fire regimes, biodiversity and ecosystems in a changing world*, pp. 51–77. CSIRO Publishing, Melbourne, Victoria, Australia.

Thomas, D. (2011). Teaching mindfulness to wildland firefighters. *Fire Management Today* **68**(2), 38–41.

Thomas, D.A., Miller, C.L., Fox, R.L. (2013). HRO-mindfulness: a natural trait of highly experienced forest fire managers. *Society and Natural Resources*, in press.

Tversky, A., Kahneman, D. (1974). Judgment under uncertainty: Heuristics and biases. *Science* **185**, 1124–1131.

Wotton, B.M., Alexander, M.E., Taylor, S.W. (2009). *Updates and revisions to the 1992 Canadian Forest Fire Behavior Prediction System.* Natural Resources Canada, Canadian Forest Service, Great Lakes Forestry Service, Sault Ste. Marie, Ontario. Information Report GLC-X-10E. 45 p.

Chapter 16

Fire management applications of wildland fire behaviour knowledge

16.1   Introduction

Wildland fire has been referred to as a 'two-edged sword' and as a 'good servant but a bad master' in recognition of the fact that, while there are many beneficial aspects, some wildland fires can turn out to be lethal and destructive (see also Part Three). In a guest editorial in the 1974 Summer issue of *Western Wildlands*, Jack S. Barrows expressed the opinion that 'there is one overriding challenge to fire management: that of maintaining full respect for the power of fire and the effects of this power on both wildland environments and the people who live and work in these environments'.

In this regard, the 1910 fires in the US Northern Rocky Mountains have come to be one of those defining moments in the history of wildland fires globally (Pyne, 2008). At least 87 firefighters and civilians met their deaths and several communities were completely destroyed, as some 1600 wildfires burned over 1.2 million hectares of forested land, most of it occurring on August 20–21.

The 1910 'big blow-up' forest fire event would have an indelible effect on those responsible for developing strategies to co-exist with wildland fires in the future.

Both fire operations personnel and fire research staff alike considered it to be a moral obligation never to let it happen again. Harry T. Gisborne exemplified the attitude well in stating his view of wildland fire research's *raison d'être* (i.e. reason or justification for existence), even though at the time of the 1910 fires he was but a 16-year-old boy (from Gisborne, 1942):

'We are not doing research for research's sake. We have a definite, decidedly practical goal, and it is still the basic, over-all goal ... stated in 1910: The first measure necessary for the successful practice of forestry is protection from forest fires. Fire research is intended to serve as directly as possible the fire-control men who must first be successful before any of the others arts or artists of forestry can function with safety.'

While there were a few initiatives following the 1910 fires, fire behaviour did not catch on as a field of forestry research until the 1930s (Figure 16.1). Interesting, while the importance of fire behaviour information in the field of forest fire protection and the associated need for research was well recognized, the term 'fire behaviour' was not included in the first

Fire on Earth: An Introduction, First Edition. Andrew C. Scott, David M.J.S. Bowman, William J. Bond, Stephen J. Pyne and Martin E. Alexander.
© 2014 John Wiley & Sons, Ltd. Published 2014 by John Wiley & Sons, Ltd.

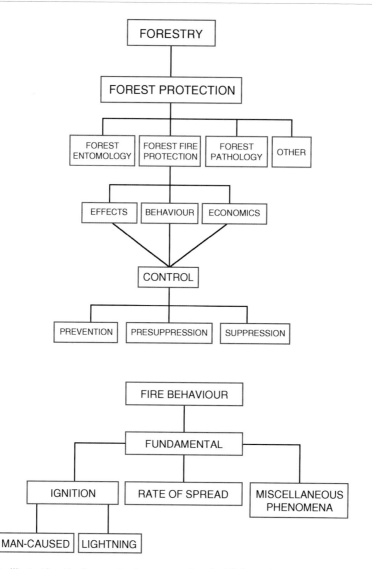

Figure 16.1 Flow charts illustrating the four main phases associated with forest fire protection (top) and four groups of studies associated with scientific study of forest fire behaviour research (bottom) as envisioned by Curry (1938). Miscellaneous phenomena include crowning and spotting, for example. Because some studies in fire behaviour apply to the whole field rather than to any particular phase, the fundamental section includes, for example, studies of the chemical and physical properties of forest fuels, classification of fuels, fuel moisture and weather relationships (after Curry, 1938).

formally published glossary of fire control terms (USDA Forest Service, 1930).

It is generally considered that nearly every aspect of wildland fire control and use requires some knowledge of fire behaviour (Figure 16.2). Operationally, this would include the following areas:

- Prevention planning (e.g. informing the public of impending fire danger, regulating access and risk

associated with public and industrial use of forest and rural areas).
- Preparedness planning (i.e. level of readiness and pre-positioning of suppression resources).
- Detection planning (e.g. lookout staffing and aircraft scheduling and routing).
- Initial attack dispatching (e.g. prioritizing of targets for air tankers and ground crews).

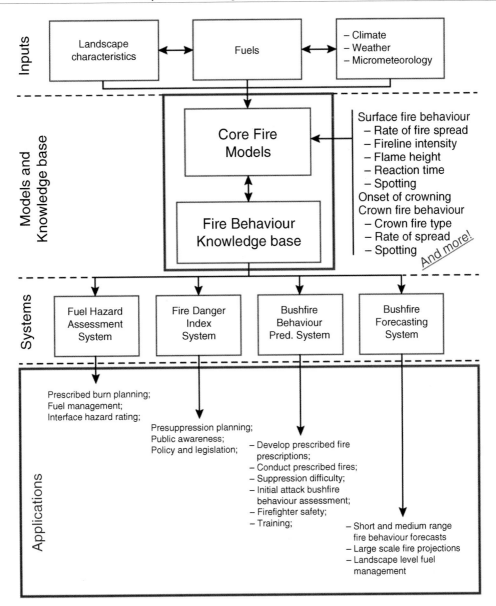

Figure 16.2 Flow chart illustrating the linkages between fire environment inputs, fire behaviour models and systems, and applications to fire and fuel management, as envisioned by Cruz and Gould (2009) in a national fire behaviour predictions system for Australia (from Cruz and Gould, 2009). Reproduced with permission of Taylor & Francis.

- Formulating suppression plans on active wildfires (including short-range predictions of fire behaviour and growth).
- Evaluating fire behaviour potential and guidelines for safe work practices for firefighters.
- Escaped fire situation analysis (including long-range projections of fire growth and behaviour).

- Prescribed fire planning and execution, including smoke management, fire and fuel management modelling and planning.
- Appraisal of impacts and effects of wildfires and prescribed fires.
- Wildland fire training.
- Wildfire fire accident investigations and reviews.
- Wildfire causal investigations.

However, since the early 1970s, the complexity of the issues requiring attention by wildland fire behaviour research has dramatically increased. For example, in Canada there is, paradoxically, a greater need for improved fire behaviour knowledge in order to manage wildfire more intelligently (e.g. no active fire suppression) in areas of low commercial timber value, or to apply prescribed fire safely in a critical habitat, than to dispatch an air tanker to a newly reported fire in an area with high values at risk. Research into the potential impacts of climatic change on fire activity and in turn fire management has steadily increased in recent years (e.g. Nitschke and Innes, 2008; Sullivan, 2010).

In this chapter, we touch on recent developments in the application of fire behaviour research and the practical knowledge gained from working on or with wildfires and prescribed fires. At the same time, though, some a historical context is provided, as well as how fire behaviour knowledge contributes to the on-the-ground application in wildland fire management.

16.2 Wildfire suppression

Fire suppression involves all the activities concerned with controlling and finally extinguishing a wildfire following its ignition (Box 16.1).

The fire triangle is commonly used to illustrate the basic principles of fire suppression (Figure 14.2a). To stop a free-burning fire, it is necessary either to:

1 remove the fuels ahead of the spreading combustion zone;
2 reduce the temperature of the burning fuels; or
3 exclude oxygen from reaching the combustion zone by smothering.

In practical terms, this means creating a physical barrier in front of the fire by removing the fuels or cooling/smothering the flames with water, covering them with mineral soil, suppressants (e.g. foam) or chemical fire retardants by various means from either the ground or the air (Alexander, 2000). If the fuel type(s) involves deep organic matter, smouldering combustion may still occur following initial containment, depending on the fuel moisture content, in which case extensive mop-up and patrol will be required before full extinguishment is achieved.

Box 16.1 Wildland fire suppression related terminology (adapted from Merrill and Alexander, 1987). Page et al. (2013a) have recently suggested a more holistic concept of resistance to control that includes fire behaviour characteristics, fire suppression operations, and firefighter safety considerations.

Difficulty of Control: the amount of effort required to contain and mop-up a fire based on its fire behaviour and persistence, as determined by the fire environment.

Resistance to Control: the relative ease of establishing and holding a fireguard and/or securing a control line, as determined by the difficulty of control and resistance to fireline or fireguard construction.

Resistance to Fireline or Fireguard Construction: the relative difficulty of constructing fireguards, as determined by fuel type characteristics (e.g., forest floor depth), effects of topography on access (e.g. slope steepness) and mineral soil type.

The forest fire behaviour research effort in the 1930s focused on estimating the rate of perimeter increase in relation to the production rates of various firefighting resources, such as ground crews, bulldozers and tractor-plough units for rapid initial attack in order to achieve containment as quickly as possible. Coupled with this emphasis was research on the early detection of initiating fires. Elapsed time since ignition, as a factor influencing fire behaviour, does not appear to have been fully appreciated in wildfire fire training, planning or operations (Box 14.2). As Alan McArthur (1968) states in his essay *The Effect of Time on Fire Behaviour and Fire Suppression Problems* (a landmark document in the field wildland fire management that should be required reading for all wildland firefighters), it is 'during the first 30 minutes

Based on an analysis of 67 fatal fires involving 222 wildland firefighter deaths in the US over a 61-year period (1926-1976), Carl Wilson (1977) identified some common features connecting these incidents. The five common denominators of fire behaviour associated with these fatal fires were:

1　Most of the incidents occurred on relatively small fires or isolated sectors of larger fires.
2　Most of the fires were innocent in appearance prior to the 'flare-ups' or 'blow-ups'. In some cases, the fatalities occurred in the mop-up stage.
3　Flare-ups occurred in deceptively light fuels.
4　Fires ran uphill in chimneys, gullies, or on steep slopes.
5　Suppression tools, such as helicopters or air tankers, can adversely modify fire behaviour (helicopter and air tanker vortices have been known to cause flare-ups).

2011) has tended to be an under-appreciated aspect of gauging fire behaviour potential. Research on fireline production rates has endeavoured to keep pace with changes in crew types and machinery over the years. Still, much of the existing information is in need of updating (Broyles, 2011). Fireline intensity and, in turn, flame length is a major determinant of the limit of effectiveness or minimum requirement for the different types of firefighting resources relative to the difficulty of control (Table 16.1).

The probability of containment will depend on sending enough resources of the right type relative to the expected fire behaviour at the time of initial attack (Figure 16.3). In order to achieve successful fire containment, the fireline production rate of the appropriate suppression resource(s) must exceed the fire's rate of perimeter increase over some specified period of time after their arrival on scene (Plucinski, 2012). Generally, there is a final target fire size in mind.

Flame length has been related directly to various measures of fire suppression over the years. Byram (1959a) recommended that, in the absence of severe spotting, the minimum width of a constructed fireline or fireguard should be 1.5 times the expected flame length. The amount of water required to extinguish wildfires has also recently been related to fireline intensity and also to flame length.

Forecasts or predictions of fire behaviour and fire growth are also needed in order to develop plans for undertaking suppression activities on large wildfires (Teie, 2005). This will take into account the method of fireline construction or attack (i.e. direct, parallel or indirect), based on safety and tactical considerations as

or so of a fire's life history, suppression forces have their greatest chance of success purely because the fire is still accelerating and has not reached its maximum rate of spread'.

The inclusion of assessments of resistance to fireline or fireguard construction (e.g. Valachovic *et al.*,

Table 16.1　The six distinct fireline intensity classes given in the Canadian Forest Fire Behavior Prediction System field guide (Taylor et al., 1997), their associated flame lengths for **surface fires**, and fire suppression related interpretations.

Fireline intensity class	Fireline intensity (kW/m)	Flame length (m)[1]	Generalized fire potential and implications for fire suppression[2]
1	<10	<0.2	New ignitions and surface fire spread unlikely. No control problems.[3]
2	10–500	0.2–1.3	Surface fires with flame heights of ≈1.0 m. Limit of control with hand tools.
3	500–2000	1.3–2.6	Vigorous surface fires with short-range spotting. Water under pressure needed.
4	2000–4000	2.6–3.5	Intermittent crowning in forests. Heavy equipment required for fireline operations.
5	4000–10 000	3.5–5.4	Onset of 'blow-ups'. Fires escape initial attack leading to major campaign incidents.
6	>10 000	>5.4	Conflagrations. A change in fuels and/or weather is needed to affect containment of the head fire.

[1]According to Byram (1959a).
[2]See Alexander and Cole (1995).
[3]Ongoing fires could still require mop-up if ground fuels are dry and plentiful.

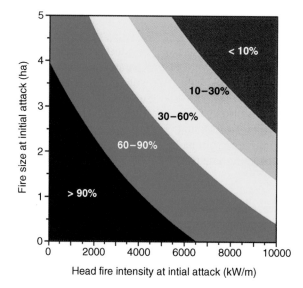

Figure 16.3 Probability of containment by a medium-sized initial attack crew (5–7 members) in the boreal spruce type of western Canada with support of water bucketing from a helicopter. This chart is based on interviews with initial attack crew leaders (from Hirsch *et al.*, 2000). Reproduced with permission of the Canadian Forest Service Publications (Natural Resources Canada).

dictated by fuel continuity and probable fire behaviour. Fire behaviour considerations have been incorporated into guides for suppression firing using ground and aerial ignition (Cooper, 1969; Burrows, 1986).

16.3 Wildland firefighter safety

Wildland fire suppression can be an especially dangerous activity (Figure 16.4). Wildland firefighter fatalities have, unfortunately, occurred from time to time, due to entrapments or burn-overs, falling trees and snags, rolling rocks, vehicle and aircraft crashes, and even heart attacks. They have occurred on both wildfires and prescribed fires.

At least 443 wildland firefighters have died as a result of being overrun or entrapped by wildfires since 1910 in the United States, up to and including the 2011 fire season (Figure 16.5). Some of these tragedies have been chronicled in book form, in addition to the official accident investigation reports. The most notable of these is Norman Maclean's (1992) epic accounting of the 1949 Mann Gulch Fire in north-western Montana, in which 12 smoke-jumpers and one fire guard perished. His son John

Maclean has since written four similar-type books, the most recent being an account of the 2006 Esperanza Fire in southern California, where a five-person engine crew died as a result of being overrun by the fire (Maclean, 2013). This incident occurred virtually on the steps of the Riverside Forest Fire Laboratory.

Current guidelines and advice on wildland firefighter safety can be traced to fire behaviour-related fatalities (e.g. Cook, 1995; Braun *et al.*, 2001; Cheney *et al.*, 2001). These include, but are not limited to, the *10 Standard Fire Orders, 18 Watch Out Situations, LCES: Lookouts – Communications – Escape routes – Safety zones* (or *LACES*, with the 'A' denoting anchor point or awareness), and the *Common Denominators of Fire Behavior on Tragedy Fires* (see Box 16.2). Aggressive initial attack on wildfires can, in fact, serve as a means of increasing wildland firefighter safety.

It is worth noting that, in discussing the findings and observation from their study of the 1994 South Canyon Fire, Butler *et al.* (1998) pointed out that no new breakthroughs in the understanding of wildland fire behaviour had occurred, but rather that *their findings support the continued need for increased understanding of the relations between the fire environment and fire behaviour* (Box 16.3). 'We can also conclude that fire managers must continue to monitor and assess both present fire behaviour and

Figure 16.4 The photo memorial of the 'faces' of fallen wildland firefighters located in the lobby of the national fire control centre of the Ministry of Environment and Forestry in Ankara, Turkey (translation of the inscription at the bottom: 'We remember with gratitude those who have lost their lives fighting forest fires') (photo Ertuğrul Bilgili, Faculty of Forestry, Karadeniz Technical University, Trabzon, Turkey).

potential future fire behaviour given the possible range of environmental factors' (Butler *et al.*, 1998).

When fire behaviour becomes threatening, wildland firefighters disengage from the fire and travel along escape routes to reach safety zones to avoid being entrapped or burnt over (Figure 16.6). In spite of the fact that the concept of escape routes and safety zones has been a formally recognized element of wildland firefighter safety since 1957, when the *10 Standard Fire Orders* first appeared, until the 1994 South Canyon Fire tragedy in Colorado, involving the deaths of 14 firefighters, there was a surprisingly paucity either of quantitative information available on firefighter travel rates using escape routes, or of scientific data describing what made for an adequate safety zone.

An extensive study of wildland firefighter travel rates carried out in west-central Alberta, Canada involving

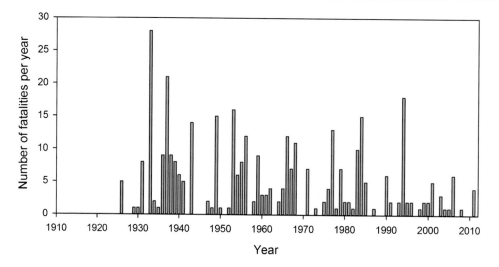

Figure 16.5 The number of reported wildland firefighter fatalities associated with burn-overs and entrapments, including heavy equipment and vehicle entrapments, by year in the United States for the years 1911–2011. A total of 78 firefighters also died as a result of the same causes during the 1910 fire season. The reporting from 1911 to the mid-1920s is undoubtedly incomplete (source: http://www.nifc.gov/safety/safety_HistFatality_report.html).

Type I, II and III wildland firefighters (Alexander *et al.*, 2005). This involved simulated runs over 250 m courses in six different fuel types/slope situations involving both 'natural' or unimproved and 'improved' routes (i.e. cleared trail and flagged), made with and without a pack (6.8 kg) and tool (fire shovel). On the basis of 360 timed runs, the following conclusions were reached:

- The fastest overall times occurred in the improved no pack/tool courses, followed by the improved pack/tool, natural no pack/tool, and then the natural pack/tool.

- The two open fuel types (i.e. grass and slash) were the easiest to travel across and the dense spruce stand was the hardest, while the mature pine stand was of intermediate difficulty.
- There was less variation in travel rates among individual crew members on improved routes.
- Trying to move upslope (a highly questionable action unless a safety zone is close by) dramatically decreases the pace a firefighter is able to attain.
- Carrying a pack and tool slows down a firefighter's rate of travel, regardless of whether they are on an

---

Box 16.3   Discussion points emanating from the fire behaviour analysis undertaken by Butler *et al.* (1998) into the 1994 South Canyon Fire, Colorado. Fourteen firefighters died as a result of being overrun by the fire.

1 Topography can dramatically influence local wind patterns.
2 Vegetation and topography can reduce a firefighter's ability to see a fire or other influencing factors.
3 Current and past fire behaviour often does not indicate the potential fire behaviour that could occur.
4 The longer a fire burns and the larger it gets, the greater the likelihood of high-intensity fire behaviour at some location around the perimeter.
5 The transition from a slow-spreading, low-intensity fire to a fast-moving, high-intensity fire often occurs rapidly. This seems to surprise firefighters most often in live fuels.
6 Escape route effectiveness should be considered in relation to potential maximum intensity fire behaviour, rather than past or present fire behaviour.
7 The under-burnt Gambel oak did not contribute to the blow-up. It was significant in that it did not provide a safety zone.
8 Smoke can significantly reduce a firefighter's abilities to sense changes in fire behaviour.

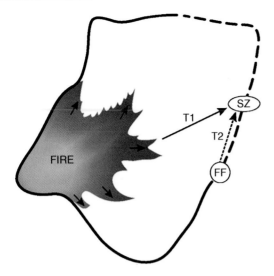

Figure 16.6 An illustration of the 'margin of safety' concept involved in wildland fire suppression (adapted from Beighely, 1995). The safety margin is a measure of the ability of wildland firefighter(s) (FF) to travel along an escape route and reach a safety zone (SZ). Mathematically, the safety margin ($\pm$) equals T1–T2, where T1 is the time for a fire to reach the SZ and T2 is the time for FF to reach the SZ. T2 depends not only on FF's rate of travel but also on their reaction time.

open improved route or in a natural standing timber fuel type. Dropping the pack and tool could allow a firefighter to increase their travel rate by up to 20% (firefighters have to be trained to perform this action – it does not come naturally).

- Firefighters can be expected to move up to 40% faster on improved routes. Simply constructing a rudimentary trail (e.g. removing or cutting through large deadfall) and flagging or marking the route in some manner can decrease the overall time taken to reach a safety zone.

Thus, by using an improved escape route and dropping the pack and tool, firefighters can travel up to two times faster than if they had attempted to travel over an unmarked/unimproved route with their pack and tool. Precious seconds gained by these actions could mean the difference between life and death on the fireline. This study also found, from simulations of rate of fire spread in relation to firefighter travel rates, that while firefighters would be able to momentarily outpace the rate of advance that most wildland fires are capable of achieving, even the most fit individuals would not be able to sustain such a pace for more than a few minutes on a moderately steep slope.

A safety zone is an area that offers refuge from the dangers associated with an approaching wildland fire. What constitutes a safety zone varies considerably amongst individuals. Ideally, it is an area completely free of readily combustible fuels. The minimum size required is determined by the characteristics of the fire environment (e.g. fuels immediately adjacent to a clearing) and the ensuing fireline intensity. Based on theoretical considerations for radiation emitted from a 'wall of flame' under idealized conditions (e.g. level terrain, steady-state fire conditions), it has been suggested, from the standpoint of flame radiation, that a safety zone should be large enough so that the distance between the human occupants and flame front should be at least four times the maximum expected flame height at the edge of the safety zone (Table 16.2). This recommendation does not directly take into account the impacts of smoke inhalation on firefighter consciousness.

How does one go about gauging the maximum flame height to be expected? Based on general field observations and measurements made of wildfires, prescribed burns and experimental fires, one can say with some degree of certainty that the maximum

Table 16.2 Minimum safety zone separation distances and sizes (circular shape) in relation to flame height for a single person (adapted from Alexander *et al.*, 2012). Reproduced with permission of Elsevier.

Safety zone size	Flame height (m)					
	2	5	10	20	40	60
Separation distance (m)[1]	8	20	40	80	160	240
Area (ha)[2]	0.02	0.13	0.5	2.0	8	18

[1]For flat topography based on the Butler and Cohen (1998) '4-times-flame-height' rule of thumb for radiant heat only (i.e. no allowance for convection, flame impingement, fire whirls or other fireline hazards (e.g. falling trees and snags, rolling rocks and logs). It is assumed that the person is standing upright and is properly clothed, including headgear and gloves.
[2]Assuming the area of a circle is equal to $(\pi \times (SD)^2) \div 10\,000$, where π equals ≈ 3.14159 and the separation distance (SD, m) is deemed to be radius of a circle. For perspective, a North American ice hockey rink is 0.16 ha in size and an American football field is 0.4 ha in size.

flame heights in grasslands, low- to medium-tall shrublands and boreal hardwood stands, for example, would range from 4–10 m, implying that a separation distance of at least 40 m should be adequate in fuel types that lack significant vertical fuel dimension. This is in contrast to crown fires in fuel complexes like mature chaparral, dry eucalypt forests and conifer stands, where flame heights in excess of 20 m and higher are commonly attained.

Past experience has shown that survival is possible in areas much smaller than that recommended for a safety zone (Alexander *et al.*, 2009). However, this is contingent on maintaining a prone position and using every means possible of protecting oneself against radiation and convective heat transfer, as the advancing flame front approaches and passes around the firefighter in the 'survival zone'.

There have been several direct applications of information from fire danger rating and fire behaviour prediction systems to enhancing situational awareness and entrapment avoidance amongst wildland firefighters. These include firefighter safety guidelines related to the real-time, diurnal and seasonal assessments of wildland fire behaviour potential (Andrews *et al.*, 1998; Beck *et al.*, 2002; Cheney *et al.*, 2001).

16.4 Community wildland fire protection

While the use of the term 'wildland-urban interface' or WUI (pronounced 'woo-ee') can be traced back some 40 years or so, in actual fact the protection of settlements or communities against the adverse impacts of wildfires has a much longer history (Cohen, 2010). As

mentioned earlier, as humans we tend to have short memories. The destructive power of wildfires in modern times was forcibly demonstrated during the 2009 Black Saturday fires in the State of Victoria in southeastern Australia, where wind-driven fires, some ignited by arsonists, led to the deaths of 173 civilians and the devastation of numerous communities. The ever increasing incidence of major WUI fires in recent years has, to a large extent, blurred the distinction between urban and wildland fire behaviour.

Advances in the understanding and prediction of wildland fire behaviour will enable fire planners and community managers to tackle the WUI fire problem more effectively. Wildland fire growth simulation models like PROMETHEUS, for example, are now being used to test containment strategies for dealing with wildfires based on fuel treatments in the context of community wildfire protection planning (Walkinshaw, 2012).

Fire weather databases have combined with fire danger rating or fire behaviour predictions system to create fire danger and fire behaviour climatologies. This has allowed for the quantification of the potential wildfire threat in simple terms that can be readily understood by non-specialists, including the general public. For example, during a public information session associated with the presentation of the information contained in Table 16.3 in March 1997, the town's emergency planner was quite surprised to learn that, on average, a fifth of the days in a fire season were capable of wildfire behaviour that would likely not be controllable should an ignition take place. Similar analyses has been undertaken in the province of Saskatchewan, Canada (Johnson *et al.*, 2005) where wildfire risk profiles have been completed for over 100 northern communities there.

Table 16.3 Average number of days in various fireline intensity classes for the black spruce forests adjacent to the town of Hay River, Northwest Territories, Canada as derived by Alexander *et al.* (2001). This compilation is based on daily fire weather records for the period 1954–1996 and using fuel type C-2 (boreal spruce) of the Canadian Forest Fire Behavior Prediction System. Reproduced with permission of Scion, New Zealand.

Fireline intensity class(es)[1]	Fireline intensity (kW/m)	Average number of days					
		May	June	July	August	September	Fire season
1 and 2	<500	14	8	10	10	16	58
3	500–2000	9	9	9	11	8	46
4	2000–4000	4	4	5	4	2	19
5 and 6	>4000	9	9	9	7	3	34

[1]The four fireline intensity class groups could be equated to low, moderate, high and extreme fire danger.

The requirement for WUI fire planners and consultants to be certified as operational fire behaviour analysts (i.e. FBANs) has not yet happened. However, FBANs assigned to incident management teams make recommendations regarding community evacuation alerts and warnings with increasing confidence, given the current state of fire behaviour science and technology (Beverly and Bothwell, 2011). For example, the second evacuation of the town of Swan Hills on May 13, 1998 during the 167 000 ha Virginia Hills Fire in central Alberta, Canada, was based on the fire's uncontrolled status in the north-eastern sector at the time, the forecasted weather, the fuel type mosaic and topography in the area between the fire and the town. This was also coupled with projections using the Canadian Forest Fire Behaviour Prediction System undertaken by the author of this chapter (MEA) while serving as the FBAN on the command team assigned to this incident. Based on interpretations of the predicted spread rate and fireline intensities, the decision was made to evacuate Swan Hills. While the forecasted weather conditions failed to materialize, the second evacuation was conducted in an orderly manner during daylight hours, in stark contrast to the haste involved in the first evacuation during the middle of the night.

Additional research and development aimed specifically at the application of fire behaviour knowledge to the WUI fire problem has been undertaken. These include the development of a model for calculating the probability of house survival that depends on estimates of fireline intensity (Wilson, 1988a), as well as the Wildland Urban Interface Evacuation (WUIVAC) Model, a spatial decision support system that can be used to evaluate public evacuation routes under emergency wildfire conditions (Cova *et al.*, 2011).

16.5   Fuels management

From a wildfire suppression standpoint, fuels management involves planned changes to living or dead wildland fuels in order to lessen fire behaviour potential and resistance to fireline construction. This increases the probability of successful containment, while minimizing adverse impacts as well as increasing the safety of wildland firefighters and the public at large. More specifically, the purpose of modifying fuels in the fire environment is to decrease the rate of fire spread and/or available fuel and, in turn, fireline intensity and fire size, as well as crowning and spot fire development. In this sense, fuels management constitutes a means of fire prevention, at least in the case of reducing the number of large fire occurrences.

The basic premise behind fuels management is that we are not capable of controlling the air mass or weather component of the fire environment (nor of modifying the weather), or reshaping the topographical features of the earth. However, we can influence the quantity and character of wildland fuels and, therefore, we can manage them to a certain extent.

The four principal means of directly managing or treating wildland fuels that are regularly applied include:

- Fuel reduction (e.g. by prescribed broadcast burning, pile burning, livestock grazing and the physical removal of fuels from the site other than by logging);
- Fuel manipulation (e.g. by pruning, pre-commercial thinning, mulching or mastication, chipping, crushing);
- Fuel conversion (i.e. a change in vegetative cover from a flammable case to a far less flammable one); and
- Fuel isolation (i.e. by the use of fuel breaks and firebreaks).

Clive Countryman's (1974) publication *Can Southern California Wildland Conflagrations Be Stopped?* represents a *tour du force* on the subject of fuels management. In the United States, the Joint Fire Science Program has supported the development of several state-of-the-art guides to fuels management practices in recent years (e.g. Jain *et al.*, 2012).

Of course, prescribed fire – the knowledgeable and controlled application of fire to a specific land area in order to accomplish planned resource management objectives (Waldrop and Goodrick, 2012) – is used for purposes other than fuels management (Figure 16.7). Such purposes may include ecosystem restoration, wildlife habitat, insect and disease control and range improvement – there are often multiple objectives for prescribed fires. General fire behaviour knowledge and predictive aids are used in developing burning prescriptions designed to accomplish the stated objectives of a particular prescribed fire. Fernandes and Botelho (2003) noted that 'Conclusive statements concerning the hazard-reduction potential of prescribed fire

Figure 16.7   View of the Blackstone Capping Unit prescribed fire in progress, south-western Alberta, May 15, 2007. This prescribed fire was carried out to create a landscape-scale fuel break, to enhance wildland habitat, to improve range conditions and also for fire crew training (photo Dennis Quintilio, D. Quintilio and Associates, Glenevis, Alberta).

application are not easily generalized, and will ultimately depend on the overall efficiency of the entire fire management process'.

Fuel reduction and fuel manipulation are often considered as a single entity – namely, fuel modification. Their resultant outcomes or effects on fire behaviour are, nevertheless, quite distinct. Fuel reduction involves fuel removal, and this correspondingly leads to a direct reduction in rate of spread and fireline intensity potential. The first published empirical proof of this concept did not appear until the late 1940s (Figure 16.8).

Fuel manipulation, on the other hand, involves only a rearrangement of the fuel. This might, for example, influence crowning potential (Figure 16.9), but the end result is that the total fuel load remains the same.

Both fuel reduction and fuel manipulation techniques may be used jointly in a fuel treatment (e.g. commercial thinning followed by prescribed under-burning). Fuel conversion involves a more or less permanent change (e.g. from a spruce stand to a deciduous or hardwood stand, by mechanical means and/or prescribed fire).

Firebreaks and fuel breaks are designed to stop outright or impede a fire's progress (Figure 16.10); they may also serve as a control line or anchor point to carry out fire suppression work from. While fuel breaks will contain burnable material (e.g. trembling aspen fuel break), firebreaks technically do not burn – and, if they do, the fuels are exceedingly sparse. A temporary firebreak might be created by ploughing a field of grass, or a permanent firebreak can be constructed by establishing and maintaining a gravel road adjacent to a community surrounded by conifer forests.

The principle of fuel isolation at a local level (e.g. a house or an entire community) certainly seems doable. At a landscape-scale, though, the task seems utterly daunting (Figure 16.11). In this respect, we need to pay attention to the lessons provided by Mother Nature (Box 16.4).

Every unwanted human-caused fire can be considered as a fire prevention failure. Even if we could

Figure 16.8 Pre-fire view of the more heavily grazed area (left of the fence) versus the more lightly grazed area (right of the fence) involved in the experimental fire study on the Nebraska National Forest, Nebraska, USA. Parallel plots, approximately 150 m by 30 m, were established on each side of the fence. Both plots were ignited so as to burn simultaneously as heading fires. The reduction in fuel (i.e. lightly grazed vs. more heavily grazed) led to a three-fold decrease in the average rate of fire spread (28 vs. 9.7 m/min) and a five-fold decrease in the average flame height (1.5 vs. 0.3 m) (from Davis, 1949).

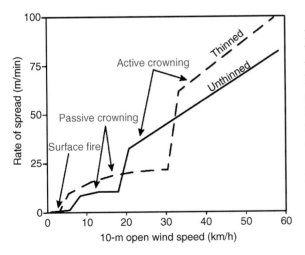

Figure 16.9 Head fire rate of spread as a function of wind speed for 12-year-old thinned (50% basal area reduction treatment) and unthinned pine plantation stands based on the Pine Plantation Pyrometrics (PPPY) fire modelling system (adapted from Cruz *et al.*, 2008). Courtesy of Taylor & Francis.

prevent all such fires, we would still have to contend with lightning-ignited fires. Is it realistic to expect that we can control all fires before they reach conflagration levels? On some days, adverse fuel, weather and topographic conditions, coupled with an ignition source, lead to instances of extreme fire behaviour for which it is impossible to effect containment until burning conditions ameliorate.

Clive Countryman considered that the best prospect for alleviation of the problem was the creation of a fuel-type mosaic that would reduce the fuel energy output. In this regard, see (for example) the case study completed by Salazar and Gonzáles-Cabán (1987) of the 1985 Wheeler Fire on the Los Padres National Forest in southern California, and the earlier analyses by Rogers (1942).

To decide how much fuel is considered acceptable requires the integration of many factors (Figure 16.12). This can be done systematically in a three-step process (Brown *et al.*, 1977; Alexander, 2007):

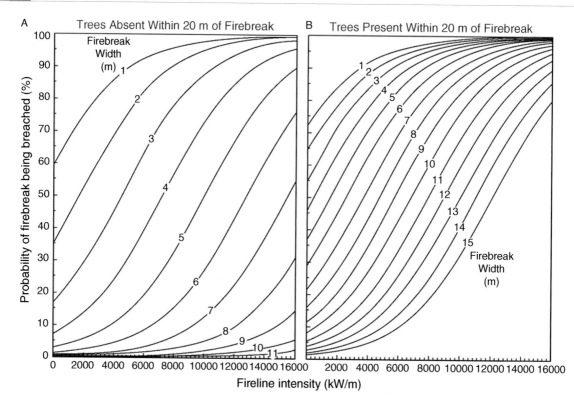

Figure 16.10   Probability of firebreak breaching by a grass fire for (A) trees absent, and (B) present within 20 m of the firebreak (adapted from Wilson, 1988b). Width of firebreak that is necessary to stop grass fires: Some field experiments. Canadian Journal of Forest Research 18, 682–687. © Canadian Science Publishing or its licensors.

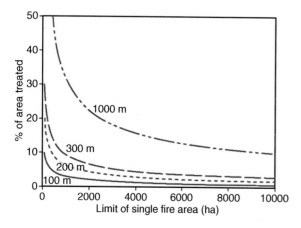

Figure 16.11   Geometrics of a proactive fuel isolation treatment. The lines are curves of the percentage of the landscape area that needs to be treated to provide a firebreak of a given width (100 m, 200 m, 300 m, 1000 m), such that the maximum fire size is limited to that shown on the x-axis. This assumes that the fuel isolation treatments are in a square grid pattern and act as perfect firebreaks in stopping further fire growth (from Amiro et al., 2001). Courtesy of CSIRO Publishing.

Step 1: Consider management objectives and values at risk. For the latter, both resource values and risk of a fire during periods of critical fire weather, and fire danger causing damage, are jointly considered.

Step 2: Appraise fuels by (a) describing fuel loads and arrangement from inventory, prediction, or ocular estimation technique, and then (b) interpreting fire behaviour and fire impact potential, such as rate of spread, intensity, flame length, crown scorch height and degree of crown fuel consumption.

Step 3: Consider other fire-related factors, such as fuel and fire behaviour potential on adjoining lands, suppression capability, frequency and severity of historical fires, and fire's ecological role.

The CFIS system mentioned in Chapter 15 allows one to evaluate the impacts of proposed fuel treatments on potential crown fire behaviour, based on its ability to manipulate three characteristics of a fuel complex (i.e. available surface fuel load, canopy base

'Logic would dictate that the chance(s) of a high-intensity crown fire occurrence would gradually increase as the size of the total plantation estate increases. The value of a dispersed pattern of relatively small to moderately sized plantations, especially in fire-prone environments exhibiting very high ignition risk coupled with an adverse fire climate, was demonstrated during the 1983 Ash Wednesday Fires in the south-eastern portion of South Australia and Victoria.

'State-owned plantations in the region managed by the Woods and Forests Department amount to approximately 80 000 ha and are comprised of a few large, more or less contiguous blocks of land. On February 16, 1983, some 21 000 ha of exotic pine plantations were burnt over in South Australia alone, most very severely, by eight fires that covered a gross area of around 120 000 ha. In contrast, private forest industry in the region, with a comparable estate of around 70 000 ha, but comprised of many smaller parcels scattered across the region, more as a result of circumstances rather than by any strategic design, suffered only minor (40 ha) wildfire losses.'

height and canopy bulk density) by silvicultural and other vegetation management techniques.

The first two research papers that used mathematical modelling to gauge the effectiveness of fuel treatments on fire potential (Anderson, 1974; Brown, 1974) were the direct result of the release of the Rothermel (1972) surface fire spread model two years earlier. The fuel appraisal process requires the coupling of mathematical or simulation modelling with experienced judgment and comparison against case study knowledge. Reliance entirely upon fire behaviour simulations based on fire modelling systems has shown to be foolhardy (e.g. Cruz and Alexander, 2010).

The number of years between fuel reduction burns in the eucalypt forests of Australia typically depends on the litter accumulation rates and the fire climatic conditions. The fuel reduction burning programme developed for the eucalypt forests in the south-western region of Western Australia, born out of the disastrous 1961 Dwellingup fires (Underwood, 2011) is unparalleled (McCaw, 2013). The interval between fuel hazard reduction burns was initially established on the basis of practical fire experience. However, it was later verified by fuel measurements and experimental fire studies carried out by fire researchers within the operational organization.

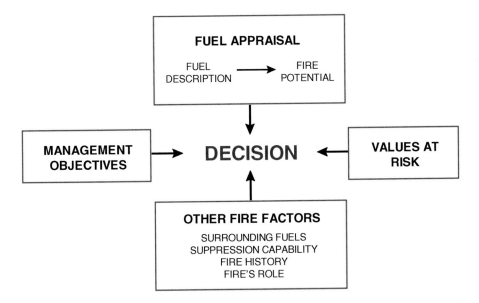

Figure 16.12   The factors considered to be important when deciding how much fuel is considered acceptable (from Brown *et al.*, 1977).

16.6   Prediction of fire effects

First order fire effects are the immediate evident consequences or impacts of a fire. Research on fire effects extends back to the late 1890s. Methven (1977) remarked, at a fire ecology workshop held in Edmonton, Alberta, Canada, that 'the most conspicuous feature of the literature on fire ecology has been the lack of reference to the nature of the fires that caused the effects being measured.' He considered this another example of the 'two solitudes in forest fire research' (see Chapter 15, section 15.3.4), as noted by Charlie Van Wagner six years earlier, suggesting that 'in the case of ecological effects, the two solitudes have been composed of fire researchers concerned with behaviour and ecologists concerned with fire effects'.

The reality is that things take time. As the science of wildland fire behaviour description and measurement began to mature in the early 1960s, so too did concerted efforts to link fire effects to fire behaviour. The end result is that today we see a whole host of fire effects modelling systems that are based on largely on information collected up to the late 1980s.

One of the first was FOFEM (First Order Fire Effects Model), a national fire effects modelling system used in the USA for predicting tree mortality, fuel consumption, emissions or smoke production, mineral soil exposure and soil heating caused by prescribed fire or wildfire (Reinhardt, 2003). FOFEM represents a collation of many separate studies. For example, the probability of post-fire tree mortality function is based on regression equations derived from empirical data on tree size and crown scorch height collected from experimental fires, prescribed fires and wildfires (Figure 16.13).

Until just recently, fire effects models have been largely statistical in nature, as opposed to process-oriented or biophysical predictive models. Like models for predicting various characteristics of fire behaviour, there are pros and cons to purely statistical models, depending on how they are formulated. The journal *Fire Ecology* recently devoted an entire special issue of eight papers on 'strengthening the foundation of wildland fire effects prediction for research and management' (Dickinson and Ryan, 2010).

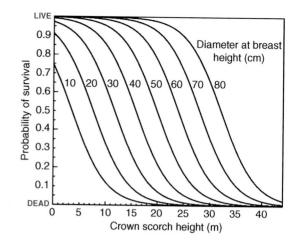

Figure 16.13 Post-fire probability of tree survival in interior Douglas-fir as a function of diameter-at-breast height and crown scorch height (adapted from Bevins, 1980). This graph is based on an equation derived from data collected on 176 trees on experimental fire plots in west central Montana, USA.

While giving the appearance of being highly robust in nature, the developers of the few biophysical fire models produced to date seem to have failed to consider some practical considerations. For example, in the software application (*CROWN SCORCH*) of the crown scorch height model developed by Michaletz and Johnson (2006), the authors did not allow for the fact that crown scorch height can also be influenced by the ignition pattern. Their model considered only a single spreading line fire and their operational software application was, thus, limited to that case.

The outputs of other biophysical modelling efforts have, in some instances, been found to match reality (e.g. Choczynska and Johnson 2009), but this is not so in other cases (in spite of the best of intentions of the modellers). It is clear that there is no substitute for 'hands-on' experience in observing experimental fires, prescribed fires, and wildfires (Figure 16.14) and their resulting aftermath, coupled with a good command of the literature, as a basis for understanding and modelling the impacts and effects of free-burning fires in relation to their behaviour (Alexander and Cruz, 2012c).

Figure 16.14 A wildfire (Red Lake 200-76) burning through snags and tree reproduction in a 1967 fire origin jack pine stand in northwestern Ontario, Canada, during the 1976 fire season. The individual with his back to the camera photographing the oncoming flame front is Brian J. Stocks of the Canadian Forest Service. (photo Martin E. Alexander).

16.7   Getting on the road towards self-improvement

Few would argue with the notion that models and modelling have become an integral component of modern day fire management practices. But there is room to wonder if we have lost sight of the idea that models were always intended simply as tools to aid us in our decisions and actions (Alexander, 2009b).

Wildland fire researchers have, indeed, provided users with a host of fire behaviour decision-making aids and knowledge designed to address specific wildland fire-related issues. There is a plethora of fire modelling systems and a mass of literature on fire behaviour subjects. Some might consider it all overkill. In spite of great expectations (e.g. USDA Forest Service, 1966), history has shown time and time again over the past hundred years or so that fire research alone will not be enough. Adopters and implementers of fire behaviour knowledge are always needed.

One can regard the products produced by wildland fire research to date as having effectively synthesized the current theory as in the 'theory and practice' of wildland fire behaviour prediction. Effective 'practice' is, however, the onus of the individual person practising fire behaviour prediction. In this regard, live fire exercises should also be considered as a means of supplementing conventional fire training using operational prescribed fires (see Figure 16.7). This would be especially valuable in fuel types with which we have limited experience (Cheney, 1994).

There is no better means of practice than experience coupled with self-study. Wildland fire icon Paul Gleason regarded it as always being a 'student of fire' (Cook and Tom, 2003) – in other words having that constant, burning desire to continue to learn about fire. The use of experienced judgment (Gisborne, 1948) in assessing fire behaviour needs to be supported by more monitoring and documentation on wildfires and prescribed fires than has been the case in the past (see Box 15.3, chapter 15, Part Four). The preparation of fire behaviour case studies should not be viewed as strictly the domain of wildland fire behaviour research. Case study preparation (Box 16.5) should be complemented by review of existing case histories, as previous events tend to have a way of repeating themselves.

Box 16.5   Suggested outline for preparing a wildland fire behaviour case study report (adapted from Alexander and Thomas, 2003b).

1 Introduction
   - Significance of the fire.
   - Regional map with fire location.
2 Fire chronology and development
   - Cause.
   - Time of origin and/or detection.
   - Initial attack action.
   - Forward spread and perimeter growth.
   - Fire characteristics, e.g. spotting distances and crowning activity.
   - Suppression strategy and tactics employed.
   - Mop-up difficulty.
   - Fire progress map, showing point of origin.
   - Final area burnt and perimeter.
   - Ground and aerial photos.
3 Details of the fire environment
   - **Topography:** review major features; include topographic map and photos.
   - **Fuels:** describe the principal fuel type(s); include a vegetation cover type map and photos.
   - **Fire weather:** describe pre-fire weather as appropriate; summarize synoptic weather features and include surface map; present daily fire weather observations; present fire danger ratings, including drought indexes, and append monthly fire weather record form; present hourly weather observations; denote location of weather station(s) on regional map or fire progress map and comment on the relevance of the readings to the fire area, including notes about the station's instrumentation.
4 Analysis of fire behaviour
   - Discuss the fire's behaviour in relation to the characteristics of the fire environment.
   - Discuss the success/failure of the suppression operations.
5 Concluding remarks
   - What, for example, did you learn about predicting fire behaviour and fire behaviour documentation?
   - What, if anything, have you been able to contribute to the general fire behaviour knowledge base?

Further reading

Agee, J.K., Bahro, B., Finney, M.A., Omi, P.N., Sapsis, D. B., Skinner, C.N., van Wagtendonk, J.W., Weatherspoon, C.P. (2000). The use of shaded fuelbreaks in landscape fire management. *Forest Ecology and Management* **127**, 55–66.

Agee, J.K., Skinner, C.N. (2005). Basic principles of forest fuel reduction treatments. *Forest Ecology and Management* **211**, 83–96.

Alexander, M.E. (2010). 'Lest we forget': Canada's major wildland fire disasters of the past, 1825–1938. In: Wade, D.D., Robinson, M.L. (eds.), *Proceedings of 3rd fire behavior and fuels conference.* International Association of Wildland Fire, Birmingham, AL. CD-ROM. 21 p.

Alexander, M.E., Baxter, G.J., Dakin, G.R. (2013). How much time does it take for a wildland firefighter to reach a safety zone? *Wildfire* **22**(4), 12–13.

Alexander, M.E., Buxton-Carr, P. (2011). Wildland fire suppression related fatalities in Canada, 1941–2010: a preliminary report. In: Fox, R.L. (ed.), *Proceedings of the 11th international wildland fire safety summit.* International Association of Wildland Fire, Missoula, MT. CD-ROM. 15 p.

Alexander, M.E., Cruz, M.G. (2011). What are the safety implications of crown fires? In: Fox, R.L. (ed.), *Proceedings of the 11th international wildland fire safety summit.* International Association of Wildland Fire, Missoula, MT. CD-ROM. 16 p.

Alexander, M.E., Heathcott, M.J., Schwanke, R.L. (2013). *Fire behaviour case study of two early winter grass fires in southern Alberta, 27 November 2011.* Partners in Protection Association, Edmonton, Alberta. 76 p.

Andrews, P.L., Bradshaw, L.S., Bunnell, D.L., Curcio, G.M. (1998). Fire danger rating pocket card for firefighter safety. Pages 67–70 In: *Preprints of second conference on fire and forest meteorology,* pp. 67–70. American Meteorological Society, Boston, Massachusetts. For further information, see http://fam.nwcg.gov/fam-web/pocketcards/default.htm)

Beverly, J.L., Bothwell, P., Conner, J.C.R., Herd, E.P.K. (2010). Assessing the exposure of the built environment to potential ignition sources generated from vegetative fires. *International Journal of Wildland Fire* **19**, 299–313.

Braun, J.W., Jones, B.L., Lee, J.S.W., Woolford, D.G., Wotton, B.M. (2010). Forest fire risk assessment: an illustrative example from Ontario, Canada. *Journal of Probability and Statistics* **2010**, Article ID 823018. 26 p.

Burrows, N.D. (1999). A soil heating index for interpreting ecological impacts of jarrah forest fires. *Australian Forestry* **62**, 320–329.

Burrows, N., McCaw, L. (2013). Prescribed burning in southwestern forests. *Frontiers in Ecology and the Environment* **11** (Online Issue 1), e25–e34.

Carlton, D.W. (1991). Fire behavior training – a look at upcoming changes. *Fire Management Notes* **52**(2): 15–19.

Cohen, J.D. (2004). Relating flame radiation to home ignition using modeling and experimental crown fires. *Canadian Journal of Forest Research* **34**, 1616–1626.

Cook, J.R. (2013). Trends in wildland fire entrapment falities...revisted. National Wildfire Coordinating Group, Boise, Idaho. 10 p. http://www.nifc.gov/wfstar/downloads/Wildland_Fire_Entrapment_Fatalities_Revisited_JC.pdf

Costa, P., Castellnou, M., Larrañaga, A., Miralles, M., Kraus, D. (2011). *Prevention of large wildfires using the fire types concept.* EU Fire Paradox Publication, Barcelona, Spain. 87 p.

De Groot, W.J. (1988). Application of fire danger rating to the wildland/urban fire problem: a case study of the Nisbet Forest, Saskatchewan. In: Fischer, W.C., Arno, S.F. (comps.) *Protecting people and homes from wildfire in the Interior West: Proceedings of the symposium and workshop*, pp. 166–170. U.S. Department of Agriculture, Forest Service, Intermountain Research Station, Ogden, Utah. General Technical Report INT-251.

Fernandes, P.M. (2009). Examining fuel treatment longevity through experimental and simulated surface fire behaviour: a maritime pine case study. *Canadian Journal of Forest Research* **39**, 2529–2535.

Hansen, R. (2012). Estimating the amount of water required to extinguish wildfires under different conditions and in various fuel types. *International Journal of Wildland Fire* **21**, 525–536, 728.

Hirsch, K.G., Martell, D.L. (1996). A review of initial attack fire crew productivity and effectiveness. *International Journal of Wildland* **6**, 199–215.

Jenkins, M.J., Page, W.G., Hebertson, E.G., Alexander, M.E. (2012). Fuels and fire behavior dynamics in bark–beetle attacked forests in western North America and implications for fire management. *Forest Ecology and Management* **275**, 23–24.

Keller, P. (ed.). (2010). *Can History Help Guide Our Fire Management Future?* A summary report on the special panel presentation featured at the International Association of Wildland Fire's Third Fire Behavior and Fuels Conference. *Beyond Fire Behavior and Fuels: Learning from the Past to Help Us in the Future.* USDA Forest Service, Wildland Fire Lessons Learned Center, Tucson, Arizona. 20 p.

Michaletz, S.T., Johnson, E.A. (2007). How forest fires kill trees: a review of the fundamental biophysical processes. *Scandinavian Journal of Forest Research* **22**, 500–515.

Murphy, K., Rich, T., Sexton, T. (2007). *An assessment of fuel treatment effects on fire behavior, suppression effectiveness, and structure ignition on the Angora Fire.* U.S. Department of Agriculture, Forest Service, Pacific Southwest Region., Vallejo, California. Technical Publication R5-TP-025. 32 p.

Mutch, R.W. (1994). New imperatives for fire behavior information. *Wildfire* **3**(3), 42–44.

National Wildfire Coordinating Group, Fire Investigation Working Team (2005). *Wildfire origin and cause determination handbook.* National Interagency Fire Center, National Fire Equipment System, Boise, Idaho. NWCG Handbook 1 – Publication NFES 1874. 111 p. http://www.nwcg.gov/pms/pubs/nfes1874/nfes1874.pdf

Stiger, E.M. "Sonny". (2012). *The sleeping giant awakens: Diary of a forester/firefighter.* Self-published by author, Helena, Montana. 77 p. For more information, see http://www.thegiantawakens.com/

Tolhurst, K.G., Cheney, N.P. (1999). *Synopsis of the knowledge used in prescribed burning in Victoria.* Department of Natural Resources and Environment, Fire Management, East Melbourne, Victoria. 97 p.

van Wilgen, B.W. (2013). Fire management in species-rich Cape fynbos shrublands. *Frontiers in Ecology and the Environment* **11** (Online Issue 1), e35–e44.

Woolley, T., Shaw, D.C., Ganio, L.M., Fitzgerald, S. (2012). A review of logistic regression models used to predict post-fire tree mortality of western North American conifers. *International Journal of Wildland Fire* **21**, 1–35.

References for part four

Abell, C.A. (1940). *Rate of initial spread of free-burning fires on the National Forests of California*. U.S. Department of Agriculture, Forest Service, California Forest and Range Experiment Station, Berkeley, CA. Research Note 24. 26 p.

Adams, M., Attiwill, P. (2011). Burning issues: sustainability and management of Australia's southern forests. CSIRO Publishing, Melbourne, Victoria, Australia.

Ager, A.A., Vaillant, N.M., Finney, M.A. (2011). Integrating fire behavior models and geospatial analysis for wildland fire risk assessment and fuel management planning. *Journal of Combustion* **2011**, Article ID 572452.

Ahrens, C.D. (2008). Essentials of meteorology: an invitation to the atmosphere. 5th edition. Thomson Brooks/Cole, Belmont, CA.

Albini, F.A. (1976). *Estimating wildfire behavior and effects*. U.S. Department of Agriculture, Forest Service, Intermountain Forest and Range Experiment Station, Ogden, UT. General Technical Report INT-30. 92 p.

Albini, F.A. (1979). *Spot-fire distance from burning trees – a predictive model*. U.S. Department of Agriculture, Forest Service, Intermountain Forest and Range Experiment Station, Ogden, UT. Research Paper INT-56. 73 p.

Albini, F.A., Alexander, M.E., Cruz, M.G. (2012). A mathematical model for predicting the maximum potential spotting distance from a crown fire. *International Journal of Wildland Fire* **21**, 609–627.

Alexander, M.E. (1982). Calculating and interpreting forest-fire intensities. *Canadian Journal of Botany* **60**, 349–357.

Alexander, M.E. (1988). Help with making crown fire hazard assessments. In: Fischer, W.C., Arno, S.F. (comps). *Protecting people and homes from wildfire in the Interior West: Proceedings of the symposium and workshop*, pp. 147–156. U.S. Department of Agriculture, Forest Service, Intermountain Research Station, Ogden, UT. General Technical Report INT-251.

Alexander, M.E. (1998). *Crown fire thresholds in exotic pine plantations of Australasia*. Australian National University, Canberra, Australian Capital Territory. Ph.D. Thesis. 228 p.

Alexander, M.E. (2000). *Fire behaviour as a factor in forest and rural fire suppression*. Forest Research, Rotorua in association with New Zealand Fire Service Commission and National Rural Fire Authority, Wellington, New Zealand. Forest Research Bulletin No. 197, Forest and Rural Fire Science and Technology Series Report No. 5. 28 p.

Alexander, M.E. (2007). Simple question; difficult answer: how much fuel is acceptable? *Fire Management Today* **67** (3), 6–11, 30.

Alexander, M.E. (2009a). Wildland fire behavior and The Course of Science flowchart: is there a connection? *Fire Management Today* **69** (3), 44–46.

Alexander, M.E. (2009b). Are we abusing our use of models and modelling in wildland fire and fuel management? *Fire Management Today* **69** (4), 24–27.

Alexander, M.E. (2010). Surface fire spread potential in trembling aspen during the summer in the boreal forest region of Canada. *Forestry Chronicle* **86**, 200–212.

Alexander, M.E., Ackerman, M.Y., Baxter, G.J. (2009). An analysis of Dodge's escape fire on the 1949 Mann Gulch Fire in terms of a survival zone for wildland firefighters. In: *Proceedings of 10th wildland fire safety summit*. International Association of Wildland Fire, Birmingham, AL. CD-ROM. 27 p.

Alexander, M.E., Andrews, P.L. (1989). Wildland fire occurrence and behavior analysis in the year 2000 and beyond. *Fire Management Notes* **50** (4), 35–37.

Alexander, M.E., Baxter, G.J., Dakin, G.R. (2005). Travel rates of Alberta wildland firefighters using escape routes. In: Butler, B.W., Alexander, M.E. (eds). *Proceedings of eighth international wildland fire safety summit: Human factors – ten years later.* International Association of Wildland Fire, Hot Springs, SD. CD-ROM. 12 p.

Alexander, M.E., Cole, F.V. (1995). Predicting and interpreting fire intensities in Alaskan black spruce forests using the Canadian system of fire danger rating. In: *Managing forests to meet people's needs, proceedings of 1994 Society of American Foresters/Canadian Institute of Forestry Convention*, pp. 185–192. Society of American Foresters, Bethesda, Maryland. SAF Publication 95-02.

Alexander, M.E., Cruz, M.G. (2012a). Interdependencies between flame length and fireline intensity in predicting crown fire initiation and crown scorch height. *International Journal of Wildland Fire* **21**, 95–113.

Alexander, M.E., Cruz, M.G. (2012b). Graphical aids for visualizing Byram's fireline intensity in relation to flame length and crown scorch height. *Forestry Chronicle* **88**, 185–190.

Alexander, M.E., Cruz, M.G. (2012c). Modelling the impacts of surface and crown fire behaviour on serotinous cone opening in jack pine and lodgepole pine forests. *International Journal of Wildland Fire* **21**, 709–721.

Alexander, M.E., Cruz, M.G. (2013a). Are the applications of wildland fire behaviour models getting ahead of their evaluation again? *Environmental Modelling & Software* **41**, 65–71.

Alexander, M.E., Cruz, M.G. (2013b). Limitations on the accuracy of model predictions of wildland fire behaviour: A state-of-the- knowledge overview. *Forestry Chronicle* **89**, 370–381.

Alexander, M.E., Cruz, M.G. (2013c). Crown fire dynamics in conifer forests. In: Werth, P.W., Potter, B.E., Alexander, M.E., *et al. Synthesis of knowledge of extreme fire behavior: Volume 2 for fire behavior specialists, researchers and meteorologists.* U.S. Department of Agriculture, Forest Service, Pacific Northwest Research Station, Portland, Oregon. General Technical Report PNW-GTR-*In press.*

Alexander, M.E., Taylor, S.W. (2010). Wildland fire behavior case studies and the 1938 Honey Fire controversy. *Fire Management Today* **70** (1), 15–25.

Alexander, M.E., Thomas, D.A. (2003a). Wildland fire behavior case studies and analyses: Value, approaches, and practical uses. *Fire Management Today* **63** (3), 4–8.

Alexander, M.E., Thomas, D.A. (2003b). Wildland fire behavior case studies and analyses: Other examples, methods, reporting standards, and some practical advice. *Fire Management Today* **63** (4), 4–12.

Alexander, M.E., Thomas, D.A. (2004). Forecasting wildland fire behavior: Aids, guides, and knowledge-based protocols. *Fire Management Today* **64** (1), 4–11.

Alexander, M.E., Thomas, D.A. (2006). Prescribed fire case studies, decision aids, and planning guides. *Fire Management Today* **66** (1), 5–20.

Alexander, M.E., Lanoville, R.A., Lowing, M., Stefner, C.N., Archibald, B. (2001). Application of the Canadian Forest Fire Danger Rating System (CFFDRS) to community fire protection in the Northwest Territories. In: Pearce, G., Lester, L. (tech coords). *Bushfire 2001 – Australasian Bushfire Conference Proceedings*, p. 317. New Zealand Forest Research Institute Ltd., Rotorua, New Zealand.

Alexander, M.E., Cruz, M.G., Lopes, A.M.G. (2006). CFIS: a software tool for simulating crown fire initiation and spread. In: Viegas, D.X. (ed). *Proceedings of 5th international conference on forest fire research.* Elsevier B.V., Amsterdam, The Netherlands. CD-ROM. 13 p.

Alexander, M.E., Lanoville, R.A. (1989). *Predicting fire behavior in the black spruce-lichen woodland fuel type of western and northern Canada.* Forestry Canada, Northern Forestry Centre, Edmonton, Alberta and Government of the Northwest Territories, Department of Renewable Resources, Territorial Forest Fire Centre, Fort Smith, Northwest Territories. Poster with text. http://cfs.nrcan.gc.ca/publications/download-pdf/23093.

Alexander, M.E., Mutch, R.W., Davis, K.M., Bucks, C.M. (2012). Wildland fires: Dangers and survival. In: Auerbach, P.S. (ed). *Wilderness medicine.* 6th edition, pp. 240–280. Elsevier, Philadelphia, PA.

Amiro, B.D., Stocks, B.J., Alexander, M.E., Flannigan, M.D., Wotton, B.M. (2001). Fire, climate change, carbon and fuel management in the Canadian boreal forest. *International Journal of Wildland Fire* **10**, 405–413.

Anderson, H.E. (1968). *The Sundance Fire: An analysis of fire phenomena.* USDA Forest Service, Intermountain Forest and Range Experiment Station, Ogden, UT. Research Paper INT-56. 39 p.

Anderson, H.E. (1974). *Appraising forest fuels: a concept.* U.S. Department of Agriculture, Forest Service, Intermountain Forest and Range Experiment Station, Ogden, UT. Research Note INT-187. 10 p.

Anderson, H.E. (1982). *Aids to determining fuel models for estimating fire behavior.* U.S. Department of Agriculture, Forest Service, Intermountain Forest and Range Experiment Station, Ogden, UT. General Technical Report INT-122. 22 p.

Andrews, P.L. (2007). BehavePlus fire modeling system: Past, present, and future. In: *Proceedings of 7th symposium on fire and forest meteorology.* American Meteorology Society, Boston, MA. 13 p. http://www.fs.fed.us/rm/pubs_other/rmrs_2007_andrews_p002.pdf.

Andrews, P.L. (2012). *Modeling wind adjustment factor and midflame wind speed for Rothermel's surface fire spread model.* U.S. Department of Agriculture, Forest Service, Rocky Mountain Research Station, Fort Collins, CO. General Technical Report RMRS-GTR-266. 39 p.

Andrews, P.L. (2013). Current status and future needs of the BehavePlus fire modeling system. *International Journal of Wildland Fire*, doi.org/10.1071/WF12167.

Andrews, P.L., Bradshaw, L.S., Bunnell, D.L., Curcio, G.M. (1998). Fire danger rating pocket card for firefighter safety. In: *Preprints of Second Conference on Fire and Forest*

Meteorology, pp. 67–70. American Meteorological Society, Boston, MA. http://fam.nwcg.gov/fam-web/pocketcards/default.htm.

Arnold, R.K., Buck, C.C. (1954). Blow-up fires – silvicultural problems or weather problems? *Journal of Forestry* **52**, 408–411.

Artsybashev, E.S. (1984). *Forest fires and their control.* A.A. Balkema, Rotterdam, Holland. Russian Translation Series 15. 160 p.

Barrow, G.J. (1945). A survey of houses affected in the Beaumaris Fire, January 14, 1944. *Journal of the Council of Scientific and Industrial Research* **18** (1), 27–37.

Barrows, J.S. (1951). *Fire behavior in Northern Rocky Mountain forests.* U.S. Department of Agriculture, Forest Service, Northern Rocky Mountain Forest and Range Experiment Station, Missoula, MT. Station Paper No. 29. 103 p.

Barrows, J.S. (1961). Natural phenomena exhibited by forest fires. In: Berg, W.G. (ed.). *Proceedings of international symposium on the use of models in fire research*, pp. 281–288. National Academy of Sciences – National Research Council, Washington, D.C. Publication 786.

Barrows, J.S. (1974). The challenges of forest fire management. *Western Wildlands* **1** (3), 3–5.

Beall, H.W. (1990). Fire research in Canada's federal forestry service – the formative years. In: Alexander, M.E., Bisgrove, G.F. (tech coords). *The art and science of fire management. Proceedings of the first Interior West Fire Council annual meeting and workshop*, pp. 14–19. Forestry Canada, Northern Forestry Centre, Edmonton, Alberta. Information Report NOR-X-309.

Beck, J.A., Alexander, M.E., Harvey, S.D., Beaver, A.K. (2002). Forecasting diurnal variation in fire intensity to enhance wildland firefighter safety. *International Journal of Wildland Fire* **11**, 173–182.

Behar, M. (2012). Burning question – why are wildfires defying long-standing computer models? *The Atlantic.* http://www.theatlantic.com/magazine/archive/2012/09/burning-question/309057/.

Beighely, M. (1995). Beyond the safety zone: creating a margin of safety. *Fire Management Notes* **55** (1), 21–24.

Beighley, M., Bishop, J. (1990). Fire behavior in high-elevation timber. *Fire Management Notes* **51** (2), 23–28.

Beverly, J.L., Bothwell, P. (2011). Wildfire evacuations in Canada 1980-2007. *Natural Hazards* **59**, 571–596.

Bevins, C.D. (1980). *Estimating survival and salvage potential of fire-scarred Douglas-fir.* U.S. Department of Agriculture, Forest Service, Intermountain Forest and Range Experiment Station, Ogden, UT. Research Note INT–187. 8 p.

Blackmarr, W.H. (1972). *Moisture content influences ignitability of slash pine litter.* U.S. Department of Forest Service, Southeastern Forest Experiment Station, Asheville, NC. Research Note SE-173. 7 p.

Braun, C.C., Gage, J., Booth, C., Rowe, A.L. (2001). Creating and evaluating alternatives to the 10 Standard Fire Orders and 18 Watch-Out Situations. *International Journal of Cognitive Ergonomics* **5**, 23–35.

Brooks, M.L., D'Antonio, C.M., Richardson, D.M., Grace, J. B., Keeley, J.E., DiTomaso, J.M., Robbs, R.J., Pellant, M., Pyke, D. (2004). Effects of invasive alien plants on fire regimes. *BioScience* **54**, 677–688.

Brown, A.A. (1937). The factors and circumstances that led to the Blackwater Fire tragedy. *Fire Control Notes* **1** (6), 384–387.

Brown, A.A. (1941). Guides to the judgment in estimating the size of a fire suppression job. *Fire Control Notes* **5** (2), 89–92.

Brown, A.A. (1959). Reliable statistics and fire research. *Fire Control Notes* **20** (4), 101–104.

Brown, A.A., Davis, K.P. (1973). *Forest fire: Control and use.* 2nd edition. McGraw-Hill, New York.

Brown, J.K. (1974). *Reducing fire potential in lodgepole pine by increasing timber utilization.* U.S. Department of Agriculture, Forest Service, Intermountain Forest and Range Experiment Station, Ogden, UT. Research Note INT-181. 6 p.

Brown, J.K., Oberheu, R.D., Johnston, C.M. (1982). *Handbook for inventorying surface fuels and biomass in the Interior West.* U.S. Department of Agriculture, Forest Service, Intermountain Forest and Range Experiment Station, Ogden, UT. General Technical Report INT-129. 48 p.

Brown, J.K., Snell, J.A.K., Bunnell, D.L. (1977). *Handbook for predicting slash weights of western conifers.* U.S. Department of Agriculture, Forest Service, Intermountain Forest and Range Experiment Station, Ogden, UT. General Technical Report INT-37. 35 p.

Broyles, G. (2011). *Fireline production rates.* USDA Forest Service, National Technology & Development Program, San Dimas Technology and Development Center, San Dimas, CA. Publication 1151 1805-SDTDC. 19 p.

Buck, C.C. (1951). Flammability of chaparral depends on how it grows. *Fire Control Notes* **12** (4), 27.

Burgan, R.E. (1988). *1988 revisions to the 1978 National Fire-Danger Rating System.* U.S. Department of Agriculture, Forest Service, Southeastern Forest Experiment Station, Asheville, NC. Research Paper SE-273. 39 p.

Burrows, N.D. (1984a). *Describing forest fires in Western Australia.* Forests Department, Perth, Western Australia. Technical Paper 9. 29 p.

Burrows, N.D. (1984b). *Predicting blow-up fires in the jarrah forest.* Forests Department, Perth, Western Australia. Technical Paper 12. 27 p.

Burrows, N.D. (1986). *Backburning in forest areas.* Department of Conservation and Land Management, Perth, Western Australia. Landnote 6. 6 p.

Burrows, N.D. (1991). *Rapid estimation of the moisture content of dead Pinus pinaster needle litter in the field. Australian Forestry* **54**, 116–119.

Burrows, N.D. (1994). *Experimental development of a fire management model for jarrah (Eucalyptus marginata Donn ex Sm.) forest.* Australian National University, Canberra, Australian Capital Territory. Ph.D. Thesis. 293 p.

Burrows, N.D. (1995). A framework for assessing acute impacts of fire in jarrah forests for ecological studies. *CALMScience* Supplement **4**, 59–66.

Burrows, N.D., Ward, B., Robinson, A. (2009). Fuel dynamics and fire spread in spinifex grasslands of the Western Desert. *Proceedings of the Royal Society of Queensland* **115**, 69–76.

Bushey, C.L., Mutch, R.W. (1990). Fire behavior service center for extreme wildfire activity. *Fire Management Notes* **51** (4), 34–42.

Butler, B.W., Bartlette, R.A., Bradshaw, L.S., Cohen, J.D., Andrews, P.L., Putnam, T., Mangan, R.J. (1998). *Fire behavior associated with the 1994 South Canyon Fire on Storm King Mountain, Colorado.* USDA Forest Service, Rocky Mountain Research Station, Ogden, UT. Research Paper RMRS-RP-9. 82 p.

Butler, B.W., Cohen, J.D. (1998). Firefighter safety zones: A theoretical model based on radiative heating. *International Journal of Wildland Fire* **8**, 73–77.

Butler, B.W., Reynolds, T.D. (1997). *Wildfire case study: Butte City Fire, southeastern Idaho, July 1, 1994.* U.S. Department of Agriculture, Forest Service, Intermountain Research Station, Ogden, UT. Research Paper INT-GTR-351. 15 p.

Byram, G.M. (1954). *Atmospheric conditions related to blowup fires.* U.S. Department of Agriculture, Forest Service, Southeastern Forest Experiment Station, Asheville, NC. Station Paper 35. 34 p.

Byram, G.M. (1959a). Combustion of forest fuels. In: Davis, K.P. (ed). *Forest fire: Control and use*, pp. 61–89, 554-555. McGraw-Hill, New York.

Byram, G.M. (1959b). Forest fire behavior. In: Davis, K.P. (ed). *Forest fire: Control and use*, pp. 90–123, 554–555. McGraw-Hill, New York.

Campbell, D. (2005). *The Campbell Prediction System.* 3rd edition. Ojai Printing and Publishing Co., Ojai, CA.

Catchpole, E.A., Alexander, M.E., Gill, A.M. (1992). Elliptical-fire perimeter- and area-intensity distributions. *Canadian Journal of Forest Research* **22**, 968–972. [errata: 23:1244. 1993; 29: 788. 1999].

Cermak, R.W. (2005). *Fire in the forest: a history of forest fire control on the national forests, 1898–1956.* U.S. Department of Agriculture, Forest Service, Pacific Southwest Region, Vallejo, CA. Publication R5-FR-003. 442 p.

Chandler, C.C., Countryman, C.M. (1959). Use of fire behavior specialists can pay off. *Fire Control Notes* **20**, 130–132.

Cheney, N.P. (1976). Bushfire disasters in Australia. *Australia Forestry* **39**, 245–268.

Cheney, N.P. (1994). Training for bushfire fighting: current trends and future needs. *Institute of Foresters of Australia – The Forester* **35** (4), 22–29.

Cheney, N.P., Gould, J.S., Catchpole, W.R. (1998). Prediction of fire spread in grasslands. *International Journal of Wildland Fire* **8**, 1–13.

Cheney, P., Gould, J., McCaw, L. (2001). The dead-man zone – a neglected area of firefighter safety. *Australian Forestry* **64**, 45–50.

Cheney, N.P., Gould, J.S., McCaw, W.L., Anderson, W.R. (2012). Predicting fire behaviour in dry eucalypt forest. *Forest Ecology and Management* **280**, 120–131.

Cheney, P., Sullivan, A. (2008). *Grassfires: Fuel, weather and fire behaviour.* CSIRO, Melbourne.

Choczynska, J.J., Johnson, E.A. (2009). A soil heat and water transfer model to predict belowground grass rhizome bud death in a grass fire. *Journal of Vegetation Science* **20**, 277–287.

Clar, C.R., Chatten, L.R. (1966). *Principles of forest fire management.* State of California, Sacramento, CA.

Cohen, J. (2010). The wildland-urban fire problem. *Fremontia* **38** (2-3) 16–22.

Cook, J. (1995). Fire environment size-up: Human limitations vs. superhuman experience. *Wildfire* **4** (4), 49–53.

Cook, J., Tom, A. (2003). Interview with Paul Gleason. *Fire Management Today* **63** (3), 91–94.

Cooper, R.W. (1969). *Preliminary guidelines for using suppression fires to control wildfires in the Southeast.* U.S. Department of Agriculture, Forest Service, Southeastern Forest Experiment Station, Asheville, NC. Research Note SE-102. 2 p.

Countryman, C.M. (1956). Old-growth conversion also converts fireclimate. *Proceedings of Society of American Foresters Meeting 1955*, 158–160.

Countryman, C.M. (1960). Fire environment and silvicultural practice. *Proceedings of Society American Foresters Meeting 1959*, 22–23.

Countryman, C.M. (1966). The concept of fire environment. *Fire Control Notes* **27** (4), 8–10.

Countryman, C.M. (1971). *This humidity business: what it is all about and its use in fire control.* U.S. Department of Agriculture, Forest Service, Forest Service, Pacific Southwest Forest and Range Experiment Station, Berkeley, CA. 15 p.

Countryman, C.M. (1972). *The fire environment concept.* U.S. Department of Agriculture, Forest Service, Forest Service, Pacific Southwest Forest and Range Experiment Station, Berkeley, CA. 12 p.

Countryman, C.M. (1974). *Can southern California conflagrations be stopped?* U.S. Department of Agriculture, Forest Service, Pacific Southwest Forest and Range Experiment Station, Berkeley, CA. General Technical Report PSW-7. 11 p.

Cova, T.J., Dennison, P.E., Drews, F.A. (2011). Modeling evacuate versus shelter-in-place decisions in wildfires. *Sustainability* **3**, 1662–1687.

Cruz, M.G. (2010). Monte Carlo-based ensemble method for prediction of grassland fire spread. *International Journal of Wildland Fire* **19**, 521–530.

Cruz, M.G., Alexander, M.E. (2010). Assessing crown fire potential in coniferous forests of western North America: a critique of current approaches and recent simulation studies. *International Journal of Wildland Fire* **19**, 377–398.

Cruz, M.G., Alexander, M.E. (2013). Uncertainty associated with model predictions of surface and crown fire rates of spread. *Environmental Modelling & Software* **47**, 16–28.

Cruz, M.G., Alexander, M.E., Wakimoto, R.H. (2003). Assessing canopy fuel stratum characteristics in crown fire prone fuel types of western North America. *International Journal of Wildland Fire* **12**, 39–50.

Cruz, M.G., Butler, B.W., Alexander, M.E. (2006a). Predicting the ignition of crown fuels above a spreading surface fire. Part II: Model behavior and evaluation. *International Journal of Wildland Fire* **15**, 61–72.

Cruz, M.G., Butler, B.W., Alexander, M.E., Forthofer, J.M., Wakimoto, R.H. (2006b). Predicting the ignition of crown fuels above a spreading surface fire. Part I: Model idealization. *International Journal of Wildland Fire* **15**, 47–60.

Cruz, M.G., Alexander, M.E., Fernandes, P.A.M. (2008). Development of a model system to predict wildfire behavior in pine plantations. *Australian Forestry* **71**, 113–121.

Cruz, M.G., Gould, J. (2009). National fire behaviour prediction system. In: *Proceedings of the biennial conference of the Institute of Foresters of Australia*, pp. 285–291. Institute of Foresters of Australia, Yarralumla, Australian Capital Territory.

Cruz, M., Gould, J. (2010). *Fire dynamics in mallee-heath*. Bushfire Cooperative Research Centre, East Melbourne, Victoria. Fire Note 66. 4 p.

Cruz, M.G., McCaw, W.L., Anderson, W.R., Gould, J.S. (2013). Fire behaviour modelling in semi-arid mallee-heath shrublands of southern Australia. *Environmental Modelling & Software* **40**, 21–34.

Cruz, M.G., Sullivan, A.L., Gould, J.S., Sims, N.C., Bannister, A.J., Hollis, J.J., Hurley, R. (2012). Anatomy of a catastrophic wildfire: the Black Saturday Kilmore East Fire. *Forest Ecology and Management* **284**, 269–285.

Cuoco, C.J., Barnett, J.K. (1996). How ICs can get maximum use of weather information. *Fire Management Notes* **56** (1), 20–24.

Curcio, G.M. (2009). *Fire danger in North Carolina: weather patterns & fuel conditions contributing to large fire growth.* North Carolina Forest Service, Raleigh, NC. NC Fire Danger Technote 1. 10 p.

Curl, A.D. (1966). The tetrahedron of fire. *Fire Fighting in Canada* **10** (6), 8–9.

Curry, J.R. (1938). The field of forest-fire protection. *Fire Control Notes* **2** (1), 1–6.

Curry, J.R., Fons, W.L. (1938). Rate of spread of surface fires in the ponderosa pine type of California. *Journal of Agricultural Research* **57**, 239–267.

Davis, J.R., Dieterich, J.H. (1976). *Predicting rate of fire spread (ROS) in Arizona oak chaparral: Field workbook.* USDA Forest Service, Rocky Mountain Forest and Range Experiment Station, Fort Collins, CO. General Technical Report RM–24. 8 p.

Davis, K.P. (ed.) (1959). *Forest fire: Control and use.* McGraw-Hill, New York.

Davis, W.S. (1949). The rate of spread – fuel density relationship. *Fire Control Notes* **10** (2), 8–9.

De Groot. WJ., Alexander, M.E. (1986). Wildfire behavior on the Canadian Shield: a case study of the 1980 Chachukew Fire, east-central Saskatchewan. In: Alexander, M. E. (ed.). *Proceedings of the third Central Region Fire Weather Committee scientific and technical seminar,* pp. 23–45. Environment Canada, Canadian Forestry Service, Northern Forestry Centre, Edmonton, Alberta.

Deeming, J.E., Burgan, R.E., Cohen, J.D. (1977). *The National Fire-Danger Rating System-1978.* U.S. Department of Agriculture, Forest Service, Intermountain Forest and Range Experiment Station, Ogden, UT. General Technical Report INT-39. 63 p.

Dickinson, M.B., Ryan, K.C. (2010). Introduction: Strengthening the foundation of wildland fire effects prediction for research and management. *Fire Ecology* **6** (1), 1–12.

Drysdale, D. (2011). *Introduction to fire dynamics.* 3rd edition. John Wiley & Sons, Ltd., Chichester.

DuBois, C. (1914). *Systematic fire protection in the California forests.* U.S. Department of Agriculture, Forest Service, Washington, DC.

Fahnestock, G.R. (1959). When will the bottom lands burn? *Forests and People* **9** (3), 18. 44.

Fahnestock, G.R. (1962). 1962 summer drought lowers moisture in living foliage. U.S. Department of Agriculture, Forest Service, Southern Forest Experiment Station, New Orleans, LA. *Southern Forestry Notes* **142**, 3.

Fernandes, P.M., Botelho, H.S. (2003). A review of prescribed burning effectiveness in fire hazard reduction. *International Journal of Wildland Fire* **12**, 117–128.

Finklin, A.I., Fischer, W.C. (1990). *Weather station handbook – an interagency guide for wildland managers.* National Interagency Fire Center, National Fire Equipment System, Boise, ID. Publication NFES 2140. 237 p.

Finney, M.A., McAllister, S.S. (2011). A review of fire interactions and mass fires. *Journal of Combustion 2011,* C Article ID 548328. 14 p.

Fons, W.L. (1946). Analysis of fire spread in light forest fuels. *Journal Agricultural Research* **72**, 93–121.

Fons, W.L. (1963). Forest fire modelling. In: BT Proceedings of American Society of Mechanical Engineers – American Society of Chemical Engineers Meeting, pp. 164–175. American Society of Mechanical Engineers, New York, NY.

Fons, W.L., Clements, H.B., George, P.M. (1963). Scale effects on propagation rate of laboratory crib fires. *Symposium (International) on Combustion* **9**, 860–866.

Forthofer, J.M., Goodrick, S.L. (2011). Review of vortices in wildland fire. *Journal of Combustion* **2011**, 1–14.

Fromm, M., Lindsey, D.T., Servranckx, R., Yue, G., Trickl, T., Sica, R., Doucet, P., Godin-Beekmann, S. (2010). The untold story of pyrocumulonimbus. *Bulletin of American Meteorological Society* **91**, 1193–1209.

Gellie, N., Mattingley, G., Gibos, K., Wells, T., Salkin, O. (2013). *Reconstructed fire spread and dynamics of the Black Saturday bushfires.* Government of Victoria, Department of Environment and Primary Industries, Melbourne, Victoria.

Gill, A.M., Burrows, N.D., Bradstock, R.A. (1995). Fire modelling and fire weather in an Australian desert. *CALMScience Supplement* **4**, 29–34.

Gill, D. (1974). Weather-induced wildfire conditions and fire-induced weather conditions in the Lower Mackenzie Valley, N.W.T. *The Musk-Ox* **14**, 62–63.

Gisborne, H.T. (1929). The complicated controls of fire behavior. *Journal of Forestry* **27**, 311–312.

Gisborne, H.T. (1942). Review of problems and accomplishments in fire control and fire research. *Fire Control Notes* **6** (2), 47–63.

Gisborne, H.T. (1948). Fundamentals of fire behavior. *Fire Control Notes* **9** (1), 13–24.

Gisborne, H.T. (1950). Tests how cigar and cigarette stubs come down hot. *Fire Control Notes* **11** (2), 28–29.

Goens, D.W. (1990). *Meteorological factors contributing to the Canyon Creek Fire blowup, September 6 and 7, 1988.* U.S. Department of Commerce, National Weather Service, Salt Lake City, UT. NOAA Technical Memorandum NWS WR-208. 21 p.

Gould, J.S., McCaw, W.L., Cheney, N.P., Ellis, P.F., Knight, I.K., Sullivan, A.L. (2007). *Project Vesta – Fire in dry eucalypt forest: Fuel structure, fuel dynamics and fire behaviour.* Ensis – CSIRO, Canberra, Australian Capital Territory and Department of Environment and Conservation, Perth, Western Australia.

Gould, J.S., McCaw, W.L., Cheney, N.P. (2011). Quantifying fine fuel dynamics and structure in dry eucalypt forest (*Eucalyptus marginata*) in Western Australia for fire management. *Forest Management and Ecology* **252**, 531–536.

Grishin, A.M. (1997). *Mathematical modeling of forest fires and new methods of fighting them.* Translated by M. Czuma, L. Chikina, and L. Smokotina. Edited by Frank Albini. Tomsk State University, Tomsk, Russia.

Haines, D.A., Kuehnast, E.L. (1970). When the Midwest burned. *Weatherwise* **23**, 112–119.

Haines, D.A., Main, W.A., Simard, A.J. (1986). *Fire-danger rating and observed wildfire behavior in the northeastern United States.* U.S. Department of Agriculture, Forest Service, North Central Forest Experiment Station, St. Paul, MN. Research Paper NC-274. 23 p.

Hardy, C.E. (1983). *The Gisborne era of forest fire research: Legacy of a pioneer.* U.S. Department of Agriculture, Forest Service, Washington, DC. Publication FS-367. 71 p.

Hester, D.A. (1952). The pinyon-juniper fuel type can really burn. *Fire Control Notes* **13** (1), 26–29.

Hirsch, K.G., Martell, D.L., Corey, P.N. (2000). *Probability of containment of medium initial attack crews in the boreal spruce fuel type.* Natural Resources Canada, Canadian Forest Service, Northern Forestry Centre, Edmonton, Alberta. Poster with text. http://cfs.nrcan.gc.ca/pubwarehouse/pdfs/18239.pdf.

Hoffman, J.V., Osborne, W.B., Jr., (1923). *Relative humidity and forest fires.* U.S. Department of Agriculture, Forest Service, Washington, DC. 12 p.

Jain, T.B., Battaglia, M.A., Han, H-S., Graham, R.T., Keyes, C.R., Fried, J.S., Sandquist, J. (2012). *A comprehensive guide to fuels management practices for dry mixed conifer forests of the Northwestern United States.* U.S. Department of Agriculture, Forest Service, Rocky Mountain Research Station, Forest Service, Fort Collins, CO. General Technical Report RMRS-GTR-292. 315 p.

Johansen, R.W. (1987). *Ignition patterns & prescribed fire behavior in Southern pine.* Georgia Forestry Commission, Research Division, Macon, GA. Georgia Forest Research Paper 72. 6 p.

Johnson, K., Maczek, P., Fremont, L. (2005). *Saskatchewan community wildfire risk assessment project: final report.* Collaboration Saskatchewan Environment – Fire Management and Forest Protection, Saskatchewan Forest Centre, and Forest Engineering Research Institute of Canada, Prince Alberta, Saskatchewan. 46 p. http://www.environment.gov.sk.ca/Default.aspx?DN=548bb226-240e-4bde-a71c-1461e72b78a6.

Kautz, J. (1997). Insulated boxes for protecting video cameras. In: Mangan, R. *Surviving fire entrapments: Comparing conditions inside vehicles and fire shelters*, pp. 39–40. U.S. Department of Agriculture, Forest Service, Missoula Technology and Development Center, Missoula, MT. Technical Report 9751-2817-MTDC.

Keeves, A., Douglas, D.R. (1983). Forest fires in South Australia on 16 February 1983 and consequent future forest management aims. *Australian Forestry* **46**, 148–162.

Kerr, J.W., Buck, C.C., Cline, W.E., Martin, S., Nelson, W.D. (1971). *Nuclear weapons effects in a forest environment – thermal and fire.* U.S. Department of Defense, Nuclear Information and Analysis Center, Santa Barbara, CA. Technical Cooperative Program Panel N-2 Report N2: TR 2-70. 268 p.

Kessell, S.R., Good, R.B., Potter, M.W. (1980). *Computer modelling in natural area management.* Commonwealth of Australia, Australian National Parks and Wildlife Service, Canberra, Australian Capital Territory. Special Publication 9. 45 p.

Kiil, A.D. (1975). Fire spread in a black spruce stand. *Canadian Forestry Service Bi-monthly Research Notes* **31**, 2–3.

Kiil, A.D., Grigel, J.E. (1969). *The May 1968 forest conflagrations in central Alberta – a review of fire weather. fuels and fire behavior.* Canada Department of Fisheries and Forestry, Forestry Branch, Forest Research Laboratory, Calgary. Alberta. Information Report A-X-24. 36 p.

Krawchuk, M.A., Moritz, M.A., Parisien, M-A., Van Dorn, J., Hayhoe, K. (2009). Global pyrogeography: the current and future distribution of wildfire. *PloS One* **4** (4), e5102.

Krueger, D.W. (1961). *Threshold values of relative humidity for large fires in Georgia.* Georgia Forest Research Council, Macon, GA. Georgia Forest Research Paper 3. 5 p.

Kuljian, H., Varner, J.M. (2010). The effects of sudden oak death on foliar moisture content and crown fire potential in tanoak. *Forest Ecology and Management* **259**, 2103–2110.

Lavoie, N. (2004). *Variation in flammability of jack pine/ black spruce forests with time since fire in the Northwest Territories, Canada.* University of Alberta, Edmonton, Alberta. Ph.D. Thesis. 332 p.

Lavoie, N., Alexander, M.E., Macdonald, E.S. (2010). *Photo guide for quantitatively assessing characteristics of forest fuels in a jack pine – black spruce chronosequence in the Northwest Territories.* Natural Resources Canada, Canadian Forest Service, Northern Forestry Centre, Edmonton, Alberta. Information Report NOR-X-419. 42 p.

Lawson, B.D. (1972). *Fire spread in lodgepole pine stands.* University of Montana, Missoula, MT. M.Sc. Thesis. 119 p.

Lawson, B.D., Armitage, O.B. (2008). *Weather guide for the Canadian Forest Fire Danger Rating System.* Natural Resources Canada, Canada Forest Service, Northern Forestry Centre, Edmonton, Alberta. 73 p.

Lindenmuth, A.W. Jr., Davis, J.R. (1973). *Predicting fire spread in Arizona's oak chaparral.* USDA Forest Service, Rocky Mountain Forest and Range Experiment Station, Fort Collins, CO. Research Paper RM-101. 11 p.

List, R.J. (1951). Smithsonian meteorological tables. Sixth revision edition. Smithsonian Institute Press, Washington, DC. 527 p.

Luke, R.H., McArthur, A.G. (1978). *Bushfires in Australia.* Australian Government Publishing Service, Canberra, Australian Capital Territory.

Maclean, J.N. (2013). *The Esperanza Fire: Arson, murder, and the agony of Engine 57.* Counterpoint, Berkeley.

Maclean, N. (1992). Young men and fire. University of Chicago Press, Chicago, IL.

McArthur, A.G. (1962). *Control burning in eucalyptus forests.* Commonwealth of Australia, Forestry and Timber Bureau, Forest Research Institute, Canberra, Australian Capital Territory. Leaflet 80. 31 p.

McArthur, A.G. (1967). *Fire behaviour in eucalypt forests.* Commonwealth of Australia, Forestry and Timber Bureau, Forest Research Institute, Canberra, Australian Capital Territory. Leaflet 107. 36 p.

McArthur, A.G. (1968). The effect of time on fire behaviour and fire suppression problems. In: *E.F.S. Manual 1968,* pp. 3–6, 8, 10-13. South Australia Emergency Fire Services, Keswick, South Australia.

McArthur, A.G., Douglas, D.R., Mitchell, L.R. (1966). *The Wandilo Fire, 5 April 1958 – fire behaviour and associated meteorological and fuel conditions.* Commonwealth of Australia, Forestry and Timber Bureau, Forest Research Institute, Canberra, Australian Capital Territory. Leaflet 98. 30 p.

McCaw, W.L. (2013). Managing forest fuels using prescribed fire – a perspective from southern Australia. *Forest Ecology and Management* **291**, 217–221.

McCaw, W.L., Gould, J.S., Cheney, N.P., Ellis, P.F.M., Anderson, W.R. (2012). Changes in behaviour of fire in dry eucalypt forest as fuel increases with age. *Forest Ecology and Management* **271**, 170–181.

Melton, M. (1989). The Keetch/Byram Drought Index: A guide to fire conditions and suppression problems. *Fire Management Notes* **50** (4), 30–34.

Merrill, D.F., Alexander, M.E. (eds.) (1987). *Glossary of forest fire management terms.* 4th edition. National Research Council of Canada, Canadian Committee Forest Fire Management, Ottawa, Ontario. Publication NRCC No. 26516. 91 p.

Methven, I.R. (1977). *Fire research at the Petawawa Forest Experiment Station: The integration of fire behaviour and forest ecology and management for management purposes.* Paper presented at the Fire Ecology and Resource Management Workshop, 6–7 December 1977, Edmonton, Alberta. 12 p. (http://cfs.nrcan.gc.ca/pubwarehouse/pdfs/33751.pdf).

Michaletz, S.T., Johnson, E.A. (2006). A heat transfer model of crown scorch in forest fires. *Canadian Journal of Forest Research* **36**, 2839–2851.

Mills, G.A. (2008). Abrupt surface drying and fire weather Part 1: Overview and case study of the South Australian fires of 11 January 2005. *Australian Meteorology Magazine* **54**, 299–309.

Mitchell, J.A., 1937. Rule of thumb for determining rate of spread. *Fire Control Notes* **1** (7), 395–396.

Moberly, H.E., Moore, J.E., Ashley, R.C., Burton, K.L., Peeples, H.C. (1979). *Planning for initial attack.* U.S. Department of Agriculture, Forest Service, Southeastern Area State and Private Forestry, Atlanta, GA. Forestry Report SA-FR 2. 41 p.

Mohr, F., Lukens, D., Terry, D. (1987). Managing confinement suppression response on the Middle Ridge and Little Granite Fires. *Fire Management Notes* **48** (3), 23–25.

Molina, D., Castellnou, M., Garcia-Marco, D., Salgueiro, A. (2010). Improving fire management success through fire behaviour specialists. In: Silva, J.S., Rego, F., Fernandes, P., Rigolot, E. (eds.), *Towards integrated fire management – outcomes of the European Project Fire Paradox,* pp. 105–133. European Forest Institute, Joensuu, Finland. Research Report 23.

Moore, T. (chief investigator) (2008). *Indians: Fire accident prevention analysis report.* U.S. Department Agriculture, Forest Service, Pacific Southwest Region, Vallejo, CA. 133 p. http://wildfirelessons.net/documents/Indians_Fire_APA_021209.pdf.

Morvan, D. (2011). Physical phenomena and length scales governing the behaviour of wildfires: a case for physical modelling. *Fire Technology* **47**, 437–460.

National Wildfire Coordinating Group (1992). *Fire behavior nomograms.* National Interagency Fire Center, National Fire Equipment System, Boise, ID. Publication NFES 2220. 28 p.

Nelson, R.M., Jr., (2001). Water relations of forest fuels. In: Johnson, E.A., Miyanishi, K. (eds), *Forest fires: Behavior and ecological effects*, pp. 79–149. Academic Press, San Diego, CA.

Nelson, R.M., Hiers, J.K. (2008). The influence of fuelbed properties on moisture drying rates and timelags of longleaf pine litter. *Canadian Journal of Forest Research* **38**, 2394–2404.

Newbould, D. (2005). *Wildland fire behavior prediction is both science and art.* http://peninsulaclarion.com/stories/040805/outdoors_0408out002.shtml.

Nitschke, C.R., Innes, J.L. (2008). Climatic change and fire potential in south-central British Columbia, *Canada. Global Change Biology* **14**, 841–855.

Noonan-Wright, E.K., Opperman, T.S., Finney, M.A., Zimmerman, G.T., Seli, R.C., Elenz, L.M., Calkin, D.E., Fiedler, J.R. (2011). Developing the US Wildland Fire Decision Support System. *Journal of Combustion* **2011**, 1–14.

Page, W.G., Alexander, M.E., Jenkins, M.J. (2013a). Wildfire's resistance to control in mountain pine-beetle attacked lodgepole pine forests. *Forestry Chronicle* **89**, in press.

Page, W.G., Jenkins, M.J., Alexander, M.E. (2013b). Foliar moisture content variations in lodgepole pine over the diurnal cycle during the red stage of mountain pine beetle attack. *Environmental Modelling & Software* **49**, 98–102.

Pearce, H.G. (2009). *Review of fire growth simulation models for application in New Zealand. New Zealand.* Scion, Rural Fire Research Group, Christchurch, New Zealand. Scion Client Report 16246. 36 p.

Pearce, H.G., Anderson, S.A.J., Clifford, V.R. (2012). *A manual for predicting fire behaviour in New Zealand fuels.* 2nd edition. Scion Rural Fire Research Group, Christchurch, New Zealand.

Pearce, H.G., Clifford, V. (2008). Fire weather and climate of New Zealand. *NZ Journal of Forestry* **53** (3), 13–18.

Pearce, H.G., Kerr, J.L., Clifford, V.R., Wakelin, H.M. (2011). *Fire climate severity across New Zealand.* Scion, Rural Fire Research Group, Christchurch, New Zealand. Scion Client Report 18264. 78 p.

Peet, G.B. (1980). *Forest fire management planning, Kenya: fire danger rating for forest areas.* Food and Agriculture Organization of the United Nations, Rome, Italy. FAO Doc. FO:DPKEN/74/024 Technical Report 1. 93 p.

Pirsko, A.R., Sergius, L.M., Hickerson, C.W. (1965). *Causes and behavior of a tornadic fire-whirlwind.* U.S. Department of Agriculture, Forest Service, Pacific Southwest Forest and Range Experiment Station, Berkeley, CA. Research Note PSW–61. 13 p.

Place, I.C.M. (2002). *75 years of research in the woods: a history of Petawawa Forest Experiment Station and Petawawa National Forestry Institute, 1918–1993.* General Store Publishing House, Burnstown, Ontario.

Plucinski, M.P. (2012). Modelling the probability of Australian grassfires escaping initial attack to aid deployment decisions. *International Journal of Wildland Fire* **21**, 219–229.

Potter, B.E. (2012). Atmospheric interactions with wildland fire behaviour – I. Basic surface interactions, vertical profiles and synoptic structures. *International Journal of Wildland Fire* **21**, 779–801.

Pyne, S.J. (2008). *Year of the fires: The story of the great fires of 1910.* Mountain Press Publishing Company, Missoula, MT.

Quintilio, D., Lawson, B.D., Walkinshaw, S., Van Nest, T. (2001). *Final documentation report – Chisholm Fire (LWF-063).* Alberta Sustainable Resource Development, Forest Protection Division, Edmonton, Alberta. Publication No. I/036. Non-paged.

Read, R.E.H. (1994). *A short history of the Fire Research Station, Borehamwood.* Building Research Establishment, Watford, UK. BRE Report 268. 157 p.

Reinhardt, E. (2003). *Using FOFEM 5.0 to estimate tree mortality, fuel consumption, smoke production and soil heating from wildland fire.* Paper presented at the 2nd International Wildland Fire Ecology and Fire Management Congress, 16-20 November 2003, Orlando, FL. 6 p. http://fire.org/downloads/fofem/5.2/FOFEM5Using.pdf.

Rogers, D.H. (1942). Measuring the efficiency of fire control in California chaparral. *Journal of Forestry* **40**, 697–703.

Rothermel, R.C. (1972). *A mathematical model for predicting fire spread in wildland fuels.* U.S. Department of Agriculture, Forest Service, Intermountain Forest and Range Experiment Station, Ogden, UT. Research Paper INT-115. 40 p.

Rothermel, R.C. (1980). Fire behavior systems for fire management. In: Martin, R.E., Edmonds, D.A., Harrington, J.B., Fuquay, D.M., Stocks, B.J., Barr, S. (eds.), *Proceedings of the sixth conference on fire and forest meteorology*, pp. 58–64. Society of American Foresters, Washington, DC.

Rothermel, R.C. (1983). *How to predict the spread and intensity of forest and range fires.* U.S. Department of Agriculture, Forest Service, Intermountain Forest and Range Experiment Station, Ogden, UT. General Technical Report INT-143. 161 p.

Rothermel, R.C. (1987). Fire behavior research: where do we go from here? In: *Conference papers of ninth conference on fire and forest meteorology*, pp. 19–22. American Meteorological Society, Boston, MA.

Rothermel, R.C. (1991a). Predicting behavior and size of crown fires in the Northern Rocky Mountains. U.S. Department of Agriculture, Forest Service, Intermountain Research Station, Ogden, UT. Research Paper INT-438. 46 p.

Rothermel, R.C. (1991b). Crown fire analysis and interpretation. In: Andrews, P.L., Potts, D. F. (eds). *Proceedings of the 11th conference on fire and forest meteorology*, pp. 253–263. Society of American Foresters, Bethesda, MD. SAF Publication 91-04.

Rothermel, R.C. (1998). Long-range fire assessment. In: Close, K., Bartlett (Hartford), R.A. (eds). *Fire management under fire (adapting to change), Proceedings of the 1994 Interior West Fire Council meeting and program*, pp. 169–179. International Association of Wildland Fire, Fairfield, WA.

Rothermel, R.C. (2000). The great western wildfires: predicting the future by looking at the past. *National Woodlands* **23** (4), 10–12.

Rothermel, R.C., Hartford, R.A., Chase, C.H. (1994). *Fire growth maps for the 1988 Greater Yellowstone Area fires.* U.S. Department of Agriculture, Forest Service, Intermountain Research Station, Ogden, UT. General Technical Report INT-304. 64 p.

Russo, J.E., Schoemaker, P.J.H. (1989). *Decision traps: Ten barriers to brilliant decision-making and how to overcome them.* Simon & Schuster Inc., New York, NY.

Ryan, K.C., Koerner, C. (2011). Fire behavior and effects: principles for archaeologists. In: Ryan, K.C., Jones, A.T., Koerner, C.L., Lee, K.M. (tech eds). *Wildland fire in ecosystems: Effects of fire on cultural resources and archaeology*, pp. 15–84. U.S. Department of Agriculture, Forest Service, Rocky Mountain Research Station, Fort Collins, CO. General Technical Report RMRS-GTR-42-volume 3.

Salazar, L.A., Gonzáles-Cabán, A. (1987). Spatial relationship of a wildfire, fuelbreaks, and recently burned areas. *Western Journal of Applied Forestry* **2**, 55–58.

Schaefer, V.J. (1957). The relationship of jet streams to forest wildfire. *Journal of Forestry* **55**, 419–425.

Schroeder, M.J., Buck, C.C. (1970). *Fire weather . . . a guide for application of meteorological information to forest fire control operations.* U.S. Department of Agriculture, Washington, DC, Agriculture Handbook 360. 229 p.

Schroeder, M.J., Countryman, C.M. (1960). Exploratory fireclimate surveys on prescribed burns. *Monthly Weather Review* **88** (4), 123–129.

Scott, J.H., Burgan, R.E. (2005). *Standard fire behavior fuel models: A comprehensive set for use with Rothermel's surface fire spread model.* U.S. Department of Agriculture, Forest Service, Rocky Mountain Research Station, Fort Collins, CO. General Technical Report RMRS-GTR-155. 72 p.

Show, S.B. (1919). Climate and forest fires in northern California. *Journal of Forestry* **17**, 965–979.

Singer, S.F. (1962). Forest fire detection from satellites. *Journal of Forestry* **60**, 860–862.

Smith, D.M. (2012). *The Missoula Fire Sciences Laboratory: A 50-year dedication to understanding wildlands and fire.* U.S. Department of Agriculture, Forest Service, Rocky Mountain Research Station, Fort Collins, CO. General Technical Report RMRS-GTR-270. 62 p.

Sneeuwjagt, R.J., Peet, G.B. (1998). *Forest fire behaviour tables for Western Australia.* 3rd edition. Department of Conservation and Land Management, Perth, Western Australia. 50 p.

Steen, H.K. (2007). *Forest fire modeling. An interview with Richard C. Rothermel, September 14–15, 2005.* Forest History Society, Inc., Durham, NC. 48 p.

Stocks, B.J. (1975). *The 1974 wildfire situation in northwestern Ontario.* Environment Canada, Canadian Forestry Service, Great Lakes Forestry Research Centre, Sault Ste. Marie, Ontario. Information Report O-X-232. 27 p.

Stocks, B.J., Alexander, M.E., Lanoville, R.A. (2004). Overview of the International Crown Fire Modelling Experiment (ICFME). *Canadian Journal of Forest Research* **34**, 1543–1547.

Stocks, B.J., Kauffman, J.B. (1997). Biomass consumption and behavior of wildland fires in boreal, temperate, and tropical ecosystems: parameters necessary to interpret historic fire regimes and future fire scenarios. In: Clark, J.S., Cachier, H., Goldammer, J.G., Stocks, B.J. (eds), *Sediment records of biomass burning and global change*, pp. 169–188. Springer-Verlag, Berlin. NATO ASI Series Volume **51**.

Sullivan, A. (2009a). Wildland fire spread modeling, 1990–2007. 1: Physical and quasi-physical models. *International Journal of Wildland Fire* **18**, 349–368.

Sullivan, A. (2009b). Wildland fire spread modeling, 1990–2007. 2: Empirical and quasi-empirical models. *International Journal of Wildland Fire* **18**, 369–386.

Sullivan, A. (2010). Grassland fire management in future climate. *Advances in Agronomy* **106**, 173–208.

Sullivan, A.L., Ellis, P.F., Knight, I.K. (2003). A review of radiant heat flux models used in bushfire applications. *International Journal of Wildland Fire* **12**, 101–110.

Sullivan, A.L., Knight, I.K., Hurley, R.J., Webber, C. (2013). A contractionless, low-turbulence wind tunnel for the study of free-burning fires. *Experimental Thermal and Fluid Science* **44**, 264–274.

Sutton, M.W. (1984). Extraordinary flame heights observed in pine trees on 16 February 1983. *Australian Forestry* **47**, 199–200.

Sweeney, B. (2012). *Final report from the Flat Top Complex Wildfire Review Committee. Alberta Environment and Sustainable Resource Development, Edmonton, Alberta.* Publication T/272. 83 p.

Taylor, S.W., Alexander, M.E. (2006). Science, technology, and human factors in fire danger rating: The Canadian experience. *International Journal of Wildland Fire* **15**, 121–135.

Taylor, S.W., Pike, R.G., Alexander, M.E. (1997). *A field guide to the Canadian Forest Fire Behavior Prediction (FBP) System.* Natural Resources Canada, Canadian Forest Service, Northern Forestry Centre, Edmonton, Alberta. Special Report 11. 60 p.

Taylor, S.W., Wotton, B.M., Alexander, M.E., Dalrymple, G.N. (2004). Variation in wind and crown fire behaviour in a northern jack pine – black spruce forest. *Canadian Journal of Forest Research* **34**, 1561–1576.

Teie, W.C. (2005). *Firefighter's handbook on wildland firefighting: Strategy, tactics and safety.* 3rd edition. Deer Valley Press, Rescue, CA.

Thomas, P.H. (1963). The size of flames from natural fires. *Symposium (International) on Combustion* **9**, 844–859.

Thorpe, D. (1999). Those really bad fire days: what makes them so dangerous? *Fire Management Today* **59** (4), 27–29.

Tolhurst, K., Shields, B., Chong, D. (2008). Phoenix: development and application of a bushfire risk management tool. *Australian Journal of Emergency Management* **23** (4), 47–54.

Trevitt, C. (1989). Ph. D. training and social responsibility in a changing climate – the roots of an educator's dilemma. *Climatic Change* **14**, 1–3.

Tymstra, C. (2014). *Blue moon, blue sun – the story of the 1950 firestorm in western Canada.* University of Alberta Press, Edmonton, Alberta, Canada.

Tymstra, C., Bryce, R.W., Wotton, B.M., Taylor, S.W., Armitage, O.B. (2010). *Development and structure of Prometheus: The Canadian wildland fire growth simulation model.* Canadian Forest Service, Northern Forestry Centre, Edmonton, Alberta, Canada. Information Report NOR-X-417. 88 p.

Underwood, R. (ed) (2011). *Tempered by fire: Stories from the firefighters and survivors of the 1961 Western Australian bushfires.* The Bushfire Front, Subiaco, Western Australia.

USDA Forest Service (1930). *Glossary of fire control terms.* U.S. Department of Agriculture, Washington, DC. Miscellaneous Publication No. 70. 22 p.

USDA Forest Service (1966). *Stopping the big ones . . . through fire research.* U.S. Department of Agriculture, Forest Service, Pacific Southwest Forest and Range Experiment Station, Berkeley, Forest Fire Laboratory, Riverside, CA. 12 p.

USDA Forest Service (1993). *Thirty-two years of Forest Service research at the Southern Forest Fire Laboratory in Macon, GA.* U.S. Department of Agriculture, Forest Service, Southeastern Forest Experiment Station, Asheville, NC. General Technical Report SE-77. 89 p.

USDI National Park Service (2003). Fire monitoring handbook. U.S. Department of Interior, National Park Service, National Interagency Fire Center, Fire Management Program Center, Boise, ID. 274 p. http://www.nps.gov/fire/download/fir_eco_FEMHandbook2003.pdf.

Valachovic, Y.S., Lee, C.A., Scanlon, H., Morgan, J.M., Glebocki, R., Graham, B.D., Rizzo, D.M. (2011). Sudden oak death-caused changes to surface fuel loading and potential fire behavior in Douglas-fir – tanoak forests. *Forest Ecology and Management* **261**, 1973–1986.

Van Wagner, C.E. (1965). Describing forest fires – old ways and new. *Forestry Chronicle* **41**, 301–305.

Van Wagner, C.E. (1969). A simple fire-growth model. *Forestry Chronicle* **45**, 103–104.

Van Wagner, C.E. (1970). Temperature gradients in duff and soil during prescribed fires. *Canada Department of Fisheries and Forestry* **26**, 42.

Van Wagner, C.E. (1971). *Two solitudes in forest fire research.* Environment Canada, Canadian Forestry Service, Petawawa Forest Experiment Station, Chalk River, Ontario. Information Report PS-X-29. 7 p.

Van Wagner, C.E. (1973). Height of crown scorch in forest fires. *Canadian Journal of Forest Research* **3**, 373–378.

Van Wagner, C.E. (1975). Convection temperatures above low intensity forest fires. *Canadian Forestry Service Bi-monthly Research Notes* **31** 21, 36.

Van Wagner, C.E. (1977a). Conditions for the start and spread of crown fire. *Canadian Journal of Forest Research* **7**, 23–34.

Van Wagner, C.E. (1977b). Effect of slope on fire spread rate. *Canadian Forestry Service Bi-monthly Research Notes* **33**, 7–8.

Van Wagner, C.E. (1978). *Metric units and conversion factors for forest fire quantities.* Environment Canada, Canadian Forestry Service, Petawawa Forest Experiment Station, Chalk River, Ontario. Information Report PS-X-71. 6 p.

Van Wagner, C.E. (1983). Fire behaviour in northern conifer forests and shrublands. In: Wein, R.W., MacLean, D.A. (eds), *The role of fire in northern circumpolar ecosystems*, pp. 65–80. John Wiley & Sons Ltd., Chichester, UK.

Van Wagner, C.E. (1985). Fire behavior modelling – how to blend art and science. In: Donoghue, L.R., Martin, R. E. (eds). *Proceedings of the eighth conference on fire and forest meteorology*, pp. 3–5. Society of American Foresters, Bethesda, MD. SAF Publication 85–04.

Van Wagner, C.E. (1987a). *Development and structure of the Canadian Forest Fire Weather Index System.* Government of Canada, Canadian Forestry Service, Ottawa, Ontario. Forestry Technical Report 35. 37 p.

Van Wagner, C.E. (1987b). Forest fire research – hindsight and foresight. In: Davis, J.B., Martin, R. E. (tech coords). *Proceedings of the symposium on wildland fire 2000*, pp. 115–120. U.S. Department of Agriculture, Forest Service, Pacific Southwest Forest and Range Experiment Station, Berkeley, CA. General Technical Report PSW-101.

Van Wagner, C.E. (1990). Six decades of forest fire science in Canada. *Forestry Chronicle* **66**, 133–137.

Van Wagner, C.E. (1993). Prediction of crown fire behavior in two stands of jack pine. *Canadian Journal of Forest Research* **23**, 442–449.

Van Wagner, C.E., Methven, I.R. (1978). Discussion: two recent articles on fire ecology. *Canadian Journal of Forest Research* **8**, 491–492.

Vigeas, D.X., Pita, L.P. (2004). Fire spread in canyons. *International Journal of Wildland Fire* **13**, 253–274.

Wade, D.D. (2011). Your fire management career–make it count! *Fire Ecology* **7**, 107–122.

Waldrop, T.A., Goodrick, S.L. (2012). Introduction to prescribed fires in Southern ecosystems. U.S. Department of Agriculture, Forest Service, Southern Research Station, Asheville, NC. Science Update SRS-054. 80 p.

Walkinshaw, S. (2012). *Greater Bragg Creek wildfire mitigation strategy*. Montane Forest Management Ltd., Canmore, Alberta. 59 p. http://www.rockyview.ca/firesmart.

Wallace, W.R., Gloe, H.L. (1938). Forest fire weather. *Australian Forestry* **3**, 28–36.

Weick, K.E. (2002). Human factors in fire behavior analysis: Reconstructing the Dude Fire. *Fire Management Today* **62** (4), 8–15.

Weir, J.R. (2004). Probability of spot fires during prescribed burns. *Fire Management Today* **64** (2), 24–26.

Weir, J.R. (2009). *Conducting prescribed fires: A comprehensive manual*. Texas A&M University Press, College Station, TX.

Werth, P., Ochoa, R. (1993). The evaluation of Idaho wildfire growth using the Haines Index. *Weather and Forecasting* **8**, 223–234.

Williams, D.E. (1954). Water table level as an indicator of drought conditions. *Forestry Chronicle* **30**, 411–416.

Williams, J.T., Rothermel, R.C. (1992). Fire dynamics in Northern Rocky Mountain stand types. U.S. Department of Agriculture, Forest Service, Intermountain Research Station, Ogden, UT. Research Note INT-405. 4 p.

Wilson, A.A.G. (1988a). A simple device for calculating the probability of a house surviving a bushfire. *Australian Forestry* **51**, 119–123.

Wilson, A.A.G. (1988b). Width of firebreak that is necessary to stop grass fires: Some field experiments. *Canadian Journal of Forest Research* **18**, 682–687.

Wilson, C.C. (1977). Fatal and near-fatal forest fires: the common denominators. *International Fire Chief* **43** (9), 9–10, 12-15.

Wilson, C.C., Davis, J.B. (1988). *Forest fire laboratory at Riverside and fire research in California: past, present, and future*. U.S. Department of Agriculture, Forest Service, Pacific Southwest Forest and Range Experiment Station, Berkeley, CA. General Technical Report PSW-105. 22 p.

Wilson, R.A., Jr., (1985). Observation of extinction and marginal burning states in free burning porous fuel beds. *Combustion Science and Technology* **44**, 179–193.

Winter, P.L., Wordell, T.A. (2009). An evaluation of the Predictive Services Program. *Fire Management Today* **69** (4), 27–32.

Wotton, B.M, Gould,J.S., McCaw, W.L., Cheney, N.P., Taylor, S.W. (2012). Flame temperature and residence time of fires in dry eucalypt forest. *International Journal of Wildland Fire* **21**, 270–281.

Index

Fire on Earth: An Introduction, First Edition. Andrew C. Scott, David M.J.S. Bowman, William J. Bond, Stephen J. Pyne and Martin E. Alexander.
© 2014 John Wiley & Sons, Ltd. Published 2014 by John Wiley & Sons, Ltd.